Genetic and Cultural Evolution of Cooperation

Goal of this Dahlem Workshop:

To elucidate the mechanisms and processes beyond kin selection that promote the emergence of cooperation from molecules to societies.

Report of the 90th Dahlem Workshop on
Genetic and Cultural Evolution of Cooperation
Berlin, June 23–28, 2002

Held and published on behalf of the President, Freie Universität Berlin: P. Gaehtgens

Scientific Advisory Board: N.S. Baer, G. Braun, E. Fischer-Lichte, F. Hucho, K. Labitzke, R. Menzel, J. Renn, W. Reutter, H.-H. Ropers, E. Sandschneider, M. Schäfer-Korting, L. Wöste

Executive Director: W. de Vivanco

Series Editor: J. Lupp

Assistant Editors: C. Rued-Engel, G. Custance

Funded by: VolkswagenStiftung

Genetic and Cultural Evolution of Cooperation

Edited by

Peter Hammerstein

Program Advisory Committee:

Peter Hammerstein, Chairperson
Samuel Bowles, Robert T. Boyd, Ernst Fehr, Olof Leimar,
Karl Sigmund, Joan B. Silk, and Eörs Szathmáry

The MIT Press

Cambridge, Massachusetts
London, England

in cooperation with Dahlem University Press

© 2003 Massachusetts Institute of Technology and Freie Universität Berlin

All rights reserved. No part of this book may be reproduced in any form
by electronic or mechanical means (including photocopying, recording, or
information storage and retrieval) without permission in writing from the publisher.

This book was set in Times by Dahlem Konferenzen.

Printed and bound in the United States of America.

Library of Congress Cataloging-in-Publication Data

Dahlem Workshop on Genetic and Cultural Evolution of Cooperation (90th : 2002 :
Berlin, Germany
Genetic and cultural evolution of cooperation / edited by Peter Hammerstein
p. cm. — (Dahlem workshop report)
"Report of the 90th Dahlem Workshop on Genetic and Cultural Evolution of
Cooperation, Berlin, June 23-28, 2002"—P. preceding t.p.
Includes bibliographic references and index.
ISBN 0-262-08326-4 (alk. paper)
1. Cooperativeness—Congresses. I. Hammerstein, Peter, 1949– II. Title.
III. Dahlem workshop reports.

BF637.H4D25 2002
302′.14—dc21

2003054052

10 9 8 7 6 5 4 3 2 1

Contents

Markets and Exploitation in Mutualism and Symbiosis

Genomic and Intercellular Cooperation: Distribution of Power and Division of Labor

Dahlem Konferenzen

History

In 1971, the initiative to establish a series of meetings in Germany to promote interdisciplinary communication among researchers originated from the scientific community. The impetus was the insight that such dialog is the basis upon which progress in research can be achieved. In subsequent discussions between the *Deutsche Forschungsgemeinschaft* and the *Stifterverband für die Deutsche Wissenschaft*, scientists were consulted to compare the needs of the scientific community with existing meeting formats (e.g., the Gordon Research Conferences and the Cold Spring Harbor Symposia). It became clear that a different approach was needed: one less focused on the state-of-the-art and more on the unknown; a form of meeting truly interdisciplinary in its approach. As a result, the *Stifterverband für die Deutsche Wissenschaft* established *Dahlem Konferenzen* in cooperation with the *Deutsche Forschungsgemeinschaft* in 1974. Silke Bernhard, formerly associated with the Schering Symposia, was engaged to lead the conference management team.

The format of the Dahlem Workshop evolved in response to the needs of science. An international scientific advisory board guided the selection and development of themes. Designed originally as a five-year project administered by the *Stifterverband, Dahlem Konferenzen* soon became firmly established in the international scientific community as an indispensable tool for the advancement of research. To secure its long-term future, *Dahlem Konferenzen* became part of the *Freie Universität Berlin* in 1990. To date, over 90 Dahlem Workshops have been held, serving a wide range of disciplines in the scientific community.

Name

Dahlem Konferenzen takes its name from a district of Berlin with strong historic connections to science. In the early 1900s, Dahlem was the seat of the Kaiser Wilhelm Institutes where, for example, Albert Einstein, Lise Meitner, Fritz Haber, and Otto Hahn conducted research.

Concept

Advancement in science is dependent upon interdisciplinary communication and innovative problem solving. For this to happen effectively, however, scientists require the space and time to interact and think creatively. A Dahlem Workshop functions as an intellectual retreat by alleviating the mundane realities that stifle creativity or limit perspectives and by providing an opportunity to focus on topics of crucial interest — topics which must be approached from an interdisciplinary perspective in order for research to progress. The overall goal is not necessarily to reach a consensus but to identify gaps in knowledge, to find new ways of approaching controversial issues, and to define priorities for future research. Workshop topics are submitted by leading scientists and approved by the Scientific Advisory Board. Themes must be problem-oriented, of high-priority interest to the disciplines involved, and timely to the advancement of science.

Dahlem Workshop Model

After a topic has been approved, a Program Advisory Committee is convened to delineate the scientific parameters of the meeting, select participants, and assign them their tasks. Participants are invited on the basis of their international scientific reputation. In addition, the integration of young German scientists is promoted through special invitations.

Dahlem Workshops are organized around four key questions, each of which is addressed by a discussion group composed of approximately ten participants. Lectures or formal presentations are taboo at Dahlem. Instead, concentrated discussion — within and between groups — is the means by which maximum communication is achieved. To facilitate this exchange during the workshop week, themes are prepared in advance of the meeting through the circulation of the "background papers," the topics and authors of which are selected by the Program Advisory Committee. These papers specifically review a particular aspect of the group's discussion topic and introduce controversies and unresolved problem areas for discussion during the workshop.

At the workshop, each group sets its own agenda to address the issues. Cross-fertilization between groups is both stressed and encouraged. By the end of the week, in a collective effort, each group has prepared a report that reflects the ideas, opinions, and contentious issues of the group, suggests directions for future research, and identifies problem areas still in need of resolution. Discussion of this kind necessarily limits the number of participants.

Dahlem Workshop Reports

The Dahlem Workshop Report series serves to complete the communication process of every workshop by disseminating the results and ideas of a workshop to the scientific community for consideration and implementation. Each volume contains the revised background papers and group reports as well as an introduction to the workshop theme. Each chapter is reviewed formally by a designated reviewer and informally by all workshop participants. The final volume is carefully edited to highlight the perspectives, controversies, gaps in knowledge, and proposed future research directions discussed during the meeting.

Dahlem Konferenzen der Freien Universität Berlin
Thielallee 50, 14195 Berlin, Germany

List of Participants with Fields of Research

CARL T. BERGSTROM Dept. of Zoology, University of Washington, P.O. Box 351800, Seattle, WA 98195–1800, U.S.A.
Evolution of communication; ecology and evolution of infectious diseases; evolutionary game theory

NEIL W. BLACKSTONE Dept. of Biological Sciences, Northern Illinois University, DeKalb, IL 60115–2861, U.S.A.
Evolution of development and complexity

SAMUEL BOWLES Santa Fe Institute, Hyde Park Road, Santa Fe, NM 87501, U.S.A.
Coevolution of individual behaviors and group-level institutions; evolution of sociality in humans

ROBERT T. BOYD Dept. of Anthropology, University of California, Los Angeles, CA 90095, U.S.A.
Evolutionary anthropology

JUDITH L. BRONSTEIN Dept. of Ecology and Evolutionary Biology, University of Arizona, P.O. Box 210088, Tucson, AZ 85721, U.S.A.
Evolutionary ecology of mutualism; implications of conflicts and concordances of interests between mutualists and the role of other species in stabilizing and disrupting their interactions

REDOUAN BSHARY Dept. of Zoology, University of Cambridge, Downing Street, Cambridge, CB2 3EJ, U.K.
Behavioral ecology; evolution of cooperation between unrelated individuals; game theoretic approach to cleaning symbiosis

TIMOTHY H. CLUTTON-BROCK Dept. of Zoology, University of Cambridge, Downing Street, Cambridge, CB2 3EJ, U.K.
Behavioral ecology; focus on the evolution of animal breeding systems and the way in which these affect population dynamics and selection (of red deer on the Isle of Rum, of Soay sheep on St Kilda, and of meerkats in the southern Kalahari)

RICHARD C. CONNOR Dept. of Biology, University of Massachusetts at Dartmouth, North Dartmouth, MA 02747, U.S.A.
Dolphin alliances and the evolution of mutualism

MARTIN DALY Dept. of Psychology, McMaster University, 1280 Main Street West, Hamilton, Ontario L8S 4KI, Canada
Risk assessment and future discounting over male and female life-courses

ERNST FEHR Institute for Empirical Research in Economics, University of Zurich, Blümlisalpstr. 10, 8006 Zurich, Switzerland
Foundations of economic behavior

DANIEL M. T. FESSLER Dept. of Anthropology, University of California, 390 Haines Hall, Box 951553, Los Angeles, CA 90095–1553, U.S.A.
Evolutionary anthropology/psychology: the evolution of emotions (especially shame, pride, anger, disgust); sex and reproduction (esp. inbreeding avoidance, pregnancy sickness, behavioral variation across the menstrual cycle); food and eating (especially meat eating, food taboos, starvation)

STEVEN A. FRANK Dept. of Ecology and Evolutionary Biology, University of California, 357 Steinhaus Hall, Irvine, CA 92697–2525, U.S.A.
Conceptual foundations of social evolution, methods of modeling kin selection and demography; evolutionary forces shaping symbioses and relation between symbiotic evolution and traditional concepts of social evolution; processes that repress reproductive conflict within groups; evolution of parasite virulence and dynamics of host–parasite coevolution with emphasis on genetic polymorphism of parasites

HERBERT GINTIS Dept. of Economics, University of Massachusetts, 15 Forbes Avenue, Northampton, MA 01060, U.S.A.
Coevolution of genes and culture in explaining human cooperation and conflict

EDWARD H. HAGEN Institute for Theoretical Biology, Humboldt-Universität zu Berlin, Invalidenstr. 43, 10115 Berlin, Germany
Evolutionary psychology; evolutionary anthropology

DAVID HAIG Dept. of Organismic and Evolutionary Biology, Harvard University, 26 Oxford Street, Cambridge, MA 02138, U.S.A.
Intragenomic conflicts; the evolution of genomic imprinting

PETER HAMMERSTEIN Institute for Theoretical Biology, Humboldt-Universität zu Berlin, Invalidenstr. 43, 10115 Berlin, Germany
Theoretical problems in evolutionary biology; game-theoretic study of conflict and cooperation; endosymbiont theory

JOSEPH HENRICH Dept. of Anthropology, Emory University, 1557 Pierce Drive, Atlanta, GA 30322, U.S.A.
Theoretical and empirical work on human social learning, with particular emphasis on social behavior, cooperation, and status; empirical work includes both field work (ethnography) and experiments

ASTRID HOPFENSITZ Centre for Research in Experimental Economics and Political Decision-Making, CREED, Roetersstraat 11, 1018 WB Amsterdam, The Netherlands
Simulations of group selection; emotions in economics

LAURENT KELLER Institute of Ecology, University of Lausanne, Bâtiment de Biologie, 1015 Lausanne, Switzerland
Kin selection; conflict & cooperation in animal societies; aging; population genetics

MICHAEL KOSFELD Institute for Empirical Research in Economics, University of Zurich, Blümlisalpstr. 10, 8006 Zurich, Switzerland
Economics; game theory; social norms

AXEL KOWALD Max Planck Institute for Molecular Genetics, Ihnestr. 73, 14195 Berlin, Germany
Modeling of complex diseases (trisomy21); population dynamics of healthy and damaged mitochondria; theory of aging

MICHAEL LACHMANN Max Planck Institute for Mathematics in the Sciences, Inselstr. 22, 04103 Leipzig, Germany
Evolution of multicellularity; evolution of signaling (differentiation control, costly signaling, and language); cellular control networks

OLOF LEIMAR Dept. of Zoology, Stockholm University, 106 91 Stockholm, Sweden
Theoretical ethology and evolutionary biology

RICHARD McELREATH Dept. of Anthropology, University of California, One Shields Avenue, Davis, CA 95616, U.S.A.
Anthropology; behavioral ecology; evolutionary theory; cultural evolution; cultural variation (East Africa)

RICHARD E. MICHOD Dept. of Ecology and Evolutionary Biology, University of Arizona, Tucson, AZ 85721, U.S.A.
Evolutionary biology; theory of evolutionary transitions and individuality with special application to the origin of multicellularity in volvocale, a green alga

MANFRED MILINSKI Dept. of Evolutionary Ecology, Max Planck Institute for Limnology, P.O. Box 165, 24302 Plön, Germany
Evolution of cooperation among egoists; sexual selection and immunogenetics; host–parasite interaction

RONALD NOË Ethologie et Ecologie comportementale des Primates, (CNRS UPR 9010), Université Louis Pasteur, 7, rue de l'Université, 67000 Strasbourg, France
Primate evolutionary ecology; primate cognition; cooperation and mutualism

MATTHIAS NÖLLENBURG VolkswagenStiftung, Kastanienallee 35, 30519 Hannover, Germany
Zoology; plant biology; environmental sciences

DAVID C. QUELLER Dept. of Ecology and Evolutionary Biology, MS 170, Rice University, P.O. Box 1892, Houston, TX 77251–1892, U.S.A.
Cooperation and conflict, especially in social insects and cellular slime molds

PETER J. RICHERSON Dept. of Environmental Science and Policy, University of California, Davis, CA 95616, U.S.A.
Cultural evolution in humans; social learning in nonhuman animals; evolution of social institutions

KARL SIGMUND Institut für Mathematik, Universität Wien, Strudlhofgasse 4, 1090 Vienna, Austria
Evolutionary games; altruism

JOAN B. SILK Dept. of Anthropology, University of California, Los Angeles, CA 90095, U.S.A.
Evolution of social behavior in primates

ERIC A. SMITH Dept. of Anthropology, University of Washington, P.O. Box, 353100, Seattle, WA 98195–3100, U.S.A.
Human behavioral ecology and evolutionary anthropology; reproductive strategies, foraging, costly signaling, collective action, resource management, environmental impacts; small-scale societies (Native American, Melanesia)

EÖRS SZATHMÁRY Dept. of Plant Taxonomy and Ecology, Eötvös University, Budapest, Hungary and Collegium Budapest, Institute for Advanced Study, 2 Szentháromság, 104 Budapest, Hungary
Evolutionary biology, major transitions; evolution of the genetic code; origin of language; astrobiology

JOHN TOOBY Center for Evolutionary Psychology, Dept.of Anthropology, University of California, Santa Barbara, CA 93106–3210, U.S.A.
Complementary interests in (1) the dynamics of the evolutionary process as revealed in certain key evolutionary problems, such as cooperation, sexual recombination, coalitions, inbreeding and its avoidance, intragenomic conflict, threat, status, and communication (the theoretical foundations of evolutionary biology); (2) combining experimentation and field investigations with the newly emerging knowledge of how natural selection operated in ancestral hunter-gatherer contexts in order to map out the design features of specialized components of the evolved human cognitive, emotional, and neural architecture (evolutionary psychology); and (3) exploring how these cognitive and neural adaptations organize social behavior, cultural processes, and economic phenomena (evolutionary social theory)

FRANZ J. WEISSING Theoretical Biology, University of Groningen, Kerklaan 30, 9751 NN Haren, The Netherlands
Evolutionary game theory; life history theory; competition theory; evolutionary consequences of sex differences; selection at different levels; emerging properties from individuals to populations; genetic vs. cultural evolution

JOHN H. WERREN Dept. of Biology, University of Rochester, River Campus, P.O. Box 270211, Rochester, NY 14627–0211, U.S.A.
Evolutionary genetics; role of cooperation and conflict in shaping evolutionary processes and genome organization; genetic conflict; endosymbiosis

MARGO I. WILSON Dept. of Psychology, McMaster University, Hamilton, ONT L 8S 4KI, Canada
Risk and decision making in young adulthood: the young male syndrome

LEWIS WOLPERT Dept. of Anatomy and Developmental Biology, University College London, Gower Street, London WC1E 6BT, U.K.
Pattern formation in development and the evolution of development; evolutional psychology of depression; biology of belief

H. PEYTON YOUNG Dept. of Economics, Johns Hopkins University, Mergenthaler Hall, 3400 North Charles Street, Baltimore, MD 21218–2685, U.S.A.
Evolution of norms and economic institutions; theories of learning in strategic environments

1

Understanding Cooperation

An Interdisciplinary Challenge

Peter Hammerstein

Institute for Theoretical Biology, Humboldt University, 10115 Berlin, Germany

INTRODUCTION

The purpose of this book is to elucidate the mechanisms and processes beyond kin selection that promote the emergence of cooperation in systems that range from molecules to societies.

In this chapter I wish to describe at a general level the prospects for interdisciplinary research on the evolution of cooperation. These prospects were discussed at the 90th Dahlem Workshop on which this book is based. The participants represented fields ranging from genetics and cell biology to evolutionary anthropology and behavioral economics. During the course of the workshop it became clear that the different fields share a variety of problems and would benefit greatly from the exchange of knowledge. In addition, a number of controversies emerged. Considering the differences in perspective from which each discipline looks at cooperation, such controversies are not unexpected and should be viewed positively as the necessary impetus to advance research on cooperation. To demonstrate the variety of perspectives, I invite the reader on a guided tour through the rich and diverse research landscape presented in this book. I begin the tour by introducing the rationale for bringing together such a wide array of scientists.

COOPERATION: ONE THEME, SEVERAL PROCESSES, VARIOUS PERSPECTIVES

The modern picture of biological evolution is one in which natural selection tends to operate more effectively at the level of individuals than at the level of groups, species, or higher units. Competition among individuals thus seems to be the key to understanding many aspects of how organisms are "designed." In this view of the living world, conflict seems very natural, and it is easy to understand, for example, why animals fight, why plants overshadow each other in the struggle for light, and why microorganisms engage in "chemical warfare." Cooperation, however, appears as a phenomenon that requires subtle explanation. Such a need for sophistication was recognized early on by William D. Hamilton (1964) and Robert Trivers (1971), who based their theories of cooperation on genetic relatedness (kin selection) and on the logic of repeated interaction (reciprocal altruism). Kin selection theory has had many successful applications and is now widely accepted in biology and increasingly in the social sciences. Yet, in its original form it does not explain cooperation between genetically unrelated individuals or between members of different species (symbiosis, mutualism). The theory of cooperation, therefore, needs to be developed far beyond the concept of kin selection.

After a long period of stagnation we currently observe major activity in this area of research. Evolutionary psychologists (Fessler and Haley, Chapter 2; Hagen, Chapter 6), primatologists (Silk, Chapter 3), and behavioral economists (Fehr and Henrich, Chapter 4) explore the important role of cognition and emotion in cooperation. Anthropologists studying how cultural evolution has shaped human societies emphasize the role of conformist decisions (Richerson, Boyd and Henrich, Chapter 19). Economists draw in a similar way on the evolutionary paradigm (Bowles and Gintis, Chapter 22). Using this paradigm, they aim to understand, for example, why social norms can be stable even when the reasons that gave rise to their existence no longer exist (Young, Chapter 20).

Looking at these research agendas, we observe that in current studies of cooperation the term "evolution" stands for several processes, including cultural transmission, learning, imitation, and, of course, natural selection acting on genotype frequencies in populations. Too often these processes are separated. There are synergistic gains, for example, from studying the coevolution of culture and genes (Richerson, Boyd and Henrich, Chapter 19; Bowles and Gintis, Chapter 22). Synergistic effects are also achievable if one explicitly represents mental mechanisms in biological studies of the evolution of cooperation (Hammerstein, Chapter 5; Hagen, Chapter 6). To include such mechanisms helps resolve the artificial separation of causal and functional analysis that has hampered progress in evolutionary research for too long.

Even within the field of biology, there is a variety of perspectives on cooperation. These perspectives relate to the multiplicity of levels at which natural selection occurs. Biologists take a deep look at the interior of organisms, for example, and ask: How much conflict needs to be overcome for genomic cooperation to evolve? The selfish gene approach promoted by Richard Dawkins (1976) suggests a strong potential for such conflict. Biologists have long known that disharmomonious chords occasionally sound in the genome. Transposable elements and parasitic B-chromosomes serve as examples that have received much attention in recent decades. In general, however, the scope for long-term evolutionary maintenance of genetic selfishness seems rather limited. The idea of a "parliament of the genes" (Leigh 1977, 1991) is sometimes used to explain this limitation. In analogy with human politics, selfish elements are "outvoted" by other genes. As tempting as this analogy may be, it can only be used with great caution, and Leigh's original concept of the parliament needs to be modified (Hoekstra, Chapter 14).

At the level of cell biology, endosymbiont evolution is an intriguing issue. This is the process by which separate organisms enter the cell and become integrated (or not) over evolutionary time. It is now well established, for example, that mitochondria originated from bacteria. These bacteria were modified into "power plants" of the cell and lost autonomy regarding their own reproduction. Mitochondria convert energy, by producing ATP, and also play a significant role in programmed cell death (apoptosis). It is worth asking whether the participation in apoptosis reflects an ancient capability of mitochondria to engage in "slave revolts" (Blackstone and Kirkwood, Chapter 17).

Whereas mitochondria can be seen as "modules of the eukaryotic cell," cells themselves are the building blocks of multicellular organisms. It seems like a simple question to ask why multicellularity evolved as a cooperative phenomenon, since it offers the potential for division of labor. We know that *cultural evolution* has produced elaborate forms of division of labor in a short time if one measures time on the evolutionary timescale. Nothing demonstrates this more impressively than the landing on the moon. But culture as such originated late in the history of life on earth. Similarly, it took *genetic evolution* an intriguingly long time to develop multicellularity from simple cell aggregations. Once in existence, multicellularity initiated a dramatic diversification of life. It seems like a great challenge to understand the origin of multicellularity (Szathmáry and Wolpert, Chapter 15; Michod, Chapter 16).

Besides handing the evolutionary paradigm over to other disciplines, biology has also learned from these disciplines how to make better use of its own paradigm. For example, when organisms from different species interact, they often need different resources. This facilitates the evolution of trade-like phenomena. Organisms that can choose between trading partners are in a situation as if they acted on a market. Addressing trade between and within species, biological market theory exploits this analogy and adopts theoretical ideas from economics (Bowles and Hammerstein, Chapter 8; Bshary and Noë, Chapter 9).

Trades between species generate surpluses and this raises the question of how they are divided. One might imagine a coevolutionary arms race "redefining" the terms of trade. It would then appear that the species that evolves faster receives a particularly favorable share of the surplus. The fastest does not always win, however. Due to its evolutionary "stubbornness," the slowly evolving species may have a strong strategic position in the "evolutionary negotiation" (Bergstrom and Lachmann, Chapter 12).

Cooperation between species (mutualism) can be undermined by individuals that take benefits but give nothing in return. It is widely believed that policing mechanisms evolved in response to the threat of exploitation. Evolution is not straightforward, however, and the origin of what is seen as policing may have little to do with its punishing effect (Bronstein, Chapter 10). For example, selective delivery of rewards to the best partners may be interpreted as a "resource conservation strategy" that helps maintaining these partners. If this is the primary effect driving the evolution of selectiveness, the punishment of exploitation would be a by-effect and there would be no need to search why it pays the punisher to punish.

Otherwise, the role of policing in animals and plants is, of course, similar to the role of sanctioning in human societies. It must be emphasized, however, that humans are very distinct from other animals in this matter. They have laid much of the sanctioning into the hands of social institutions — a key to the understanding of human sociality (Bowles and Gintis, Chapter 22). Perhaps the only natural analogies with human institutions are mechanisms that operate within organisms and help maintain the internal cooperation and integrity (Hoekstra, Chapter 14).

Mutualism may look more cooperative than it deserves. In pseudo-reciprocity, for example, one individual "pays a price" and gives aid to another individual to induce a response that *happens to be* a beneficial return but did not evolve for this reason. Leimar and Connor (Chapter 11) illustrate pseudoreciprocity by a tale, where a bird drops its feces next to a bush that benefits from fertilization. This helps the bush grow and shade the bird's nest. The plant's tendency to transform resources into growth, however, did not evolve as an act of "gratitude." Biologists and social scientists can learn from this tale how important it is to identify correctly the relevant processes that generate a cooperative phenomenon.

THE WORKSHOP

An interdisciplinary effort will help elucidate the various mechanisms and processes that promote the emergence of cooperation. The Dahlem Workshop on the "Genetic and Cultural Evolution of Cooperation" was convened with this goal in mind. For a week of intense "brain storming," the workshop united scientists from a wide array of disciplines. Discourse began two months before the workshop with the circulation of background papers. The book contains these revised background papers as well as reports of the discussion groups. Some of the background papers were substantially revised in response to the workshop (and referee reports). The changes reflect interdisciplinary convergence of ideas as well as divergence based on interesting debates and controversies.

CONTROVERSIES

A variety of conceptual approaches to cooperation is reflected in the reports on group discussions (McElreath et al. Chapter 7; Bergstrom et al. Chapter 13;

Lachmann et al., Chapter 18; Henrich et al., Chapter 23). This conceptual variety was bound to provoke controversies. Many of them simply relate to differences in perspective and seem easily resolvable by making perspectives more visible. Other controversies require major research efforts for their resolution. I conclude the chapter with two examples.

Evolutionary psychologists (Fessler and Haley, Chapter 2; Hagen, Chapter 6) frequently invoke the notion of adaptation, as do most of the disciplines that participated in the workshop. Human behavioral ecologists (Smith, Chapter 21), for example, define their discipline as the study of phenotypic adaptation to varying social and ecological conditions. So far, this seems like full harmony. Yet, depending on the assumed underlying processes, the term *adaptation* has very different meanings. For example, we know that many cooperative phenotypes cannot evolve *genetically* in a given population because this would require group selection and selection is often too weak at the group level. The same phenotypes might evolve by group selection in *cultural* evolution, however, if some conformist transmission is involved (Richerson, Boyd and Henrich, Chapter 19). These cooperative phenotypes would be adaptive with respect to cultural evolution but not with respect to genetic evolution.

Social learning mechanisms thus create a second system of inheritance that differs substantially from genetic inheritance. How is human cooperation affected by *both* of these inheritance systems? That is the question and food for controversial thought. Instead of seeing their own field diminished, however, evolutionary psychologists have good reasons to participate actively in the study of cultural evolution: Social learning takes place in a structured mind, not in a wax-like material. We also know that emotions relevant to cooperation have been found across many different cultures, with the possible exception of "guilt" (Fessler and Haley, Chapter 2). This speaks strongly for a role of genetic evolution in the origin of emotions. Still, culturally evolved institutions have shaped much of the environment to which emotions seem to have adapted — a coevolutionary feedback loop (Bowles and Gintis, Chapter 22).

Let us now switch to the next controversy. In the beginning it must be emphasized that — contrary to its reputation — economics has more to offer than just theory. The field of behavioral economics makes major empirical efforts to identify regularities in human decision making. The regularities found so far change the picture of economics dramatically and pose a challenge to evolutionary research in biology and evolutionary psychology. One of the major controversies at the workshop arose from the behavioral economists' observation of "strong reciprocity" (Fehr and Henrich, Chapter 4). Strong reciprocity means, in particular, that people punish violation of cooperation even in anonymous one-shot encounters with strangers. If punishment is costly, this can be seen as an altruistic act. Theoretical biologists would *not* predict such altruism by just sitting at their desks, analyzing simple models of evolution.

So, what are the evolutionary circumstances that promote the emergence of altruistic punishment? I think that evolutionary psychologists, with their interest in mental mechanisms, should provide an answer to this question. The reason is that rather than "writing on a blank sheet of paper," evolution leaves its traces in nature by modifying the existing mechanisms. This can lead to more cooperation than one would otherwise expect (Hammerstein, Chapter 5). It is worth mentioning in this context that human friendship seems to be maintained by processes that do not quite reflect the simple logic of reciprocity (Silk, Chapter 3).

If we really want to understand the broad scope for human cooperation, mental mechanisms and the underlying structural organization of the mental machinery will tell us much of what we need to know.

ACKNOWLEDGMENTS

In the name of all participants I wish to thank the *VolkswagenStiftung* for generously funding the entire workshop. I am also grateful to the Scientific Advisory Board of the Dahlem Workshops and the *Freie Universität Berlin* for offering us the opportunity to convene such a prestigious conference. Furthermore, I thank the *Wissenschaftskolleg zu Berlin*, the Santa Fe Institute in New Mexico, and the *Deutsche Forschungsgemeinschaft* (SFB 618 *Theoretische Biologie*) for helping some of us prepare for the workshop and reflect thereafter on the intriguing insights generated.

I am deeply indepted to Julia Lupp who has managed this Dahlem Workshop in an incredibly creative, professional, and warmhearted way. I would also like to extend many thanks to Julia's highly motivated and experienced staff members, Caroline Rued-Engel, Angela Daberkow, and Gloria Custance. With all your help and encouragement, it appeared as "mere child's play" to prepare and conduct the workshop, and to edit this book. Unlike me, the Dahlem staff had to work miracles, especially when there was the need to help us scientists emotionally survive the intellectual battles that sometimes went beyond play.

I wish to express my gratitude to Sam Bowles, Rob Boyd, Ernst Fehr, Ole Leimar, Karl Sigmund, Joan Silk, and Eörs Szathmáry for their active participation in the planning of this meeting, to the moderators for their support, and to all participants for their contributions to the workshop and subsequent reviews of the chapters. I appreciate the hard work that Richard McElreath, Carl Bergstrom, Michael Lachmann, and Joe Henrich invested in the group reports. My special thanks go to Ed Hagen with whom I have had many discussions about this book and to Rolf Hoekstra for his perceptive chapter.

REFERENCES

Dawkins, R. 1976. The Selfish Gene. Oxford: Oxford Univ. Press.

Hamilton, W.D. 1964. The genetical evolution of social behaviour. I, II. *J. Theor. Biol.* **7**:1–16, 17–52.

Leigh, E.G.J. 1977. How does selection reconcile individual advantage with the good of the group? *Proc. Natl. Acad. Sci. USA* **74**:4542–4546.

Leigh, E.G.J. 1991. Genes, bees and ecosystems: The evolution of common interest among individuals. *Trends Ecol. Evol.* **6**:257–262.

Trivers, R.L. 1971. The evolution of reciprocal altruism. *Qtly. Rev. Biol.* **46**:35–57.

2

The Strategy of Affect

Emotions in Human Cooperation

Daniel M. T. Fessler and Kevin J. Haley

Department of Anthropology, University of California,
Los Angeles, CA 90095–1553, U.S.A.

ABSTRACT

Emotions appear to be a key determinant of behavior in cooperative relationships. Emotions affect behavior both directly, by motivating action, and indirectly, as actors anticipate others' emotional responses. The influence of emotions is understandable once it is recognized that (a) the ability to benefit from cooperative relationships has been a key determinant of biological fitness throughout our species' history, and (b) panhuman emotions are adaptations crafted by natural selection. Different emotions affect cooperative behavior in different ways: some emotions lead actors to forego the temptation to defect, some lead them to reciprocate harm suffered or benefits provided, and some lead them to repair damaged relationships. An important class of emotions influences cooperative behavior in part by motivating conformity to norms and/or punishment of norm violators. This chapter discusses thirteen emotions that seem to have the greatest impact on cooperation. In addition to reviewing empirical evidence of the role of emotions in cooperation, the chapter presents a variety of explanatory hypotheses and provides a number of discrete testable predictions.

INTRODUCTION

In the Bengkulu fishing village in Sumatra where one of us (Dan Fessler) conducted ethnographic fieldwork, ceremonies are communal affairs. The atmosphere is festive and people are happy. However, if someone appears not to be working hard or, even worse, fails to help at all, people scowl, make disparaging remarks about the shirker, and may even sever social relations, forgoing future opportunities to interact with him and benefit from his hospitality. When asked why they work so hard, and why they are willing to ostracize shirkers even when it is costly to themselves, people answer in one of two ways: They make reference either to past social interactions (e.g., "I'm cooking for her wedding because she helped at my father's funeral"), or to emotions (e.g., "I'd be ashamed not to help out when everyone else is working so hard"). Moreover, even when people only make reference to past events, they often do so in a highly emotional fashion.

We believe that, far from reflecting a parochial culture, the patterns described above illustrate universal aspects of human psychology and behavior. This chapter is premised on the claim that human cooperation is profoundly shaped by, and perhaps only possible because of, emotions. We will examine the manner in which different emotions shape behavior in cooperative contexts; we include under the rubric of "emotion" additional subjective experiences, such as *sympathy,* which have strong affective connotations. Although framed within an evolutionary psychological perspective, our goal is not to present definitive evidence of the validity of this particular approach, but rather to spur future investigations of the role of emotions in cooperation. Toward that end, on an emotion-by-emotion basis we will both briefly describe a variety of existing findings and present a number of hypotheses, specifying discrete, testable predictions whenever possible.

Theoretical Background

To stimulate debate and prompt additional research, we adopt here an extremely broad conception of cooperation. Whether defined in terms of absolute or relative payoff structures (cf. Dugatkin 1990) or merely with regard to the intentions of the actors involved, human cooperation encompasses an enormous range of contexts and behaviors. Cooperation may involve either simultaneous or sequential actions. The behaviors of cooperators may be relatively independent, or they may be tightly coordinated and synchronized. The number of cooperators may range from two to several million. Cooperative action may take place over timescales ranging from minutes to generations and may involve direct or indirect reciprocity (Trivers 1971; Alexander 1987). Such activities can consist of either coordinated collective action or individual contributions to public goods. Moreover, the line between cooperative and noncooperative action is often blurry. Since (a) it is often empirically difficult to discern whether activities such as resource sharing constitute genuine cooperation or merely a coerced compromise and (b) the latter is sometimes a precursor to true cooperation, we cast a broad net, including in our discussion a class of interactions that we term "pseudocooperation," that is, superficially harmonious yet not truly cooperative social behavior.

Today cooperation is arguably one of the most important determinants of human survival and success, and this is likely to have been even truer for that vast majority of our species' history when we lived as nomadic hunter-gatherers (Boehm 2000). This suggests that natural selection will have favored psychological attributes that enhance the individual's ability to engage in, and profit from, cooperative enterprises. This observation interlocks with a growing movement in the psychological sciences, wherein emotions are viewed as discrete mechanisms crafted by evolutionary processes in order to shape behavior in ways that enhanced biological fitness (i.e., survival and reproduction) under ancestral conditions (Frank 1988; Nesse 1990; Cosmides and Tooby 2000).

Identifying and defining emotions is a complex (and contested) enterprise. Although facial expressions and other display behaviors can be a useful index (Ekman and Friesen 1971), we believe that the most productive approach is that which seeks to describe the generic eliciting conditions and outcome behaviors associated with a given emotion (see Lazarus 1991; Russell 1991; Fessler 1999). In this view, each emotion is associated with a logically distinct class of events, and each emotion shapes the organism's resulting behavior in a broadly predictable fashion. This approach is congruent with a theoretical framework wherein emotions are viewed as adaptations produced by natural selection, as emotions seem to parse the world into distinct fitness-relevant tasks, directing attention and memory resources to the given task, heightening the salience of particular courses of action, and reweighting the assessed costs and benefits of different courses of action in a fashion that is adaptive in the environment in which the organism evolved (Izard 1977; Nesse 1990; Cosmides and Tooby 2000). For example, the emotion *fear* is elicited by the threat of imminent harm; it channels attention selectively to the source of the threat, highlights information relevant to the threatening situation in memory, foregrounds behavior relevant to avoidance or escape, and diminishes the perceived costs associated with self protection. As Darwin (1872) recognized, emotions such as fear likely possess a deep phylogeny, as a wide variety of mammals respond to the same class of events in much the same fashion, and exhibit similar display behaviors. However, in contrast to fear, complex emotions such as *shame* and *moral outrage*, though constructed upon pan-mammalian foundations, nevertheless appear to be unique to humans, a conclusion that is consistent with both our elaborate cognitive capacities and our extreme reliance upon socially transmitted information and cooperation among nonkin (cf. Fessler 1999).

Some emotions are explicable primarily in terms of the influences that they exercise in potential cooperative contexts; others function in a wide range of contexts, including situations of potential cooperation and pseudocooperation. Some emotions operate primarily in dyadic interactions, whereas others have their strongest effects in collective action contexts, a distinction that serves to organize the discussion which follows.

EMOTIONS THAT OPERATE PRIMARILY IN DYADIC RELATIONSHIPS

Romantic Love

A barrier to cooperation is the impulse to defect in the short term for immediate gains, behavior which destroys ongoing mutual trust relationships. This impulse may stem from foreshortened time horizons, steep time discounting, or inaccurate assessments of the likelihood that others will learn of one's actions.[1] However, when the benefits of mutual trust relationships exceed the gains to be reaped by defection, individuals are well served by the possession of mechanisms that counteract this impulse and lead them to forego defection. A number of investigators (Hirshleifer 1987; Frank 1988, 2001; Fiske 2003) have suggested that some emotions can be understood as mechanisms designed to commit people to behavior that yields long-term payoffs, thus overcoming the temptation for short-term defection. *Romantic love*, a universal human emotion that underpins pair bonding (Jankowiak and Fischer 1998; Harris 1995), appears to be such a mechanism.[2]

As a consequence of the prolonged period of juvenile dependence characteristic of our species, both males and females can benefit from long-term cooperative mating relationships, as the increased survivorship resulting from biparental care can outweigh the benefits of more promiscuous mating patterns. However, both parties nevertheless face strong temptations to defect. On the one hand, because a man's minimum obligate investment in reproduction is small, following copulation men may abandon their mates in favor of other women, thus foisting all of the costs of childrearing onto the mother. On the other hand, because women can reap both genetic and material benefits from extra-pair copulation, women may cuckold their mates, leading men to invest mistakenly in other men's offspring. Romantic love appears to be part of a suite of mechanisms designed to prevent mates from defecting on the relationship and, importantly, signaling this commitment to their partners.

During the initial or *limerant* phase of romantic love, individuals focus all of their mating-related thoughts and actions on a single person. Later, following sustained emotional and physical consummation of the relationship, this obsessive focus fades away (Tennov 1998). Although the birth of children and the creation of similar joint investments likely reduces the temptation to defect later due to a common interest (a condition indexed subjectively via the emotion *companionate love*), existing accounts of love-as-commitment-device (e.g., Frank 1988, 2001) have overlooked the time-limited nature of the limerant

---

1 It is unclear why natural selection did not simply eliminate these handicaps rather than constructing compensatory mechanisms; nevertheless, an understanding of the advantages of circumventing these pervasive limitations sheds considerable light on the adaptive utility of a variety of emotions (Frank 1988, 2001; Bowles and Gintis 2002; Fiske 2003).

2 In many traditional societies, individuals often do not choose their spouses, as marriages are arranged by kin. However, it is probable that, in ancestral hunter-gatherer societies, the combination of economic self-sufficiency and considerable physical mobility were such that individual preferences could nevertheless substantially influence mate selection (for a contemporary example, see Shostak [1981]; for additional discussion, see Harris [1995]).

phase. It is precisely this initial phase that is particularly important given that the risk of defection is very high in the early part of a relationship. Elizabeth Pillsworth (pers. comm.) has hypothesized that, during the initial phase of a relationship, the obsessive aspect of romantic love may influence behavior in a fashion that effectively spans the ensuing interval, as individuals who are "crazy in love" may "burn their bridges" by openly eschewing alternative mating opportunities and weakening or severing other valuable social ties. Having limited their mating options and social contacts, individuals are thus strategically committed to maintaining the chosen relationship; likewise, because the "bridge-burning" behavior is public, both the chosen partner and any potential rivals are able to assess the level of commitment, i.e., the behavior constitutes an honest signal. In this model, romantic love functions to enhance commitment in different ways at different stages of the relationship. Initially, limerance leads to a single-minded focus on one partner. As limerance fades, commitment is maintained by the social consequences of behavior during the limerant phase. Finally, the creation of joint investments, a state subjectively marked by companionate love, solidifies commitment once more.

In the above account, companionate love serves to mark a valuable relationship, highlighting it in a fashion that decreases the likelihood of defection. This function seems to be achieved through the combination of a number of subjective components. Partners experience satisfaction and security in one another's company and distress at prolonged separation, emotions that motivate the actors to preserve the relationship. Importantly, actors also experience a sympathetic orientation toward the partner wherein the prospect of harm befalling the partner is cause for distress; the desire to avoid inflicting harm then motivates abstention from defection (see Frank 1988, 2001). It is likely that the same features characterize subjective experiences attending friendships, relationships which, like mateships, both present the opportunity for defection and, in the event that defection can be avoided, hold the promise of substantial long-term benefits (see Silk, this volume).

Gratitude

Companionate love, sympathy, and affiliative "liking" all address overarching features of a given social relationship, i.e., how an actor feels about *somebody.* In contrast, a second class of emotions relevant to cooperation addresses how an actor feels about *something somebody has done*. For example, though remarkably understudied (Haidt 2003b), *gratitude* likely plays an important role in fostering and maintaining cooperation (Trivers 1971). Gratitude focuses both attention and a positive, affiliative orientation on a party who has supplied the actor with a substantial benefit. In the context of its initial elicitation, gratitude seems to prompt the actor to recognize a valuable interaction partner and subsequently signal a willingness to reciprocate, thus either (a) establishing the grounds for a new relationship, or (b) reassuring a long-standing partner that the debt has been registered. Most important of all, whereas the duration and intensity of gratitude are likely functions of both the perceived size of the benefit and a variety of individual attributes, to the extent that gratitude endures, in conjunction with affiliative attitudes it motivates a desire to reciprocate and to defend the interests of the benefactor (Trivers 1971). Subjects in behavioral economics games often violate the assumptions of traditional rational actor models by demonstrating a willingness to incur monetary costs in order to reward partners for perceived cooperative or altruistic behavior (Berg et al. 1995; Andreoni, Harbaugh, and Vesterlund, unpublished). Experimental games conducted by Andreoni et al., for instance, revealed that even in one-shot dyadic proposer-responder games, rewards increased when exogenous factors compelling generosity became unavailable (i.e., when opportunities to punish were absent). The pattern of results obtained by Andreoni et al. suggests that increased willingness to reward was linked to responders' evaluation of high offers that were uncompelled by the threat of punishment as being more generous. Although such experiments do not directly examine the role of emotions, these results are in keeping with the possibility that subjects respond with gratitude to uncompelled acts of generosity and thus feel subjectively motivated to reciprocate in kind. In sum, whereas it is possible to experience gratitude in anonymous or transitory interactions, the emotion appears to be designed to prolong potentially beneficial cooperative relationships between known actors.

Anger

If gratitude is elicited by receipt of a benefit, its opposite is anger, elicited by actual or attempted exploitation or harm (Izard 1977). More formally, anger is the response to the infliction of a cost. In addition to showing an "irrational" willingness to reward generosity, subjects in behavioral economics experiments also show an eagerness to punish uncooperative partners (Roth 1995; Fehr and Gächter 2000; Andreoni et al., unpublished). In the ultimatum game, where partners in the role of respondent can "spitefully" prevent both partners in the dyad from receiving a payoff by rejecting the offer of the proposer, respondents demonstrate reluctance to accept low offers and a willingness to punish by rejecting such offers (Roth 1995; Fehr and Gächter 2000; Andreoni et al., unpublished). Exploring the relationship between perceptions of fairness, emotions, and choices, Pillutla and Murnighan (1996) conducted an ultimatum game employing outside options, varying information, and varying common knowledge, then asked respondents to report their feelings. Anger, elicited by perceived unfairness, was commonly associated with rejections and was particularly frequent when respondents rejected offers that exceeded their outside options. Bosman and van Winden (2002) conducted power-to-take games in which players could only reduce the amount that others could appropriate by destroying their own endowment. Using self-report measures of emotional response to appropriation, the authors found that irritation (strongly correlated with anger) and contempt (see below) were linked with the spiteful elimination of a player's own earnings. Similar patterns occur in public goods games, as initially cooperative subjects eventually spitefully reduce their contributions in order to strike out at low contributors whose free riding angers them (Andreoni 1995). Together, these results clearly demonstrate that, even within the confines of finite anonymous games, angry individuals often place paramount importance on harming the transgressor, and are willing to incur substantial costs in order to do so.

Though sometimes destructive to cooperation, anger can also be eminently functional. Focusing attention on the transgressor to the exclusion of other facets of the world, anger motivates actors to strike out at those who transgress against them, thus inflicting costs on the transgressor which then reduce the attractiveness of future attempts at transgression (see also Trivers 1971; McGuire and Troisi 1990; Edwards 1999, pp. 140–141). The stronger the response to transgression, the greater the deterrent effect (Daly and Wilson 1988): this goal is effected in anger by simultaneously enhancing the subjective value of retribution and reducing the salience of costs entailed therein, thus sometimes leading to punishments that seem out of proportion with the offense (cf. Trivers 1971). Consistent with this, experiments demonstrate that angry subjects make optimistic risk estimates and show more risk-seeking behavior (Lerner and Keltner 2001). These psychological changes are likely enhanced by the presence of an audience since, when news of the "irrational" strength of the actor's response spreads, others who might have contemplated transgressing against the actor will also be deterred (Daly and Wilson 1988).

Anger, with its universally recognized and largely involuntary facial display (Ekman 1992), can be seen as yet another example of the manner in which natural selection has used emotions as a means of overcoming the consequences of the tendency to both discount the future and underestimate the extent of others' knowledge of one's actions (cf. Fiske 2003). Time discounting alone would reduce the incentive for responding to transgression since the costs of reacting are paid in the present but the benefits of deterrence are reaped in the future (Frank 1988, 2001). Similarly, misjudging the extent of others' knowledge would reduce the incentive for responding since the reputational aspects of deterrence may be underestimated.

Although anger is not limited to contexts relevant to cooperation because exploitation is the antithesis of cooperative interaction, knowledge of others' propensity to experience anger promotes cooperation by reducing the temptation to exploit actual or potential cooperative partners (cf. Hirshleifer 1987; Frank 1988). Empirical results in economics experiments again confirm this view. When asked about behavior in the context of economic games, subjects report (a) expecting to feel angry toward free riders, (b) expecting to be the target of others' anger if the subject herself free-rides (Fehr and Gächter 2000), and (c) being motivated to pursue cooperative strategies by the anticipation of others' anger (Fehr and Gächter 2002; see also Prasnikar and Roth 1992). Overall, punishment or the threat thereof proves more salient than does reward as an incentive for cooperation or generosity in experimental economics games (Andreoni et al., unpublished), and anger seems to be a key factor in the willingness to punish.

Although cooperation is not isomorphic with the equitable distribution of resources, the two are linked in that absence of the latter may interfere with the former, an interaction that is importantly mediated by emotions. Howsoever "equitable distribution" is locally defined (cf. Henrich et al. 2001), it seems to involve a sense of entitlement such that, when distributions are seen as inequitable, the less-benefited party often experiences the inequality as a transgression. The detection of transgression then triggers anger, resulting in the infliction of costs on the other party. Awareness of the possibility that the recipient of a smaller share will react with anger thus leads actors to increase the equity of distributions; whether this behavior constitutes true cooperation or merely pseudocooperation, by preserving the peace, a recognition of others' potential for anger facilitates continued interaction, a prerequisite for future cooperation.

Envy

When actors identify a sizeable disparity between parties in the possession of, or access to, valued goods or opportunities, those having less often wish to obtain more. The propensity to experience such covetous desire is understandable given that, under ancestral conditions, resources were likely a principal determinant of reproductive success. However, in addition to a simple desire to obtain more resources, humans (and possibly other social mammals) experience a more complex emotion, namely *envy*.[3] In contrast to the simple desire to obtain that which others possess, envy also includes a measure of hostility toward the more fortunate party. Around the world, beliefs such as the "evil eye" concretize the observation that envious individuals are positively dangerous to those they envy (Dundes 1992; Schoeck 1969). Under more controlled conditions, in an anonymous economic game, Zizzo and Oswald (2001) found that a majority of participants were willing to give up large portions of their real-money stakes in order to destroy portions of the winnings of more successful individuals; although no psychological data were collected, the authors interpret this behavior as driven by envy.

Behavioral ecologists have suggested that much apparent sharing of resources (a form of cooperation) is actually tolerated theft (a form of

---

[3] Envy and jealousy are often treated as synonyms. However, although they partially overlap, the two are logically distinct; whereas envy is elicited solely by a disparity in possession of a valued good, jealousy is elicited by a third party's (actual or potential) disruption of an (actual or desired) dyadic social relationship. Jealousy experienced by the partner within such a relationship is thus akin to anger (a response to transgression), whereas jealousy experienced by the nonpartner desirous of such a relationship is akin to envy (a wish to displace the individual who possesses the coveted resource).

pseudocooperation), since resource possessors should not exclude others when the costs of defense exceed the costs of allowing others access (Blurton Jones 1987; see Smith, this volume). In one-shot interactions, those who do not possess the given resource should be willing to incur costs approaching, but always less than, the benefits that they would reap by gaining access; whenever incurring such costs allows the actor to inflict costs on the resource possessor that exceed the costs to the latter of sharing, access should be granted. It is therefore noteworthy that envious individuals seem to be willing to incur huge costs in order to inflict costs on those envied—presumably, it is this willingness that generates the widespread fear of envious individuals. The disparity between the benefits of obtaining access to a resource and the costs that envious individuals seem to be willing to incur suggests that the strategic benefits of envy derive from iterated rather than one-shot interactions.

In a social environment consisting only of one-shot interactions, a resource possessor cannot be sure of the costs that excluded actors are willing to pay in order to gain access, hence the latter will have to incur costs in order to demonstrate such willingness, and this behavior must be repeated in each one-shot interaction. However, in a world of repeated interactions, a few dramatic examples of an actor's willingness to pay substantial costs suffice to indicate to resource possessors that failure to share with the actor in the future will be very expensive. Moreover, as in the case of anger, reputational effects greatly augment this pattern, as third parties who witness or otherwise learn of an actor's envious behavior are informed of the potential costs of not sharing with the actor. Hence, by paying the large up-front costs of initially inflicting excessive harm on those who do not share, the envious actor may avoid paying the individually smaller but highly iterated costs of repeatedly forcing tolerated theft access. In this view, envy, like anger, serves to instantiate in the present the dynamics of potential future transactions in a world of repeated interactions. Like anger, envy thus overcomes time preferences that would otherwise lead actors to forego paying high costs in the present in order to avoid even higher costs in the future. This account of envy entails several testable predictions. First, in contrast to simple covetous desire, envy should reliably be accompanied by a willingness to incur substantial costs, including costs in excess of the value of the benefits associated with the resource at issue. Second, individuals should generally be envious of (i.e., both covet the resources of and be hostile toward) only those with whom they interact on a repeated basis — individuals may desire the resources of strangers and may even attempt to take such resources by force, but they should not feel true envy toward them. More broadly, just as knowledge that others may experience anger should promote (overtly) harmonious iterated social interaction by increasing the equity of distributions, knowledge that others may experience envy should foster peaceful coexistence by enhancing generosity in general.

Guilt

Emotions such as anger can promote cooperation because they motivate actors to inflict costs on selfish individuals. However, inflicting costs on individuals who are not selfish is corrosive to both the establishment and the maintenance of cooperation, whether such actions constitute intentional exploitation or accidental harm. Interestingly, such behavior can evoke a discrete emotion: Although *guilt* can be elicited by a variety of events (including simple norm violations), the central elicitor is the infliction of harm on another, whether intentional or unintentional (Hoffman 1982; Keltner and Buswell 1996), prototypically within a communal relationship characterized by expectations of mutual concern (Baumeister et al. 1994). Deliberate defection or careless mistakes can elicit guilt, just as initial gratitude can segue into guilt when failure to reciprocate becomes perceived as defection. Guilt focuses attention on the action and the harm that has been done to the other party, inflicts subjective discomfort on the actor via its strongly aversive valence, and motivates the actor to make amends by aiding or otherwise compensating the victim (Izard 1977; Baumeister et al. 1994; Tangney 1998). The functioning of guilt is thus precisely tuned to identify and reverse the damage done to a cooperative relationship. Furthermore, just as anticipation of another's anger often leads actors to refrain from intentionally transgressing, anticipation of their own guilt often leads them to refrain from intentionally defecting. Hence, via multiple avenues, guilt can enhance cooperation (Trivers 1971; Frank 1988).

Using an iterated Prisoner's Dilemma game, Ketelaar and Tung Au (2003) found that inducing guilt increased cooperativeness among previously uncooperative players. Using an iterated ultimatum game in which affect was measured after the first round, the authors also found that individuals who made selfish offers and reported experiencing guilt subsequently made generous offers one week later. In both cases, comparisons (with those who had already played fairly, with controls who did not undergo guilt induction, and with individuals who did not report guilt after making selfish offers) indicated that guilt was a key factor in increasing cooperativeness among uncooperative actors.

The fit between the demands of maintaining cooperative relationships and the functioning of guilt is remarkable, and results such as Ketelaar and Au's provide compelling evidence of guilt's efficacy in promoting cooperation. Given these facts, and given our argument that natural selection has favored emotions that enhance the ability to benefit from cooperative relationships, it seems logical that guilt would be a universal human emotion. However, a word of caution is in order. Unlike many of the emotions discussed thus far, it is questionable as to whether guilt is experienced in all or nearly all societies. Although one large study reports evidence of guilt in many cultures (Scherer and Wallbott 1994), our own unsystematic survey of the ethnographic literature suggests that guilt is not lexically marked in many cultures, a somewhat surprising finding given guilt's obvious utility in maintaining social order and enforcing norm adherence. Furthermore, given that a key function of guilt is the repair of damage done to relationships, we might expect that natural selection would have crafted an emotion display, particularly one containing elements that are outside of volitional control, to accompany the subjective and cognitive aspects of guilt. The flood of protestations of regret, apologies, and so on that often attend guilt testify to the importance of communicating to the harmed party that reparations are sincerely intended, harm was inflicted accidentally, etc. If guilt is an evolved panhuman emotion, why is there no universal involuntary display associated with guilt (cf. Keltner and Buswell 1996)?[4] One possibility is that guilt is distinctly different from, say, romantic love or anger in that, rather than being a product of biological evolution, guilt may be the result of cultural evolution, cobbled together from evolved emotions and dispositions such as regret, sympathy, and so on. In this scenario, concepts of guilt and ways of inculcating this emotion developed only in cultures in which the social structure, means of subsistence, etc. were most compatible with a highly autonomous mode of behavior regulation (in contrast, e.g., to shame as will be discussed below). We believe that both biological and cultural accounts of the origins of guilt are sufficiently coherent as to justify a concerted research effort to determine which is correct.

Righteousness

The core elicitor for guilt is the infliction of harm. However, norm violation alone can potentially elicit guilt. This observation draws our attention to the emotion that is arguably the opposite of guilt, an understudied affect that we refer to as *righteousness*. Although guilt is an aversive state experienced as a consequence of rule violation, righteousness is a rewarding state experienced as a consequence of rule adherence. We claim that humans feel a distinct positive emotion when they "do the right thing" (cf. Thomas Aquinas' *Summa Theologiae*). People feel good when they help a friend, provide gifts for a mate, or comfort a child. Whereas these feelings likely stem in part from sympathy, as the actor empathically experiences the benefit obtained by the recipient (cf. Trivers 1971; Frank 2001), there seems to be an additional component to this subjective state. Righteousness, a distinctive subjective reward, may be experienced when actors behave in a fashion that promotes the formation and maintenance of social relationships, reflecting the recognition that the actor has become valuable to others or has earned social credit. More broadly, righteousness may play an important role in motivating general norm adherence, behavior which, as we will discuss at length, may benefit the actor in part through subsequent enhanced recruitment into cooperative ventures (cf. Bowles and Gintis

4 Some advocates of the position that guilt is an evolved emotion argue that dynamic whole-body displays accompany guilt (T. Ketelaar, pers. comm.). However, Wallbott (1998) found that, in contrast to emotions such as shame, trained actors were unable to use dynamic whole-body displays to convey guilt to coders.

2002). Hence, via a number of pathways, righteousness may play an important role in cooperation, one that is deserving of further investigation.

Contempt

We have suggested that guilt and righteousness facilitate the formation and preservation of cooperative relationships. However, not all cooperative relationships are worthwhile. In some cases, the benefits of defection exceed the benefits of cooperation. In a world without emotions that function to preserve cooperative relationships, steep time discounting alone would lead to high rates of defection. However, the existence of relationship-preserving emotions creates a situation in which it may be advantageous to mark explicitly individuals who have little of value to offer the actor. We suggest that *contempt* is the emotion accompanying exactly such an evaluation. By highlighting the low value of the other individual, contempt predisposes the actor to either (a) avoid establishing a relationship, (b) establish a relationship on highly unequal (i.e., exploitative) grounds, or (c) defect on an existing relationship. Consistent with the low valuation of the other, contempt seems to preclude the experience of prosocial emotions in the event that the actor is able to exploit the partner, apparently by framing the harm as merited.

Because contempt is highly corrosive to the formation and maintenance of cooperative relationships, actors can be expected to be highly sensitive to any indications that a prospective or current partner experiences contempt toward them. When an actor concludes that an other's contempt accurately reflects disparities in their relationship caused by either (a) the actor's own failure to adhere to a social norm (cf. Rozin et al. 1999) or (b) an overarching inequality in status, the recipient of contempt often feels *shame* (see below), an emotion that motivates either appeasement or avoidance (Izard 1977; Gilbert 1997; Elster 1998; Fessler 1999). Appeasement serves to maintain the relationship despite the disparity, whereas avoidance serves to minimize exploitation. In contrast, when an actor concludes that an other's contempt exceeds what is merited by any disparities in their relationship, the recipient of contempt often feels anger (Fessler 1999; see also Bowles and Gintis 2002), a response that serves to preempt exploitation through the demonstration of a willingness to incur high costs in order to inflict harm on transgressors.

EMOTIONS THAT IMPORTANTLY OPERATE IN COLLECTIVE CONTEXTS

Shame and Pride

Although there is substantial evidence supporting the universality of contempt (Biehl et al. 1997), as in the case of guilt, the status of righteousness as a panhuman emotion is unclear. Investigating these emotions in Bengkulu, Dan

Fessler found that informants rarely discussed anything resembling guilt, frequently only providing accounts of *regret* (e.g., "I wish that I hadn't cheated because it caused so many problems"). Likewise, attempts to elicit accounts of righteousness often led merely to reports of sympathy (e.g., "When I give her things she is happy, and that makes me happy too"). However, whereas guilt and righteousness were essentially absent from informants' discourse, reports of shame and its opposite, *pride,* were pervasive.[5]

Shame is the negative emotion experienced when an actor knows that others are aware that the actor has behaved in a blameworthy fashion, whereas pride is the positive emotion experienced when the actor knows that other parties are aware that the actor has behaved in a commendable fashion (Gilbert 1997; Fessler 1999; Katz 1999).[6] Shame thus constitutes a subjective penalty for norm violation and pride constitutes a subjective payoff for norm adherence. Conformity to norms is fundamental to myriad forms of human cooperation for at least three reasons:

1. Many norms directly address cooperative behavior (i.e., the need to reciprocate, etc.) (Cooter and Eisenberg 2001; Henrich et al. 2001).
2. Many norms structure interactions in a fashion that precludes negotiation and conflict (cf. Young, this volume).
3. Complex cooperation is often contingent on the precisely timed coordination of behavior, an objective that is best achieved through the sharing of understandings regarding the nature of the appropriate actions at hand (cf. McElreath et al. 2003).

Although an optimal strategy might therefore be to adopt a Machiavellian approach to norms, conforming only when the benefits (including those mediated by reputation) exceed the costs, the cognitive constraints discussed earlier are such as to make it likely that Machiavellian actors will err, eventually destroying reputations that have enormous value over the long run. In contrast, such errors will not be committed by actors who (a) have internalized norms, and hence see them as self-evidently valid, and (b) experience others' assessments of norm-conforming or norm-violating behavior as intrinsically rewarding or

5 Recently Fontaine et al. (2002), seeking to compare the semantic domains of emotion in Indonesian and Dutch, argued that the Indonesian term *bersalah* is equivalent to guilt. However, *bersalah*, which literally means "to be in the wrong" or "to have committed a wrong," is, as the authors themselves demonstrate, semantically associated primarily with fear (cf. Heider 1991). There is thus little evidence that *bersalah* is equivalent to guilt. Even holding aside differences in subjective experience (or "qualia"), behavioral outcomes suffice to reveal the lack of similarity—whereas guilt motivates reparations and/or self-punishment, fear motivates avoidance.

6 Note that these eliciting conditions depend upon the ability to understand others' mental states. Reflecting their roots in the emotions shared with other primates, shame and pride can also be elicited by cognitively simpler conditions of subordinance or dominance, respectively (Fessler 1999; Gilbert et al. 1994).

punishing (Cooter and Eisenberg 2001; Bowles and Gintis 2002). Shame and pride, for which there is substantial evidence of universality (Scherer and Wallbott 1994; Fessler 1999), perform precisely the latter function (Fessler 1999; Bowles and Gintis 2002; Bowles and Gintis, this volume). Hence, it appears that, together with the existence of external punishment (discussed below), the benefits to be derived from cooperation may have been a significant factor favoring the evolution of an affective system that promotes norm adherence (Fessler 1999; Bowles and Gintis, this volume).

Shame and pride can promote cooperation in purely dyadic interactions, as the actor can feel shame if she defects and the partner knows about, or is likely to learn of, her defection, whereas she can feel pride if she fulfills her reciprocal responsibilities and the partner knows about, or is likely to learn of, her actions. However, although these emotions, or the anticipation thereof, can influence choices even within encapsulated dyads, it is in the greater social arena that shame and pride most profoundly affect behavior: Informants in both Bengkulu and California made it abundantly clear that the larger the audience that is privy to one's actions, and the more prestigious the members of that audience, the more intense the emotions that attend failure or success, norm violation or norm fulfillment. The influence of an audience is understandable if the function of shame and pride is to promote conformity to social standards in order to both avoid punishment and gain access to cooperative enterprises, as (a) the larger the audience, the larger the number of prospective punishers and prospective collaborators and (b) the more prestigious the audience, the greater the value of prospective collaborators.

The impact of audience awareness on the intensity of shame and pride likely influences cooperative behavior in two important ways. First, gossip networks raise both the costs of defection and the benefits of cooperation in dyadic relationships, as the reputational consequences of actions will determine both the actor's access to additional opportunities for cooperation and the actor's exposure to third-order punishments or rewards. Actors who are concerned with the prospect of experiencing shame or pride are more likely to consider publicity when weighing possible courses of action; hence these emotions often shape behavior in ways that benefit from information transmission. Second, many of the most important human cooperative ventures are communal rather than dyadic.

Communal cooperative enterprises are substantially more complex than dyadic cooperation: In communal enterprises, the large number of actors makes it more difficult to keep track of individuals' actions, thus enhancing opportunities for free riding and other forms of defection (e.g., it is easier to get away with not pulling one's weight when there are 30 people on a rope than when there are two). In addition, communal enterprises often require more elaborate coordination of behavior, including more extensive role specialization and more multiplex synchronization, than is true in dyadic interactions (e.g., raising a barn is vastly more complicated than paddling a two-person canoe). Importantly, the large number of actors involved in communal enterprises means that a sizeable and interested audience is readily at hand should it come to light that an actor has behaved in a blameworthy or praiseworthy fashion. As a consequence, the potential immediate intensity of shame and pride is greater in communal undertakings than in dyadic interactions. Taken together, the above factors make it plausible that, as our ancestors developed more elaborate forms of communal cooperation, the selective pressures favoring motivational mechanisms that would enhance inclusion in such ventures increased, thus reinforcing the evolution of the capacity for, and propensity to experience, shame and pride.

The influence of the size and composition of the audience on the experience of shame and pride also shapes behavior in communal cooperative relationships that do not involve collective action, chief among which are common goods contexts. Not involving the difficulties of coordination inherent in collective action, common goods contexts are often even more vulnerable to cheating than communal activities since, in contrast to the latter, individuals often use, and sometimes contribute to, the commons when others are not present (e.g., the cleanliness of departmental microwave ovens deteriorates because these shared resources are accessed individually). Thus, common goods contexts sometimes pose problems of oversight; they share with communal activities the fact that they involve a built-in audience with an interest in information pertaining to individuals' performances. This provides potential leverage over actors in part via shame and pride. In Bengkulu, the names of those who failed to contribute to the maintenance of the village drainage ditches were read over the mosque's loudspeaker, as were the names of those who made contributions to the upkeep of the mosque. This elicited intense shame and pride in the respective individuals, with noticeable consequences for ensuing behavior. More formally, Bowles and Gintis (2002) discuss public goods games with costly punishment in which low contributors increase their contributions in response to punishment despite the fact that doing so is not an optimal response within the payoff structure of the game; the authors infer that low contributors interpret punishment as simultaneously delineating the norm for cooperation and expressing disapproval at its violation, conditions that elicit shame, resulting in increased prosociality.

Moral Outrage and Moral Approbation as Solutions to Common Goods Problems

Publicizing the identities of cheaters and cooperators are forms of active punishment and reward in the cooperative context, categories of behavior that include ostracism and violence, on the one hand, and recruitment and gifts of resources on the other. The utility of such actions in dyadic cooperation is clear, and this behavior is demonstrably important in fostering communal cooperation (Fehr and Gächter 2002). Communal cooperation involves the added difficulty that the actions of punishing cheaters and rewarding cooperators are themselves both costly and a form of common good. As a consequence, actors will be tempted to free ride by letting others pay the costs of punishing and rewarding, sharing in the benefits of enhanced cooperation that result. The solution to such second-order cheating is to institute third-order punishment, that is, to punish individuals who fail to punish individuals who cheat (and, less commonly, to punish individuals who fail to reward individuals who cooperate in an exemplary fashion) (Boyd and Richerson 1992).

Models suggest that systems of third-order punishment are stable once a sizeable number of punishers exist (Boyd and Richerson 1992; Henrich and Boyd 2001). Likewise, experimental results underline the importance of "altruistic punishment" in sustaining cooperation in public goods games (Yamagishi and Sato 1986; Fehr and Gächter 2002). However, the self-reinforcing nature of such systems does not explain the forces needed to create the initial critical mass of individuals inclined to punish second-order cheaters (Boyd and Richerson 1992). We propose that this issue is resolved once it is recognized that enforcing norms serves a communicative function. All else being equal, present behavior is a reasonable predictor of future behavior (cf. Shoda et al. 1990; Eron and Huesmann 1990). Moreover, the more costly the present behavior, the more likely this is to be true, as cheap actions, being easily engaged in, are less revealing of stable underlying dispositions. By incurring costs in policing norm adherence, third-order punishers advertise to the community their support for, and conformity to, shared standards for behavior. Accordingly, active, costly, and highly public pursuit of norm violators, including second-order cheaters, indicates a high likelihood that the actor will herself conform to norms in the future (cf. Smith, this volume, on resource sharing). Because conformity to norms is closely linked to cooperation, advertising one's conformity increases one's attractiveness as a potential cooperative partner. When long-term benefits are taken into consideration, punishing norm violators can thus be seen as a self-interested act performed in pursuit of the benefits of cooperation.[7] In Bengkulu, Dan Fessler witnessed a mob's attempt to "tar and feather" a prostitute. Bystanders commented approvingly on the vociferous outrage expressed by the mob's young male leaders, including outrage directed at other young men who hung back, failing to take an active role in the enterprise. When young men were later recruited for participation in communal ceremonies, the leaders of the mob were prominent in the group selected. As this example illustrates, because the reputational benefits obtained by punishing norm violators are independent of gains reaped directly from the cooperation-enhancing effects of punishment, actors can profit by inflicting costs on individuals who violate any of a wide

[7] Note that our position shares some of the signaling features outlined by Bowles and Gintis (this volume) whereas our account of prosocial behavior as ultimately wholly self-interested stands in contrast to the proposal that some form of group selection has favored a predisposition for "strong reciprocity" (cf. Bowles and Gintis, this volume; Richerson et al., this volume; Fehr et al. 2002; Gintis 2000; Boehm 1997, 2000). While our perspective does not rule out the possibility of group selection, if correct, it may lessen or even obviate the need for it to have occurred.

variety of norms, including norms that pertain to cooperative ventures of which the actor is not herself a part or, more broadly, norms that do not directly pertain to cooperation at all.

Because the benefits entailed by punishing norm violators only accrue over the long-term, issues of short time horizons and steep discounting of the future would often lead actors to forgo punishing norm violators. Similarly, the problem of accurately assessing the likelihood that others would quickly learn of one's actions would lead individuals to often erroneously refrain from punishment, thereby risking becoming the targets of punishment themselves. Yet again, the solution hit upon by natural selection appears to be to employ emotions in shaping propensities and behaviors. Moreover, in this case the answer to the problem was readily at hand, for selection needed only to exapt (i.e., put to a new purpose) an existing emotion. *Moral outrage*, an emotion subjectively indistinguishable from simple anger, is that state which occurs when norm violations are experienced *as if* they were transgressions against the self. It thus appears that, with the coevolution of complex forms of cooperation and shared standards for behavior, selective pressure favored individuals who possessed a motivational system that would lead them to punish norm violators spontaneously; this was achieved by subjectively linking transgressions against norms to transgressions against the self, thus recruiting the pan-mammalian emotion anger to the uniquely human job of advertising one's own norm adherence.

The above account of moral outrage leads to specific predictions regarding contextual and demographic features of the experience of this emotion. First, genuine, spontaneous moral outrage (as opposed to faked versions thereof) is likely to be stronger when a norm violation is witnessed or communicated in front of an audience than when no audience is present. Second, the makeup of the audience should influence the intensity of the moral outrage experienced—moral outrage should be maximal when the audience consists of attractive potential collaborators (e.g., persons having high prestige, valued skills, strong social networks) and/or individuals who constitute an avenue for disseminating information to such potential collaborators, and minimal when the audience consists of individuals who are neither attractive potential collaborators nor an avenue for disseminating information to potential collaborators (e.g., outgroup members). Third, the intensity of moral outrage and the costliness of the ensuing behavior, should be highest in individuals who have the greatest need to advertise their attractiveness as cooperative partners (e.g., young men who are entering the political arena for the first time), and lowest in individuals who have the least need to advertise their attractiveness as cooperative partners (e.g., well-established high-prestige senior men). (Note that this prediction runs counter to a common-sensical assessment of norm-policing, as high-ranking individuals are typically seen as the arbiters of norms, hence one might expect that they are the ones most outraged by norm violations.)

Like the punishing of norm violators, the rewarding of those who fulfill norms in an exemplary fashion simultaneously generates a common good and serves as a means of advertising the individual's own norm adherence. Hence, in part because of the value of inclusion in cooperative relationships, actors benefit by contributing to the costs of supplying both the stick and the carrot used to promote conformity. We therefore hypothesize that, for reasons similar to those obtaining in the case of moral outrage, natural selection has produced an emotion, which we term *moral approbation,* that leads individuals to be positively inclined toward, and seek to reward, virtuous actors who behave in a model fashion. Although in moral outrage the actor experiences another individual's blameworthy norm violation as if it had inflicted a cost on the actor, in moral approbation the actor experiences another individual's praiseworthy norm adherence as if it had provided a benefit for the actor. Accordingly, just as moral outrage exapts anger from the domain of dyadic interactions to the domain of norm compliance, moral approbation exapts gratitude in a similar fashion. The hypothesized benefits to the individual of moral approbation are the same as those proposed for moral outrage. In both cases, because conformists are more reliable than nonconformists, the actor profits from the conspicuous advertisement of his or her endorsement of norms. Accordingly, the same predictions described for moral outrage regarding audiences, demography, and so on also apply to moral approbation. Additionally, because both punishing norm violations and rewarding norm adherence reinforce norms, these actions can be further motivated by the experience of righteousness, the emotion that serves as a proxy for the benefits of conformity. Finally, by increasing both the costs accompanying norm violation and the benefits to be reaped from norm adherence, the presence of actors with a propensity to experience moral outrage, moral approbation, and righteousness increases the attractiveness of norm adherence. In turn, increased norm adherence furthers cooperation, both directly (via conformity to norms concerning reciprocity, equitable divisions, etc.) and indirectly (via the facilitation of coordination across participants).

Admiration and Elevation

Our informal observations suggest that moral approbation overlaps with the emotion *admiration* but is not isomorphic with it. Admiration occurs in contexts in which the admirer lacks some or all of the traits of the admired individual and wishes to acquire them (Henrich and Gil-White 2001). In contrast, the desire to reward the virtuous actor that is central to moral approbation is independent of the individual's own attributes or stature—a war hero may feel moved to praise a Boy Scout who saves a baby from a burning building despite the fact that the former exceeds the latter in bravery, altruism, and so on. As this example suggests, in direct contrast to admiration, in moral approbation superiors may be inclined to bestow benefits on inferiors.

Moral approbation also appears to overlap substantially with the emotion that Haidt (2003a) terms *elevation*. Haidt describes elevation as a positive emotion experienced upon witnessing a good deed. However, whereas the motivational component of moral approbation focuses on rewarding the praiseworthy individual, in elevation the motivational component focuses on carrying out similar deeds oneself. Many of the examples that Haidt collected in Japan, India, and the U.S. revolve around providing a benefit to others. Although Haidt does not offer an account of how such an emotion could have evolved, we believe that this question is amenable to the same form of explanation as that which we applied to moral outrage and moral approbation, namely that a seemingly altruistic act in fact contains a hidden benefit for the actor in the form of advertising the actor's norm adherence, an action which increases the actor's attractiveness as a partner in future cooperative enterprises (see Smith, this volume, on resource sharing; also, cf. Gintis et al. 2001).

To a large extent, assessment of norm adherence is relative—behavior is often judged as praiseworthy or blameworthy through a process of comparison with others' recent actions (hence the common justification offered in defense of rule violations, "everybody does it"). This means that whenever people publicly behave in a praiseworthy fashion, they incrementally reinforce, or even raise, the standard for appropriate action. If praiseworthy actions are an avenue whereby individuals gain access to valuable cooperative opportunities, and if such opportunities are limited, then public praiseworthy behavior on the part of one individual can constitute a threat to others who are competing for the same opportunities. In effect, public praiseworthy behavior throws down a gauntlet, challenging others to live up to the same standard or else lose out in the race for inclusion in the most valuable cooperative ventures. Haidt specifies that elevation motivates individuals to perform praiseworthy, often altruistic acts in the immediate aftermath of witnessing such behavior; however, both the feeling and the motivation fade after a short time. This is exactly what we might expect if elevation is in fact prompting competitively prosocial behavior. Actors ought not be any more prosocial than they need to be in order to secure coveted cooperative opportunities. Accordingly, when others evince extensive prosocial behavior, actors ought to respond in kind, and, conversely, when others limit their prosocial behavior, actors too should scale back their efforts—to avoid either repeatedly overbidding or underbidding in the game of costly norm adherence, actors ought not adhere to a fixed level of prosociality, hence the influence of having observed others' actions should be time-limited. Finally, if this account of elevation is correct, factors similar to those detailed above for moral outrage ought to influence the intensity of the emotion and the costliness of the resulting behavior, that is, (a) the size and composition of an audience present ought to play a role, and (b) the actor's relative need for recruitment into coalitions ought to contribute. Finally, note that there are parallels between our position and good genes costly signaling explanations of prosocial behavior (e.g., Smith and Bliege Bird 2000; also Gintis et al. 2001), because our account focuses not merely on ability but, moreover, on conformism and predictability as well, our position is not susceptible to the criticism that, if show-off altruism signals genetic quality, it ought to occur in many species (Gil-White and Richerson 2002). Likewise, the importance of reputation in the generation of benefits via conformism (see Smith, this volume) explains why show-off altruism does not occur even in those nonhuman species capable of developing rudimentary behavioral traditions (cf. Fragaszy and Perry 2003), since the absence of symbolic communication constrains the reputational benefits of conformism among nonhuman animals.[8]

Mirth

By constituting subjective proxies for the fitness consequences that potentially attend others' assessments of the actor's behavior, shame, pride, moral outrage, moral approbation, and elevation all shape behavior so as to increase the likelihood of inclusion in beneficial coalitions. Once such initial inclusion has been achieved, a second class of emotions come into play, emotions that both index the value of the relationship for the actor and motivate signaling behavior that reinforces the relationship by conveying that valuation. Although there may be a number of such emotions, we are particularly struck by the importance of *mirth.* Building on the work of previous investigators, Flamson (2002) has developed a theory exploring the origins and function of mirth. Noting that, ontogenetically, mirth first appears in response to tickling, Flamson argues that mirth serves to index the fact that safe intimate contact is occurring, that is, the given interaction, though potentially agonistic, is actually affiliative. Laughter, the behavioral expression of mirth, serves to signal that the vulnerable actor trusts the affiliative intent of the other; the recipient of this signal then frequently provides additional affiliative overtures, reconfirming her benign intent.

Mirth subjectively rewards the achievement of affiliation and shapes signaling that reinforces the relationship. In adulthood, these paired functions are frequently evident during the linguistically mediated establishment of dyadic alliances such as friendships and mateships (cf. Grammer 1990). However, it is in the larger social arena that this emotion seems particularly important, as mirth and laughter appear to underlie a considerable amount of solidarity-building in multiperson coalitions. Consistent with the claim that the core event consists of recognizing and signaling solidarity, mirth is often elicited by statements which, from the perspective of the outside observer, lack humorous content (Provine 1993). However, the solidarity-building aspects of mirth can be enhanced through several types of strategic utterances. First, it seems that speakers often employ information that is indexical of in-group membership. Second, it appears that speakers often derogate out-groups. In both cases, mirthful response confirms both the speaker's and the listener's statuses as in-group members,

[8] Although Gintis et al. (2001) make reference to examples of food sharing in nonhuman animals, we are not convinced that these are comparable to human behavior in which the possessor of the resource often ends up with little or none of the shared item.

often prompting additional utterances from other parties, and sometimes resulting in what seems to be an almost orgiastic spiraling of solidarity-building.

It is not difficult to observe the patterns of behavior described above. Somewhat surprisingly, given both its evident frequency and its potential importance, mirth has received relatively little scientific attention to date. Nevertheless, consistent with Flamson's general account, investigations of prejudice suggest that humorous derogation of the out-group plays an important role in the generation and maintenance of discrimination (Terrion and Ashforth 2002), and organizational studies report higher efficiency in work groups that laugh together compared to those that do not (Pollio and Bainum 1983). More formally, participants in a gift-exchange game behaved more altruistically after viewing a humorous movie clip (Kirchsteiger et al. 2001), and negotiators in a bargaining simulation behaved more cooperatively after examining humorous cartoons (Carnevale and Isen 1986), patterns that are consistent with the premise that, because it normally indexes the existence of a cooperative relationship, mirth motivates prosocial behavior.[9]

In addition to the need to investigate systematically the influence of mirth and laughter on cooperation, many intriguing questions remain unexplored, including the possibility that humorous individuals provide a public good by catalyzing solidarity enhancement, a cost which they might recoup via a number of avenues, including (a) thereby increasing their attractiveness as a coalition member, (b) attracting admiring clients (cf. Henrich and Gil-White 2001), or (c) signaling their high mate value (cf. Miller 2000). Alternately, coalition members might compensate humorous individuals for their services by accepting lower contributions of other currencies, a pattern which, if it occurs, raises questions as to the management of second-order free riding.

Corporate Emotions and Cooperation

Throughout the above discussions we have repeatedly emphasized that cooperation is facilitated in part through the promotion and maintenance of particular forms of social relationships. Emotions play critical roles in these processes both by promoting prosocial behavior and by raising the costs of antisocial behavior. However, in addition to these functions, emotions can play a key role in cooperation by virtue of the fact that they can be experienced in a *corporate fashion.* By this we mean that anger, shame, pride, gratitude, and so on can be elicited by actions that affect some part of a group in which the actor is a member, even though the actor was not directly involved in the interaction (cf.

[9] Both Kirchsteiger et al. (2001) and Carnevale and Isen (1986) suppose that behavior is a function of moods. This premise led both sets of investigators to employ manipulations (respectively, viewing a clip from Schindler's List, intended to evoke negative mood; giving a free notepad, intended to evoke positive mood), the influence of which on discrete emotions is unclear.

Richerson et al., this volume). For example, intervillage violence occurs not infrequently in Bengkulu, most commonly when a young man from one village insults someone from another village — the action is experienced as a transgression by all members of the second village, leading to widespread anger and calls for retribution. We believe that the experience of corporate emotions is a consequence of the interaction between those mental mechanisms responsible for producing the various emotions and a separate mental mechanism, one which defines the boundaries of the individual as an interest party. Close kin, buddies in an army squad, residents of the same village, or occupants of a single lifeboat — in each case the interests of the individual are often aligned with the interests of the group. Apparently as a consequence of the recognition of this alignment, the emotions that normally respond to directly experienced interindividual behavior can come to respond to any information that pertains to the fate of the larger interest party.

The ability to experience corporate emotions interdigitates with many of the aspects of cooperative behavior discussed thus far. For example, experiencing anger at transgressions committed against a village mate, and incurring costs as a consequence in order to harm the transgressor, actively demonstrates to other members of the community that the actor aligns his interests with theirs. Likewise, such actions show that the actor adheres to norms such as those dictating community solidarity, mutual defense, and so on. As a result, individuals who experience corporate anger and advertise and act on that experience constitute attractive partners for future cooperative ventures. We believe that this explains why people go far out of their way not only to advertise their affiliations to various groups, but also to demonstrably express corporate emotions — we have seen fans of a winning sports team leave their television sets, rush out of their homes, and frantically search the streets for anyone with whom they can express pride in the (spuriously corporate, in this case) group's achievements. In short, corporate emotions not only function to promote the individual's interests by leading actors to act in the group's interests, they also function to promote the individual's interests by shaping relationships with fellow group members.

THE IMPACT OF CULTURE ON THE ROLE OF EMOTIONS IN COOPERATION

The identity and boundaries of interest groups are often culturally defined.[10] This is simply one of the innumerable ways in which culture influences

[10] We suspect that some of the cooperative behavior observed in economics experiments reflects the Western liberal cultural concept that all humans are members of a single social category and are thus deserving of equal treatment. In stark contrast to this idea, most cultural value systems are premised on the existence of concentric social worlds, with out-group members not being seen as deserving of the same consideration as in-group members (Johnson and Earle 2000, p. 25).

cooperative behavior, notably including the delineation of the appropriate levels of generosity, reactivity to transgression, participation in communal ventures, and so on (cf. Henrich et al. 2001). Of particular relevance for the present discussion is the influence of culture on the subjective salience, motivational power, and moral valence of emotions. Cultures differentially elaborate on or ignore various emotions. Whereas the absence of lexical labels for cultural schemas about, and socialization practices concerning a given emotion does not preclude the ability to experience that emotion, these conditions do reduce the salience of the emotion, the extent to which it shapes behavior, and the frequency with which it is elicited (Levy 1973). Conversely, elaborate cultural marking of an emotion can greatly enhance its subjective salience; how much the emotion shapes behavior, and how frequently it is experienced, are in turn partially a function of the moral valence assigned to the emotion in the relevant cultural schemas (Levy 1973; Briggs 1970; Fessler 2003). Together, these observations indicate that, even for those many emotions which, being the product of our shared phylogeny, are panhuman, the influence of any given emotion on cooperative behavior can nevertheless be expected to differ substantially across cultures. Investigators interested in exploring the role of emotions in cooperation must therefore attend closely to the cultural backgrounds of participants in a given venture. Perhaps even more important, diplomats must be attuned to the salience and valence of particular emotions in specific cultures if they are to mediate international conflicts effectively and foster large-scale cooperation.

DISCUSSION

Humans often behave in ways that contradict predictions derived from economists' traditional rational actor models of behavior. The principal weakness of such models is their failure to recognize fully both the proximate and the ultimate determinants of utility. Once it is understood that (a) emotions change the subjective importance of costs and benefits and (b) actors take account of the influence of emotions on others' behavior, then many observable strategies are "rational" in the sense that they serve to maximize subjective utility. Whereas recent efforts by economists to capture the constituents of subjective utility more accurately using the concept of "social preferences" (cf. Fehr et al. 2002) are an important step forward, we believe that if investigators are to understand human cooperation fully, they must not be satisfied merely with characterizations of the proximate determinants of subjective utility but rather must also investigate the ultimate factors responsible for those determinants. Idiosyncratic factors unquestionably play a role in subjective experience; however, we are impressed by the underlying similarities evident across diverse individuals and disparate cultures, similarities best explained using an evolutionary perspective.

Natural selection produces mechanisms that shape behavior in the service of maximizing a single ultimate utility, biological fitness. Importantly, such mechanisms are reliant upon the presence of features of the environment that reliably occurred over the course of the mechanism's evolution (Tooby and Cosmides 1992). To the extent that it is shaped by evolved mechanisms, contemporary behavior can thus be expected to reflect strategies that would have been biologically rational under ancestral conditions. It is highly plausible that our ancestral social environment, the world in which humans are "designed" to operate, consisted of relatively stable small-scale acephalous social groups in which cooperation generated critical benefits (Boehm 2000). Natural selection has thus produced a suite of emotions which, when operating in such a setting, effectively mitigate both the temptation of short-term defection and the danger that others will be similarly tempted. Likewise, because of the importance of conformity to shared standards for behavior in human cooperation, selection has crafted emotions that enhance both norm adherence and the punishing of nonconformity. Complementing these prosocial emotions, natural selection has also produced emotions such as envy and contempt, which maximize individual benefit extraction; in turn, awareness of these emotions in others elicits compensatory responses.

The above perspective suggests that although contemporary actors may behave in ways that maximize their subjective utilities, *the more that a contemporary setting deviates from the ancestral environment, the less likely it is that such actions will be rational from a biological perspective* (Tooby and Cosmides 1992). Consider first the case of "road rage" on Los Angeles freeways. Providing redundant stereotypical exemplars of the dynamics of anger, drivers who suffer transgressions are often furious at, and agress against, their transgressors (cf. Katz 1999, pp. 18–83). We have argued that the capacity to experience anger evolved because of the benefits of deterring future transgressions, benefits that only accrue when interactions are iterated. All but the most dim-witted Los Angeles drivers surely understand that they are unlikely to ever interact again with either the targets of their anger or the other drivers who observe their actions, yet conscious awareness that freeways are populated by anonymous hordes does not suffice to preclude aggressive response to transgression. This is presumably because the autonomous mental mechanisms at issue are for some reason mistaking the ephemeral social world of the freeway for the stable social world of the village.

The proximate causes of this type of (ultimately erroneous) elicitation are clearly evident in a second case, the intense emotions exhibited by modern sports fans. A few simple symbols and a short period of spectating often evoke active, at times violent, solidarity in sports fans. We suggest that this occurs not because the fans are so foolish as to think that they themselves will be the beneficiaries of the many material and social rewards awarded to a winning team, but rather because culturally evolved cues (cf. Richerson et al., this volume) elicit corporate pride and corporate shame in spite of the fans' overt knowledge that they are not really players in the game. The fact that an overweight nearsighted middle-aged man with a heart condition can joyously scream "We won!" sitting on a couch watching world-class athletes on television is understandable once it is recognized that discrete cues of affiliation (banners, replicas of team jerseys, and the identification of a team with a particular locale) activate autonomous psychological mechanisms responsible for producing corporate pride, mechanisms that evolved in an environment within which both local affiliation and overt symbolic markers committed individuals to in-group coalitions, a world in which there were substantive rewards for advertising one's membership in cooperative ventures (McElreath et al. 2003; Richerson et al., this volume; Bowles and Gintis, this volume).

Note that, in contrast to broader descriptions of social preferences, an evolutionary approach directs the investigator's attention to the specific features of the environment that activate the particular emotions observed. We might predict, for example, that, on a per mile basis, road rage will be experienced more frequently when following one's daily commute than when navigating unfamiliar freeways, since it is plausible that familiarity with the setting is one criterion used by evolved mechanisms in evaluating the likelihood of iterated interactions. Similarly, if, as we suspect, sports teams evoke corporate emotions using cues detected by mechanisms designed to operate in a world of local coalitions, there should be considerable resistance to changing the name or mascot of a team, teams should suffer a drop in popularity whenever they relocate, and fans should be particularly devoted to athletes who refuse lucrative offers to join other teams.

Fehr and Henrich (this volume) question the notion that contemporary humans' propensity to incur costs to punish noncooperators and reward cooperators is a maladaptation. While the absence of citations makes it unclear with whom they are arguing, their extensive treatment of emotions and frequent mention of "evolutionary psychologists" suggests that they presume this to be our position. However, we in no way suppose that so-called strong reciprocity is intrinsically maladaptive, nor do we claim that people are unable to distinguish friend from stranger or are insensitive to factors affecting the opportunity for reputation formation (indeed, we make explicit predictions to the contrary).

As Fehr and Henrich note, there is substantial experimental evidence that pure strong reciprocity (i.e., devoid of reputational gain) can be elicited in diverse cultures. Importantly, Henrich and colleagues' work on the subject (Henrich et al. 2001) reveals considerable cultural variation in many aspects of such behavior. This implies that cultures differentially exploit the underlying emotional architecture that makes strong reciprocity possible. As accounts of road rage suggest, emotions can be elicited when novel environments present features that resemble those found in the ancestral past, yet differ in their informational value regarding the costs and benefits of various actions.[11] Professional sports exemplifies the manner in which cultures take advantage of such deviations, creating work-arounds (Richerson and Boyd 1999; Richerson et al., this volume) that rely on emotions. Many cultural work-arounds promote cooperative bahavior. In some instances, such behavior is individually maladaptive, as when "heroes" die in battle upholding the symbols and social structures of enormous modern armies (Richerson and Boyd 1999).[12]

Participants in economics experiments display a range of behaviors, including both those that are rational within the confines of the game and those that are not, but would be under other conditions (Fehr et al. 2002; Fehr and Gächter 2000). The perspective developed here suggests that, while conscious calculations often alter behavior in response to changes in game parameters, a sizeable portion of inter-individual and inter-cultural variation may nonetheless be explicable in terms of differences in participants' emotional experiences, and, relatedly, their interpretations of the meaning of the game behavior within their culturally constituted systems of values and beliefs.

CONCLUSION

We believe that the long-standing Western tradition of viewing emotions as interfering with rational decision making stems from a twofold error: (a) a failure to recognize the nature of the currency (biological fitness) that psychological mechanisms are intended to maximize, and (b) a failure to recognize the consequences of evolutionary disequilibrium, the disjunction between many contemporary circumstances and the environment in which our species evolved. When viewed in the context for which they were designed, our emotions, long disparaged as both a reflection of our animality and the source of our irrationality, are thus exactly the opposite, namely, the keys to our complexity, efficacy, and remarkable ability to cooperate.

---

[11] On the basis of speculation and results from relatively low-stakes laboratory experiments, Fehr and Henrich claim that the cognitive impenetrability of emotions (see discussion in McElreath et al., this volume) has been overstated. However, consider the television program *Fear Factor* in which, for a prize of $50,000, six competitors are challenged to engage in stunts such as high-wire acrobatics or eating bovine recta. Despite knowing that they are secured by a safety rope, that the meat has been cooked, etc., visibly shaken participants are often unable to even attempt a stunt. This is understandable once it is recognized that emotions evolved in a world in which cues (e.g., visual cliff, fecal contact, etc.) reliably indicated the potential costs or benefits of actions — overt knowledge to the contrary may fail to alter the behavior of even highly motivated individuals.

[12] Because culture can evolve much more rapidly than can human psychology, the existence of group-functional, individual-dysfunctional institutions does not demand that genetic group selection has taken place.

ACKNOWLEDGMENTS

We thank Robert Kurzban, H. Clark Barrett, Alan P. Fiske, Jack Hirshleifer, Timothy Ketelaar, Jennifer Fessler, and our fellow Dahlem participants, particularly Ed Hagen and Richard McElreath, for valuable discussions and feedback.

REFERENCES

Alexander, R.D. 1987. The Biology of Moral Systems. Hawthorne, NY: de Gruyter.

Andreoni, J. 1995. Cooperation in public-goods experiments: Kindness or confusion. *Am. Econ. Rev.* **85**:891–904.

Baumeister, R.F., A.M. Stillwell, and T.F. Heatherton. 1994. Guilt: An interpersonal approach. *Psych. Bull.* **115**:243–267.

Berg, J., J. Dickhaut, and K. McCabe. 1995. Trust, reciprocity, and social norms. *Games Econ. Behav.* **10**:122–142.

Biehl, M., D. Matsumoto, P. Ekman, and V. Hearn. 1997. Matsumoto and Ekman's Japanese and Caucasian Facial Expressions of Emotion (JACFEE): Reliability data and cross-national differences. *J. Nonverb. Behav.* **21**:3–21.

Blurton Jones, N.G. 1987. Tolerated theft: Suggestions about the ecology and evolution of sharing, hoarding, and scrounging. *Soc. Sci. Info.* **26**:31–54.

Boehm, C. 1997. Impact of the human egalitarian syndrome on Darwinian selection mechanics. *Am. Nat.* **150**:100–121.

Boehm, C. 2000. Conflict and the evolution of social control. *J. Consc. St.* **7**:79–101.

Bosman, R., and F. van Winden. 2002. Emotional hazard in a power-to-take game experiment. *Econ. J.* **112**:147–169.

Bowles, S., and H. Gintis. 2002. Prosocial emotions. Santa Fe Institute Working Paper 02-07-028.

Boyd, R., and P.J. Richerson. 1992. Punishment allows the evolution of cooperation (or anything else) in sizable groups. *Ethol. Sociobiol.* **13**:171–195.

Briggs, J.L. 1970. Never in Anger: Portrait of an Eskimo Family. Cambridge, MA: Harvard Univ. Press.

Carnevale, P.J., and A.M. Isen. 1986. The influence of positive affect and visual access on the discovery of integrative solutions in bilateral negotiation. *Org. Behav. Hum. Dec. Proc.* **37**:1–13.

Cooter, R., and M.A. Eisenberg. 2001. Fairness, character, and efficiency in firms. *Univ. Penn. Law Rev.* **149**:1717–1733.

Cosmides, L., and J. Tooby. 2000. Evolutionary psychology and the emotions. In: Handbook of Emotions, 2d ed., ed. M. Lewis and J.M. Haviland-Jones, pp. 91–115. New York: Guilford.

Daly, M., and M. Wilson. 1988. Homicide. New York: Aldine de Gruyter.

Darwin, C. 1872. The Expression of the Emotions in Man and Animals. London: Murray.

Dugatkin, L.A. 1990. N-person games and the evolution of cooperation: A model based on predator inspection in fish. *J. Theor. Biol.* **142**:123–136.

Dundes, A., ed. 1992. The Evil Eye: A Casebook. Madison: Univ. of Wisconsin Press.

Edwards, D.C. 1999. Motivation and Emotion: Evolutionary, Physiological, Cognitive, and Social Influences. Thousand Oaks, CA: Sage.

Ekman, P. 1992. An argument for basic emotions. *Cog. Emot.* **6**:169–200.

Ekman, P., and W.V. Friesen. 1971. Constants across cultures in the face and emotion. *J. Pers. Soc. Psych.* **17**:124–129.

Elster, J. 1998. Emotions and economic theory. *J. Econ. Lit.* **36**:47–74.

Eron, L.D., and L.R. Huesmann. 1990. The stability of aggressive behavior: Even unto the third generation. In: Handbook of Developmental Psychopathology: Perspectives in Developmental Psychology, ed. M. Lewis and S.M. Miller, pp. 147–156. New York: Plenum.

Fehr, E., U. Fischbacher, and S. Gächter. 2002. Strong reciprocity, human cooperation, and the enforcement of social norms. *Hum. Nat.* **13**:1–25.

Fehr, E., and S. Gächter. 2000. Cooperation and punishment in public goods experiments. *Am. Econ. Rev.* **90**:980–994.

Fehr, E., and S. Gächter. 2002. Altruistic punishment in humans. *Nature* **415**:137–140.

Fessler, D.M.T. 1999. Toward an understanding of the universality of second order emotions. In: Biocultural Approaches to the Emotions: Publications of the Society for Psychological Anthropology, ed. A.L. Hinton, pp. 75–116. New York: Cambridge Univ. Press.

Fessler, D.M.T. 2003. The male flash of anger: Violent response to transgression as an example of the intersection of evolved psychology and culture. In: Missing the Revolution: Darwinism for Social Scientists, ed. J.H. Barkow. New York: Oxford Univ. Press, in press.

Fiske, A.P. 2003. Moral emotions provide the self-control needed to sustain social relationships. *Self and Emotion*, in press.

Flamson, T.J. 2002. The Evolution of Humor and Laughter. M.A thesis, Dept. of Anthropology, Univ. of California, Los Angeles.

Fontaine, J.R.J., Y.H. Poortinga, B. Setiadi, and S. Markam. 2002. Cognitive structure of emotion terms in Indonesia and The Netherlands. *Cog. Emot.* **16**:61–86.

Fragaszy, D., and S. Perry, eds. 2003. The Biology of Traditions: Models and Evidence. New York: Cambridge Univ. Press, in press.

Frank, R.H. 1988. Passions within Reason: The Strategic Role of the Emotions. New York: Norton.

Frank, R.H. 2001. Cooperation through emotional commitment. In: Evolution and the Capacity for Commitment, ed. R.M. Nesse, pp. 57–76. New York: Russell Sage.

Gilbert, P. 1997. The evolution of social attractiveness and its role in shame, humiliation, guilt and therapy. *Brit. J. Med. Psych.* **70**:113–147.

Gilbert, P., J. Pehl, and S. Allan. 1994. The phenomenology of shame and guilt: An empirical investigation. *Brit. J. Med. Psych.* **67**:23–36.

Gil-White, F.J., and P.J. Richerson. 2002. Large-scale human cooperation and conflict. In: Encyclopedia of Cognitive Science. London: MacMillan.

Gintis, H. 2000. Strong reciprocity and human sociality. *J. Theor. Biol.* **206**:169–179.

Gintis, H., E.A. Smith, and S. Bowles. 2001. Costly signaling and cooperation. *J. Theor. Biol.* **213**:103–119.

Grammer, K. 1990. Strangers meet: Laughter and nonverbal signs of interest in opposite-sex encounters. *J. Nonverb. Behav.* **14**:209–236.

Haidt, J. 2003a. Elevation and the positive psychology of morality. In: Flourishing: Positive Psychology and the Life Well-lived, ed. C.L.M. Keyes and J. Haidt, pp. 275–289. Washington, D.C.: American Psychological Assn.

Haidt, J. 2003b. Moral emotions. In: Handbook of Affective Sciences, ed. R.J. Davidson, K.R. Scherer, and H.H. Goldsmith, pp. 852–870. Oxford: Oxford Univ. Press.

Harris, H. 1995. Human Nature and the Nature of Romantic Love. Ph.D. diss., Univ. of California, Santa Barbara.

Heider, K.G. 1991. Landscapes of Emotion: Mapping Three Cultures of Emotion in Indonesia. New York: Cambridge Univ. Press.

Henrich, J., and R. Boyd. 2001. Why people punish defectors: Weak conformist transmission can stabilize costly enforcement of norms in cooperative dilemmas. *J. Theor. Biol.* **208**:79–89.

Henrich, J., R. Boyd, S. Bowles et al. 2001. Cooperation, reciprocity and punishment in fifteen small-scale societies. *Am. Econ. Rev.* **91**:73–78.

Henrich, J., and F.J. Gil-White. 2001. The evolution of prestige: Freely conferred status as a mechanism for enhancing the benefits of cultural transmission. *Evol. Hum. Behav.* **22**:165–196.

Hirshleifer, J. 1987. On the emotions as guarantors of threats and promises. In: The Latest on the Best: Essays on Evolution and Optimality, ed. J. Dupré, pp. 307–326. Cambridge, MA: MIT Press.

Hoffman, M.L. 1982. Affect and moral development. *New Dir. Child Dev.* **16**:83–103.

Izard, C.E. 1977. Human Emotions. New York: Plenum.

Jankowiak, W.R., and E.F. Fischer. 1998. A cross-cultural perspective on romantic love. In: Human Emotions: A Reader, ed. J.M. Jenkins, K. Oatley, and N. Stein, pp. 55–62. Malden, MA: Blackwell.

Johnson, A.W., and T. Earle. 2000. The Evolution of Human Societies: From Foraging Group to Agrarian State, 2d ed. Stanford, CA: Stanford Univ. Press.

Katz, J. 1999. How Emotions Work. Chicago: Univ. of Chicago Press.

Keltner, D., and B.N. Buswell. 1996. Evidence for the distinctness of embarrassment, shame, and guilt: A study of recalled antecedents and facial expressions of emotion. *Cog. Emot.* **10**:155–171.

Ketelaar, T., and W. Tung Au. 2003. The effects of feeling guilt on the behaviour of uncooperative individuals in repeated social bargaining games: An affect-as information interpretation of the role of emotion in social interaction. *Cog. Emot.*, in press.

Kirchsteiger, G., L. Rigotti, and A. Rustichini. 2001. Your morals are your moods. Working Paper E01-294. Economics Dept., Univ. of California, Berkeley. http://repositories.cdlib.org/iber/econ/E01-294.

Lazarus, R.S. 1991. Emotion and Adaptation. New York: Oxford Univ. Press.

Lerner, J., and D. Keltner. 2001. Fear, anger, and risk. *J. Pers. Soc. Psych., Spec. Iss.* **81(1)**:146–159.

Levy, R.I. 1973. Tahitians: Mind and Experience in the Society Islands. Chicago: Univ. of Chicago Press.

McElreath, R., R. Boyd, and P.J. Richerson. 2003. Shared norms and the evolution of ethnic markers. *Curr. Anthro.* **44**:122–129.

McGuire, M.T., and A. Troisi. 1990. Anger: An evolutionary view. In: Emotion, Psychopathology, and Psychotherapy, ed. R. Plutchik and H. Kellerman, pp. 43–57. San Diego: Academic.

Miller, G.F. 2000. The Mating Mind: How Sexual Choice Shaped the Evolution of Human Nature. New York: Doubleday.

Nesse, R.M. 1990. Evolutionary explanations of emotions. *Hum. Nat.* **1**:261–289.

Pillutla, M.M., and J.K. Murnighan. 1996. Unfairness, anger, and spite: Emotional rejections of ultimatum offers. *Org. Behav. Hum. Dec. Proc.* **68**:208–224.

Pollio, H.R., and C.K. Bainum. 1983. Are funny groups good at solving problems? A methodological evaluation and some preliminary results. *Small Group Behav.* **14**: 379–404.

Prasnikar, V., and A.E. Roth. 1992. Considerations of fairness and strategy: Experimental data from sequential games. *Qtly. J. Econ.* **107**:865–888.

Provine, R.R. 1993. Laughter punctuates speech: Linguistic, social and gender contexts of laughter. *Ethology* **95**:291–298.

Richerson, P.J., and R. Boyd. 1999. The evolutionary dynamics of a crude superorganism. *Hum. Nat.* **10**:253–289.

Roth, A.E. 1995. Bargaining experiments. In: Handbook of Experimental Economics, ed. J.H. Kagel and A.E. Roth, pp. 253–248. Princeton, NJ: Princeton Univ. Press.

Rozin, P., L. Lowery, S. Imada, and J. Haidt. 1999. The CAD triad hypothesis: A mapping between three moral emotions (contempt, anger, disgust) and three moral codes (community, autonomy, divinity). *J. Pers. Soc. Psych.* **76**:574–586.

Russell, J.A. 1991. Culture and the categorization of emotions. *Psych. Bull.* **110**: 426–450.

Scherer, K.R., and H.G. Wallbott. 1994. Evidence for universality and cultural variation of differential emotion response patterning. *J. Pers. Soc. Psych.* **66**:310–328.

Schoeck, H. 1969. Envy: A theory of social behaviour (trans. M. Glenny and B. Ross). New York: Harcourt Brace and World.

Shoda, Y., W. Mischel, and P.K. Peake. 1990. Predicting adolescent cognitive and self-regulatory competencies from preschool delay of gratification: Identifying diagnostic conditions. *Dev. Psych.* **26**:978–986.

Shostak, M. 1981. Nisa, the Life and Words of a !Kung woman. Cambridge, MA: Harvard Univ. Press.

Smith, E.A., and R.L. Bliege Bird. 2000. Turtle hunting and tombstone opening: Public generosity as costly signaling. *Evol. Hum. Behav.* **21**:245–261.

Tangney, J.P. 1998. How does guilt differ from shame? In: Guilt and Children, ed. J. Bybee, pp. 1–17. San Diego: Academic.

Tennov, D. 1998. Love madness. In: Romantic Love and Sexual Behavior: Perspectives from the Social Sciences, ed. V.C. de Munck, pp.77–88. Westport, CT: Praeger.

Terrion, J.L., and B.E. Ashforth. 2002. From "I" to "we": The role of putdown humor and identity in the development of a temporary group. *Hum. Rel.* **55**:55–88.

Tooby, J., and L. Cosmides. 1992. The psychological foundations of culture. In: The Adapted Mind: Evolutionary Psychology and the Generation of Culture, ed. J.H. Barkow, L. Cosmides, and J. Tooby, pp. 19–136. New York: Oxford Univ. Press.

Trivers, R.L. 1971. The evolution of reciprocal altruism. *Qtly. Rev. Biol.* **46**:35–57.

Wallbott, H.G. 1998. Bodily expression of emotion. *Eur. J. Soc. Psych.* **28**:879–896.

Yamagishi, T., and K. Sato. 1986. Motivational bases of the public-goods problem. *J. Pers. Soc. Psych.* **50**:67–73.

Zizzo, D.J., and A. Oswald. 2001. Are people willing to pay to reduce others' incomes? *Annales d'Economie et de Statistique* **63–64**:39–65.

3

Cooperation without Counting

The Puzzle of Friendship

Joan B. Silk

Department of Anthropology, University of California,
Los Angeles, CA 90095, U.S.A.

ABSTRACT

Cooperative relationships, which involve the exchange of altruistic behaviors that are costly to the actor and beneficial to the recipient, are thought to be the product of kin selection or reciprocal altruism. Humans form close, enduring, cooperative relationships with nonrelatives. In these relationships, which we call friendships, both emotional and material support are exchanged. If these relationships are shaped by the adaptive logic of Tit-For-Tat reciprocal altruism, then we would expect people to keep track of benefits given to and received from friends, and for there to be contingencies between favors given now and favors received in the past. However, the social science literature suggests that Tit-for-Tat reciprocity is characteristic of relationships among casual acquaintances and strangers, not among friends. A considerable body of empirical work indicates that people value balanced reciprocity in their relationships with friends, but avoid keeping careful count of benefits given and received, and are offended when friends reciprocate immediately and directly. Thus, the dynamic of friendship does not fit the logic of models of reciprocity and presents a puzzle for evolutionary analysis.

INTRODUCTION

Friendship is a common, perhaps universal, feature of human societies. One of the defining features of friendship is that it involves the exchange of costly favors and services, including both material help and emotional support. Evolutionary theory predicts that altruistic interactions will be shaped by kin selection or reciprocal altruism. Since costly help is often extended to nonrelatives, and does not benefit the actor directly, evolutionary theory predicts that friendship will conform to the logic of reciprocity. The social science literature indicates that reciprocity and equity are important among friends, but Tit-for-Tat reciprocity is antithetical to the formation and maintenance of close friendship. If these seemingly contradictory claims are correct, then friendship presents a puzzle for evolutionary analysis. The goal of this chapter is to lay out the pieces of this puzzle and try to see how they fit together.

I begin by considering the phylogenetic history of cooperative relationships in the primate order. This is an important place to begin because it is possible that friendship is a derived feature of human societies, one that appears after humans diverged from their last common ancestor with other primates five to ten million years ago. If so, then the evolution of friendship may be linked to emergent features of human societies which produced the capacity for collective action, strong norms of fairness, a willingness to inflict costly punishment on strangers, and other forms of highly cooperative behavior (Richerson and Boyd 1998; Fehr and Gächter 2001). Primatologists, however, have recently begun to use the term *friendship* to describe affiliative social bonds among nonhuman primates. If nonhuman primates (or other animals) form relationships that embody the essential features of human friendships, then these relationships may be ancestral traits that evolved before the other highly cooperative features of modern human societies emerged. Thus, it is important to examine the mechanisms that underlie cooperation in nonhuman primates and to consider the phylogenetic roots of friendship in the primate order.

Next, I examine empirical evidence about reciprocity in relationships with friends and strangers. There is a broad consensus in the social science literature that short-term, Tit-for-Tat reciprocity is not a feature of close friendships, but concerns about equity and reciprocity are nonetheless important among friends. These seemingly contradictory claims are supported by empirical studies that demonstrate that people tend to obscure contributions to joint tasks completed with friends, but not strangers, but are disturbed about inequities in their relationships with others. Despite this evidence, most evolutionary analyses of friendship in humans assume that friendship evolves through Tit-for-Tat reciprocal altruism. If the empirical claims made by social scientists are correct, then evolutionary explanations based on reciprocal altruism need to be amended.

RECIPROCITY IN COOPERATIVE RELATIONSHIPS IN PRIMATES

In nonhuman primates, as in other animals, evidence for reciprocal altruism is much more limited than evidence for kin selection (Dugatkin 1997; Hammerstein, Chapter 5, this volume). This is somewhat surprising because primates are good candidates for reciprocal altruism. All monkeys and apes, except for orangutans, live in stable social groups of known individuals and have many opportunities to interact. They have good memories and are able to solve complex social problems. For example, they keep track of their own kinship, dominance, and affiliative relationships with other group members, and know something about the nature of kinship, dominance, and affiliative relationships among others (Tomasello and Call 1997).

A number of naturalistic studies document exchanges of altruistic behaviors within pairs of individuals and measure the statistical significance of the associations between behaviors initiated and received. In many of these studies, positive correlations between various types of friendly behaviors, such as grooming and proximity, can be detected.

An example of this kind of work that seems relevant to the notion of friendship comes from recent work on chimpanzees at Ngogo, in the Kibale Forest of Uganda. Male chimpanzees form close and well-differentiated social relationships. These kinds of relationships are uncommon among nonhuman primate males. This is probably related to the fact that males in most species are the dispersing sex and consequently live in groups composed mainly of nonkin. In addition, males compete with one another for resources that cannot be shared equitably, namely receptive females. This limits the potential benefits derived from cooperation among males, and relationships among adult males typically range from indifferent to hostile. In chimpanzees, however, males are the philopatric sex and males form close ties with other males. Chimpanzee males groom one another, hunt in groups, share meat with other males, support one another in conflicts, jointly patrol the borders of their territories, participate in hostile intergroup encounters, and guard access to receptive females. Careful analyses of the patterning of these activities at Ngogo indicate that males groom, share meat, and support one another reciprocally (reviewed in Watts 2002). Males apparently exchange grooming for support. Moreover, males tend to hunt with the same males that they groom, support in conflicts, and accompany on border patrols. Present data (Mitani et al. 2002) suggests that males do not associate preferentially with their maternal kin. These data, and data from other chimpanzee communities, suggest that reciprocity plays an important part in the lives of chimpanzees.

However, even the most comprehensive correlational studies provide an unsatisfying foundation for studying reciprocity for several reasons. First, it is notoriously difficult to draw causal deductions from correlational data. In this case, it is important to make sure that correlations between one form of cooperation and another are not the product of third variable, such as kinship or dominance rank. Second, correlational analyses do not address the mechanisms underlying behavioral exchanges, although reciprocal altruism relies on the ability of animals to detect defection and terminate relationships when partners cheat. Third, correlational studies do not account for the possibility that different processes may shape interactions in different dyads. Females might unilaterally support their offspring, trade grooming for support from males, and balance grooming with nonrelatives of adjacent rank. Fourth, it is very difficult to specify the relevant behavioral and temporal domains in which exchanges might take place.

Better evidence for contingent exchanges comes from detailed studies of turn taking during grooming bouts. In some cases, one monkey grooms its partner for a short period, then they switch roles (Barrett and Henzi 2001; Cords 2002). Not all grooming bouts involve turn taking, and there is no evidence that primates "raise the stakes" by extending the duration of grooming in each successive round (Barrett et al. 2000). Nonetheless, these data suggest that grooming is parcelled into short, low-cost units and exchanged on a contingent basis.

Several experimental studies provide further evidence that nonhuman primates adopt contingent strategies in the deployment of altruism to nonrelatives. Using tape-recorded vocalizations of females' screams, which signal distress and are often used to recruit support, Seyfarth and Cheney (1984) showed that free-ranging vervet females were more attentive to screams of other unrelated group members if they had been groomed by the screamer shortly before they heard the scream than if they had not been groomed by the same individual. This experiment demonstrates that monkeys' responses are contingent on prior interactions, a key component of the tactics of reciprocal altruism. However, because the conflicts were simulated, there was no opportunity for monkeys to intervene, leaving some doubt about the meaning of their responses. This shortcoming was remedied in a study conducted on captive long-tailed macaques by Hemelrijk (1994). She artificially induced fights among familiar, unrelated macaques housed temporarily in groups of three. When fights between two females occurred, aggressors sometimes received support from the third member of the trio. Supporters were more likely to intervene on behalf of females who had previously groomed them.

These experimental studies must be weighed against naturalistic studies of the association between grooming and support among nonrelatives. Schino (2001) has found consistent support for a number predictions about the distribution of grooming derived from Seyfarth's hypothesis; however, evidence of direct associations between grooming and support among nonkin is quite limited. Schino (2001) suggests that it may not be possible to find statistically significant correlations between grooming and support because alliances are rare, whereas Henzi and Barrett (1999) interpret the absence of such correlations as evidence that monkeys do not exchange grooming for support.

Chimpanzees sometimes share plant foods and meat and use specialized "begging" gestures to solicit food from others. In a group of captive chimpanzees, de Waal (1997a) assessed the relationship between grooming and subsequent food sharing. He and his colleagues observed chimpanzees for several hours before and after they were provisioned with leafy branches. He found that the chimpanzees were more likely to share with individuals who had previously groomed them than with individuals who had not groomed them in the past few hours. Moreover, if there had been no grooming before provisioning, possessors were more likely to respond aggressively to efforts to share. Interestingly, the magnitude of the effect of prior grooming was influenced by the nature of the relationship between the two individuals: for pairs that rarely groomed, sharing was strongly contingent on recent grooming, whereas for pairs that groomed at higher rates, recent grooming had a smaller impact on sharing.

The dynamics of food sharing in capuchin monkeys has also been studied by de Waal in the laboratory. In this setting, capuchins are strongly motivated to sit close together and are very sloppy eaters. When they are given food, they frequently carry the food back toward other group members and allow them to take pieces of food that have dropped to the floor of the cage. De Waal (1997b) took advantage of the capuchins' tolerance to examine the patterning of food exchanges within dyads. In one set of experiments, a pair of familiar monkeys were placed in adjacent cages separated by wire mesh. The holes in the mesh were large enough to allow the monkeys to reach into the adjacent cage and take food items. In the first phase of the experiment, one member of the dyad was given food and all exchanges of food were monitored. In the second phase of the experiment, the other monkey was given food and exchanges were monitored again. In this experimental situation, the vast majority of exchanges occurred when one monkey reached through the mesh and helped itself to scraps of food dropped by the owner; owners tolerated these initiatives but did not actively donate food to their partners. Among females, the number of transfers from the owner to her partner in the first phase of the experiment was correlated with the rate of transfer when their roles were reversed in the second phase of the experiment. Dyads that tended to associate frequently and fight little had higher transfer rates than dyads that associated less often and fought more frequently.

The primate data are important for several reasons. First, they demonstrate that cooperation is (sometimes) contingent on prior interactions. Second, some types of exchanges involve potentially high cost forms of behavior, coalitionary support, or access to mates. Third, the experiments reveal that the dynamics of reciprocity differ across dyads. Fourth, the data span a broad spectrum of the monkeys and apes, including New World monkeys, Old World monkeys, and apes. This suggests that the capacity for Tit-for-Tat reciprocal altruism may have deep roots in the primate order.

THE PHYLOGENY OF FRIENDSHIP

Observers of savanna baboons were the first to use the word friendship to describe close ties between certain pairs of adult males and females. Smuts' book, *Sex and Friendship in Baboons* (1985), made friendship a respectable topic for primatological analysis, and the word began to appear with greater frequency in the literature. Friendship is sometimes used as a synonym for close, affiliative bonds, which are thought to involve high levels of nonaggressive behaviors, such as grooming and proximity, tolerance and mutual attraction, and reciprocity (reviewed by Silk 2002).

In baboon groups, pairs of adult males and females sometimes form close relationships. In East African baboon groups, these relationships are characterized by high frequencies of proximity (mainly maintained by the female), grooming (mainly performed by the female), and support (mainly performed by the male on behalf of the female and her offspring). Typically, each female has just one close male associate, spending very little time with other males. These pairs are labeled as "friends." Smuts (1985) hypothesized that males and females both benefit from these relationships. Females obtain protection for themselves and their offspring, whereas males gain future mating advantages and access to infants that they can use in triadic interactions with other males.

In baboon groups in the Moremi Reserve in Botswana, these relationships look much the same, vis-à-vis proximity maintenance and grooming, but differ in their function. There, immigrant males often rise quickly to the top-ranking position within the group and then kill unweaned infants (Palombit et al. 2000). The death of these infants causes females to resume cycling much sooner than they would otherwise. Because top-ranking males monopolize access to high-ranking females, infanticidal males also gain mating opportunities. In Moremi, mothers of new infants form close ties with familiar males, often former mating partners and likely fathers of their infants (Palombit et al. 1997). Males are attentive to these females and their infants, and rush to their defense when they are distressed. Males often hold infants and carry them in confrontations with new immigrants. Infants provide the pivotal link in these relationships. If the infant dies or disappears, males soon lose interest in their partners' welfare. In this case, male-female relationships seem to be a form of parental investment in the welfare of their joint offspring.

Thus, male-female relationships in baboons seem to be a form of mating effort or joint parental investment in the welfare of offspring. I have argued elsewhere that these relationships are different than close friendships among humans because they hinge on the presence of a third party, are often asymmetric and relatively short-lived, and have instrumental functions (Silk 2002).

Empirical support for the existence of friendships, aside from male-female friendships in baboon groups, is still quite limited. There is good evidence that social relationships are frequently differentiated — not all dyads interact with the same frequency or in the same contexts. However, we know little about the behavioral repertoire of friendship — do grooming partners also protect each other from aggression or predators, sit together, tolerate attempts to handle their infants, or share food with one another? Also, we do not know how long these relationships last. Barrett and Henzi (2002) detected frequent changes in preferred grooming partners among female baboons, suggesting that stable long-term relationships may not be common in these animals. Is this true of other groups and species? We know even less about the emotional tenor of affiliative relationships. Are primates more relaxed in the presence of close associates?

Although friendship is often linked to reciprocity, some primatologists have begun to question whether monkeys and apes have the cognitive ability to keep track of costs incurred and benefits received across long periods of time and different currencies (Barrett and Henzi 2002). Most cooperation among nonkin may be based on short-term objectives, such as getting groomed or obtaining access to infants. In these cases, the costs involved in exchanges may be low and the time frame over which accounts must be kept may be quite short. De Waal (2000) also doubts whether monkeys are capable of managing relationships that require careful record keeping. He suggests that balanced exchanges might simply arise from mutual tolerance or high rates of association between partners. On the other hand, de Waal (1992) has suggested that chimpanzees may hold grudges against group members for long periods, suggesting that there may be taxonomic differences in the form of reciprocal relationships among nonhuman primate species.

HOW DOES HUMAN FRIENDSHIP WORK?

Friendships in contemporary Western societies are voluntary, intimate, supportive, reciprocal relationships between equals (Hinde 2002). Companionship, trust, self-disclosure, loyalty, commitment, affection, acceptance, empathy, and mutual regard are important elements of close friendships (Hinde 1997). Time spent together is an important relational currency, but friendships can endure long separations and infrequent contact. Compatability is an important element of friendship, although friendships can weather some degree of tension and conflict (Bleiszner and Adams 1992). Even though people gain both material and emotional support from their friends, emotional support seems to be particularly important in the satisfaction that people derive from their friends and in the benefits that people derive from friendship.

There is some dispute about whether this notion of friendship is a universal feature of human societies. Some social scientists believe that our contemporary notion of friendship as an intimate, private, noninstrumental relationship among nonrelatives is specific to contemporary Western societies and emerged with the rise of commercial societies during the eighteenth century (Adams and Allan 1998; Allan 2001; Bell and Coleman 1999; Pahl 2000; Silver 1990). They point out that in some times and places, social networks are almost entirely limited to close kin; there are also societies in which friendships are institutionalized and lose something of their voluntary and private character. Others contend that friendship is a ubiquitous feature of human societies (Argyle and Henderson 1984), and point to ethnographic descriptions of friendships based on sentiments of affection, intimacy, and empathy. Some evolutionary psychologists hypothesize that there is a universal psychology of friendship (Bleske and Shackelford 2001; Bleske-Rechek and Buss 2001). Here, I focus primarily on the contemporary Western notion of friendships as voluntary, intimate, and private relationships that provide both material and emotional support.

There is some dispute among psychologists about the processes that sustain friendship in contemporary Western societies. Equity theorists contend that inequality in relationships produces dissatisfaction and distress (Walster and Walster 1975). According to this theory, people are equally unhappy when they give more than they receive and when they receive more than they give, and the same processes govern all kinds of close relationships. However, the evidence suggests that although people do value equality in their relationships, they have different expectations about different kinds of relationships (e.g., Bar-Tal et al. 1977; Rook 1987; Winn et al. 1991).

Building on work by Goffman (1961), who distinguished between relationships based on social exchange and economic exchange, Clark and Mills (1979) drew a distinction between exchange relationships and communal relationships. In exchange relationships, benefits are given with the expectation that they will be reciprocated. When one party receives a benefit, she incurs an obligation to return the benefit, and both parties are principally concerned with equity. In evolutionary terms, exchange relationships rely on Tit-for-Tat reciprocal altruism. In communal relationships, benefits are given according to the other's need, and receiving a benefit does not create an obligation to reciprocate. Exchange relationships are thought to characterize relationships among strangers and casual acquaintances, whereas communal relationships are thought to characterize relationships among close friends and kin. Very similar kinds of distinctions are drawn in the sociological and anthropological literature. For example, Wolf (1966) distinguished between instrumental and expressive relationships, and Reisman (1981) distinguished between associative (casual), reciprocal (close), and receptive (asymmetric) friendships.

There is broad consensus in the social science literature that close friendship is independent of short-term, Tit-for-Tat reciprocity (Argyle and Henderson 1984; Hinde 2002; O'Connor 1992). Even Adam Smith recognized the fundamental difference between market exchanges among strangers and transactions among friends. In *The Theory of Moral Sentiments*, he wrote: "The actions required by friendship, humanity, hospitality, generosity are vague and indeterminate."

The communal-exchange distinction articulated by Clark and Mills would be of little interest if it was not reflected in the behavior of people in everyday life. However, the results of several experiments suggest that this distinction maps onto the behavior of people in consistent ways.

In one experiment, subjects were asked to read a short account of a series of interactions between two people (Clark 1981). In these accounts, one person asked another person for a favor, such as a ride to work. In half the accounts, the recipient of the favor subsequently provided the same benefit to the other person (i.e., if they were given a ride to work, they offered the other person a ride to work), and in half the accounts the recipient of the favor subsequently provided a different kind of benefit to the other person (i.e., if they were given a ride to work, they offered to buy the other lunch). Subjects were asked to evaluate the quality of the friendship between the two individuals after they read these accounts. Subjects reported that individuals who exchanged comparable benefits were less close than individuals who exchanged benefits of different types.

Asked why they made these assessments, subjects said that they interpreted the exchange of comparable benefits as a form of repayment, something that they evidently did not associate with close friendship.

Similarly, Shackelford and Buss (1996) examined the effects of immediate reciprocity on relationships between committed mates, close friends, and coalition partners. In this experiment, coalition partners were described as people who worked together to accomplish specific objectives, but were not close friends. Subjects were asked how strongly they thought someone would feel betrayed if immediate reciprocity was offered or demanded by close friends or coalition partners. The results indicate that immediate reciprocity elicited stronger feelings of betrayal among mates and close friends, who are expected to have communal relationships, than coalition partners, who are expected to have exchange relationships.

Boster et al. (1995) examined the effects of "pre-giving" on subsequent compliance with requests from close friends and strangers. Their experiment builds on previous evidence that the receipt of a favor or gift makes recipients more likely to feel obligated to reciprocate, perhaps because pre-giving elicits a norm of reciprocity. In these experiments, subjects requested close friends or strangers to purchase $1 raffle tickets from them. In one treatment, the subject gave a soda to their partner before making the request, and in one treatment, no soda was given. When subjects were paired with strangers, pre-giving nearly doubled the number of raffle tickets purchased. When subjects were paired with friends, pre-giving had no effect, though close friends in both conditions purchased more raffle tickets than strangers.

Clark and her colleagues have conducted a series of experiments investigating contributions to joint tasks (described in Mills and Clark 1994). In one experiment, subjects were assigned a joint task on which they would be rewarded on the basis of their performance. They were required to complete the task in ink and were provided with pens by the experimenters. One subject began the task, and shortly later the other subject was asked to join in the task in a separate room. When the two subjects were strangers, the second subject nearly always used a different color pen than the first subject, but when the two subjects were friends, they were more likely to use the same color pen. The differences between friends and strangers were more exaggerated when the subjects were asked to do the task at the same time face to face.

In another experiment, experimenters monitored subjects' attention to a light that flashed when their partner needed help or when their partner had made a substantial contribution to a joint task. When the signal indicated that help was needed, friends looked at the light more often than strangers. When the signal indicated that their partner had made a contribution to a joint task, strangers monitored the light more often than friends. In a similar experiment, the subjects were more likely to monitor others' needs for help (even when they were unable to provide actual support) when a communal relationship was desired than when an exchange relationship was desired.

Taken together these experiments provide empirical support for the distinction between exchange and communal relationships. More importantly, they support the hypothesis that communal relationships are not based on strict Tit-for-Tat reciprocity. People use Tit-for-Tat reciprocity as a diagnostic criteria for the existence of close friendships; when benefits are balanced directly, relationships are assumed to be casual and ephemeral. People seem to make concerted efforts to obscure the accounting of costs and benefits among their friends — in joint tasks, they hide their own contributions and avoiding monitoring their friends' contributions.

It is important to emphasize that the exchange-communal distinction does not imply that people do not care about the cost-benefit balance in close relationships. In fact, people are unsatisfied when they perceive relationships with close friends to be unbalanced in either direction, and they become resentful when their requests are not granted or when they feel that they are being asked to do too much (Allan 1998; Rook 1987; Walker 1995; Winn et al. 1991). The failure to provide help when requested or needed produces a sense of betrayal and can lead to the dissolution of friendships (O'Connor 1992; Walker 1995).

Mills and Clark believe that the exchange-communal distinction implies that the process that preserves the balance in these two different kind of relationships differs. In exchange relationships, help is given with the explicit expectation that it will be reciprocated. In communal relationships, help is given because it is needed or desired; when both partners have the same communal orientation, benefits will flow back and forth, but they will not be strictly contingent on expectations of future benefits.

HOW DID FRIENDSHIP EVOLVE?

Most researchers interested in the evolution of human social relationships have been preoccupied with kin relations, parenting decisions, and mate choice, giving little attention to the problem of human friendship. When friendship is mentioned, it is usually assumed to be the product of kin selection, which is misdirected toward nonkin or Tit-for-Tat reciprocity.

The argument that friendship is derived from kin selection relies on the logic that our altruistic dispositions were shaped during the millions of years in which people lived in conditions like those of modern foragers. In these societies, people interacted mainly with close relatives and had no need to distinguish between kin and nonkin, or between reciprocators and nonreciprocators. We continue to treat close associates like kin because our ancestors had few opportunities to interact with strangers and had little need to discriminate between kin and nonkin. Accordingly, we form friendships because we have a long history of nepotistic associations (e.g., Alexander 1979; Kenrick and Trost 2000).

I find this hypothesis unconvincing because it assumes that people are less flexible in their behavior than other primates. In many nonhuman primate groups, the average degree of relatedness among females is relatively high. Nonetheless, they clearly discriminate among potential partners, interacting selectively with close kin and reciprocating partners. Even in small foraging societies, people interact regularly with both relatives and nonrelatives, and have opportunities to discriminate between close kin and distant kin, between relatives with high reproductive value and low reproductive value, and between reliable and unreliable reciprocators.

Others have hypothesized that friendship is the product of reciprocal altruism (e.g., Kenrick and Trost 2000; Hewlett 2001). Shackelford and Buss (1996, p. 1153; italics in original) wrote, "One of the most important characteristics of close relationships is a reciprocity of time, resources, and effort expended by one relationship members for the benefit of the other. This exchange of costs and benefits between relationship parties has been termed *reciprocal altruism.*" Humans are good candidates for reciprocal altruism because natural selection seems to have equipped humans with well-tuned mental mechanisms to detect violations of social contracts (Cosmides and Tooby 1992), and these mechanisms could operate in the context of friendship.

Shackelford and Buss (1996) suggest that the difference in the dynamics of reciprocity in communal and exchange relationships reflects differences in the timescale over which accounting is done. According to their view, in coalitions and exchange relationships, the shadow of the future is short, and immediate reciprocity is required to prevent exploitation and cheating. In communal relationships (such as close friendships), the shadow of the future is extended, and there is more tolerance of short-term imbalances in relationship accounts. In such cases, insistence on immediate reciprocity signals uncertainty about the continuation of the relationship, and this elicits feelings of concern, distress, or betrayal. They hypothesize that the difference in responses to requests for immediate reciprocation by close friends and coalition partners described earlier arises because a demand for immediate reciprocity implies that future interactions are unlikely to occur. This is more disturbing for close friends, and elicits stronger feelings of betrayal, than for coalition partners. Although this explanation might explain why friends avoid Tit-for-Tat reciprocity, it does not explain why they obscure their contributions to joint tasks with friends.

FRIENDSHIP IS NOT MUTUALISM

It is possible that friendship is a form of mutualism, a relationship in which each party benefits directly from the things that they do for each other. There is growing interest in the role of mutualism and pseudoreciprocity in nature (Leimar and Connor, this volume). Clutton-Brock (2002) argues persuasively that mutualism plays an important role in the evolution of cooperation in cooperative breeders.

Tooby and Cosmides (1996) emphasize the importance of mutualistic processes in friendship. They begin by challenging the relevance of the conventional definition of altruism, which is based on costs to the actor and benefits to the recipient. They point out that there are many situations in which benefits can be provided at little cost to the actor. For example, if you own a television, it costs you nothing to let others watch with you. This is roughly analogous to what is called by-product mutualism (Dugatkin 1997). To understand the evolution of friendship, they argue, we need to understand how evolution shapes mechanisms that are designed to deliver benefits to others.

Tooby and Cosmides note that when we need help the most, we are often least able to reciprocate. They call this the banker's paradox, likening it to the banker's problem in deciding who to loan money to — those who need it most are often the worst credit risks. Tooby and Cosmides suggest that the solution lies in choosing the right friends. The most reliable sources of support will be those who consider their friends to be unique and irreplacable, because they will be most motivated to preserve the relationship. Thus, if you are the only person in the neighborhood who owns a television, you will be much sought after as a friend. However, it is also important to distinguish between sincere and loyal friends and "fair-weather" friends, because only the former will be willing to help when your needs are greatest. This may be why help received in times of great need is particularly memorable.

Tooby and Cosmides suggest that it is important to be selective in choosing friends because there are practical constraints on the number of friends that a person can have. Thus, when we choose friends it is important (a) to consider how many friends we already have, recruiting friends when we have few friends, discouraging new friendships when we have many; (b) to evaluate the qualities of potential friends, preferring those who possess positive externalities (qualities such as strength, wealth, prestige, and power) that provide benefits with no obligation to repay; and (c) to select those who are able to read your mind and thus anticipate your needs and desires, who consider you to be irreplacable, and who want what you want.

Tooby and Cosmides's verbal model reflects some important features of the psychology of friendship, focusing on the many ways in which friendship increases the benefits that we gain from our relationships with others (Blieszner and Adams 1992). For example, by forming friendships with people who share our interests and understand our needs we can increase the net value of benefits that we derive. (Thus, you might like me because I let you watch my television, but you will derive little benefit from the experience if you are a Star Trek fan and I only watch BBC nature documentaries. Trekkies should seek other Trekkies as friends.)

Tooby and Cosmides also emphasize the importance of choosing the right partners. This may mean choosing partners with positive externalities who can provide copious benefits, or choosing partners who will provide help when you need it. De Vos and Zeggelink (1997) show that the tendency to request support selectively from previous supporters facilitates the evolution of cooperation in small groups living under harsh conditions.

I find it difficult, however, to understand how Tooby and Cosmides's scheme avoids the underlying logic of reciprocity completely. The metaphor of the banker's paradox is based on the implicit assumption that reciprocity matters. If bankers were unworried about being repaid, they would loan money to anyone who asked. Tooby and Cosmides argue that the mechanisms for obtaining benefits matter more than mechanisms that focus on contingent exchange of benefits and costs, but their argument implicitly assumes that costs limit peoples' willingness to provide benefits to others. The banker's paradox is not resolved by ignoring costs and obligations to reciprocate, but by choosing friends for whom the cost-benefit balance is most favorable. It may be that it is easier to inflate the benefit side of the equation (maximizing the benefits that others derive from their association with you), than to deflate the cost side; however, this does not mean that costs are irrelevant.

Finally, I do not think that the Tooby and Cosmides model gives sufficient weight to the fact that close friendship sometimes involves real costs. Such costs may be necessary for friendship: "By definition all friendship must be both sentimental in inspiration and instrumental in effects since there is no other way to demonstrate one's sentiments than through those actions which speak louder than words" (Pitt-Rivers 1973, p. 97). Friendship involves material investments of time, energy, and resources (O'Connor 1992). Moreover, friends may put themselves at risk because same-sex friendships increase vulnerability to sexual rivalry (Bleske and Schackelford 2001) and jealousy (Argyle and Henderson 1984). Although we may be best off choosing friends so that we minimize costs to our friends and maximize benefits to ourselves, friends are valued because they are the ones who are willing to provide help even when it is costly to themselves. Thus, you would be more appreciative if a friend gives you the shirt off his back than if he gives you one of two dozen shirts he has stacked in his closet. The benefit is the same, but the cost to your friend is different. Moral sentiments that we attach to acts of altruism are particularly sensitive to the costs paid.

Thus, I would argue that close friendship is not a form of mutualism. This is not to say that mutualism plays no role in human affairs. We may derive some direct benefits from associations with other people, and mutualistic payoffs may be relevant in those relationships. In some cases, we may even invest in others in order to receive by-product benefits (or pseudoreciprocity, sensu Connor 1986). Thus, it makes sense for me to strike up a relationship with someone who has a big screen television as the World Cup final approaches, even to contribute something to the cost of the television, as long as I get to watch the game. However, this does not provide an adequate description of close friendships. We provide costly favors, services, and support to our friends, and we do not benefit directly when we do so. We only benefit to the extent that our friends provide us with similar benefits.

COOPERATION WITHOUT COUNTING

Friendship is friendship, but accounts must be kept. *(Chinese proverb)*

Friendship in contemporary Western societies seems to be based on two fundamentally incompatible rules. The first rule is that it is inappropriate to keep careful and accurate track of benefits given and received from friends, or to help friends with the explicit expectation of being repaid. This is not just rhetoric; in the laboratory, people obscure their own contributions to joint tasks with friends and avoid keeping track other of their friends' contributions. The second rule is that costs and benefits should be balanced in relationships with friends. Friendships are expected to be based on equality, and people seem to be dissatisfied with relationships in which the benefit-cost balance is tipped in favor of themselves or in favor of their partners.

The existence of these two rules implies that people value reciprocity in relationships with friends and strangers, and rely on the mechanisms of Tit-for-Tat reciprocal altruism to regulate their behavior toward strangers, but not toward friends. We have no models of the evolution of reciprocity that can accommodate both these rules. Theoretical work on reciprocity generally suggests that natural selection will favor strategies that are highly sensitive to recent interactions and require contingent (but not necessarily equal) distribution of benefits. The psychology of friendship contradicts the logic of these models.

The rules that govern exchanges among friends seem to facilitate systematic exploitation. By consistently giving just a little less than she receives, an unscrupulous individual could take advantage of an uncalculating friend. The perceptions of equity in relationships provide some protection against exploitation, but if accounting is imprecise, there may be considerable opportunity for cheating. Moral sentiments, which produce guilt when we cheat our friends and resentment and anger when we think we are being cheated (Hinde 2002), may be effective when asymmetries are detected, but what will trigger these emotions at appropriate times if we do not keep careful cost-benefit accounts? Cheater detection mechanisms seem well designed to catch single transgressions of social contracts (Cosmides and Tooby 1992), but it is not clear that we are equipped to deal with kinds of accounting problems that long-term relationships create.

Although the threat of exploitation seems very real, the practical difficulties of keeping track of costs and benefits seem intractable. How could people keep track of long-term patterns of exchange in multiple currencies with many different partners? In theory, this is necessary to sustain friendship; in practice, it does not seem feasible. It is possible that people only keep track of acts that have substantial costs, and make little effort to monitor the many small exchanges with their friends. It is also possible that people take stock of their relationships periodically, conducting random mental audits of their friendships (Pillsworth, pers. comm.). Other shortcuts for accounting might be used, though we lack evidence on this point. We still need to explain why people deny that they keep track of accounts with close friends and why accounting interferes with friendship.

One obvious solution to this puzzle is to assume that the empirical evidence is wrong. Experiments conducted on undergraduates in the laboratory involve trivial stakes and extremely unnatural settings; they may tell us little about the real psychology of friendship. However, the experiments are consistent with more qualitative descriptions of the motivations of people toward their friends. The congruence of these results may simply mean that people consistently misrepresent their own motivations to themselves and to experimenters in different experimental settings. Subjects may deny that they monitor benefits given and received from their friends and act accordingly when they are asked to perform cooperative tasks in the laboratory, but behave differently outside of these artificial experimental environments. Still, it seems unreasonable to simply ignore these data because they do not fit our theoretical preconceptions. Doubts about the credibility of these kind of laboratory experiments must be addressed by collecting relevant data in more realistic settings.

Thus, the puzzle remains unresolved. People establish close cooperative relationships with nonrelatives, care about reciprocity, but avoid keeping careful count of benefits given and received. None of our models of reciprocity can accommodate the psychology of human friendship. As always, we need more data and better models.

ACKNOWLEDGMENTS

I thank Peter Hammerstein for his invitation to participate in this Dahlem Workshop and the participants from the Dahlem Workshop for stimulating discussions of some of the ideas discussed here. My work on friendship is part of a larger project on the adaptive significance of social relationships in female baboons and work is supported by the National Science Foundation (BCS–0003245), the LSB Leakey Foundation, National Geographic Society, and the UCLA Division of Social Sciences. This paper was written while I was on leave in Berlin; I thank the *Wissenschaftskolleg zu Berlin* for generous support of my research activities.

REFERENCES

Adams, R.G., and G. Allan, eds. 1998. Placing Friendship in Perspective. Cambridge: Cambridge Univ. Press.

Alexander, R.D. 1979. Natural selection and social exchange. In: Social Exchange in Developing Relationships, ed. R.L. Burgess and T.L. Huston, pp. 197–221. New York: Academic.

Allan, G. 1998. Friendship and the private sphere. In: Placing Friendship in Perspective, ed. R.G. Adams and G. Allan, pp. 71–91. Cambridge: Cambridge Univ. Press.

Allan, G. 2001. Personal relationships in late modernity. *Pers. Rel.* **8**:325–339.

Argyle, M., and M. Henderson. 1984. The rules of friendship. *J. Soc. Pers. Relats.* **1**:211–237.

Barrett, L., and S.P. Henzi. 2001. The utility of grooming in baboon groups. In: Economics in Nature, ed. R. Noë, J.A.R.A.M. van Hooff, and P. Hammerstein, pp. 119–145. Cambridge: Cambridge Univ. Press.
Barrett, L., and S.P. Henzi. 2002. Constraints on relationship formation among female primates. *Behaviour* **139**:263–289.
Barrett, L., S.P. Henzi, T. Weingrill et al. 2000. Female baboons do not raise the stakes, but they give as good as they get. *Anim. Behav.* **59**:763–770.
Bar-Tal, D., Y. Bar-Zohar, M.S. Greenberg, and M. Hermon. 1977. Reciprocity behavior in the relationship between donor and recipient and between harm-doer and victim. *Sociometry* **40(3)**:293–298.
Bell, S., and S. Coleman, eds. 1999. The Anthropology of Friendship. Oxford: Berg.
Bleiszner, R., and R.G. Adams. 1992. Adult Friendship. New York: Sage.
Bleske, A.L., and T.K. Shackelford. 2001. Poaching, promiscuity, and deceit: Combating mating rivalry in same-sex friendships. *Pers. Rel.* **8**:407–424.
Bleske-Rechek, A.L., and D.M. Buss. 2001. Opposite-sex friendship: Sex differences and similarities in initiation, selection, and dissolution. *Pers. Soc. Psych. Bull.* **27(10)**:1310–1321.
Boster, F.J., J.I. Rodríguez, M.G. Cruz, and L. Marshall. 1995. The relative effectiveness of a direct request message and a pregiving message on friends and strangers. *Comm. Res.* **22**:475–484.
Clark, M.S. 1981. Noncomparability of benefits given and received: A cue to the existence of friendship. *Soc. Psych. Qtly.* **44**:375–381.
Clark, M.S., and J. Mills. 1979. Interpersonal attraction in exchange and communal relationships. *J. Pers. Soc. Pysch.* **37**:2–24.
Clutton-Brock, T.H. 2002. Breeding together: Kin selection and mutualism in cooperative societies. *Science* **296**:69–72.
Connor, R. 1986. Pseudo-reciprocity: Investing in mutualism. *Anim. Behav.* **34**:1562–1584.
Cords, M. 2002. Friendship among adult female blue monkeys. *Behaviour* **139(2–3)**:291–314.
Cosmides, L., and J. Tooby. 1992. Cognitive adaptations for social exchange. In: The Adapted Mind, ed. J.H. Barkow, L. Cosmides, and J. Tooby, pp. 163–228. Oxford: Oxford Univ. Press.
De Vos, H., and E. Zeggelink. 1997. Reciprocal altruism in human social evolution: The viability of reciprocal altruism with a preference for "old-helping-partners." *Evol. Hum. Behav.* **18**:261–278.
de Waal, F.B.M. 1992. Chimpanzee Politics. London: Cape.
de Waal, F.B.M. 1997a. The chimpanzee's service economy: Food for grooming. *Evol. Hum. Behav.* **18**:375–386.
de Waal, F.B.M. 1997b. Food transfers through mesh in brown capuchins. *J. Comp. Psych.* **111**:370–378.
de Waal, F.B.M. 2000. Attitudinal reciprocity in food sharing among brown capuchin monkeys. *Anim. Behav.* **60**:253–261.
Dugatkin, L.A. 1997. Cooperation among Animals: An Evolutionary Perspective. Oxford: Oxford Univ. Press.
Fehr, E., and S. Gächter. 2001. Altruistic punishment in humans. *Nature* **415**:137–140.
Goffman, E. 1961. Encounters: Two Studies in the Sociology of Interaction. Indianapolis: Bobbs-Merrill.
Hemelrijk, C.K. 1994. Support for being groomed in long-tailed macaques, *Macaca fasicularis*. *Anim. Behav.* **48**:479–481.

Henzi, S.P., and L. Barrett. 1999. The value of grooming to female primates. *Primates* **40**:47–59.

Hewlett, B. 2001. Neoevolutionary approaches to human kinship. In: New Directions in Anthropological Kinship, ed. L. Stone, pp. 93–108. Lanham, MD: Rowman and Littlefield.

Hinde, R.A. 1997. Relationships: A Dialectical Perspective. Brighton: Psychology Press.

Hinde, R.A. 2002. Why Good Is Good: The Sources of Morality. London: Routledge.

Kenrick, D.T., and M.R. Trost. 2000. An evolutionary perspective on human relationships. In: The Social Psychology of Personal Relationships, ed. W. Ickes and S. Duck, pp. 9–35. New York: Wiley.

Mills, J., and M.S. Clark. 1994. Communal and exchange relationships: Controversies and research. In: Theoretical Frameworks for Personal Relationships, ed. R. Erber and R. Gilmour, pp. 29–42. Hillsdale, NJ: Erlbaum.

Mitani, J.C., D.P. Watts, J.W. Pepper, and D.A. Merriwether. 2002. Demographic and social constraints on male chimpanzee behaviour. *Anim. Behav.* **64**:727–737.

O'Connor, P. 1992. Friendships between Women: A Critical Review. New York: Guilford.

Pahl, R. 2000. On Friendship. Cambridge: Polity.

Palombit, R.A., D.L. Cheney, J. Fischer et al. 2000. Male infanticide and defense of infants in chacma baboons. In: Male Infanticide and Its Implications, ed. C.P. van Schaik and C.H. Janson, pp. 123–151. Cambridge: Cambridge Univ. Press.

Palombit, R.A., R.M. Seyfarth, and D.L. Cheney. 1997. The adaptive value of "friendships" to female baboons: Experimental and observational evidence. *Anim. Behav.* **54**:599–614.

Pitt-Rivers, J. 1973. The kith and the kin. In: The Character of Kinship, ed. J. Goody, pp. 85–109. Cambridge: Cambridge Univ. Press.

Reisman, J.M. 1981. Adult friendships. In: Personal Relationships. 2. Developing Personal Relationships, ed. S. Duck and R. Gilmour, pp. 205–230. New York: Academic.

Richerson, P., and R. Boyd. 1998. The evolution of human ultra-sociality. In: Ideology, Warfare, and Indoctrinability, ed. I. Eibl-Eibesfeldt and F. Salter, pp. 71–95. Oxford: Berghahn.

Rook, K.S. 1987. Reciprocity of social exchange and social satisfaction among older women. *J. Pers. Soc. Psych.* **52**:145–154.

Schino, G. 2001. Grooming, competition, and social rank among female primates: A meta-analysis. *Anim. Behav.* **62**:265–271.

Seyfarth, R.M., and D.L. Cheney. 1984. Grooming, alliances, and reciprocal altruism in vervet monkeys. *Nature* **308**:541–543.

Shackelford, T.K., and D.M. Buss. 1996. Betrayal in mateships, friendships, and coalitions. *Pers. Soc. Psych. Bull.* **22**:1151–1164.

Silk, J.B. 2002. Using the "F" word in primatology. *Behaviour* **139**:421–446.

Silver, A. 1990. Friendship in commercial society: Eighteenth-century social theory and modern sociology. *Am. J. Sociol.* **95**:1474–1504.

Smuts, B.B. 1985. Sex and Friendship in Baboons. New York: Aldine de Gruyter.

Tomasello, M., and J. Call. 1997. Primate Cognition. Oxford: Oxford Univ. Press.

Tooby, J., and L. Cosmides. 1996. Friendship and the banker's paradox: Other pathways to the evolution of adaptations for altruism. *Proc. Brit. Acad.* **88**:119–143.

Walker, K. 1995. "Always there for me": Friendship patterns and expectations among middle- and working-class men and women. *Sociol. Forum* **10**:273–296.

Walster, E., and G. Walster. 1975. Equity and social justice. *J. Soc. Issues* **31**:21–43.

Watts, D.P. 2002. Reciprocity and interchange in the social relationships of wild male chimpanzees. *Behaviour* **139**:343–370.

Winn, K.I., D.W. Crawford, and J. Fischer. 1991. Equity and commitment in romance versus friendship. *J. Soc. Behav. Pers.* **6**:301–314.

Wolf, E. 1966. Kinship, friendship and patron–client relationships in complex societies. In: The Social Anthropology of Complex Societies, ed. M. Blanton, pp. 1–21. London: Tavistock.

4

Is Strong Reciprocity a Maladaptation?

On the Evolutionary Foundations of Human Altruism

Ernst Fehr[1] and Joseph Henrich[2]

[1]Institute for Empirical Research in Economics, University of Zurich, Switzerland
[2]Department of Anthropology, Emory University, Atlanta, GA 30322, U.S.A.

ABSTRACT

In recent years, a large number of experimental studies have documented the existence of strong reciprocity among humans. Strong reciprocity means that people willingly repay gifts and punish the violation of cooperation and fairness norms even in anonymous one-shot encounters with genetically unrelated strangers. This chapter provides ethnographic and experimental evidence suggesting that ultimate theories of kin selection, reciprocal altruism, costly signaling, and indirect reciprocity do not provide satisfactory evolutionary explanations of strong reciprocity. The problem with these theories is that they can rationalize strong reciprocity only if it is viewed as maladaptive behavior, whereas the evidence suggests that it is an adaptive trait. Thus, alternative evolutionary approaches are needed to provide ultimate accounts of strong reciprocity.

INTRODUCTION

A large body of evidence has emerged in recent years from lab experiments indicating that a substantial fraction of people willingly repay gifts and punish the violation of cooperation and fairness norms, even in anonymous one-shot encounters with genetically unrelated strangers (see, e.g., Fehr and Gächter 1998a, b; Henrich et al. 2001; McCabe et al. 1998; Fehr, Fischbacher, and Gächter 2002). This behavioral propensity has been termed strong reciprocity (Gintis 2000; Bowles and Gintis 2001; Fehr, Fischbacher, and Gächter 2002).

In this chapter we discuss the evidence bearing on the question of whether strong reciprocity represents adaptive or maladaptive behavior. When the above-mentioned evidence is presented to biologists, zoologists, primate researchers, or evolutionary psychologists, they often spontaneously provide an ultimate account of strong reciprocity as a maladaptation (e.g., Johnson et al. 2003). They argue that in the evolutionarily relevant past, humans evolved in small groups with frequent repeated interactions and strong reputation mechanisms. In the presence of repeated interactions, or when people's reputation was at stake, they faced a strong fitness incentive to cooperate because, in response to their noncooperative behavior, other group members may have refused to engage in profitable future interactions with them, or they may even have punished them directly. Moreover, so the argument goes, because anonymous one-shot interactions have been rare in the past, human psychology tends to "misfire" in the anonymous one-shot encounters that characterize modern life and laboratory experiments. That is, humans misapply behavioral rules, which make adaptive sense in repeated interactions or when their reputation is at stake, to one-shot situations — that is, they "mistakenly" cooperate and punish in one-shot situations. In this view, strong reciprocity represents a behavioral trait that can only exist in evolutionary disequilibrium. If one-shot encounters become frequent, as is undoubtedly the case in modern societies, natural selection is expected to reduce the frequency of strongly reciprocal individuals because these individuals will do worse than individuals who do not bear the costs of cooperation and punishment in one-shot encounters.

Despite the superficial plausibility of this maladaptation argument, we believe that there is little evidence in favor of this view and fairly strong evidence against it. We show that there is a lot of laboratory evidence as well as field evidence from small-scale societies that contradicts the maladaptation hypothesis. To provide the basis for our discussion, we first define strong reciprocity more precisely and present some of the evidence on strong reciprocity that is relevant for our discussion. Then we take a look at the evolutionary history of humans to see whether it is indeed the case that encounters with no or a low probability of future interactions have been rare.

It is important to stress that this chapter deals with the *ultimate* sources of strong reciprocity. Thus, we do not discuss the importance of strong reciprocity for the functioning of friendships, neighborhoods, markets, organizations, and the political economy. Elsewhere it has been shown that strong reciprocity has decisive effects for the functioning of many aspects of modern societies (Fehr and Fischbacher 2002b; Fehr, Fischbacher, and Gächter 2002). This role of strong reciprocity remains true irrespective of whether the maladaptation hypothesis is valid or not.

Our critique of the maladaptation hypothesis does not imply that we consider other ultimate sources of human altruism — kin selection, reciprocal altruism, and reputational forces — as unimportant. There is good evidence indicating that they are important. However, our critique means that these other ultimate sources of human altruism do not provide an explanation of strong reciprocity.

They provide, therefore, a rather incomplete picture of the evolutionary forces that shaped human cooperation and altruism.

Before plunging into the details, we will briefly describe the typical circumstances under which the experiments that provide much of the evidence for strong reciprocity take place. In these experiments, researchers meticulously ensure that all interactions between the subjects take place anonymously, so that neither before, during, nor after the experiments the subjects were informed of the identities of their interaction partners. There are at least two reasons for the anonymity requirements. First, anonymous interactions provide an interesting baseline case, which, when compared with different types of nonanonymous interactions, allow us to measure the impact on nonanonymity on behavior. Second, if helping or punishing behavior shows up in nonanonymous interactions, one can always argue that because the subjects know each other they might somehow engage in repeated interactions after the experiment. Thus, altruistic helping and punishing behavior cannot be identified in a clean way in nonanonymous interactions. In this large body of experiments, subjects also had to incur real monetary costs when repaying gifts or punishing others, with stakes sometimes approaching three months' income (Cameron 1999; Fehr, Tougareva, and Fischbacher 2002). Several of the experiments also ensured that not even the experimenters could observe the individual actions of the subjects (and this was made transparent to subjects). In these experiments, the experimenter could only observe the aggregate statistical results, not the behavior of specific individuals. Thus, even when monetary stakes are high, there are no repeated interactions, and subjects can be sure that nobody else knows their behavior (so that reputational factors are ruled out), many subjects exhibited strongly reciprocal responses.

PROXIMATE PATTERNS OF STRONG RECIPROCITY

A person is a strong reciprocator if she is willing (a) to sacrifice resources to bestow benefits on those who have bestowed benefits (= strong positive reciprocity) and (b) to sacrifice resources to punish those who are not bestowing benefits in accordance with some social norm (= strong negative reciprocity). The essential feature of strong reciprocity is a willingness to sacrifice resources in both rewarding fair behavior and punishing unfair behavior, *even if this is costly and provides neither present nor future economic rewards for the reciprocator.* Whether an action is perceived as fair or unfair depends on the distributional consequences of the action relative to a neutral reference action (Rabin 1993; Falk and Fischbacher 1999).

It is important to distinguish strong reciprocity from "reciprocal altruism" (Trivers 1971) and from "indirect reciprocity" (Alexander 1987; Nowak and Sigmund 1998). A reciprocally altruistic actor will incur short-run costs to help other individuals only when the actor expects to recoup some long-term net benefits from helping. An indirect reciprocator may also be willing to help if the act of helping can be credibly communicated to others, so that the others are more likely to exhibit cooperative behavior toward the indirect reciprocator. However, in the absence of repeated interactions or the possibility to gain a favorable reputation, these actors never help. This contrasts sharply with a strong reciprocator, who is willing to help another person in response to a kind behavior of this person even in the absence of repeated interactions and opportunities to gain a reputation. The distinction between strong reciprocity, indirect reciprocity, and reciprocal altruism can most easily be illustrated in the context of a *sequential* Prisoner's Dilemma (PD) that is played *only once*. Moreover, the game is played under complete anonymity so that any kind of reputation formation can be ruled out. In a sequential PD, player A first decides whether to defect or to cooperate. Then player B observes player A's action after which she decides to defect or to cooperate. To be specific, let material payoffs for (A, B) be (5, 5) if both cooperate, (2, 2) if both defect, (0, 7) if A cooperates and B defects, and (7, 0) if A defects and B cooperates. If player B is a strong reciprocator, she defects if A defected and cooperates if A cooperated because she is willing to sacrifice resources to reward a behavior that is perceived as fair. A cooperative act by player B, despite the economic incentive to cheat, is a prime example of such fairness. The cooperation of a strong reciprocator is thus *conditional* on the perceived fairness of the other player. In contrast, reciprocal altruists or indirect reciprocators, when in the role of player B, will always defect in an anonymously played sequential *one-shot* PD because in this game there are no future interactions nor is it possible to gain a reputation. This, of course, assumes that players have the ability to comprehend a one-shot game.

The structure of a sequential PD neatly captures the problem of economic and social exchanges under circumstances in which the quality of the goods exchanged is not enforced by third parties, such as an impartial police and impartial courts. Fehr and Gächter (1998a, b) and Fehr, Fischbacher, and Gächter (2002) describe the results of many slightly more general sequential PDs (often called gift exchange experiments or trust experiments) in which the parties are not constrained to pure "cooperate" or "defect" choices but can also choose several different intermediate cooperation levels. The upshot of these experiments is that there is a strong positive correlation between the level of cooperation of player A and the level of cooperation of player B. Depending on the details of the parameters, between 40% and 60% of the B-players typically respond in a strongly reciprocal manner to the choice of player A: the more A gives/cooperates, the more B gives/cooperates. If player A chooses zero cooperation, then strongly reciprocal B-players also choose zero cooperation. However, there are often also between 40% and 60% of second movers who *always* choose zero cooperation irrespective of what player A does. These players thus exhibit purely selfish behavior.

There is an interesting extension of the generalized sequential PD if player A is given the additional option to punish or reward player B after observing the action of player B. In Fehr and Gächter (1998b), player A could invest money to reward or punish player B in this way. Every dollar invested into rewarding increased player B's earnings by 2.5 dollars and every dollar invested into punishment of B reduced player B's earnings by 2.5 dollars. Since after the reward and punishment stage the game is over, a selfish player A will never reward or sanction in this experiment. In fact, however, many A-players rewarded player B for high cooperation and punished low cooperation. Moreover, subjects in the role of player B expected to be rewarded for high cooperation and punished for low cooperation; consequently, the cooperation rate of player B was much higher in the presence of a reward and punishment opportunity. Thus, it is not only the case that many B-players exhibit strongly reciprocal responses in the sequential PD, it is also the case that — in the extended version of the sequential PD in which A can punish or reward — the B-players expect the A-players to exhibit strongly reciprocal behavior. This expectation, in turn, causes a large rise in the cooperation of the B-players relative to the situation in which the A-players have no reward and punishment opportunity.

There are many real-life examples of the desire to take revenge and to retaliate in response to harmful and unfair acts. One important example is that people frequently break off bargaining with opponents who try to squeeze them. This example can be nicely illustrated by so-called ultimatum bargaining experiments (Güth et al. 1982; Camerer and Thaler 1995; Roth 1995). In the ultimatum game, two players have to agree on the division of a fixed sum of money. Person A, the proposer, moves first and makes exactly one proposal of how to divide the amount. Then person B, the responder, can accept or reject the proposed division. In the case of B's rejection, both players receive nothing, whereas in the case of acceptance the proposal is implemented. In populations from industrialized societies, the results robustly show that proposals that would leave the responder *positive* shares below 20% of the available sum are rejected with a very high probability. This indicates that responders do not behave in a self-interest maximizing manner. In general, the motive indicated for the rejection of positive, yet "low," offers is that subjects view them as unfair. As in the case of strong positive reciprocity, it is worthwhile to mention that strong negative reciprocity is observed in a wide variety of cultures, and that rather high monetary stakes do not change or have only a minor impact on these experimental results. By now there are literally hundreds of studies of one-shot ultimatum games. Rejections of positive offers are observed in Israel, Japan, many European countries, Russia, Indonesia, and the United States. For an early comparison across countries, see Roth et al. (1991). In the study of Cameron (1999), the amount to be divided by the Indonesian subjects represented the income of three months for them. Other studies with relatively high stakes are Hoffman et al. (1996) where $100 had to be divided by U.S. students, and Slonim and Roth (1998).

Strong reciprocity also plays a decisive role in *n*-person situations that involve the production of a public good. Human history is full of public goods situations, such as cooperative hunting, food sharing, collective warfare, and the like. An essential characteristic of a public good is that it is difficult, impossible, or not desirable to exclude those from the consumption of the good that did not contribute to producing it. This then raises the question why anybody should contribute to the provision of the good if the nonproviders can also consume the good. In a recent paper, Fehr and Gächter (2002) showed that altruistic punishment provides a proximate solution to this problem. A substantial fraction of the subjects in public goods experiments are willing to cooperate *and* to punish the defectors, if given the chance to do so. In these situations, the punishment threat provides an incentive for potential defectors to cooperate. Whereas in the absence of targeted punishment opportunities, cooperation typically breaks down, cooperation flourished when targeted punishment of defectors was possible.

HOW PLAUSIBLE ARE MALADAPTATION ACCOUNTS OF STRONG RECIPROCITY?

An important challenge for the maladaptation account of strong reciprocity is the evidence from nonhuman primates (Boyd and Richerson 2003). Many extant nonhuman primate species live in small groups very much like those presumed for early humans. In all primate species the members of at least one sex leave their natal groups and join other groups where, in many cases, their only relatives will be their own offspring. It is also well known that primates are able to distinguish kin from nonkin. In most primate groups, there is ample opportunity for repeated interactions among unrelated individuals as well as for reputation formation. However, cooperation among unrelated individuals in primate groups is far less developed than among humans, and no behaviors that come close to strong reciprocity have been observed among them (Silk 2003). Therefore, the maladaptation account of strong reciprocity must explain why repeated interactions among humans lead to strongly reciprocal behavior; among primates, it does not. We conclude that the maladaptation account of strong reciprocity is, at least, incomplete.

Furthermore, in zoos and research facilities, provisioned primates often live in much larger social groups than in their natural habitats. However, despite being such "unnatural" social environments, nonhuman primates do not exhibit any of the "mistaken" cooperation that is attributed to human living in larger social groups that characterizes modern society. For the same reason that humans mistakenly cooperate in the modern context, the maladaptation hypothesis predicts that nonhuman primates should "mistakenly" cooperate in such novel social environments. If nonhuman primates are not fooled by such unnatural social environments, it seems unlikely that humans would be.

The maladaptation account is based on the idea that no ultimate explanation beyond kin selection (Hamilton 1964), reciprocal altruism (Trivers 1971), indirect reciprocity (Alexander 1987; Nowak and Sigmund 1998), or costly signaling (Zahavi and Zahavi 1997; Gintis et al. 2000) is necessary to explain strong reciprocity. In a sense, strong reciprocity is viewed as a by-product of one of these other ultimate accounts of human cooperation. Proximate mechanisms that have been caused by one of these other ultimate forces are held responsible for the existence of strong reciprocity. These proximate mechanisms, so the idea goes, have been shaped by natural selection but are not sufficiently fine-tuned to the modern human condition, where lots of one-shot interactions occur. In addition, they are, in particular, not fine-tuned to the laboratory world of anonymous one-shot experiments. This argument implies that the behavioral rules of humans that produce cooperation and punishment should not be fine-tuned to the distinction between low and high frequency of future encounters. In other words, humans should exhibit roughly the same behavior in encounters with a high and a low probability of future encounters — this should be especially true in the conditions of the experimental laboratory. As we will show below, this prediction is strongly contradicted by experimental evidence.

We do not doubt that the other ultimate forces explain important aspects of human cooperation. There is, in fact, persuasive evidence that the nepotistic motives associated with kin selection and the (long-term) selfish motives associated with reciprocal altruism and indirect reciprocity have powerful effects on human cooperation (Silk 1980, 1987; Daly and Wilson 1988; Gächter and Falk 2002; Keser and van Winden 2000; Milinski et al. 2002). However, these theories do not provide good ultimate explanations of strong reciprocity. Kin selection could, in principle, account for cooperation in one-shot encounters if humans were driven by rules that do not distinguish between kin or nonkin. However, humans, like other primates, cognitively and behaviorally distinguish kin from nonkin (Tomasello and Call 1997). Furthermore, people generally feel stronger emotions toward kin than toward nonkin. Parents, for example, have no trouble differentially bestowing benefits on their own offspring, even when their offspring are intermixed with the offspring of others in the same household (Daly and Wilson 1998; Case et al. 2001).

Reciprocal Altruism and Strong Reciprocity

Reciprocal altruism could account for cooperation and punishment in one-shot encounters if the behavioral rules of humans did not depend on the probability of future interactions with potential opponents. However, humans are well capable of distinguishing "partners," with whom they are likely to have many future interactions, from "strangers," with whom future interactions are less likely. There is ample evidence that humans cooperate much more if they expect frequent future interactions than if future interactions are rare or absent (Gächter and Falk 2002; Keser and van Winden 2000; Fehr and Gächter 2000). Gächter and Falk, for instance, conducted sequential PD experiments with many intermediate cooperation possibilities. They implemented a pure one-shot condition in which every player A met ten different B-players (ten different one-shots). In addition, they did a repeated interaction condition in which a pair of A- and B-players interacted for ten periods. The results in the Gächter and Falk study show that the B-players behave much more cooperatively in the repeated condition than in the one-shot condition. Similarly, Keser and van Winden (2000) conducted public goods experiments in a one-shot condition and in a repeated condition. Again, cooperation rates were much higher in the repeated condition. Likewise, Fehr and Gächter (2000) also conducted public goods experiments under one-shot and repeated conditions. Their results also show that cooperation is much higher in the repeated condition, irrespective of whether opportunities exist for targeted punishment. For example, in the repeated interaction condition, when it is possible to punish specific other group members directly, subjects contribute 95% of their endowment to the public good, whereas in the one-shot situation with punishment, subjects invest "only" between 60% and 70% of their endowment to the public good.

This evidence strongly suggests that laboratory subjects have no problems in understanding the difference between one-shot and repeated interactions. The same researchers who spontaneously put forth the "maladaptation hypothesis" would also likely explain the acuity with which subjects distinguish one-shot from repeated encounters as a result of the "cognitive architecture" shaped by selective processes favoring reciprocal altruism. In fact, it would be quite surprising if subjects did not understand that the probability of being cheated by a stranger in a foreign town is orders of magnitude bigger than the probability of being cheated by a close friend or a business partner or a colleague at the work place. One of us (Ernst Fehr) has often conducted sequential one-shot PDs in the laboratory. After the experiment, subjects were often disappointed because they failed to exhaust large parts of the potential gains from cooperation. As a consequence, they often complained about the one-shot rules of these experiments saying that it is difficult to establish trust and cooperation with somebody with whom one interacts only once. This all indicates that subjects have no *cognitive* problems in grasping the difference between low- and high-frequency encounters. Perhaps their *emotions* are not fine-tuned to the differences across these two kinds of encounters. It is plausible that emotions, such as shame or anger, enhance the willingness to cooperate and to punish, and if these emotions show up regardless of whether we face a one-shot encounter or a repeated encounter, they might be responsible for the existence of strongly reciprocal behavior.

Emotions and Strong Reciprocity

We doubt that our emotions cannot discriminate between, for instance, being cheated by a long-term interaction partner (e.g., a friend) or a short-term interaction partner (e.g., a stranger in a foreign town). Most of us probably feel much stronger negative emotions if we have been cheated by a friend. To check whether this intuition is correct, we conducted a questionnaire among students ($n = 172$). We asked them whether they felt angrier when a long-term partner had cheated them compared with when a stranger had cheated them. Roughly 80%

indicated overwhelmingly stronger feelings of anger when the long-term partner cheated them. If anger is indeed the proximate force underlying the punishment of noncooperators, then these answers suggest that the emotional impulse to punish the partner is much stronger.

Sometimes it is also argued that emotions are cognitively impenetrable, suggesting that they are overwhelming determinants of human behavior. To support this claim, some members of this Dahlem Workshop group (see McElreath et al., this volume) refer to experiments conducted by Paul Rozin and colleagues. Rozin, Millman, and Nemeroff (1986) have performed experiments in which an experimenter gives a subject fudge and then asks the subject (in a between subjects design) if they would be willing to eat more of the same fudge in (a) the shape of a disc or (b) in the shape of feces. Even though the subject knew consciously that the substance is the same fudge they have already eaten (because they could see how the experimenter produced the fudge in the form of discs or feces), most subjects refused to eat the fudge in the shape of feces. In our view, these experiments do not really show that emotions are cognitively impenetrable determinants of human behavior. They only show that subjects who are nearly indifferent between eating a further piece of fudge or stopping to eat are affected by the emotion of disgust. We suspect that if subjects were paid for eating the feces-shaped fudge, they would eat the fudge for a relatively small amount of money (e.g., $20). More importantly, it seems highly likely that the minimum amount of money (or hunger) needed to persuade individuals to eat the feces-shaped fudge would be substantially less than the amount of money that would be necessary to get them to eat real feces in the same shape (we suspect the amount of difference will be infinite). This suggests that the behavioral impulse of emotions is far from being overwhelming or cognitively impenetrable, because if the costs of not eating the fudge become sufficiently high, subjects will eat the fudge. Thus, although the existence of emotions affects our tastes, humans seem to weigh the costs and benefits of different courses of action cognitively, irrespective of whether the action involves emotions or not.

If this argument is correct and if emotions like guilt, shame, and anger are driving forces of strong reciprocity, strongly reciprocal behavior patterns should quickly respond to changes in the costs and "benefits." Experiments strongly confirm this argument. Recall the generalized sequential PDs described above, in which player A had the option to reward and punish player B after observing player B's choice. Fehr, Gächter, and Kirchsteiger (1997) have conducted experiments in which they increased the cost of rewarding and punishing player B by a factor of five. If A-players' emotions are "penetrable" (i.e., if subjects are capable of understanding costs and benefits in novel situations) as we argue, then they will take into account this cost increase. Therefore, they will reward and punish less. In addition, B-players who understand this should expect less rewarding and punishment and, consequently, they should cooperate less. Moreover, if the A-players know that — in the high cost condition — any increase in their own cooperation will elicit a smaller increase in the cooperation of the B-players, then A-players should also reduce their cooperation. Note that this argument requires a very subtle chain of reasoning and an understanding of the impact of higher costs, not only on one's own (emotion-driven) behavior but also on the (emotion-driven) behavior of the other player. Nevertheless, the behavioral evidence powerfully supports the conclusion that subjects quickly respond to cost changes by adapting their own behavior and anticipating adaptations in the behavior of their opponent: the A-players immediately (before any trial and error learning could occur) punish and reward less in the high cost condition, the B-players immediately expect less punishment and rewarding and, hence, cooperate less. In response, the A-players also reduce their cooperation significantly.

Another example of instantaneous behavioral changes in response to changes in the benefits is provided by the experiments of Fischbacher, Fong, and Fehr (2002). These authors conducted an ultimatum game with competition among responders. Instead of only one responder there were two and five responders. As in the bilateral case, the (single) proposer made one offer. Then, the responders simultaneously accepted or rejected the proposal. If all responders rejected, all players earned zero. If more than one responder accepted the offer, one of the accepting responders was randomly allocated the proposed amount of money, the proposer received the remaining money, and the rejecting responders received nothing. If strong reciprocity were just blind emotion-driven (impenetrable) revenge that is not tailored to the subtleties of the circumstances, one would expect that players in the responder competition condition behave similarly to the bilateral case. If the players' emotions do not understand the difference between one-shot and repeated play, why should they understand the much subtler distinction between the bilateral case and the responder competition case? From the viewpoint of the evolutionary history of humans, the bilateral one-shot ultimatum game is as artificial as the one-shot game with responder competition. Both games were probably rarely played throughout human evolution, and there is thus no reason why human behavior should be well adapted to the differences across these two games. If, however, there are adaptive reasons for why humans want to punish unfair behavior, that is, if there is an adaptive account for strong reciprocity, then the prediction is different. In the bilateral case, the responder basically has a property right in punishment. By rejecting a greedy offer, the responder can ensure with certainty that the proposer is punished. The situation is dramatically different in the case of responder competition. Here the responder can no longer unilaterally ensure the punishment of the proposer. In fact, in the two-responder case, a rejection by responder 1 only ensures the punishment of the proposer *if* responder 2 rejects with certainty, too. Thus, if the adaptive goal of rejections is to punish the proposer, we should observe that the responders punish much less in the responder competition condition. Moreover, the reduction in the willingness to punish should be driven by the responders' beliefs about the likelihood that all other responders will punish as well, because only in this case can the punishment of the proposer be ensured. Finally, rational proposers who anticipate that the responders will reject less in the competitive condition will make much greedier offers in this condition. Note that these arguments are again quite subtle and require considerable sophistication.

A great variety of experimental results strongly confirm the strong reciprocity prediction. The responders reject much less in the competitive condition and the proposers respond accordingly. The rejection rate is highest in the bilateral case, much lower in the two-responder case, and even lower in the five-responder case. For instance, whereas in the bilateral case offers of 20% of the bargaining cake are rejected with probability 0.8, the same offers are only rejected with probability 0.15 in the five-responder case. As a consequence, the average share of the bargaining cake that goes to the responders declines from roughly 40% in the bilateral case to 20% in the two-responder case, and further to roughly 15% in the five-responder case. Moreover, the decline in the rejection rate across conditions is *exclusively* driven by the responders' beliefs about the other responders' rejection behavior. If the responders in the competitive condition believe that all other responders also punish a greedy offer, the probability of rejection in the competitive case is as high as in the bilateral case. However, if the responders believe that some of the other responders accept a greedy offer, their willingness to reject the offer becomes much lower — as predicted by the strong reciprocity approach. Thus, it is not the case that because of the competitive situation the responders *somehow* exhibited more competitive preferences, which induced them to reject less often. Instead, the reduction in the willingness to reject is exclusively the result of a change in beliefs — as predicted by the strong reciprocity approach.

Reputation Formation and Strong Reciprocity

Indirect reciprocity can only account for cooperation in one-shot encounters if our behavioral rules are not contingent upon the likelihood that our actions will be observed by others. Again, there is strong evidence to the contrary. The experiments of Milinski et al. (2002) show, for instance, that cooperation in public goods games breaks down if the possibility to gain a favorable individual reputation is removed, whereas if subjects can gain individual reputations, cooperation flourishes. People are well calibrated to observe opportunities for reputation formation, even in novel laboratory environments; they are not "fooled" into thinking reputation is always important, as the maladaptation hypothesis would propose. Incidentally, this breakdown of cooperation is regularly observed in repeated anonymous public goods experiments where reputation formation and punishment are ruled out. This breakdown is observed even if the same group of people can stay together for the whole experiment. Thus, in this context, the problem is not to explain why humans cooperate but why humans cannot maintain cooperation. It can be shown that the peculiar patterns of strong positive reciprocity provide a plausible explanation of the breakdown of cooperation.

The peculiarities of strong positive reciprocity in public goods games have been examined by Fischbacher, Gächter, and Fehr (2001). In their experiment a self-interested subject is predicted to defect completely, irrespective of how much the other group members contribute to the public good. However, only a minority of subjects behave in this way. About 50% of the subjects are willing to contribute to the public good if the other group members contribute as well. Moreover, these subjects contribute more to the public good when others are expected to increase their contributions, indicating a strongly reciprocal cooperation pattern. Only 20% of these subjects, however, are willing to match the average contribution of the other group members whereas 80% of the strongly reciprocal types contribute less than the average contribution of the other group members. Roughly 30% of the subjects behave in a fully selfish manner and always defect, irrespective of how much they expect others to contribute. The remaining 20% exhibit other patterns.

Based on the patterns of reciprocal behavior observed by Fischbacher, Gächter, and Fehr (2001), the breakdown of cooperation in repeated public goods experiments can be neatly explained by the dynamics of the interaction between strongly reciprocal strategies and selfish strategies. For any given *expected* average contribution of the other group members in period t, strong reciprocators either match this average contribution or contribute somewhat less than the expected average contribution. Moreover, the selfish types contribute nothing. Thus, the *actual* average contribution in period t clearly falls short of the average contribution that has been expected for period t, inducing the subjects to reduce their expectations about the other members' contributions in period $t + 1$. Due to the presence of reciprocal types, however, the lower expected average contributions in period $t + 1$ cause a further decrease in the actual contributions in $t + 1$. This process repeats itself over time until very low contribution levels are reached.

It has also been shown that the punishment behavior of laboratory subjects is contingent on the possibility of building a reputation. Fehr and Fischbacher (2002a) conducted a series of ten ultimatum games in two different conditions. Under both conditions, subjects played against a different opponent in each of the ten periods. In each period of the baseline condition, proposers knew nothing about the past behavior of their current responders. Thus, the responders could not build up a reputation for being "tough" in this condition. By contrast, in the reputation condition, proposers knew the full history of behavior of their current responder, i.e., the responders could build up a reputation for being "tough." In the reputation condition, a reputation for rejecting low offers is, of course, valuable because it increases the likelihood to receive high offers from the proposers in future periods.

If the responders cognitively and emotionally understand that there is a pecuniary payoff from rejecting low offers in the reputation condition, one should observe higher acceptance thresholds in this condition. If, in contrast, subjects do not understand the logic of reputation formation and apply the same habits or cognitive heuristics to both conditions one should observe no systematic differences in responder behavior across conditions. Since the subjects participated in both conditions, it was possible to observe behavioral changes at the individual level. It turns out that the vast majority (82%) of the responders increase their acceptance thresholds in the reputation condition relative to the baseline condition. Moreover, the increase in acceptance thresholds occurs *immediately* after the reputation condition is introduced. There was not a single subject that reduced the acceptance thresholds in the reputation condition relative to the baseline in a statistically significant way. Again, this contradicts the hypothesis that the subjects do not understand the difference between anonymous and nonanonymous play. Subjects seem keenly attuned to the differences in reputation building opportunities.

It is sometimes argued (e.g., during discussions with members of Group 1; cf. McElreath et al., this volume) that the application of different behavioral rules across one-shot and repeated interactions as well as across anonymous and nonanonymous interactions do not refute the maladaptation account. Perhaps subjects have a kind of baseline belief such that when they are put in an anonymous one-shot experiment they actually believe with *positive* probability that their actions will not remain anonymous or that they do in fact play a repeated game. Then, if one changes from the anonymous one-shot situation to a nonanonymous situation or to a situation with repeated play, their assessments of the probability of repeated interaction increase. This change in beliefs could then be responsible for the different behaviors. Fehr and Henrich have tested this argument by asking subjects after each session of a series of anonymous (4-person) one-shot public goods experiments whether they believed the experimenters' claims in the instructions regarding anonymity and one-shot play. The following introductory statement was provided to our questions: "Unfortunately, it happens from time to time that our experimental instructions are not sufficiently precise, or that participants forget or do not believe certain aspects of the instructions. To improve our instructions for future experiments we ask you to answer the following questions." After this introduction, there was the following statement regarding one-shot play: "In the instructions we told you that in every period of the experiment you will be matched with three new persons. However, we do not know whether — during the experiment — you also perceived this in this way." Then the subjects had to indicate YES or NO for the following statement: "During the experiment I assumed that in every period I will be matched with three new persons." After subjects had answered the first question, the following statement regarding anonymity was given: "In the instructions we told you that the other group members will never be informed that you have been together with them in a group, that is, the other group members do not receive information about your personal identity. We do not know whether — during the experiment — you believed this." Then they had to indicate YES or NO to the following statement: "I assumed that the other participants will not receive any information about my personal identity."

Ninety-six percent of the subjects ($n = 120$) indicated that they believed that they would be matched with new persons in every period and that their anonymity would be preserved. This suggests that carefully written instructions in combination with the experimenters' credibility to always tell the truth (which is unfortunately often not the case in psychology experiments) induce subjects to believe the anonymous one-shot character of experiments. Sometimes it is argued that the belief that there is a positive probability that one's identity will be revealed or that one plays in fact a repeated game may be driven by unconscious mechanisms. It is, however, often not clear what these unconscious mechanisms could be and how they work. In the absence of more concrete specifications, this argument remains elusive, and it is difficult to refute because if one provides evidence challenging argument A it is always possible to say, "Oh, I didn't mean argument A but argument B."[1] In fact, to our knowledge, the maladaptation account has never been carefully formulated in terms of empirically refutable predictions. According to our experience, it often comes in the form of vague intuitions based on imprecise and questionable generalizations about human evolutionary history.

If indeed unconscious mechanisms are the reason for helping and punishing responses, why do subjects respond *so quickly* to changes in the cost of helping or punishing? Likewise, why do subjects *instantaneously* change their behavior when repeated interactions or the possibility of reputation formation are introduced? These quick behavioral changes are almost certainly mediated by sophisticated evaluations of the costs and benefits of different courses of action that are available to them.[2] We find it hard to reconcile subjects' quick responses to treatment changes, which almost surely are mediated by sophisticated, conscious, cognitive acts, with the view that a cognitively inaccessible mechanism drives the baseline pattern of reciprocal responses.

Another problem with the "greater than zero baseline probability of repeated interaction" argument described above is that it can only arise from a fundamental misunderstanding of evolutionary models of repeated interaction. The canonical models of the evolution of cooperation via repeated interactions specify

1 If emotions are thought to be the mechanism, the claim is doubtful because emotions such as gratefulness, which enhance cooperative responses to cooperative acts, or anger, which provides a basis for strong negative reciprocity, are not cognitively impenetrable.

2 For example, when one introduces a cost change within an experimental session, subjects have to understand consciously what higher costs of punishment mean for their earnings. Subjects in the experiments are always paid at the end of the whole session. Thus, they do not directly experience these higher costs during the session because the gains and losses accruing during the session are just documented on their decision sheets or in a computer file. Only at the end do they experience the higher costs in terms of lower cash earnings.

that the selection of a cooperative strategy depends on the probability of future interaction and the fitness costs and benefits. If the ratio of costs to benefits is sufficiently large, no number of repeated interactions (given finite lifetimes) will favor reciprocating strategies. Given that life in ancestral environments provides a full range of costs and benefits of different kinds (from giving one's life for a nonrelative in warding off a predator to providing an ounce of meat from a large kill), evolutionary psychology should predict that individuals are able to switch from a zero baseline (full defection) to full cooperation, depending on the detail. Similarly, if groups are fluid and migration is common, the expected number of future interactions will vary on a case by case basis, so individuals should be geared up to withdraw substantial forms of cooperation (high cost-benefit ratio) completely from ephemeral individuals. Again, evolutionary psychology should predict that individuals are fully capable of grasping a zero baseline.

Using the maladaptation account, it is also difficult to explain the fundamental heterogeneity in subjects' behavior. In most experiments there is a large group of subjects that indicate strongly reciprocal responses *as well as* a large group that behave in a completely selfish manner. The relative size of these groups depends on the economic costs and benefits of strongly reciprocal actions. In fact, it is almost surely possible to wipe out any reciprocal responses by making them sufficiently expensive. However, if the costs are not too high many subjects reciprocate. How can the maladaptation account explain the existence of completely self-interested behavior? If the maladaptation account is correct, why do we not observe everybody engaging in strongly reciprocal behavior?

The Frequency of "One-shot" Interactions

Laboratory evidence as well as casual observations from everyday life suggest that humans have fine-tuned behavioral repertoires which take into account whether they face kin or nonkin, partners or strangers, and whether they can or cannot gain an individual reputation. This suggests that humans who exhibited these behavioral traits had an evolutionary advantage over those humans who exhibited more blunt behaviors. The likely reason for this advantage is, contrary to common mythology, that humans faced many interactions where the probability of future interactions was sufficiently low to make defection worthwhile. In addition, the costs of mistakenly treating unrelated individuals as kin, or treating strangers as partners, were very high. That is, if one were to "reverse engineer" from the available empirical evidence back to the ancestral environment (environment of evolutionary adaptiveness, EEA) that characterized human evolution, one would predict that our ancestors had frequent encounters with strangers, and that these encounters had substantial fitness consequences. As we illustrate with ethnographic data from small-scale societies, a lack of vigilance in interactions with unfamiliar individuals often had deadly consequences. Using data from both the primatological and ethnographic records, we find no support for the idea that the EEA lacked fitness-relevant interactions with strangers (by "strangers" we mean individuals who were neither kin nor long-term interactants). To the contrary, data from small-scale foraging societies and chimpanzees clearly show that interactions with strangers were likely common and highly fitness relevant. Before we sketch the evidence, it is important to discard the false dichotomy between completely anonymous one-shot interactions on one side and nonanonymous repeated interactions on the other.

The real question is not whether 100% nonrepeated interactions (completely anonymous) have been frequent in evolutionary history or not — those unfamiliar with the mathematical details of the relevant theory often couch the argument in these terms. Instead, the really important distinction is between encounters with a sufficiently low probability of future encounters, such that defection was the fitness-maximizing strategy, and encounters with a sufficiently high probability of future encounters, such that cooperation was the fitness-maximizing strategy. If early humans faced a mix of situations such that sometimes the best strategy was defection and sometimes the best strategy was cooperation, then there is an a priori case for the selection of strategies that do *distinguish* between these two situations (this also implies that individuals should be fully capable of taking a "zero-baseline" of cooperation). Under these circumstances, the logic of the maladaptation argument implies that selection should have favored strategies that can distinguish encounters that have a sufficiently low probability of repeated interaction, and generate noncooperating, nonpunishing behavior in one-shot experiments. Thus, in these circumstances, the maladaptation argument cannot account for strongly reciprocal responses in one-shot encounters.

To be precise, assume that if the probability of future encounters in a simultaneously played PD is below $p = 0.7$, the fitness-maximizing strategy is to defect, whereas if the probability is above 0.7, the fitness-maximizing strategy is to cooperate.[3] Then individuals who defect in the first and cooperate in the second situation have a selection advantage over those individuals who cooperate in both situations. Thus, selection should favor the individuals who discriminate between the two situations. Yet, if an individual is capable of understanding that defection is the fitness-maximizing strategy in case of a probability smaller than $p = 0.7$, why should this individual then be unable to understand that defection is also the best thing when the continuation probability is zero? In fact, if the continuation probability is zero, it is clear that there are no future gains from

[3] Here we abstract from the multiple equilibrium problem that is inherent in any repeated PD or public goods game with a sufficiently high continuation probability. The maladaptation account typically forgets that repeated interactions are by no means sufficient for cooperation, even if the probability of future encounters is one, because there are typically infinitely many equilibria implying less than full cooperation. Thus, even if it were true that humans never faced encounters with a low probability of future interactions, it is not guaranteed that evolution favors strongly reciprocal behavior patterns. The reason is simply that there are also many equilibria in repeated games with noncooperative outcomes.

cooperation; however, if the continuation probability is, for example 0.5, the situation is more ambiguous, and the costs and benefits from different actions are more difficult to assess. The upshot of this argument is that we believe that interior continuation probabilities were common in evolutionary history and that evolution has, therefore, favored discriminating strategies. If humans have been selected to apply discriminating strategies in the case of interior probabilities, that is, they defect even when the continuation probability is positive but insufficiently high, then the discriminating strategy also induces them to defect in the case of a zero continuation probability. Therefore, the maladaptation view cannot account for the existence of strongly reciprocal behavior.

How plausible is the assumption that interior continuation probabilities have been common in our evolutionary history? To get a first grip on this question, consider a simultaneously played PD in which players A and B receive the payoff (c, c) if both cooperate, (d, d) if both defect, $(c + t, c - s)$ if A defects and B cooperates, and vice versa in the opposite case. To ensure that the conditions of the PD hold the payoffs obey $c + t > c > d > c - s$. The gains of the defector arising from a deviation from joint cooperation are measured by t, and s measures the loss of the cooperator if the other player defects. It can be shown that in a cooperative equilibrium obtained by so-called trigger strategies, the continuation probability has to exceed a critical level p_0, which is given by $p_0 = t/(c - d + t)$.[4] Thus, the higher the temptation t, the higher the critical threshold p_0; the higher the increase in gains arising from mutual cooperation $(c - d)$, the lower the

4 Consider a cooperative equilibrium in which the joint cooperation outcome is supported by so-called trigger strategies. In a trigger-strategy equilibrium, both players start with cooperation. They cooperate in period $\tau > 1$ as long as nobody has defected in the past. When one of the players has defected in the past, both defect in period τ. Thus, defection by the other player is punished with future noncooperation, but the defector also does not cooperate in the future because he expects the cheated player to defect in the future. A trigger strategy is thus unforgiving and imposes large costs on the defector, implying that even for a relatively low continuation probability punishment can be sufficiently high to ensure cooperation. Thus, our assumption of a trigger-strategy equilibrium favors the maladaptation view because we allow for large punishments, which render cooperation already an equilibrium at relatively low continuation probabilities. The lower the continuation probability necessary to sustain a cooperative equilibrium, the more likely human cooperation is achieved in repeated interactions, and the more likely the maladaptation argument applies. How large is the threshold continuation probability that ensures the existence of a cooperative equilibrium sustained by trigger strategies? To compute this probability, we first derive the expected payoff (viewed from period τ onwards) of an individual in the joint cooperation equilibrium. With probability p the individual receives in each future period c implying that the expected payoff is given by the discounted sum $\Sigma p^{\tau} c$, which is equal to $c/(1 - p)$. If, instead, an individual defects in period τ, it receives $c + t$ in this period, but from the next period on, the payoff is only d. Thus, in case of defection in period τ the expected payoff from τ on is given by $(c + t) + p\Sigma p^{\tau} d$, which is equal to $(c + t) + (pd)/(1 - p)$. If the expected payoff from joint cooperation, $c/(1 - p)$, is higher than from defection, $(c + t) + (pd)/(1 - p)$, it is rational for a selfish individual to cooperate. Simple manipulations of this inequality show that this is the case if $p > [t/(c - d + t)]$.

critical threshold. If temptation becomes very large and approaches plus infinity, the critical threshold approaches 1, rendering cooperation among selfish individuals highly unlikely. In times where the survival of an individual is at stake, the temptation value can be plausibly set equal to "plus infinity." Note also that if $c - d$ becomes very small, the critical threshold again approaches $p_0 = 1$.

All the characteristics of repeated encounters that affect $c - d$ and t affect the critical threshold. It is plausible that throughout evolutionary history humans have faced a wide variety of different conditions shifting the threshold up and down; the threshold, however, was always in the interior because $t > 0$ prevailed. Therefore, for sufficiently low continuation probabilities (i.e., if $p < p_0$) defection was the fitness-maximizing strategy. Moreover, due to variation in conditions over individuals' lifetimes and across different contexts, ancestral humans probably faced many situations in which defection was worthwhile, even when the probability of future encounters was relatively high (e.g., during times of scarcity, families could offer their children as food to be cooked and consumed by unrelated individuals who would be expected to offer their children in the next scarcity). Surely, selection would favor a "defect" strategy in this context, as the benefit-cost ratio is too low.

To assemble an understanding of the relevant characteristics of the EEA, we marshal evidence from both contemporary foraging populations (and other small-scale societies) and extant nonhuman primates to show that ephemeral contacts with strangers (low repeat interactants) were likely to be both frequent and to carry substantial fitness consequences. We think that this evidence should lead evolutionarily minded scholars to predict that humans should be equipped with specialized cognitive machinery capable of distinguishing low-frequency interactants (when $p < p_0$) from long-term repeat interactants. Above, we summarized a substantial body of empirical evidence showing that people do seem fully capable of adapting their behavior in response to information about the likelihood of future interaction, even when that information comes in the form of "novel" laboratory cues. Thus, we think that much of the laboratory-observed behavior results from adaptive processes acting on human psychology over the course of hominid evolution. In what follows, we sketch the basic pattern of evidence, but since the facts are spread over a large number of ethnographies and primatological sources, we cannot provide a full picture here.

Hunter-gatherers vary widely in settlement patterns, social structure, economic integration, institutional forms, population density, and resource use (Kelly 1995; Arnold 1996). For our purposes, we focus on the nomadic and semi-nomadic foraging populations that are *assumed* to give the most insight into Palaeolithic lifestyles.[5] Typically, ethnolinguistic groups of hunter-

[5] This assumption, however, deserves substantially more scrutiny than it typically receives. Most of the foragers used as "models" of EEA societies live in the remote marginal environments left over after 10,000 years of the agricultural revolution. The few historical and archaeological cases we have of hunter-gatherers living in rich nonmarginal environments show societies that are substantially denser, larger, and

gatherers number at least in the low thousands, but populations between 10,000 and 15,000 are not uncommon. At the local level, many nomadic foragers live in mobile bands of approximately 25 individuals. The membership of these bands is fluid and constantly shifting as individuals move from one band to another for various reasons. These local groups often aggregate around centralized resources during certain seasons (e.g., waterholes, pinon nuts, annual herds). In these recurrent circumstances, social life becomes very intense: Mates are found, ancestors glorified, alliances formed, and scores settled. Once every twelve years or so, Shoshone families would aggregate in larger groups, consisting of 75 people or more, for "antelope drives" (Johnson and Earle 2000).

More irregularly, periodic environmental fluctuations (floods, windstorms, plagues, droughts, hurricanes, volcanic eruptions) often (in evolutionary terms) bring together substantial numbers of strangers during fitness-critical times, as individuals and groups travel over great distances in search of water, caribou, and other resources. In the Kalahari, for example, the most severe droughts hit on average about two per hundred years, which means that the average !Kung experiences one in his lifetime (Lee 1998, pp. 79–83). Under these conditions many of the "permanent waterholes" failed, and bands had to travel great distances. Of the five permanent waterholes in the 64,000 km^2 Dobe area, only two have never failed in living memory. During these kinds of aggregations, people will encounter many strangers that they are not likely to encounter again in their lifetime. This was actually observed by J. Marshall during the severe drought of 1952, when seven San groups converged on a waterhole that many of them had not used in living memory and were unlikely to use again in their lifetimes (Lee 1998, p. 86). Obviously, these events had substantial fitness consequences as large aggregations of strangers attempted to share water, game, and monogongo nuts. Wiessner (1982) described how the combination of high winds (which destroyed the mongongo nut crop) and a plague (which decimated the meager domestic stocks) caused the local population to scatter itself across numerous camps all over an area. Similar drought-related patterns exist throughout Australian and American ethnographic and historical record (e.g., Cane 1990, p. 157; Aschmann 1959, p. 96; Peterson 1975, 1978; Peterson and Long 1986). If nothing else, environmental shocks would have guaranteed that strangers encountered one another during fitness critical times and had to divide resources (or fight) in some fashion. The reader should keep in mind that environments were likely substantially more variable in the Palaeolithic than the Holocene (now), so such population mixes driven by shocks would have likely been more common.

more complex than any of the "standard models." The marginal environments inhabited by these remnant foragers are those few places on Earth that resisted plant domestication. The inhabitants of these marginal environments, the smallest-scale, nomadic foragers known, provide the "worst cases" for our argument. If we used the full spectrum of foraging societies, with their high densities, slaves, classes, and chiefs, the maladaptation hypothesis would do even worse.

The !Kung institution of *hxaro* (dyadic, long-term trading partnerships) provided a risk-managing means of coping with both spatial and temporal variation in food and water. In times of acute stress, individuals in families often "activated" a trading partnership by traveling 200 km to visit one of their trading partners and staying for several weeks. The diffuse networks of *hxaro* trading relationships combined with both temporal and spatial variation, guaranteed a well-mixed population, which means that individuals were likely to encounter strangers in the camps of their *hxaro* partners. It is often forgotten that while an individual had direct, long-term, reciprocal ties with the *hxaro* partner, they did not have such ties with all the other members of their partners' encampments, with whom they, nevertheless, shared water, game, meat, and other local resources as if they were a member of the group. Further, in managing the *hxaro* relationship, of which the average adult with mature children had around 24 (that is 48 per couple), the average !Kung had to travel an area of at least 10,000 km,[2] wherein they could encounter easily more than thousand people. However, 10% of *hxaro* partnerships (i.e., 2.4 of each person's 24 partners) ranged over roughly 140,000 km^2, where they could meet up to 14,000 inhabitants. It is hard to believe that these individuals did not have many cooperation opportunities that were associated with low frequencies of future interactions.

Again, using Kalahari data, Harpending calculated the average distance between the birthplaces of spouses and the birthplaces of parents and offspring. The average distance between parent and offspring is 66.5 km; over 10% of the population had distances between 100 km and 220 km. The average distance between the birthplaces of mates is 70 km (Harpending 1998). This means that in searching for a mate, the average individual covered a territory of 15,000 km^2, an area in which he or she was likely to encounter 1,500 people. It is essentially impossible that any individual maintained long-term repeated interactions with so many people.

Foragers, sometimes alone or in small groups, are known to have traveled extensively over large regions for a variety of reasons, often to obtain specific resources such as a particular kind of wood or ochre. Australia is perhaps the best place on Earth to look for evidence on this matter because, until European conquest in the late 19th and early 20th century, much of Australia was inhabited entirely by full-time foragers. (Agriculture was first introduced in Australia by the European settlers.) In the Western Australian Desert, particular areas attracted a constant influx of thousands of individuals, who had traveled hundreds of miles to obtain a particular type of acacia wood used in crafting fine spears. Myers (1986) recorded life histories from dozens of Pintupi Australians who had not seen a white person until their later years. These life histories tell of substantial traveling and constant encounters with complete strangers. Sometimes these encounters were friendly and an exchange of goods or information occurred, whereas at other times hostilities ensued. There seems little doubt that many of these were one-shot interactions with strangers in the middle of the desert, hundreds of miles from one's home territory.

Further, diverse hunter-gatherer groups have rituals for bringing strangers into the camp. In several cases, these rituals were observed, or experienced, by the earliest European explorers (and recorded in detail by later ethnographers). On the continent of hunter-gatherers (Australia), ethnographers and explorers found elaborate rituals that were *specifically* used for bringing *strangers* peacefully into camp (Thomson 1932, pp. 163–164). Strangers not performing these rituals were treated as having hostile intentions and usually killed (note the fitness effects, the ability to "treat" strangers differently). If foraging life does not involve encountering strangers, why would nearly all aboriginal groups have elaborate culturally evolved rituals specific for admitting strangers? Moreover, this basic ritualized interaction managed to diffuse over a vast region, so it is hard to believe this was a newly developed practice. On the other side of the world, inhabiting the fierce climate on the southern tip of South America (Tierra del Fuego), two foraging groups — the Ona and Yaghan — maintain a similar (though less elaborated) ritual process for bringing strangers into camp. At first European contact, several of Magellan's crew, in search of provision, approached a small encampment of these foragers during their trip through the Strait of Magellan. Unfortunately, these men did not perform the proper ritual and were immediately killed. These lonely foragers immediately spotted these strangers and reduced their fitness substantially.

Far to the north, arctic foragers traveled extensively to maintain knowledge of geography and ecology over vast regions. The Nunamiut maintained knowledge of nearly 250,000 km^2 (Binford 1983), while typically only using about 25,000 km^2. During occasional recognizance travels, they may have encountered any of the 5,000 inhabitants. It seems unlikely that each hunter knew and personally maintained long-term relationships with these thousands of adults.

Another line of evidence comes from archaeological and ethnohistorical evidence of warfare and long-distance raiding. As we showed for Australia, foragers often traveled long distances to trade or obtain resources from distant groups of strangers, and they encountered many strangers along the way. Human groups, including foragers, have for as far as we can see back in the archaeological record had violent interactions with other groups (Keely 1996). As far back as the Upper Palaeolithic, we see evidence of large settlements behind defensive walls (Johnson and Earle 2000). For our case, the most persuasive evidence comes from long-distance raiding. Here war parties traveled hundreds of miles to raid, pillage, and steal wives from strangers. North American ethnohistorical data show that groups like the Tlingit, from the Alaskan panhandle, raided as far south as Puget Sound, and the Mohave raided groups on the Californian coast. During the historic period, the Iroquois raided Delaware, the Great Lakes, and the Mississippi Valley. This kind of raiding, if present in the EEA (there is no reason to think it was not), would have provided consistent selection pressure to distinguish strangers (possible raiders) from friendly locals. The idea that human psychology should have evolved to cause individuals to "hedge their bets" by assuming that any particular stranger might be a long-term interactant seems mistaken. Combining this raiding data with the above evidence suggests that people may well have routinely encountered both strangers who were friendly and interested in trading, as well as strangers who were dangerous and bent on raiding, stealing, and murder. What we do not see in the ethnographic data is any hint that people rarely encountered strangers in fitness-relevant circumstances.

Nonhuman primates provide another line of evidence that should be used to develop an understanding of the EEA. In this case, the evidence is stark. The mating patterns of nonhuman primates means that at least one sex leaves their natal group to find and join another social group. Upon arrival at their new group (which may or may not contain any kin), the existing group members do not confuse the individual with their kin or coalition partners. The individual has to build her own coalitions gradually, find a mate or mates, and produce some kin. In nonhuman primate societies, even in the unnaturally large and provisioned ones in zoos and research centers, animals do not "get confused" and mistakenly treat strangers as long-term partners and kin.

Chimpanzees, our closest genetic relatives, provide an instructive example. Chimpanzee social groups are known to go on "patrols" along the border of their territories. If they encounter a smaller group of "stranger chimpanzees" (from another chimpanzee group), they attack and kill (or drive off) the strangers (Manson and Wrangham 1991).[6] Chimpanzees do not get confused and treat strangers like long-term coalition partners, other group members, or kin. Chimpanzees apparently have the ability to distinguish group members from strangers and treat them differentially. Chimpanzees do not have the "nonzero baseline" of mistaken prosociality toward other chimpanzees, despite their long history in small-scale societies based on reciprocity and kinship. From this perspective, it seems odd to argue that humans carry around a nonzero baseline of mistaken prosociality for strangers, whereas chimpanzees do not.

The above account necessarily lacks the kind of systematic rigor and evidence with which one would like to address the EEA question because such evidence simply does not exist. Nevertheless, we believe that it demonstrates that the widely held view of ancestral human societies as isolated groups, which did not mix or interact with surrounding social groups or strangers, has little, if any, empirical support.

CAN STRONG RECIPROCITY SURVIVE IN EVOLUTIONARY EQUILIBRIUM?

Bad theories often survive if no other theories are available to replace them. Therefore, if strong reciprocity is unlikely to be a maladaptive trait, it is important to develop an adaptive account of strong reciprocity. Recently Price et al.

---

6 Female strangers may, however, be incorporated into the raider's group.

(2002) argued that the punishment of noncooperators in public goods situations evolved because the punishers can reduce or overturn the payoff differences between themselves and the punished defectors. Unfortunately, this argument is theoretically and empirically invalid. It is theoretically invalid because it does not solve the core problem of why nonpunishing cooperators do not replace the punishing cooperators. This question is decisive because nonpunishing cooperators will reap the benefits created by the presence of punishing cooperators without paying the cost. On the empirical side, experiments conducted by Falk et al. (2001) show that if the punisher cannot reduce the payoff differences — because every dollar invested into punishment only reduces the payoff of the punished by one dollar — the willingness to incur costs to punish defectors is still very high. Thus, a lot of punishment occurs even if payoff differences cannot be affected.

On the positive side, several theoretically rigorous models have been proposed that provide an adaptive evolutionary foundation for strong reciprocity. Taking advantage of the fact that humans, unlike other primates, are heavily dependent on certain kinds of imitation and other forms of social learning, Henrich and Boyd (2001) show that an arbitrarily small amount of conformist transmission makes cooperate–punish a stable culturally evolved equilibrium in one-shot n-person interactions. They go on to show that, once this equilibrium spreads (a process modeled in Boyd and Richerson 2003), within-group individual selection will favor prosocial genes that allow individuals to avoid the costs of being punished — these are genes that would otherwise not be favored without the interaction of genes and culture (see also Henrich et al., this volume; Richerson et al., this volume; Henrich 2003).

More recently, Boyd, Gintis, Bowles and Richerson (2003) used a simulation to show that a model of cultural group selection is able to explain altruistic cooperation as well as altruistic punishment in large groups. They calibrate the parameters of their model to mimic the likely evolutionary conditions of humans. The idea behind their model is that in the vicinity of an evolutionary equilibrium, where altruistic cooperators and altruistic punishers are frequent, within-group selection against altruistic punishers is very weak because noncooperation rarely occurs and, hence, few punishment costs have to be born by the altruistic punishers (a logic first pointed out in Henrich and Boyd 2001). They show that there is an important asymmetry between altruistic cooperation and altruistic punishment because, in the absence of altruistic punishment, within-group selection against altruistic cooperation is always strong. Thus, cultural group selection cannot sustain altruistic cooperation without altruistic punishment.

Despite their formal rigor, cultural group selection models are often treated with skepticism because of the long running controversy about genetic group selection within biology (Sober and Wilson 1998). However, this skepticism, where it is found, is typically based on overgeneralized suspicions of anything using the word "cultural" or "group selection," rather than an in-depth understanding of the details of the *differences* between cultural evolution (and culture–gene coevolution) and genetic evolution. Criticisms that apply to genetic group selection and make it an unlikely force in human evolution, do *not* apply to cultural and culture–gene coevolutionary models discussed above (Boyd and Richerson, in press; Henrich and Boyd 2001; Henrich 2003). As we found at this Dahlem Workshop, skeptics are unable to come up with specific criticisms, and when pressed, they voice only an untargeted, vague distrust.

Gintis (2000) also developed an evolutionary model showing how strong reciprocity can evolve and persist in evolutionary equilibrium.[7] His model is based on the plausible idea that in the relevant evolutionary environment human groups faced extinction threats (wars, famines, environmental catastrophes) with a positive probability. When groups face such extinction threats, neither reciprocal altruism nor indirect reciprocity can sustain the necessary cooperation that helps the groups to survive the situation because the shadow of the future is too weak. Kin selection also does not work here because in most human groups membership is not restricted to relatives but is also open to nonkin members. However, groups with disproportionately many strong reciprocators are better able to survive these threats. Hence, within-group selection creates evolutionary pressures against strong reciprocity because strong reciprocators engage in individually costly behaviors that benefit the whole group. In contrast, between-group selection favors strong reciprocity because groups with disproportionately many strong reciprocators are better able to survive. The consequence of these two evolutionary forces is that in equilibrium, strong reciprocators and purely selfish humans coexist. This logic applies to genes, cultural traits, or both in an interactive process. Thus, this approach provides a logically rigorous argument as to why we observe heterogeneous responses in laboratory experiments.

SUMMARY

Our main purpose in this chapter was to address the question of whether strong reciprocity results from the maladaptive operation of a psychology that evolved in ancestral human environments under processes described by the canonical models of cooperation (reciprocal altruism, indirect reciprocity, and reputation). The weight of the evidence suggests that strong reciprocity is unlikely to be accounted for by the "maladaptationist approach." This means that the prevailing evolutionary accounts, which often ignore population structure and our second system of inheritance (culture), cannot explain strong reciprocity. Cultural group selection models and culture–gene coevolutionary models are capable of providing ultimate equilibrium explanations of strong reciprocity. However, we

---

7 For other evolutionary models of strong reciprocity, see Bowles and Gintis (2001) and Sethi and Somananthan (2001, 2003).

do not believe that these models hold the last word in the debate about the ultimate causes of strong reciprocity. More empirical and theoretical work is necessary to evaluate the plausibility of these models and to discriminate between them and possibly other forthcoming accounts of strong reciprocity. Because of our limited empirical knowledge about human evolutionary history and the general lack of systematic attempts by evolutionary theorists to generate sharp testable predictions, our conclusions necessarily have to be preliminary. In the future, we strongly encourage theoreticians to generate sharply focused, testable hypotheses that allow discrimination between alternative ultimate explanations. Empirical work should not just aim at providing evidence that is consistent with one of the prevailing approaches but should also aim at discriminating between competing approaches. There is thus ample room for further theoretical and empirical investigations.

ACKNOWLEDGMENTS

This paper is part of a research project on strong reciprocity financed by the Network on Economic Environments and the Evolution of Individual Preferences and Social Norms of the MacArthur Foundation.

REFERENCES

Alexander, R.D. 1987. The Biology of Moral Systems. New York: de Gruyter.

Arnold, J.E. 1996. The archaeology of complex hunter-gatherers. *J. Archaeol. Meth. Theory* **3(2)**:77–126.

Aschmann, H. 1959. The Central Desert of Baja California. Ibero-Americana 42. Berkeley: Univ. of California.

Binford, L. 1983. In Pursuit of the Past. London: Thames and Hudson.

Bowles, S., and H. Gintis. 2001. The evolution of strong reciprocity. Discussion Paper, Univ. of Massachusetts at Amherst.

Boyd, R., H. Gintis, S. Bowles, and P. Richerson. 2003. The evolution of altruistic punishment. *Proc. Natl. Acad. Sci. USA* **100**:3531–3535.

Boyd, R., and P.J. Richerson. 2003. Solving the puzzle of human cooperation. In: Evolution and Culture, ed. S. Levinson. Cambridge MA: MIT Press, in press.

Camerer, C.F., and R.H. Thaler. 1995. Ultimatums, dictators and manners. *J. Econ. Persp.* **9**:209–219.

Cameron, L.A. 1999. Raising the stakes in the ultimatum game: Experimental evidence from Indonesia. *Econ. Inq.* **37(1)**:47–59.

Cane, S. 1990. Desert demograhy: A case study of pre-contact aboriginal densities. In: Hunter-Gatherers Demography: Past and Present, ed. B. Meehan and N. White, pp. 149–59. Oceania Monograph 19. Sydney: Univ. of Sydney.

Case, A., I. Lin, and S. McLanahan. 2001. Educational attainment of siblings in stepfamilies. *Evol. Hum. Behav.* **22**:269–289.

Daly, M., and M. Wilson. 1988. Homicide. New York: de Gruyter.

Falk, A., E. Fehr, and U. Fischbacher. 2001. Driving forces of informal sanctions. Working Paper 59. Institute for Empirical Research in Economics, Univ. of Zurich.

Falk, A., and U. Fischbacher. 1999. A theory of reciprocity. Working Paper 6. Institute for Empirical Research in Economics, Univ. of Zurich.

Fehr, E., and U. Fischbacher. 2002a. Retaliation and reputation. Institute for Empirical Research in Economics, Univ. of Zurich, mimeo.

Fehr, E., and U. Fischbacher. 2002b. Why social preferences matter: The impact of non-selfish motives on competition, cooperation and incentives. *Econ. J.* **112**:C1–C33.

Fehr, E., U. Fischbacher, and S. Gächter. 2002. Strong reciprocity, human cooperation, and the enforcement of social norms. *Hum. Nat.* **13**:1–25.

Fehr, E., and S. Gächter. 1998a. How effective are trust- and reciprocity-based incentives. In: Economics, Values and Organization, ed. A. Ben-Ner and L. Putterman, pp. 337–363. Cambridge: Cambridge Univ. Press.

Fehr, E., and S. Gächter. 1998b. Reciprocity and economics: The economic implications of homo reciprocans. *Eur. Econ. Rev.* **42**:845–859.

Fehr, E., and S. Gächter. 2000. Cooperation and punishment in public goods experiments. *Am. Econ. Rev.* **90**:980–994.

Fehr, E., and S. Gächter. 2002. Altruistic punishment in humans. *Nature* **415**:137–140.

Fehr, E., S. Gächter, and G. Kirchsteiger. 1997. Reciprocity as a contract enforcement device. *Econometrica* **65**:833–860.

Fehr, E., E. Tougareva, and U. Fischbacher. 2002. Do high stakes and competition undermine fairness? Working Paper 125. Institute for Empirical Economic Research, Univ. of Zurich.

Fischbacher, U., C. Fong, and E. Fehr. 2002. Fairness and the power of competition. Working paper 133. Institute for Empirical Research in Economics, Univ. of Zurich.

Fischbacher, U., S. Gächter, and E. Fehr. 2001. Are people conditionally cooperative? Evidence from a public goods experiment. *Econ. Lett.* **71**:297–404.

Gächter S., and A. Falk. 2002. Reputation and reciprocity: Consequences for the labor relation. *Scand. J. Econ.* **104**:1–25.

Gintis, H. 2000. Strong reciprocity and human sociality. *J. Theor. Biol.* **206**:169–179.

Gintis, H., E. Smith, and S. Bowles. 2000. Costly signaling and cooperation. *J. Theor. Biol.* **213**:103–119.

Güth, W., R. Schmittberger, and B. Schwarze. 1982. An experimental analysis of ultimatium bargaining. *J. Econ. Behav. Org.* **3**:367–388.

Hamilton, W.D. 1964. Genetical evolution of social behavior I, II. *J. Theor. Biol.* **7(1)**:1–52.

Harpending, H. 1998. Regional variations in !Kung popuations. In: Kalahari Hunter-Gatherers: Studies of the !Kung San and Their Neighbors, ed. R.B. Lee and I. DeVore, pp. 153–165. Cambridge, MA: Harvard Univ. Press.

Henrich, J. 2003. Cultural group selection, coevoutionary processes, and large-scale cooperation. *J. Econ. Behav. Org.*, in press.

Henrich, J., and R. Boyd. 2001. Why people punish defectors: Weak conformist transmission can stabilize costly enforcement of norms in cooperative dilemmas. *J. Theor. Biol.* **208**:79–89.

Henrich, J., R. Boyd, S. Bowles et al. 2001. In search of *homo economicus*: Behavioral experiments in 15 small-scale societies. *Am. Econ. Rev.* **91**:73–78.

Hoffman, E., K. McCabe, and V. Smith. 1996. On expectations and monetary stakes in ultimatum games. *Intl. J. Game Theory* **25**:289–301.

Johnson, A.W., and T.K. Earle. 2000. The Evolution of Human Societies: From Foraging Group to Agrarian State. 2d ed. Stanford: Stanford Univ. Press.

Johnson, D.P., P. Stopka, and S. Knights. 2003. The puzzle of human cooperation. *Nature* **421**:911–912.

Keeley, L.H. 1996. War before Civilization. New York: Oxford Univ. Press.

Kelly, R. 1995. The Foraging Spectrum. Washington, D.C.: Smithsonian Institution Press.

Keser, C., and F. van Winden. 2000. Conditional cooperation and voluntary contributions to public goods. *Scand. J. Econ.* **102**:23–39.

Lee, R.B. 1998. !Kung spatial organization: An ecological and historical perspective. In: Kalahari Hunter-Gatherers: Studies of the !Kung San and Their Neighbors, ed. R.B. Lee and I. DeVore, pp. 74–97. Cambridge, MA: Harvard Univ. Press.

Manson, J.H., and R.W. Wrangham. 1991. Intergroup aggression in chimpanzees and humans. *Curr. Anthro.* **32**:369–390.

McCabe, K.A., S.J. Rassenti, and V.L. Smith. 1998. Reciprocity, trust, and payoff privacy in extensive form bargaining. *Games Econ. Behav.* **24**:10–24.

Milinski, M., D. Semmann, and H.-J. Krambeck. 2002. Reputation helps solve the "tragedy of the commons." *Nature* **415**:424–426.

Myers, F.R. 1986. Pintupi Country, Pintupi Self: Sentiment, Place and Politics among Western Desert Aborigines. Washington, D.C.: Smithsonian Institution Press.

Nowak, M.A., and K. Sigmund. 1998. Evolution of indirect reciprocity by image scoring. *Nature* **393**:573–577.

Peterson, N. 1975. Hunter-gatherer territoriality: The perspective from Australia. *Am. Anthro*. **77**:53–68.

Peterson, N. 1978. The importance of women in determining the composition of residential groups in aboriginal Australia. In: Woman's Role in Aboriginal Society, ed. F. Gale, pp. 16–27. Canberra: Australian Institute of Aboriginal Studies.

Peterson, N., and J. Long. 1986. Australian Territorial Organization. Oceania Monograph 30. Sydney: Univ. of Sydney.

Price, M.E., L. Cosmides, and J. Tooby. 2002. Punitive sentiment as an anti-free rider psychological device. *Evol. Hum. Behav.* **23**:203–231.

Rabin, M. 1993. Incorporating fairness into game theory and economics. *Am. Econ. Rev.* **83**:1281–1302.

Roth, A.E. 1995. Bargaining experiments. In: Handbook of Experimental Economics, ed. J. Kagel and A. Roth, pp. 253–348. Princeton, NJ: Princeton Univ. Press.

Roth, A.E., V. Prasnikar, M. Okuno-Fujiwara, and S. Zamir. 1991. Bargaining and market behavior in Jerusalem, Ljubljana, Pittsburgh, and Tokyo: An experimental study. *Am. Econ. Rev.* **81**:1068–1095.

Rozin, P., L. Millman, and C. Nemeroff. 1986. Operation of the laws of sympathetic magic in disgust and other domains. *J. Pers. Soc. Psych.* **50**:703–712.

Sethi, R., and E. Somananthan. 2001. Preference evolution and reciprocity. *J. Econ. Theory* **97**:273–297.

Sethi, R., and E. Somananthan. 2003. Understanding reciprocity. *J. Econ. Behav. Org.* **50(1)**:1–27.

Silk, J.B. 1980. Adoption and kinship in Oceania. *Am. Anthro.* **82**:799–820.

Silk, J.B. 1987. Adoption and fosterage in human societies: Adaptations or enigmas? *Cult. Anthro*. **2**:39–49.

Silk, J.B. 2003. The evolution of cooperation in primate groups. In: The Moral Sentiments: Origins, Evidence, and Policy, ed. H. Gintis, S. Bowles, R. Boyd, and E. Fehr. Princeton, NJ: Princeton Univ. Press, in press.

Slonim, R., and A.E. Roth. 1998. Financial incentives and learning in ultimatum and market games: An experiment in the Slovak Republic. *Econometrica* **65**:569–596.

Sober, E., and D.S. Wilson. 1998. Unto Others: The Evolution and Psychology of Unselfish Behavior. Cambridge, MA: Harvard Univ. Press.

Thomson, D.F. 1932. Ceremonial presentation of fire in North Queensland: A preliminary note on the place of fire in primitive ritual. *Man* **32**:162–166.

Tomasello, M., and J. Call. 1997. Primate Cognition. Oxford: Oxford Univ. Press.

Trivers, R. 1971. The evolution of reciprocal altruism. *Qtly. J. Biol.* **46**:35–57.

Wiessner, P. 1982. Risk, reciprocity and social influences on !Kung San economics. In: Politics and History in Band Societies, ed. E. Leacock and R. Lee, pp. 61–84. New York: Cambridge Univ. Press.

Zahavi, A., and A. Zahavi. 1997. The Handicap Principle: A Missing Piece of Darwin's Puzzle. New York: Oxford Univ. Press.

5

Why Is Reciprocity So Rare in Social Animals?

A Protestant Appeal

Peter Hammerstein

Institute for Theoretical Biology, Humboldt University, 10115 Berlin, Germany

ABSTRACT

After three decades of worldwide research on reciprocal altruism and related phenomena, no more than a modest number of animal examples have been identified. Even in primates, evidence for reciprocity is surprisingly scarce. In contrast to the shortage of support, reciprocal altruism and Tit-for-Tat-like behavior have been used as the prime explanation for cooperation among nonkin. From models based on this line of reasoning, one easily gets the impression that reciprocity should be widespread among social animals. Why is there such a discrepancy between theory and facts? A look at the best-known examples of reciprocity shows that simple models of repeated games do not properly reflect the natural circumstances under which evolution takes place. Most repeated animal interactions do not even correspond to repeated games. Partner switching and mobility often counteract the evolutionary stability of reciprocal altruism. Moreover, if learning is involved in mental implementation, then the timescale in which reciprocity can occur is often dramatically shortened. In the few known examples, quick reciprocation seems to be the rule, yet standard game theory fails to account for this empirical finding. More generally, it must be emphasized that mental mechanisms shape the evolution of reciprocity. An impressive mental machinery is required for nontrivial examples of reciprocity, as illustrated by the attribution problem (i.e., the problem of classifying other individuals' actions as cooperative, intentionally uncooperative, or unintentionally uncooperative). Emotions may play a role in the machinery underlying cooperation, but current game theory is conceptually not designed to account for the role emotions play. Collectively, this shows that many obstacles can impede the evolution of reciprocity and that evolutionary game theory needs new conceptual tools to understand these obstacles adequately.

THESES

Some theoretical ideas appear to be so compelling that the lack of supporting evidence is indulged by major parts of the scientific community. This criticism applies to current thought in evolutionary biology regarding cooperation in repeated interactions. Thus, to provoke a change, I am "tacking up" some theses for public display. My aim is not to create an entirely new theory but rather to steer its course closer toward reality. It is in this light that I ask the reader to approach the following discourse.

In its simplest form the biological theory of reciprocity aims to explain apparently altruistic behavior by revealing its nonaltruistic nature. It strongly resembles the theory of repeated games. A repeated game, or supergame, consists of a series of interactions between the same two (or more) players. In each interaction, they play the same game. This game forms the building block upon which the supergame is built. After each round, the game is repeated with some probability so that the decision about continuation is externalized. Long before evolutionary game theory was born, it has been known (Luce and Raiffa 1957) that more cooperation is possible in a repeated game scenario than in a one-shot encounter.

The so-called folk theorems capture this popular wisdom in a mathematically rigorous way (e.g., Fudenberg and Tirole 1991). In essence, they state that if *any* one-shot game has a solution (Nash equilibrium) that does not fully exploit the scope for cooperation, then higher degrees of cooperation can be observed in appropriate solutions of the supergame, provided the expected number of future interactions is sufficiently large. The idea of supergame cooperation thus applies to a wide range of very different scenarios, and it helps to understand problems far beyond reciprocity, altruism, and the Prisoner's Dilemma. In this generality lies the temptation to overestimate the explanatory power of repeated games.

The drawback is that the theory of supergame cooperation is based on a very narrow picture of the long-term interaction pattern. Strategy spaces of repeated games do not include the option to end the sequence of interactions. Therefore, the decision to leave a partner and interact instead with other individuals is not permitted in the formal structure of a repeated game. As has been pointed out by Friedman and Hammerstein (1991), biological examples of reciprocity require different modeling approaches, and biologists often talk about something fundamentally different from repeated games when they discuss reciprocity (see also Connor 1992). Let us sum this up in a thesis.

Thesis 1: The assumption of forced interactions severely limits the applicability of repeated games.

The theory of repeated games applies to a large class of games and is therefore broad. Concurrently, it is very narrow in that players are treated as if they were attached to each other by some "magic glue."

There are, of course, examples of animal and especially human interaction patterns (see also Hagen, this volume) for which it seems reasonable to idealize them as supergames. For example, two human neighbors in a residential area or two neighboring territory owners of an animal species may have to deal with each other for quite some time due to the transaction cost associated with moving to another house or territory.

In contrast, car drivers are not attached to their gas stations. If it saves them money, many will switch from one station to another. This creates a market which is crucial for the understanding of why drivers shop at a particular place. To some extent, the car driver logic seems to apply to the mutualistic relationship between a cleaner fish and its clients. As Bshary and Noë discuss in this volume, clients that come from a long distance to the "cleaning station" seem to have a tendency to switch stations (cleaners) according to their offer. Even with such a mechanism at work, one may observe repeated visits of the same station. These repeats are driven by partner choice, however, and not by the magic glue of a repeated game. This leads to Thesis 2.

Thesis 2: Repeated interactions, as such, are not evidence for repeated games.

Repeated interactions alone are not sufficient evidence for a repeated game. Repetition can and will almost always result from strategic benefits that interacting animals incur by deliberately continuing to interact with their partners. When casting biological examples into the form of a repeated game, one excludes a potential incentive for noncooperative behavior from the analysis. This incentive is to take benefits and then leave the partner behind without giving anything in return. If one aims at explaining cooperation, it is necessary to understand why the scope for such exploitation is limited.

To illustrate this thesis, consider the famous egg-trading procedure in a fish called the black Hamlet (*Hypoplectrus nigricans*). This fish is a simultaneous hermaphrodite, i.e., it produces both eggs and sperm. According to Fischer (1980, 1981), the black Hamlet typically has more sperm than needed to fertilize all the eggs of a mating partner. Eggs are, therefore, a precious commodity at the mating site where fish congregate that are ready to spawn during a given afternoon. In principle, it would be beneficial for an individual to fertilize the eggs of more than one partner. In practice, however, this is difficult to achieve, because eggs are "parceled" so that individuals cannot fertilize all the eggs of a mate at once. Egg parcels are never exchanged simultaneously and, to a large extent, mating takes place in an alternating sequence of giving (eggs) and taking (fertilizing). Fischer interpreted this as a Tit-for-Tat strategy in a repeated game. However, in a repeated game, two partners are forced to stay together and there would be no incentive for defection for the following reason: both partners need their eggs to be fertilized before nightfall and both gain a benefit from fertilizing the other's eggs. Partner control in a Tit-for-Tat-like fashion would seem unnecessary in this particular model[1] (which, of course, does not capture what Fischer really had in mind).

The Hamlet egg-trading problem, in other words, does not meet the assumptions of a formal repeated game. Instead, it may better be considered in the context of markets. The Hamlet spawning ground shares an interesting feature with illegal drug markets. When two dealers are about to trade a large amount of heroin for a large amount of money, the one with the heroin may abscond with the money before handing over the heroin. The following strategic offer from the buyer, however, can create an incentive to participate *deliberately* in a repetitive interaction: "Give me the first portion of your heroin and I will pay for it; give me the next portion, and I will pay again, etc." Once the dealers enter this interaction sequence with heroin parceling, a cheater cannot run away with more than one unpaid portion of heroin. Now, if the cost of switching trading partners is high enough, then cheating would not be worth the free portion.

In the drug market, partner switching is costly because police are watching and violence is involved. Although the Hamlet fish have no guns and do not have to contend with authority, switching may nevertheless be costly. The cost can be estimated from how long it takes to find a new partner, at what time the mating market will close in the evening, how many eggs will be traded with the new partner, etc. Using Fischer's data, Friedman and Hammerstein (1991) made an attempt to show in a model that there is no incentive for cheating given the observed practice of parceling. After a spawning bout, a mate still possesses enough eggs to remain the preferred trading partner. This keeps the interaction going and we see how market arguments, rather than thoughts about repeated games, can enlighten the study of reciprocity. This leads me, therefore, to state the following thesis.

Thesis 3: Partner switching and partner markets are important but often neglected issues in the study of reciprocity.

A vital strategic element in maintaining cooperation is to make it unprofitable for a social partner to switch. Therefore, the investigation of cooperation typically requires consideration of partner markets.

In opposition to Thesis 3, one might be tempted to negate it on the premise that switching is rare or seldom observed in a number of animal examples. However, this would be a mistake since even then the unprofitability of switching is one of the main keys to the understanding of cooperation. Friedman and Hammerstein (1991) demonstrated this in their model for the egg-trading of the Hamlet fish. Conversely, Enquist and Leimar (1993) emphasized the profitability of switching and made the general point that mobility seriously restricts the evolution of cooperation in many animals. They argued, however, that the effect of mobility

[1] In general, partner control is an important issue even in market theory.

might be counteracted to some extent by behavioral propensities, such as suspiciousness toward strangers and gossiping. Vehrencamp (1983) expressed the idea that dominant animals in nonegalitarian animal groups cannot "exploit" other group members to extremes because this would create an incentive for subordinates to migrate and search for a group in which exploitation is less severe. In her theory of reproductive skew, Vehrencamp made an important conceptual step toward thinking about the role of partner markets in cooperation.

Whereas Vehrencamp focused her attention on the partner choice exerted by subordinate animals, Noë and Hammerstein (1994) showed that social partner choice exerted by dominant individuals can increase "exploitation" of subordinates by forcing them to be more cooperative than they would be in the absence of this choice. The degree of the subordinate's cooperation then depends on the "animal labor market" (see also Bowles and Hammerstein, this volume).

At this point it must be emphasized that we would "throw the baby out with the bathwater" if we claimed that functional analysis of reciprocity pertains only to partner choice and switching. Partners, even preferred ones, may have to be checked. So, why is some partner control necessary even when both partners know that in principle they are a perfect match? To answer this question, let us return to the gas station example. We approach our preferred station with good reason to believe that someone will sell us gas. Occasionally, we encounter a defective pump and have not been forewarned through a sign. When this happens, would we stand at the pump and wait forever? Fortunately, routine processes protect us from pursuing such unsolvable tasks for too long. It is very likely that these processes are not special adaptations to cheating in social interactions, since the propensity to change goals when tasks are unsolvable is crucial to the management of many problems an animal faces. Returning to the Hamlet example, if a fish fails to receive eggs from a partner for a long time, this resembles the situation where the pump is out of order, and a similar logic can be applied. This leads us to the next thesis.

Thesis 4: Partner control can result from general mental processes that are not specific tools against cheating in social interactions.

Animals cannot waste their time on unsolvable tasks. If a partner fails to provide an expected "commodity," routine task switching may cause the animal either to search elsewhere for this commodity or to end the search. This kind of task control is likely to be a rather general feature of the mental machinery — one that protects the animal against locking itself into endless waiting states. Therefore, the behavioral contingency in reciprocal cooperation may not be a specific adaptation to cheating.

Thesis 4 implies that general mental processes can be seen as preadaptations to reciprocal altruism. An animal that does not reciprocate would risk losing a cooperating partner due to that partner's task control. To obtain reciprocity, however, much more needs to be implemented.

If learning is at least partially involved in the mental implementation of reciprocal altruism, then the following kind of problem arises. We learn to associate a stomach problem with the fish that we ate just prior to the onset of this problem but not with the steak that we ate a week ago. In a similar spirit, it seems plausible that learning would not allow animals to develop reciprocity when there is a significant temporal or contextual gap between the situations for giving and taking, or when rewards from cooperation are delayed relative to the rewards from noncooperation. This is nicely illustrated by the following experiments.

Clements and Stephens (1995) exposed captive blue jays to the repeated Prisoner's Dilemma. In their experimental setup the birds obtained immediate rewards from noncooperative behavior, whereas the rewards from cooperative action were slightly delayed. The authors reported that the jays were unable to learn sustained cooperation. Stephens et al. (2002) repeated this experiment using a modified setup where rewards were not directly given to the birds but accumulated in a transparent plastic box. Here, the jays could see their food gains but not consume them until a flap was finally opened. In this experiment, there was no delay between the rewards from cooperative and noncooperative action. The birds actually did learn sustained cooperation. Taken together, the two experiments demonstrate nicely that timescale considerations are important in explaining the facts, which brings us to the next thesis.

Thesis 5: There is surprisingly little evidence for reciprocity in nonhuman animals, and the known examples seem to be largely restricted to reciprocation on short timescales.

Ever since Trivers (1971) wrote his seminal paper on reciprocal altruism, models of cooperation in repeated games have preoccupied and entertained the scientific community. However, as far as convincing data are concerned, the harvest has been very modest. There are few animal examples outside the primate world. Even in primates, the evidence for reciprocity is scarce (see Silk, this volume). The most typical form of reciprocity is the reciprocal grooming found in ungulates and some primates. Here, the effort is often parceled like in the egg trading of the Hamlet fish. The reward for a grooming act is often instantaneous. As explained above, the quick succession of giving and taking facilitates the implementation of reciprocity by learning.

To challenge the message of Thesis 5, let us look at an example in which the timescale for reciprocation is not short. Wilkinson (1984) conducted a fascinating empirical study in which he describes blood donations among female vampire bats. The females roost in groups. Every day they fly out in search of blood. If a female fails to obtain a blood meal for two days in succession, her risk of starvation becomes very high. When females return from an unsuccessful foraging excursion, they solicit a blood donation from other females. In an experiment with genetically unrelated individuals, Wilkinson demonstrated that if a donation takes place, the donator is more likely to be a female that has already received a donation from the soliciting bat than a female that had not been a beneficiary of her help.

Admittedly, the bat example looks very much like a repeated game, and it has a strong flavor of reciprocity. But, Wilkinson did not expose the bats to the crucial contingency test. From his experiment, we cannot conclude that a female bat would be less inclined to cooperate in the future if another female refused to donate blood to her. Wilkinson's data could, for example, result from a tendency of the bats to like some females more than others on the basis of characteristics such as smell. This idea is not far-fetched because under natural circumstances the communally roosting females are actually kin groups. The explanation behind the blood donations may ultimately lie in the genetic relatedness of helpers and receivers, despite the fact that Wilkinson used unrelated individuals in his experiment. The reason is that kin selection may produce mechanisms that appear like reciprocity in the experiement with unrelated individuals: a kin recognition mechanism may be operating that needs calibration based on the concrete group in which it is used. Such a mechanism could produce friendly affinities in groups of unrelated individuals.

The more convincing examples of reciprocal altruism are indeed characterized by short timescales on which reciprocation occurs. Quick exchanges of altruistic acts are typical for reciprocal grooming, as it occurs in ungulates. Impala serve as an impressive example. They possess teeth that are adapted to grooming, often referred to as the "antelope comb." This comb is used to remove ectoparasites such as ticks. Much of the removal is done by self-grooming but, for obvious reasons, this does not include the head and neck. Females and nonterritorial males engage in interaction sequences where partners alternate in grooming one another. Newborn impala develop this capability during the first weeks of their life after which they groom with adults other than their mother and with other lambs (Mooring 1999). Adult grooming is not restricted to relatives. The amount of this activity that is given in an interaction sequence matches approximately the amount received (Hart and Hart 1992). Grooming is costly and the benefits from tick removal are substantial. Connor (1995) compares the impala system with the egg parceling found in the Hamlet fish. According to Connor, partner switching is not difficult given the impala's gregariousness but switching seems to be sufficiently costly because the initiator of an interaction has to groom first.

Now, if most reciprocity occurs on short timescales, this poses a serious problem to evolutionary game theory since standard models from game theory are "blind" to this restriction and would not predict it. Any passionate game theorist, like myself, should consider this as an alarm signaling the need for conceptual change.

Thesis 6: Models from standard game theory fail to explain the short timescales on which animal reciprocation takes place.

Typical reciprocity models from evolutionary game theory are designed in such a way that they predict reciprocal altruism regardless of whether the time distance between giving and taking is short or long.

Thesis 6 is not meant to convey the notion that just by introducing time into the models, one can achieve progress in understanding cooperation based on reciprocity. The crux of the issue is to invoke the actual mechanisms upon which evolution operates.

Thesis 7: Mechanisms shape the evolution of cooperation.

The evolution of cooperation in repeated interactions depends strongly on assumptions about how much simple learning is involved in decision making as opposed to higher forms of cognition and hard-wired strategies. It is thus impossible to disentangle causal and functional analysis in this field of research.

To illustrate this claim, let us take a short detour and consider an experiment conducted by Selten and Stoecker (1986) in their laboratory of behavioral economics. Randomly paired subjects played a fixed number of rounds of the repeated Prisoner's Dilemma. After this, partners were exchanged and the same repeated game was played with other randomly chosen partners. Many such exchanges took place in succession. The experiment, therefore, simulated the process of social learning without free partner choice in a finite population.

Game theorists would maintain that rational players should never cooperate in the repeated Prisoner's Dilemma of Selten and Stoecker's experiment, since the end of the imposed interaction sequence is fixed. Rational players then have to treat the last round as if they were operating in a one-shot game. Therefore, it pays to defect in the last round. Using the cognitive tool of backward induction, a similar argument can be made for all rounds, leading to all-out defection. Evolution acting on hard-wired strategies for this game would mimic the backward induction, provided there is enough genetic variation. Furthermore, it is possible to describe learning rules that would equally please the classical game theorist. With these parallels between advanced cognition, evolution, and simple learning in mind, one might wish to dismiss Thesis 7. The facts, however, teach us another lesson: real learning in humans produces a qualitatively different outcome, as we will now see.

In the series of encounters, subjects first developed or were predisposed with a cooperative attitude. Subsequently they discovered the benefit from defecting toward the end, the so-called *end effect*. However, the learning population did *not* mimic much of the backward induction. After many repeated games had been played in the population of subjects, the typical onset of defection failed to take place anywhere near the first round, and a lot of cooperation still occurred.

Selten and Stoecker tried to interpret this empirical finding as follows. After gaining some initial experience with the game, a learning process guides an individual's behavior. The subjects are prepared to cooperate first, but switch to noncooperation if either the partner has started to defect or a personal limit, say round *r*, is reached. If the current opponent switches to defection in a round earlier than *r*, an individual will have a tendency to move the switching threshold down in the next repeated game with a new opponent. Conversely, if the current opponent is still cooperating in round *r*, the opposite tendency will occur, namely to shift *r* to a higher value. Simulations of this population learning process showed that the onset of defection moves a little from the end to earlier periods of the game but does not move all the way down to zero. A lot of cooperation is maintained.

Without taking this proximate learning mechanism[2] and its structural properties into account, it would have been difficult to understand the experimental findings under discussion. The learning mechanism as such was probably favored by natural selection because it is a *robust* method of dealing with a variety of problems. Evolution does not design a new mental tool for every problem that animals face, and it always operates by *modification of existing mechanisms*, not by selecting a strategy from an abstract strategy space. This explains why mechanisms can shape the evolution of cooperation, as expressed in Thesis 7.

Let us now move to another aspect of proximate causation, namely the mental bookkeeping involved in partner control. We have already seen that the necessity of this bookkeeping may have been overestimated in models that ignore the partner market. Still, we expect it to play some role in many social interactions. For a *long* time experimentalists have been challenged to demonstrate this bookkeeping. Seyfarth and Cheney (1984), for example, succeeded in providing some evidence for bookkeeping in vervet monkeys, but the interpretation of their well-known empirical results is difficult (Hammerstein 2001).

Silk (this volume) reviews the psychological literature and concludes that certain rigid forms of bookkeeping, like "I gave you this much, you owe me that much," seem to be counterproductive for the maintenance of human friendships. Beyond the primate world, our knowledge of bookkeeping is still extremely limited (see McElreath et al. [this volume] for a discussion of the stickleback predator inspection trips described by Milinski [1987]; see also Dugatkin [1997]).

The issue of mental bookkeeping becomes particularly interesting if one studies indirect forms of reciprocity. In indirect reciprocity (Alexander 1979, 1987; Sugden 1986), the return from a social investment is expected from someone other than the beneficiary of aid. The investment increases the investor's *reputation* in the social group where others have a propensity to help those with a good reputation. This idea raises the empirical and theoretical concern of how updating of reputation can be organized and what role strategic gossiping plays in this context. Some basic problems with the updating of reputation are discussed by Leimar and Hammerstein (2001) and McElreath et al. (this volume).

2 Since Selten and Stoecker's experiment, economists (including Selten) have elaborated on the process by which human subjects adjust their behavior and beliefs.

Thesis 8: Constraints on basic mental abilities of animals restrict considerably the evolution of reciprocity.

When strategies such as Tit-for-Tat and reciprocal altruism are discussed, it is often said that they require very little mental machinery because all the animal has to do is to remember what happened during the last round of an interaction. Such statements reflect the simplicity of models rather than account for the problems real animals would face if they engaged in reciprocal altruism beyond grooming.

Mental machinery has to perform complex tasks in order to achieve subtle reciprocity based on partner control. If the partner, for example, fails to exhibit a cooperative act, this poses the attribution problem to determine whether the observed behavior really belongs to the class of noncooperative moves. The mental updating machinery must solve this problem. Routine learning may interfere with its information processing, which may be costly and error prone. Following Fessler and Haley (this volume), emotions probably play an important role and thus should be reflected in the model. This would require a radical change from the Bayesian approach that governs thinking and modeling in classical game theory. By assumption, a Bayesian decision maker is forced to use all available information for updating his decisions. In contrast, a strategic aspect of emotions may consist in shutting off or distorting input channels. To give an example, one cannot easily argue with an angry person; the door for communication is temporarily closed. This strategic stubbornness may add credibility to the threat that defection from cooperation would have negative consequences.

* * *

Whether or not these theses will spur the hoped-for reformation remains to be seen. Most certainly, if we invested the same amount of energy in the resolution of all problems raised in this discourse, as we do in the publishing of toy models with limited applicability, we would be further along in our understanding of cooperation. No protest(ant) would then be necessary.

ACKNOWLEDGMENTS

I wish to thank Ed Hagen, Joan Silk, Rob Boyd, Ernst Fehr, and Jack Werren for their ecumenical advice.

REFERENCES

Alexander, R.D. 1979. Darwinism and Human Affairs. Seattle: Univ. of Washington Press.

Alexander, R.D. 1987. The Biology of Moral Systems. New York: Aldine de Gruyter.

Clements, K.C., and D.W. Stephens. 1995. Testing models of non-kin cooperation: Mutualism and the Prisoner's Dilemma. *Anim. Behav.* **50**:527–535.

Connor, R.C. 1992. Egg-trading in simultaneous hermaphrodites: An alternative to Tit for Tat. *J. Evol. Biol.* **5**:523–528.

Connor, R.C. 1995. Impala allogrooming and the parcelling model of reciprocity. *Anim. Behav.* **49**:528–530.

Dugatkin, L.A. 1997. Cooperation among Animals: An Evolutionary Perspective. New York: Oxford Univ. Press.

Enquist, M., and O. Leimar. 1993. The evolution of cooperation in mobile organisms. *Anim. Behav.* **45**:747–757.

Fischer, E.A. 1980. The relationship between mating system and simultaneous hermaphroditism in the coral reef fish *Hypoplectrus nigricans* (Serranidae). *Anim. Behav.* **28**:620–633.

Fischer, E.A. 1981. Sexual allocation in a simultaneously hermaphroditic coral reef fish. *Am. Nat.* **117**:64–82.

Friedman, J.W., and P. Hammerstein. 1991. To trade or not to trade; that is the question. In: Game Equilibrium Models. I. Evolution and Game Dynamics, ed. R. Selten, pp. 257–275. Berlin: Springer.

Fudenberg, D., and J. Tirole. 1991. Game Theory. Cambridge, MA: MIT Press.

Hammerstein, P. 2001. Games and markets: Economic behaviour in humans and other animals. In: Economics in Nature, ed. R. Noë, L.A.R.A.M. van Hooff, and P. Hammerstein, pp. 1–19. Cambridge: Cambridge Univ. Press.

Hart, B.L., and L.A. Hart. 1992. Reciprocal allogrooming in impala, *Aepyceros melampus*. *Anim. Behav.* **44**:1073–1083.

Leimar, O., and P. Hammerstein. 2001. Evolution of cooperation through indirect reciprocity. *Proc. Roy. Soc. Lond. B* **268**:745–753.

Luce, R.D., and H. Raiffa. 1957. Games and Decisions. New York: Wiley.

Milinski, M. 1987. Tit for Tat and the evolution of cooperation in sticklebacks. *Nature* **325**:433–435.

Mooring, M. 1999. Impala: The living fossil. *Afr. Env. Wildlife* **7**:52–61.

Noë, R., and P. Hammerstein. 1994. Biological markets: Supply and demand determine the effect of partner choice in cooperation, mutualism and mating. *Behav. Ecol. Sociobiol*. **35**:1–11.

Selten, R., and R. Stoecker. 1986. End behaviour in sequences of finite Prisoner's Dilemma supergames. *J. Econ. Behav. Org.* **7**:47–70.

Seyfarth, R.M., and D.L. Cheney. 1984. Grooming, alliances and reciprocal altruism in vervet monkeys. *Nature* **308**:541–543.

Stephens, D.W., C.M. McLinn, and J.R. Stevens. 2002. Discounting and reciprocity in an iterated Prisoner's Dilemma. *Science* **298**:2216–2218.

Sugden, R. 1986. The Economics of Rights, Co-operation and Welfare. Oxford: Blackwell.

Trivers, R. 1971. The evolution of reciprocal altruism. *Qtly. Rev. Biol.* **46**:35–57.

Vehrencamp, S.L. 1983. Optimal degree of skew in cooperative societies. *Am. Zool.* **23**:327–335.

Wilkinson, G.S. 1984. Reciprocal food sharing in the vampire bat. *Nature* **308**:181–184.

6

The Bargaining Model of Depression

Edward H. Hagen

Institute for Theoretical Biology, Humboldt University, 10115 Berlin, Germany

ABSTRACT

Minor depression — low mood often accompanied by a loss of motivation — is almost certainly an adaptation to circumstances that, in ancestral environments, imposed a fitness cost. It is, in other words, the psychic equivalent of physical pain. Major depression is characterized by additional symptoms — such as loss of interest in virtually all activities and suicidality — that have no obvious utility. The frequent association of these severe and disabling symptoms with apparently functional symptoms, like sadness and low mood, challenges both dysfunctional and functional accounts of depression. Given that the principal cause of major unipolar depression is a significant negative life event, and that its characteristic symptom is a loss of interest in virtually all activities, it is possible that this syndrome functions somewhat like a labor strike. When powerful others are benefiting from an individual's efforts, but the individual herself is not benefiting, she can, by reducing her productivity, put her value to them at risk to compel their consent and assistance in renegotiating the social contract so that it will yield net fitness benefits for her. In partial support of this hypothesis, depression is associated with the receipt of considerable social benefits despite the negative reaction it causes in others.

DEPRESSION IS STILL A MYSTERY

After more than a century of inquiry, unipolar depression remains a profound scientific mystery. Like people working on a large and difficult jigsaw puzzle, researchers in genetics, biochemistry, cognitive psychology, social psychology, and psychodynamics have pieced together detailed accounts of depression from their various theoretical vantage points, but these disparate views have yet to be integrated into a single, coherent whole. Just as an unfinished puzzle often reveals itself in parts that give little clue of the final picture, depression is well understood in aspects, yet no one can answer the question: What, ultimately, is depression? Recent personal problems are clearly implicated in its onset, and psychotherapy — talking about these problems — has been shown to be about as effective in reducing depressive symptoms as the latest antidepressants (U.S. Department of Health and Human Services 1993). These facets of depression must be reconciled with the equally significant genetic and biochemical correlates. As the editor of a recent volume on depression concluded, "Despite a great deal of thorough research there is no agreement concerning etiology, symptomatology, and treatment methods" (Wolman 1990, preface). The editor's choice of terminology reflects what is perhaps the single point of agreement among depression researchers: major depression is an illness.

With no consensus on causes, symptoms, or treatment, little-to-no evidence that depression in general is caused by infections, toxins, or physical injury to the brain, excellent evidence that depression is caused by social circumstances that would have occurred repeatedly in the environment of evolutionary adaptedness (EEA) — often dangerous social circumstances in which a genuine cognitive impairment would have been disastrous — and given that most sufferers of depression experience a complete recovery often in association with (and possibly caused by) major life improvements like getting a better job or relationship, one wonders why there is such conviction that depression is a mental illness. Several unpleasant experiences such as physical pain and nausea are in fact adaptations designed to protect the sufferer from harm.

In the first part of this chapter I will argue that there was a selection pressure for the evolution of a bargaining strategy in humans; in the second, I will argue that clinical, unipolar depression may be just such a strategy.

THE INDIVIDUAL VERSUS SOCIETY IN THE EEA

In the EEA, costly conflicts between individuals and groups were probably common, particularly in the wake of individual social losses and failures. Individuals suffering social losses or failures often need assistance, additional material or social resources, or to renegotiate their relationships with group members. Social partners could not be expected to provide help or make changes immediately, however, particularly when they were benefiting from the status quo.

Conflicts between individuals are common in many species and often result in physical aggression because injuring, or threatening to injure, others is an effective means of influencing or deterring their actions (Clutton-Brock and Parker 1995; McElreath 2003). In humans, this strategy is closely identified with the emotion anger (Fessler 2003). There is, however, a key limitation to an aggressive strategy. In the EEA, it would have been difficult for a single individual to use aggression when one's opponent was physically more formidable, or when one was opposed by a group. If one needed to influence the behavior of a single powerful individual or a group, physical threats (especially by a female) would rarely have been effective: even two people can almost always overpower one.

Persuasion was also an option, but if an individual's claims were difficult to verify and/or if inherent conflicts of interest existed among the parties, persuasion was likely to fail. Consequently, an effective strategy to compel assistance or change would have provided substantial fitness benefits. The solution proposed here is that one could have efficiently imposed costs on powerful others, thereby influencing them, by withholding benefits that one provided them until desired changes were forthcoming. In other words, individuals could bargain.

BARGAINING

Social Conditions in Which Bargaining Can Be Effective

Bargaining, the withholding of benefits to compel changes by others, can only work, and is only necessary, in particular social circumstances — circumstances that were likely to have been ubiquitous in the EEA.

Viscous Social Markets and Monopoly Power

When there are many resource providers, i.e., when there is a market instead of a monopoly, one has little need to pay a cost to influence others because one can always obtain the necessary resources elsewhere (resource costs are then determined by the supply and demand curves of standard economic theory).[1] In the EEA, however, it may frequently have been the case that there was little-to-no market; all parties often had effective monopolies on resources that were crucial to other group members. Kin- and family-based social organization, high levels of biparental care, low population densities, ethnicity, and occasional intergroup aggression meant that switching social partners was difficult.

It would have been difficult, for example, for mothers to raise offspring without help from the father and/or other family members; conversely, the fitness of the father, parents, and other family members depended critically on the mother successfully raising offspring. Abandonment of one party by another would have entailed a significant fitness cost to all (for further details, see Hagen 1999). In another typical example, political alliances between families may have often depended on an arranged marriage between a man from one family and a woman from the other, as is commonly seen in contemporary hunter-gatherer groups (Rodseth et al. 1991). If so, important political relationships between families depended critically on sons and daughters; conversely, sons' or daughters' relationships with their families depended critically on their willingness to participate in the arranged marriage.

---

[1] If there is a market on only one side, the single seller (monopoly) or single buyer (monopsony) sets the price unilaterally.

Contract Enforcement

Partners can also maintain an effective monopoly on resources they provide, thereby ensuring their personal importance to others, when they can exclude competitors or when they can easily punish defection, both likely important aspects of ancestral social environments. Punishing defectors, in particular, is increasingly recognized as an important social strategy (e.g., Boyd and Richerson 1988, 1992). The ability to impose costs efficiently on defectors raises the specter that individuals who do not benefit from a cooperative venture could nonetheless be forced to participate despite the fitness costs they might suffer.

In sum, the market for certain kinds of social partners in the EEA may often have been anything but fluid. Given this high degree of interdependence in foraging bands (see also Boehm 1996), individuals who withheld benefits would have imposed significant costs on other band members.

When to Bargain

Individuals should attempt to compel assistance when they suffer high costs that can be alleviated by others. Such costs can have many causes but can frequently come in the wake of social losses and failures; when critical social strategies fail, the benefits one is receiving plummet. Increased benefits may be possible, however, if others are willing to provide assistance or make major social changes. In the EEA, individuals could have suffered social losses and failures in numerous ways. Important social partners such as mates and allies could have died or severed relations, forcing one to abandon the current strategy; social strategies could have failed to realize fitness benefits, such as when efforts to increase or maintain social status failed, or when a mateship yielded a low viability infant; competitors could have blocked access to critical resources, including key social relationships; one could have been coerced by powerful others; one could have been betrayed by social partners; one could have been prevented from pursuing new, more profitable opportunities; or one could simply have chosen the wrong strategy or executed it poorly.

In many such cases, individuals could have unilaterally pursued an alternative strategy, like finding a new mate after the death of a spouse. If evidence from contemporary small-scale societies is any guide, however, in many other cases, individuals often required the consent and/or cooperation of group members to mitigate the costs of social failures. If a husband were abandoned by his wife, for example, physical threats might have secured her return, but they might also have been counterproductive (Figueredo et al. 2001). If the husband could have convinced group members to spend political capital in securing the return of the wife or procuring another, chances of success would have been far greater. Unfortunately, there often could have been conflicts of interest between the individual and the group. Group members might not want to spend their political capital securing another mate for someone who had one, but lost her due to his abusive behavior, or because the group preferred using its capital to secure a mate for a higher status individual. In another example, arranged marriages are frequently made with little regard for the personal preferences of those to be married. Those betrothed to an undesirable mate often face formidable opposition from their families and other group members, however, if they resist the marriage (e.g., Shostak 1981). This opposition could exist because there is a genuine conflict of interest between the parties, or because the family and group members simply have little reliable information about the relative quality of the mates (and thus would not want to make costly changes for an unknown benefit).

Given that even relatively high degrees of relatedness, although important, appear insufficient to sustain cooperation in foraging bands, given the high mutual interdependence of individuals in these bands, and given that small cooperative groups of foragers only had the time and resources to achieve limited goals, which might not meet the needs of all members, conflicts between individual members and the group were inevitable. This was especially so when one member was suffering costs that others were not. Individuals therefore needed a strategy to pressure other group members to alleviate these costs despite conflicts of interest or concern about their legitimacy.

As a consequence of viscous social markets, enforcement of social contracts, and conflicts of interest, there was a strong selection pressure among humans to evolve bargaining strategies to compel assistance and/or modification of social arrangements that were no longer profitable. Bargaining is necessary and effective when (a) at least one participant is not benefiting from the current social contract, (b) others are benefiting from the social contract, and (c) participants have a monopoly or near monopoly on the resources they provide — otherwise, disaffected parties could simply choose to cooperate with someone else (for a review, see Kennan and Wilson 1993).

Private Information and Credible Signaling: The Function of Delay

When the value of cooperation decreases with time, withholding benefits can also credibly signal that one is truly suffering costs to those who might not otherwise recognize those costs. It is difficult for group members to assess the costs and benefits incurred by their social partners accurately: she claims she is not benefiting from a relationship, but perhaps she really is and just wants more than her fair share; her true valuation is private information.

The discount factor, δ, is the fraction of cooperative benefits still available after each round of bargaining and is thus a measure of delay costs due to multiple rounds of bargaining. Kennan and Wilson (1993) argue that quick agreements are usually possible in most models of bargaining where valuations and discount factors are common knowledge (i.e., no private information). Informally, if each participant knows what the other participants know, each will come to the same conclusions about how any sequence of bargaining rounds will proceed; each participant will also come to the same conclusions about the "optimal" outcome for other participants, and thus this outcome can be offered in the first round. In a simple game of alternating offers by a buyer and seller, if $0 < \delta < 1$, then the maximum benefit decreases as δ^t, where t represents the number of rounds. The seller must make an offer just sufficiently generous such that the buyer cannot do better by waiting another round — when delay is costly, each party has an incentive to minimize the number of rounds of bargaining in order to maximize benefits. It can be shown that if the seller makes the first offer, she will offer a price that gives her $1/(1 + \delta)$ of the benefits, which the buyer accepts immediately (Rubinstein 1982).

If, however, participants in a cooperative venture do not know how other participants value the potential benefits or the costs they will suffer from delays, as was often likely in the EEA, it will be impossible for all participants to reach the same conclusion about the "optimal" agreement. If participants could credibly signal to other participants their true valuations and discount factors, then an agreement could be reached. Kennan and Wilson (1993) argue that the willingness of a participant to suffer the costs of multiple rounds of bargaining (due to discount factors less than one), coupled with the sizes of the offers made each round, represents credible information about that participant's true valuation — a greater willingness to delay signals lower valuations (because the more valuable the potential benefits from cooperation are to a participant, the less she can afford to delay). Once each participant acquires a relative level of certainty about the other participants' private valuation by observing their willingness to incur delays, the bargaining game becomes equivalent to one where valuations and discount factors are public knowledge, and an agreement can be quickly reached.

I argue that the costly symptoms of depression have a function, and that function is to impose costs efficiently on other group members by withholding critical benefits, credibly signaling to them that one is suffering costs (Watson and Andrews 2002), and compelling them to provide assistance or make changes. According to this view, depression is an (unconscious) social manipulation strategy that is triggered when individuals perceive that they are suffering costs that can only be alleviated by the actions of fellow group members (Hagen 1996, 1999, 2002; MacKey and Immerman 2000; Watson and Andrews 2002). Much as striking workers withhold benefits to impose costs on management, in the hope of inducing an increase of wages, a depressed individual may be strategically reducing productivity to impose costs on fellow group members, hopefully inducing them to act in ways more beneficial to her. To paraphrase Clausewitz, depression is the continuation of personal politics by other means.

MAJOR UNIPOLAR DEPRESSION: AN OVERVIEW

The two major classification systems of psychiatric disorders, ICD-10 and DSM-IV, both recognize that in typical depressive episodes, the individual

Table 6.1 Symptoms of a major depressive episode according to DSM-IV (APA 1994) and their hypothesized functions in the bargaining model. Bracketed text indicates functions that require additional assumptions. The diagnostic criteria for a major depressive episode are that an individual experiences either symptom one or symptom two, and at least four of the remaining seven symptoms nearly every day for a period of not less than two weeks.

Symptoms of Major Depression	Hypothesized Functions according to the Bargaining Model
1. Sad or depressed affect	Information to the sufferer that the current social strategy or circumstance is imposing a net fitness cost
2. Marked loss of interest in virtually all activities	a) Reduce investment in the costly strategy (minor depression b) Reduce investment in oneself and others (major depression)
3. Significant weight loss or gain	Loss: reduce investment in oneself (Gain: store resources for tough times ahead. Weight gain was probably difficult in the EEA)
4. Hypersomnia or insomnia	Hypersomnia: reduce productivity (Insomnia: allocate additional cognitive resources toward finding a profitable resolution to the current crisis)
5. Psychomotor retardation or agitation	Retardation: reduce productivity (Agitation: comorbid anxiety. Conflicts with social partners are often dangerous.)
6. Fatigue or loss of energy	Reduce productivity
7. Feelings of worthlessness or guilt	Worthlessness: contributions undervalued by others Guilt: defecting from social contracts imposes costs on others
8. Diminished ability to think or concentrate	Reduce productivity (and, more importantly, divert cognitive resources to renegotiating the current venture or toward finding more profitable alternatives)
9. Recurrent thoughts of death	Threaten to put future productivity at risk

suffers from depressed or sad mood, loss of interest and enjoyment, reduced energy, and diminished activity. This suite of emotions and behaviors has been observed in virtually all human societies (Patel 2001).[2] Table 6.1 provides an overview of the symptoms and their hypothesized functions (note: bipolar depression will not be discussed).

2 Asians may be more willing to report somatic symptoms relative to cognitive or affective symptoms, but it appears that they are just as likely to experience cognitive and affective symptoms as are Westerners; similarly, somatic symptoms are the most commonly reported by Westerners as well (Patel 2001).

Any theoretical explanation of depression must account for the following characteristics of depression: low mood and loss of interest in virtually all activities, a significant reduction in productivity, suicidality, a possible negative impact on health, a cross-culturally robust 2:1 female bias, a relatively high worldwide annual prevalence rate of around 5–10% (WHO 2001; rates vary widely by country), the substantial evidence that depression is closely associated with chronic activation of the hypothalamic-pituitary-adrenal axis (HPA) (e.g, Nemeroff 1996) — which prepares the body for fight-or-flight — and the fact that the most significant known cause is a major, negative life event. More on each of these below.

LIMITATIONS OR PROBLEMS WITH PREVIOUS ADAPTATIONIST HYPOTHESES FOR DEPRESSION

The challenge for an evolutionary account of depression is to reconcile the close association of plausibly functional symptoms (e.g., sadness and loss of interest in some activities) with its many costly symptoms (e.g., suicidality). The most theoretically coherent and empirically supported hypothesis for minor depression (a much less severe form of major depression) is the "psychological pain" hypothesis (Alexander 1986; Hagen 1999; Nesse 1991; Suarez and Gallup 1985; Thornhill and Thornhill 1989; Tooby and Cosmides 1990; Watson and Andrews 2002). Whereas physical pain functions to inform individuals that they have suffered a physical injury — motivating them to cease activities that would exacerbate this injury, as well as to avoid similar future situations which would also likely result in such an injury — psychological pain informs individuals that their current social strategy or circumstance is imposing a fitness cost, motivating them to cease activities that would exacerbate this cost, as well as to avoid similar future situations which would also likely result in a fitness cost. Such circumstances include, e.g., the death of children and relatives, loss of status, loss of a mate.

The "social competition" or "social yielding" hypothesis similarly proposes that short-term depression is an adaptation to force the loser of a social conflict involving status or rank (a) to stop competing with the winner, (b) to accept the fact that s/he has lost, and (c) to signal submission, thereby avoiding further conflict with the winner (Price et al. 1994). The "yielding" hypothesis obviously has much in common with the "psychic pain" hypothesis and is probably best considered an important, special case of the latter — loss of a social competition is certainly a prime example of a social circumstance that imposes a fitness cost, and the pain of depression could quite plausibly motivate losers to cease competing, thus avoiding the costs of continuing a futile competition. The yielding hypothesis cannot be a complete explanation for even minor depression, however, because loss of a social competition is not its only cause — having a baby with temperament problems (C.T. Beck 1996) is but one well-documented cause of depression that does not involve losing a status competition.

Neither the yielding hypothesis nor the psychic pain hypothesis accounts for major depression, and comments by proponents of these theories suggest that they are not intended to. Losers of social competitions should yield quickly, so proponents of the yielding hypothesis (Price et al. 1994) logically argue (from their perspective) that severe and prolonged depression is maladaptive (a major depressive episode can typically persist for months). Similarly, Nesse (1999, p. 356), a proponent of the psychic pain hypothesis, suggests that "sadness is almost certainly adaptive, but depression may arise from dysregulated sadness or from an entirely separate mechanism." A pronounced and sustained loss of interest and enjoyment in virtually all activities, loss of energy, and diminished activity are core features of major depression. Some psychic pain theorists (Tooby and Cosmides 1990, 2000; Nesse 2000) have cogently argued that, in the face of a major social failure, one should take pause. Immediately pursuing another social strategy without first evaluating the recent failure would likely only lead to another, costly failure. A distinction must be made, however, between a short-term reluctance to pursue one's social strategies, which often would have been wise in such circumstances, and long-term reduced self-care, which does not improve analysis of social failures or ability to unilaterally respond to social opportunities. Except when faced with an immediate threat, individuals simply analyzing a social failure should never stop eating, bathing, and grooming; individuals who did so in the EEA would have found that their health deteriorated rapidly, with (under this hypothesis) no compensating benefits.

Not only does depression have a significant, long-term negative impact on productivity, there is, as will be briefly discussed below, legitimate concern that the lack of self care accompanying depression may cause increased mortality, even in populations with ready access to resources and sophisticated medical care. Suicidality is also a very common symptom of major depression, yet there is no reason for an individual who has suffered a severe fitness cost, such as losing a social competition, to contemplate imposing additional costs on herself—especially the ultimate cost of death!

Energy conservation is another commonly proposed function for depression (e.g., A. Beck 1996). Although energy conservation was certainly an important reproductive problem in the EEA, depression does not show evidence of having been well designed by natural selection to solve it. Depression has some features that would reduce energy consumption, such as psychomotor retardation, but it has many features that have nothing to do with energy conservation, such as the intensely negative emotions that are the hallmark of depression. Neither fatigue nor sleep, two recognized energy-conserving adaptations, are associated with such intensely negative emotions in nondepressed individuals. Similarly, why would depression often be associated with loss of appetite when food is available? If it were an adaptation to resource-poor conditions, the opposite should always be the case. Why would depression be associated with insomnia, intense social rumination, or psychomotor agitation, which increase energy consumption? Why would it often be associated with feelings of guilt or anxiety? In sum, the symptoms of depression would have added nothing to, and would often have subtracted much from, the efficacy of fatigue and sleep as energy-conserving adaptations.

A common and reasonably compelling hypothesis is that depression is an evolved signal of social need (Lewis 1934; Henderson 1974). Many human emotions are closely associated with facial expressions and other types of signaling such as laughing and crying. Could the symptoms of depression, including suicide threats, simply be costly and therefore credible signals of need? However theoretically attractive this hypothesis, it is not supported by the evidence. Research has clearly shown that individuals react negatively to people who are depressed or exhibit symptoms of depression (Segrin and Dillard 1992), precisely opposite the desired reaction if depression were merely a generic signal of social need.[3]

In general, the symptoms of major depression seem designed to prevent the acquisition of benefits. A marked loss of interest in virtually all activities, significant weight loss, psychomotor disturbances, fatigue or loss of energy, and suicidal ideation would all have impeded ancestral humans from engaging in critical, beneficial activities, such as food gathering and consumption, buffering food shortages, personal hygiene, avoiding environmental hazards, information gathering, helping relatives and friends, etc. An adaptationist account of major depression must incorporate, not avoid or reinterpret, its costly symptoms.

MAJOR UNIPOLAR DEPRESSION AS A BARGAINING STRATEGY

Social Losses, Failures, and Other Causes of Depression

Numerous studies have shown that circumstances in which individuals may need to compel social assistance — adverse life events — are a potent cause of depression (Kendler et al. 1995; Mazure et al. 2000). Kendler et al.'s (1993) etiologic model of depression among female twins captures the essentials as well as any. In a longitudinal study of 680 female-female twin pairs, Kendler et al. found that the strongest predictors of a major depressive episode were, in descending order, (a) recent stressful life events, (b) genetic factors, (c) previous history of major depressive disorder, and (d) neuroticism. Their full, nine-variable model explained 50.1% of the variance in liability to depression (see also Kendler et al. 2002). For illustration, the four adverse life events which predicted onset of major depression in women with an odds ratio of 10 in a study by Kendler et al. (1995) were death of a close relative, assault, serious marital

[3] Note that, despite the negative feelings engendered by depression, actual rejection would have been difficult for most group members in the EEA if the depressed person had a monopoly on benefits they provided to the group.

problems, and divorce/breakup. Cross-culturally, depression case rates strongly covary with rates of adverse life events (Brown 1998).

Decades of research have shown that postpartum depression (PPD) is similar, if not identical, to depression in general (e.g., Whiffen and Gotlib 1993), and is therefore a good model for depression.[4] Human mothers should not automatically invest in offspring, but rather should weigh the decision carefully based on infant viability, levels of social support, access to resources, negative consequences for their other children, etc. (Trivers 1972; Clutton-Brock 1991). There is excellent evidence that lack of social support is a cause of PPD and substantial evidence that problems with the pregnancy, delivery, or infant, lack of resources, and concern about their ability to care for their other children are also closely associated with PPD (Hagen 1999). Childrearing costs that others could mitigate appear to cause PPD.

Social Constraints: Viscous Social Markets and Monopoly Power

A key prediction of the bargaining model is that depression should be caused not simply by loss, failure, and other social costs, but also by circumstances where individuals cannot unilaterally alleviate these costs. There is considerable evidence that this is the case. A perceived inability to control events — variously termed external locus-of-control (e.g., Rotter 1966), helplessness or hopelessness (Abramson et al. 1989), or entrapment (Brown 1998) — is clearly implicated in depression.

Meta-analyses of nearly 100 studies (Benassi et al. 1988; Presson and Benassi 1996) found that external locus-of-control and depression were significantly related, that the relation was moderately strong, and that it was consistent across studies; in addition, a belief that events were controlled by powerful others and chance was associated with higher levels of depression. Under the bargaining model, depression is a strategy to redress the causes of helplessness/hopelessness/lack-of-control/entrapment and that is why depression is expected to be associated with them. In the EEA, even seemingly irredeemable losses, such as the abandonment by or death of a spouse, could often have been readily addressed by powerful individuals in one's social group.

Studies of PPD also support the contention that constraints on unilateral action are associated with depression. Hagen (2002) found that for mothers in general, there was no correlation between social constraints on abortion and their PPD levels, nor should there have been. A social constraint on abortion is inconsequential for mothers who want the new child. The depression scores of mothers with unwanted or unplanned pregnancies, however, significantly positively correlated with their perception that having an abortion would damage their

4 PPD has a number of methodological advantages in the study of depression; e.g., it is easy to identify a population — pregnant women — who will be experiencing a stressful and potentially costly life event — birth — at a predictable point in time.

relationship with their spouse (there was, however, an interesting nonlinearity). Because mothers' perceptions could have been biased by their depression levels, fathers were also asked to report how much damage a wife's abortion would have caused their relationship. Fathers' perceptions of damage also correlated with mothers' depression levels, suggesting that actual, and not merely perceived, social constraints on reproductive decision making are associated with PPD.

Men's reproductive decisions are also constrained. Men, but not women, can substantially increase their reproductive success by mating with multiple partners. Hence, the opportunity cost of socially imposed monogamy is predicted to be much higher for men, especially during the postpartum period when their nursing wives are infertile, encumbered with a new infant, and therefore significantly hindered from finding other mates. This cost, however, will only be borne by men who have additional mating opportunities. Hagen (2002) found exactly this. Men with more sexual opportunities were more depressed postpartum, whereas women with more sexual opportunities were not. About one half the effect for men was found to be due to relationship problems, whereas the other half was due simply to sexual opportunities.

Conflicts of Interest and Private Information in the EEA

In the bargaining model, a need to influence others plus the inability to act unilaterally are necessary, but not sufficient, to cause major depression. There must also be a conflict of interest between group members and the individual, a conflict that can arise, in part, from private information (if there were no conflict, group members would simply provide the needed benefits). Note that this conflict need not be overt nor even consciously recognized by those involved. Although the evidence presented above certainly suggests a conflict with others, there is also considerable direct evidence that social conflict is involved. In a meta-analysis of 48 studies, Finch et al. (1999) found that social negativity had a significant correlation with depression in the expected direction, and results of longitudinal studies suggest a causal influence of negative social interactions on subsequent depression (e.g., Vinokur and van Ryn 1993); depression may, in turn, exacerbate social negativity (e.g., Coyne 1976). A follow-up study by Finch et al. (1999) suggests that interference/hindrance, anger, and insensitivity are the three aspects of social negativity that are most salient as predictors of depression. Each seems relevant to the bargaining model.

Because changing social relations within a group can be a difficult and costly affair, most group members will resist such a change without clear evidence that it is necessary (Watson and Andrews 2002); otherwise group members could easily be exploited by deceptive individuals. If the individual has information that she is suffering a cost, but the other group members do not, the individual must credibly communicate this private information to others. Because this is a novel aspect of the bargaining model, there is no evidence (yet) that private information is associated with depression. It is very likely, however, that individuals often had private information about their costs and benefits in important cooperative ventures. Childrearing provides a nice example. The mother, having carried the child for nine months, may have considerable information about its health that is unavailable to either the father or other family members, or she may have information about her own health that necessitates changing her levels of investment.

Withholding Benefits and the Costs of Delay

Withholding benefits until better terms are forthcoming (asserting monopoly power) is the essential feature of any bargaining strategy and is one of the central functions of depression proposed here. In addition, the willingness of a depressed individual to delay investment in a cooperative venture is a credible signal to her social partners that the endeavor is unprofitable (Watson and Andrews 2002). Conversely, the degree of reluctance of other participants to increase the benefits they are providing is an equally credible signal of their true valuation of the venture: the longer they are willing to delay, the less they value the venture. It is important to note that depression is not simply a costly and therefore honest signal of social need (Spence 1974; Zahavi 1975). First, in the classic theory of costly signaling, the recipient of the signal does not incur a cost, only the sender (ignoring the relatively small costs of signal detection). This is not the case in the bargaining model of depression. The recipients of the signal (group members) may incur substantial costs; this, in fact, is a principal objective of the strategy.

Second, although it is widely assumed that costs guarantee the honesty of a signal, it is not the costs of a signal per se but rather that inherent aspects of the signal necessarily distinguish between individuals in different states. Here, the sender, — the depressed person — may incur little or no fitness cost when sending a credible signal. Consider, for example, the extreme case of a worker who is paid nothing, but whose boss profits handsomely from her labor. Because she has no wages to lose, it costs her nothing to go on strike, but it costs her boss plenty. Her willingness to delay working indefinitely is a credible (but not costly) signal of her low valuation of her current salary. Similarly, there would have been little fitness difference between an indissoluble marriage to an infertile mate and a complete cessation of all activities, including feeding and self care. The "message" of depression is that, for the sufferer, there is little difference in the fitness benefits obtained from investing heavily in her current social strategy or investing little. Depression is a credible signal because individuals who are profiting from their social strategies cannot afford the delay required to send it. Depression is a relatively affordable (and therefore sendable) signal only for those senders whose social circumstances are imposing significant opportunity costs.

Consistent with the bargaining model's requirement that depression cause a reduction in benefits generated by the afflicted individual, a loss of interest in virtually all activities is a prominent symptom of major depression, and depression has a very significant, negative impact on productivity. Worldwide, it is the leading cause of disability as measured by YLDs, and the fourth leading contributor to the global burden of disease (DALYs) in 2000.[5] Depression is the second most significant cause of DALYs in the age category 15–44 years for both sexes combined (WHO 2001). Wells et al. (1989) found that the poor functioning uniquely associated with depressive symptoms (with or without depressive disorder) was comparable to, or worse than, that uniquely associated with eight major, chronic medical conditions. For example, the unique association of days in bed with depressive symptoms was significantly greater than the comparable association with arthritis, diabetes, and hypertension.[6]

Further, numerous studies have found a significant impact of depression on mortality rates, suggesting that either depression itself, or the poor self-care caused by its symptoms, or both, might have an important negative impact on health. Unfortunately, most of these studies did not sufficiently control for important associated health risks like smoking and alcohol use. As a recent systematic review of the mortality of depression concluded (Wulsin et al. 1999, p. 15):

> The existing body of studies, so rich with mixed findings and so lean in the numbers of well-controlled comparable studies, suggests a substantial effect of depression on mortality in some populations, but to estimate the true size and the source of this effect (whether it is a direct result of the pathophysiology of depression or the indirect result of poor self-care) will require more rigorous study.

In the postpartum model, mothers with PPD should (a) experience a loss of interest in the infant and (b) actually reduce their investment in the infant. As predicted, loss of interest in the infant is a major symptom of PPD. In addition, mothers with PPD unequivocally reduce their investment in the new offspring along virtually every dimension at the same time that they appear to have reached a negative assessment of the childrearing venture. In the EEA, such reduced care would have had a serious negative impact on infant, and therefore family members' fitness (Hagen 1999).

---

5 "DALYs for a disease are the sum of the years of life lost due to premature mortality (YLL) in the population and the years lost due to disability (YLD) for incident cases of the health condition. One DALY can be thought of as one lost year of 'healthy' life." (WHO 2001, p. 25)

6 Depression is also associated with serious physical illness. There is a well-documented elevated risk of acute coronary syndromes in persons with major depression, which may be caused by the increased platelet reactivity/aggregability that has been observed in depressed patients (because these increase risk of intra-arterial thrombus formation, i.e., clotting) (e.g., Shimbo et al. 2002). An obvious interpretation of these findings is that, in the EEA, a social threat of the type hypothesized to cause depression frequently resulted in physical injury; thus, blood clotting system is on "high alert."

Given the time-sensitive nature of most human cooperative activities (e.g., foraging, territorial defense, and parenting), the withholding in the EEA of the benefits documented above would have certainly imposed the costs of delay on others required by the bargaining model. Even if an individual did not receive increased investment as a consequence of bargaining, she would have credibly signaled her low valuation of this cooperative venture to her social partners and would have received credible information from her social partners regarding their valuation of the venture. This information would have been of considerable utility for her future strategic decision making.

Does Depression Elicit Benefits?

Critical to the bargaining hypothesis is evidence that depression can improve one's social environment (or would have in the EEA). Just as management would react negatively to a labor strike but still be forced to provide benefits, depression should cause negative reactions in others yet still elicit benefits from them. The substantial evidence that depression causes negative reactions in others (Segrin and Dillard 1992) implies conflict. Does depression nonetheless elicit benefits? For much of the last century in the West, researchers have viewed depression as an illness, so studies investigating its power to work deep, and ultimately positive, long-term changes in the lives of those afflicted have been few-to-none. However, accounts of depression's transformative capabilities are frequently found in the penetrating autobiographies of those who have known the "black dog" (e.g., Jeffery Smith's *Where the Roots Reach for Water*).

In comparison to the current lack of objective evidence for long-term benefits, there is solid evidence that depression elicits short-term benefits. Before presenting the evidence for the benefits that are obvious predictions of the bargaining model, the rationale for an additional benefit — reduced risk of punishment — will be developed. Unilateral defection from a cooperative relationship, as occurs in the bargaining model, invites punishment for cheating (e.g., Axelrod and Dion 1988). If those choosing to withhold benefits could convince others that, despite not providing benefits, they were not taking benefits either, they might be able to avoid punishment for cheating, at least in the short term. The behavioral "shutdown" that characterizes major depression effectively prevents individuals not only from providing benefits, but also from taking benefits provided by others. It is important to have a thorough behavioral shutdown. Theoretical treatments of punishment and the evolution of cooperation make clear that error rates can be a critical parameter (e.g., Boyd and Richerson 1992). If group members mistakenly perceive that an individual is taking benefits but not reciprocating, they might impose devastating costs. A marked loss of interest in virtually all activities can significantly decrease the odds that the depressed individual will be perceived by anyone to be taking benefits.

A number of behavioral studies have demonstrated that although depression in one family member prompts negative feelings from other family members, it nonetheless appears to deter their aggressive behavior and to cause an increase in their tendency to offer solutions to problems in a positive or neutral tone and an increase in their solicitous behavior (e.g., caring statements), consistent with the bargaining model.[7] In the short term, depression has also been shown to elicit help and support from nonfamily members (i.e., roommates) in naturally occurring as well as laboratory situations, although longer-term studies indicate high levels of hostility and a progressive decline in social contact and satisfaction with the depressed person. In non-EEA social settings where social partners such as roommates often do not have the power to make major social changes and are not dependent on the depressed, it is not surprising that depression continues unabated and that social partners elect to reduce social contact. For a review of this literature, see Sheeber et al. (2001). Behavioral studies thus confirm that depression causes an increase in provisioning of social benefits and a decrease in aggressive responses, as predicted.

Similarly, the spouses of individuals experiencing PPD should report increasing their investment in parenting, and in fact they do. Depression scores for one spouse were positively correlated with reports of increasing investment in childcare by the other spouse (Hagen 2002). High levels of help from spouses and better interactions with infants in one study were also the only variables associated with remission of PPD (Campbell et al. 1992).

Major life improvements are associated with remission of depression and may even play a causal role (citations in text omitted):

> Even more thought-provoking was the investigation of the "meaning" of those fresh start experiences which, more often than not, preceded depressive remission.... Although all these data were collected retrospectively, the time order between these and remission, and the high proportion of such events which were independent of the subject's agency, lent plausibility to this being the effect of the environment on pathology. It seemed fresh starts were the mirror image of those producing the generalised hopelessness of Beck's depressive cognitive triad.... They either involved events like starting a new job after months unemployed, starting a course after years as a housewife, establishing a regular relationship with a new boy friend/girl friend after many months single, or the reduction of a severe difficulty, usually with interpersonal relationships, housing or finance. They seemed to embody the promise of new hope against a background of deprivation. It was notable that even for women who continued to experience difficulties of a depressogenic severity in one life domain such as marriage, a fresh start in another life domain — starting an access course — often seemed to tip the balance and set them on course for remission. (Harris 2001, p. 19)

It is not yet apparent whether depression symptoms themselves help enable "fresh starts" (or would have in the EEA), but this is, of course, precisely the proposed function of depression. It is therefore encouraging that "fresh starts" are closely associated with the remission of depression and may even cause it.

---

[7] Oddly, such responses seem to be viewed negatively by researchers in this field because they are seen as "facilitating" or "reinforcing" depressive behavior.

Depression in the Ethnographic Record

In small-scale, kin-based societies, which most closely resemble ancestral human communities, what little evidence exists suggests that depression occurs for the reasons predicted by the bargaining model, and that it has the predicted effects on the group. Among the Kaluli of the tropical forest in Papua New Guinea, for example, emotions (in general) and depression (in particular) must be understood for the roles they play in the system of reciprocity upon which Kalulian society is based (Schieffelin 1985). Emotions like grief and anger are appeals or demands to redress losses. If grief is an appeal to satisfy a "legitimate" claim, depression is an appeal to satisfy an "illegitimate" claim. Scheiffelin argues (p. 117) that depression should "arise in circumstances where an individual was placed unwillingly into a long-term life situation in which his or her assertive moves were regularly rebuffed or frustrated and in which there were no socially acceptable grounds for expressing anger or feeling owed." Thus, according to both Schieffelin and the bargaining model, grief should occur when there is a loss but little conflict between the individual and powerful others, and depression should occur when there is loss (more accurately, an opportunity cost) but a significant conflict between the individual and powerful others.

A careful study of an indigenous Quechuan malady, *pena*, which closely resembles depression (Tousignant and Maldonado 1989), also illustrates the impact of depressive symptoms on others in a small, kin-based society. Like major depression, severe cases of *pena* are characterized by a lack of concern for personal hygiene, loss of appetite often resulting in serious weight loss and dehydration, sleep disturbances, an inability to enjoy life, and a wish to die. Also, like major depression, *pena* is invariably associated with some kind of loss. Tousignant and Maldonado argue that *pena* functions to restore the balance of reciprocity upset by the loss and that "restitution of some form or another is the goal of the emotional strategy" (1989, p. 901). The impact of *pena* on the community closely matches the predictions of the bargaining model:

> [L]ong periods of sadness in a woman will attract the attention of kin. They will investigate with whom the fault lies, usually suspecting the husband, and see in what way the situation can be corrected. In case of failure, the eldest adults of the community will get involved and, if discussions fail, more stringent admonitions and punishments, even flogging, may be applied. As was pointed out by McKee [unpublished ms], guilt is not the core element of punishment. The goal of the intervention is not to make the abuser ashamed but to facilitate reparation. (Tousignant and Maldonado 1989, p. 900)

Both Schieffelin (1985) and Tousignant and Maldonado (1989) argue that the meaning and social consequences of depression among the Kaluli and the Quechua can only be understood in the context of the central organizing principle of these societies: reciprocity. Given the ubiquitous importance of reciprocity in contemporary hunter-gatherer groups, depression may well have had the same meaning and social consequences among ancestral human foragers.

The conceptualization of depression in some larger-scale traditional societies is also quite similar to the bargaining model. The Bengali illness concept *mathar golmal* (disturbance of the head), which appears to include depression, is an example. It is caused by "shock" such as the death of a loved one, business or career failures, or rejection by a lover (Bhattacharyya 1981, p. 153).

> [T]hese emotional states all seem to point to frustration as a key cause. This frustration may be economic (money worries), academic (failure in exams), career (lack of advancement), or emotional (unrequited love). As several respondents have noted, being unable to obtain what is deeply desired is the source of frustration. The most extreme example of such frustration and the one most frequently cited is [intense grief] where the death of a loved one prevents the fulfillment of one's desires. Thus, the primary attribute of "shock" is an emotional response to an intensely frustrating situation. The gratification of desires is prevented because of some obstacle which makes the desired outcome beyond one's control, thus rendering one's own efforts totally ineffectual. (Bhattacharyya 1981, p. 201)

Consistent with the bargaining model, informants believe that the affliction "can be cured if the desires of the individual are met." Examples include obtaining a spouse or securing the return of a boyfriend (Bhattacharyya 1981, p. 203).

Suicidality

Depression and suicidality are deeply intertwined (see Table 6.2). Suicidality is a diagnostic symptom for major depression (Table 6.1), and depression is the most common mental disorder leading to suicide, although substance abuse and schizophrenia are also major contributors (WHO 2001). A successful theory of depression must explain suicidality, and the bargaining model, building on the work of Giddens (1964), Brown (1986), and Watson and Andrews (2002), does.

Suicide permanently removes oneself as a source of valuable benefits for the group. Suicide threats are therefore threats to impose substantial costs on group members and can be viewed as a means to signal cheaply and efficiently to a large social group that it may suffer such costs if assistance or change is not

Table 6.2 The close association of depression and suicide.

Percent of suicides who had a mood disorder*	60%	NIMH (2000)
Percent of severely depressed (inpatient population treated for depression) who commit suicide	4.4%	Bostwick and Pankratz (2000)
Percent of less severely depressed (mixed inpatient/ outpatient population treated for depression) who commit suicide	2.2%	
Percent of those treated for nondepression illness who commit suicide	< 0.5%	

* Major depression, bipolar disorder, dysthymia.

forthcoming. Suicide attempts are necessary to underwrite the credibility of suicide threats and must therefore entail a genuine risk of serious injury or death. Failed attempts resulting in injury can still impose costs on group members and indicate the seriousness of future attempts. Completed suicides are the cost of maintaining a credible threat. A suicidal signaling/bargaining strategy could evolve if it involved warning others beforehand (allowing them to respond to the suicidal person's needs), if the rate of threats were much higher than the rate of attempts, and if the rates of attempts were much higher than the rate of completions. Under these circumstances, the average benefits received over many generations by genes coding for this strategy, when group members were successfully influenced, could exceed the average costs suffered by those genes when suicide attempts succeeded.[8]

In depression-related suicidality, individuals do commonly warn others of their intentions and frequently choose unreliable methods (Kreitman 1977; Stengel 1974). Major depression has been found to be by far the greatest risk factor for suicidal ideation, and the lifetime prevalence of suicidal ideation and attempts is several hundred times greater than the annual suicide rate (Table 6.3).

Across numerous studies, five psychological constructs have consistently been associated with suicide: impulsivity/aggression, depression, anxiety, hopelessness, and self-consciousness/social disengagement (Conner et al. 2001). Most of these are consistent with the bargaining model in obvious ways.

Previous research suggests that both clinicians (Bancroft et al. 1979; Hawton et al. 1982) and families (James and Hawton 1985) tend to attribute

Table 6.3 Rates of suicidal ideation and attempts are high compared to suicide rates.

Annual suicide rate (age standardized)[1]	0.015%	WHO (2001)
Two-week prevalence of suicidal ideation	2.6%	Goldney et al. (2003)
Lifetime prevalence of suicidal ideation[2]	10–18%	Weissmann et al. (1999)
Lifetime prevalence of suicide attempts[2]	3.5%	Weissmann et al. (1999)
Suicide attempts per completion[3]	8–25	NIMH (2000); Platt et al. (1992)

[1] Worldwide rate for 1996; approximately four times as many men (0.024%) as women (0.0068%) commit suicide, a bias that is probably due to men choosing more lethal methods.
[2] Cross-cultural study based on self-report.
[3] Based on conservative criteria such as suicide attempt-related hospitalizations.

[8] It may have been adaptive for very elderly or infirm individuals who were burdening their close kin to kill themselves reliably and without warning (deCatanzaro 1981). This does not account, however, for the large number of healthy, productive people who attempt suicide. (Suicide is among the three leading causes of death among young people 15–34 years of age [WHO 2001].) Healthy individuals who are suffering negative fitness due to costs imposed on their kin should simply leave the group.

manipulative motives to suicide attempters, consistent with the bargaining model. Although studies of adolescents' stated reasons for suicide indicate that few mention a manipulative motive (e.g., in a study by Boergers et al. [1998], only 18% did so), numerous data from small, kin-based societies confirm that suicide threats are used by individuals for exactly the political purposes proposed here. Giddens's 1964 article on the cross-cultural sociology of suicide is worth quoting at length (citations in text omitted):

> An example [of suicide as part of a wider social system of punishment and sanction in some societies] was given by Malinowski, in what has been recently described as "the best-known suicide in the ethnographic literature".... This was the case of a youth who committed suicide after he had been publicly accused of incest. This action, says Malinowski, served to expiate his crime. The suicide, by means of his act, "declares that he has been badly treated"...; the probability that a wronged or humiliated individual would kill himself serves as "a permanent damper on any violence of language or behavior, or any deviation from custom or tradition, which might hurt or offend another".... Suicide thus functions to facilitate social order; suicide, or the possibility of suicide, serves as a sanction in situations of controversy or dispute. A similar conclusion is reached by Berndt in a recent discussion of suicide.... Jefferys has collected together a number of examples of what he calls "revenge" suicide: in these examples, again, suicide functions as a form of social sanction against those towards whom the individual has a grievance.... Such suicide usually has ritualized elements in it — the suicide method, for example, is often standardized.
>
> Attempted suicide and verbal threats of suicide, can also be seen in some societies to be part of a recognized social pattern. In Tikopia, for example, according to Firth, the suicidal threat is recognized as an appropriate response in certain types of situations. Verbal suicide threats are used as a form of social pressure in the judicial process. The announcement of intention to commit suicide draws public attention to the individual who believes himself wronged, and provides an indictment of the wrongdoer.... A similar mechanism involving "a threat of suicide dramatically announced" operates, according to Honigman, among the Kashka Indians.... In Ovimbuandu, in central Angola, suicide threats are similarly used to put pressure on others in disputes; the suicidal threat is also recognized as an important form of social sanction among the Fulani.... Other examples are not hard to find. In all of these cases, suicide threats are part of a defined social pattern relating to the settlement of disputes.
>
> Attempted suicide, of course, often simply represents a suicide which fails through technical reasons. But this is by no means always the case. Malinowski, for example, notes that, in the Trobriands, there are two "serious" methods used in suicide — these virtually always produce death; there is also a "milder" method, from which the individual usually recovers. The "milder" method is usually the one used in matrimonial quarrels and other relatively minor disputes.... Among the Kuma of New Guinea, suicide attempts are "expected" of women when they are contractually married. The suicide attempt is always by drowning. The attempt only occasionally results in the death of the individual. The suicide attempt is an accepted method of protest against the relatives who have brought about the undesired match.... Fortune describes various cases of attempted suicide in Dobu. Here

> attempted suicide is mainly associated with matrimonial disputes. The suicide attempt is typically made in the spouse's village, and serves as a means of registering protest, in front of relatives, against the conduct of the spouse.... Gorer remarks upon similar instances among the Lepchas of the Southern Himalaya. An individual who believes himself wronged may attempt to commit suicide; this serves both to affirm his own innocence in the matter in question, and as a public indictment of the transgressor. The individual attempts suicide, but the attempt is made "in such situations that he is bound to be saved"....
>
> In all of these examples, the suicidal act is a recognized type of social mechanism, an accepted method of bringing pressure to bear upon others. (Giddens 1964, pp. 115–116)

Brown's (1986) detailed analysis of suicide among the Jivaroan Aguaruna, a group of hunter-horticulturalists who live in the rugged uplands of the Amazon in northern Peru, similarly reveals that the social etiology of suicides among this group is precisely that predicted by the bargaining model — suicide is used by individuals to impose costs on group members with whom they have a conflict:

> Some segments of Aguaruna society — specifically, *women and young men who are unable to organise collective responses to conflict* — use solitary acts of violence directed against the self to express anger and grief, as well as to punish social antagonists. (Brown [1986], p. 311; emphasis added)

Sex Bias

Women are about twice as likely to suffer from a major depressive episode as men, a finding that, cross-culturally, is quite robust (e.g., Ustun and Sartorius 1995). Matching men and women by social role variables (e.g., employment, marriage status, and number of children) within cultures appears to reduce the female bias by about 50% (Maier et al. 1999); the remaining bias has yet to be explained.

Under the bargaining model, women are expected to have higher rates of depression because (a) it was more often a better strategy for them, and (b) they had more conflicts with powerful others (cf. Wenegret 1995; Watson and Andrews 2002; MacKey and Immerman 2000). Women should have a lower threshold for, and higher rates of, depression than men because, in the EEA:

1. Physical aggression was a less-effective strategy for females in intersexual conflict.
2. Patrilocality[9] meant that females, more often than males, were living with nonkin and were thus more likely to have conflicts with the group (e.g., Rodseth et al. 1991; see also Hess and Hagen, unpublished).
3. Female reproductive capacity was a scarce resource, so females were, more than males, victims of social manipulation by powerful others.

---

[9] Males live with kin, females transfer — the modal pattern for humans.

4. Most females could put scarce reproductive and childcare investment capacities at risk, whereas only some males had, for example, valuable hunting or military benefits to put at risk (i.e., there was less variability in female reproductive value relative to male reproductive value).

Biochemistry

The monoamine hypothesis of depression proposes that the physiological basis for depression is a deficiency of central noradrenergic and/or serotonergic systems, and that rectifying such deficiencies with an antidepressant would reduce or eliminate depression. Consistent with this hypothesis, the symptoms of depression can be alleviated by agents that, via several mechanisms, increase synaptic concentrations of monoamines, like serotonin and norepinephrine. This hypothesis has a number of problems, however, including the fact that it usually takes weeks or months of antidepressant treatment before depressive symptoms lift, even though antidepressants increase availability of the target neurotransmitters immediately. The hypothesis also fails to explain why depletion of serotonin does not cause depression in nondepressed subjects, nor does it exacerbate symptoms in depressed subjects (for review, see Bell et al. 2001). In addition, not all drugs which enhance serotonergic or noradrenergic transmission effectively treat depression. These and other deficiencies of the monoamine hypothesis are widely recognized, although it has by no means been abandoned (for a review, see Hirschfeld 2000).

According to the bargaining model, individuals should experience depression when they have potential conflicts with powerful others and cannot act unilaterally. Such circumstances would obviously induce long-term stress. Hundreds of studies have demonstrated increased levels of the stress hormone cortisol in depressed patients, and there is rapidly accumulating evidence that chronic activation of the HPA axis, the hormonal system that regulates the "fight-or-flight" (i.e., stress) response, is a proximate cause of depression. Pariante and Miller (2001, p. 391) summarize these findings in their review of the role of glucocorticoid receptors and stress hormones in major depression:

> Hyperactivity of the HPA axis in patients with major depression is one of the most consistent findings in biological psychiatry. Specifically, patients with major depression have been shown to exhibit increased concentrations of [the stress hormone] cortisol in plasma, urine, and cerebrospinal fluid (CSF); an exaggerated cortisol response to adrenocorticotropic hormone (ACTH); and an enlargement of both the pituitary and the adrenal glands.... These HPA axis alterations are believed to be secondary to hypersecretion of corticotropin-releasing hormone (CRH), which has behavioral effects in animals that are similar to those seen in depressed patients, including alterations in activity, appetite, and sleep....

Elevated levels of stress hormones among depressives were recognized even before antidepressants were discovered, but these changes were seen as epiphenomena of the stressful experience of depression. A vast amount of evidence has since accumulated that altered stress hormone secretions in depression are not epiphenomenal but are causally involved in its development and course. Further, there is evidence that traditional antidepressants may function by effecting changes in corticosteroid receptors; thus in the HPA axis, changes which then lead to clinical recovery (Holsboer 2000; Pariante and Miller 2001). In sum, considerable biochemical evidence is consistent with the bargaining model's prediction that certain kinds of social stress cause depression.[10]

Other Etiological Factors and Findings

Three factors that are important in the etiology of depression — genetic background, prior episodes of depression, and personality — do not clearly support the bargaining model, yet they are not inconsistent with it either. That there is a significant heritable component to unipolar depression is perhaps the strongest evidence against it being viewed as an adaptation.[11] However, just as there could be heritable differences in thresholds for physical pain (which clearly is an adaptation), there could be heritable differences in depression thresholds or heritable differences in the likelihood of experiencing depressogenic events. Kendler and colleagues have found just this: A significant fraction of the heritable component of depression consists of heritable differences in the sensitivity to the environmental stimuli that trigger depression, and heritable differences in the likelihood of selecting oneself into environments that cause depression. That is, the genetic effects, at least in part, act on the environmental pathways to depression (e.g., Kendler et al. 2002).

Prior episodes of depression appear to be, in and of themselves, a cause of current episodes. Evidence is also accumulating that with each depressive episode, the association between stressful life events and a depressive episode decreases. Although early episodes are strongly correlated with stressful life events, later episodes onset with little apparent provocation (Kendler et al. 2000, 2001). This effect was strongest for those at low genetic risk. This "kindling" effect is probably responsible for the clinical observation that some cases of depression are not clearly related to life stressors. One functional interpretation of this effect is that defensive strategies become increasingly "hair-triggered." Much as the immune system becomes sensitized to specific antigens in order to

[10] Of course, all biomedical researchers investigating depression assume that differences in biochemistry between depressed and nondepressed individuals reflect pathology. This appears, however, to be little more than an assumption.

[11] Although studies consistently find that individual environmental factors play the largest role in the development of unipolar depression, various twin studies have found modest degrees of heritability, ranging from approximately 0.30–0.50. There is also some evidence that depression might be more heritable in women than men (see, e.g., Bierut et al. 1999).

respond with maximum speed and efficiency when it encounters them again, so, too, may social defense strategies become sensitized to social circumstances that are likely to reoccur and require a rapid and perhaps even preemptive response. It is also possible, of course, that the kindling effect is simply a by-product of the neurological changes that are associated with chronic stress.

Vulnerability factors, such as having a "neurotic" personality, also account for some of the variability in depression and are good predictors of future episodes. Although the origins of such personality factors are still obscure, they may be based on genetic background, experiences during childhood, and long-term exposure to particular social circumstances (e.g., Goldberg 2001). Given that an anxious disposition is a central feature of neuroticism, the vulnerability factor most reliably associated with depression, it is reasonable that "high-n" individuals believe themselves to be facing, or vulnerable to, social threats. If so, then neuroticism, whatever its origins, is understandably a "risk factor" for depression under the bargaining model.

A number of differences in cognitive performance between depressed and nondepressed individuals, typically involving memory, attention, and executive functions, have been well established (for a brief review, see Austin et al. 2001). These differences are widely interpreted as "deficits" indicative of an underlying neurological pathology. If depression is an adaptation, a number of cognitive differences along with their associated neuronal differences would also be expected between depressed and nondepressed individuals. The mere fact of differences is not, in and of itself, evidence that depression is a pathology, and it is possible that the documented differences are in fact related to adaptive functions of depression. Specific pathological models will have to be tested, both against functional models and against each other, to determine the best interpretation of these and the other data on depression.

CONCLUSION

Although effective in many circumstances, aggression and persuasion are poorly suited to resolve genuine conflicts between an individual and powerful others. Given the limited ability of ancestral groups to meet all the needs of all members, such conflicts would have been common, especially when most group members' social strategies were yielding benefits, but one individual's were not. If the individual had a monopoly, or near monopoly, on the benefits she was providing to the group, she could put these benefits at risk, forcing group members to provide assistance or bargain over the terms of the social contract. This strategy might have been particularly effective for women.

In the context of social conflict and feelings of entrapment/lack-of-control/helplessness, a severe negative life event frequently causes depression, especially among women. Depressive symptoms, such as sad or depressed mood, a loss of interest in virtually all activities, and suicidality, cause productivity to plummet. Despite negative feelings about the depressed, family members and other social partners consequently provide a surprising number of benefits, including increased concern, offers of advice, childcare and other forms of support, and decreased aggression.

Depression and suicidality in at least some of the small-scale, kin-based societies, which most closely resemble ancestral communities, are seen to be caused by loss, socially unacceptable anger, or "frustrated desires." Further, they are understood to redress losses and elicit help and concern from community members. Given the high degree of interdependence and reliance on reciprocity in these societies, it is difficult to imagine that depressive symptoms would not have such effects. Depression remits in association with fresh-start experiences and increased social support. Numerous biochemical investigations indicate that depression may be caused, not by neurotransmitter deficits per se, but by chronic stress.

The hypothesis that depression is an adaptation triggered by social costs that functions to compel social investment and change is supported by much of what is known about depression; however, finer-grained longitudinal studies will be required to determine adequately if depression can, in fact, cause meaningful and ultimately beneficial changes in social circumstances, or could have in the EEA. If so, then non-Western conceptualizations of depression, such as the Quechuan view of depression as an emotional strategy to restore the balance of reciprocity upset by loss, are largely correct, whereas the Western conceptualization of depression as a mental illness is largely incorrect.

ACKNOWLEDGMENTS

Many thanks to Leda Cosmides, Peter Hammerstein, Nicole Hess, Andy Thomson, John Tooby, Paul Watson, Aaron Sell, and members of the Center for Evolutionary Psychology lab group and Institute for Theoretical Biology for numerous comments and suggestions.

REFERENCES

Abramson, L.Y., G.I. Metalsky, and L.B. Alloy. 1989. Hopelessness depression: A theory-based subtype of depression. *Psychol. Rev.* **96**:358–372.

Alexander, R.D. 1986. Ostracism and indirect reciprocity: The reproductive significance of humor. *Ethol. Sociobiol.* **7**:253–270.

APA (American Psychiatric Association). 1994. Diagnostic and Statistical Manual of Mental Disorders. Washington, D.C.: American Psychiatric Assn.

Austin, M., P. Mitchell, and G.M. Goodwin. 2001. Cognitive deficits in depression. *Brit. J. Psychiatry* **178**:200–206.

Axelrod, R., and D. Dion. 1988. The further evolution of cooperation. *Science* **242**: 1385–1390.

Bancroft, J., K. Hawton, S. Simkin et al. 1979. The reasons people give for taking overdoses: A further inquiry. *Brit. J. Med. Psychol.* **52**:353–365.

Beck, A.T. 1996. Depression as an evolutionary strategy. Presented at Annual Meeting of the Human Behavior and Evolution Society, June 27, Evanston, IL. Abstract. http://www.hbes.com/HBES/abst96.htm.

Beck, C.T. 1996. A meta-analysis of the relationship between postpartum depression and infant temperament. *Nursing Res.* **45**:225–230.

Bell, C., J. Abrams, and D. Nutt. 2001. Tryptophan depletion and its implications for psychiatry. *Brit. J. Psychiatry* **178**:399–405.

Benassi, V.A., P.D. Sweeney, and C.L. Dufour. 1988. Is there a relation between locus of control orientation and depression? *J. Abnorm. Psychol.* **97**:357–367.

Bhattacharyya, D.P. 1981. Bengali Conceptions of Mental Illness. Ph.D. diss., Univ. of Michigan, Ann Arbor.

Bierut, L.J., A.C. Heath, K.K. Bucholz et al. 1999. Major depressive disorder in a community-based twin sample: Are there different genetic and environmental contributions for men and women? *Arch. Gen. Psychiatry* **56**:557–563.

Boehm, C. 1996. Emergency decisions, cultural-selection mechanics, and group selection. *Curr. Anthro.* **37**:763–793.

Boergers, J., A. Spirito, and D. Donaldson. 1998. Reasons for adolescent suicide attempts: Associations with psychological functioning. *J. Am. Acad. Child Adoles. Psychiat.* **37**:1287–1293.

Bostwick, J.M., and V.S. Pankratz. 2000. Affective disorders and suicide risk: A reexamination. *Am. J. Psychiatry* **157**:1925–1932.

Boyd, R., and P.J. Richerson. 1988. The evolution of reciprocity in sizable groups. *J. Theor. Biol.* **132**:337–356.

Boyd, R., and P.J. Richerson. 1992. Punishment allows the evolution of cooperation (or anything else) in sizable groups. *Ethol. Sociobiol.* **13**:171–195.

Brown, M.F. 1986. Power, gender, and the social meaning of Aguaruna suicide. *Man New Series* **21**:311–328.

Brown, G.W. 1998. Genetic and population perspectives on life events and depression. *Soc. Psychiat. Epidemiol.* **33**:363–372.

Campbell, S.B., J.F. Cohn, C. Flanagan et al. 1992. Course and correlates of postpartum depression during the transition to parenthood. *Dev. Psychopathol.* **4**:29–47.

Clutton-Brock, T.H. 1991. The Evolution of Parental Care. Princeton, NJ: Princeton Univ. Press.

Clutton-Brock, T.H., and G.A. Parker. 1995. Punishment in animal societies. *Nature* **373**:209–216.

Conner, K.R., P.R. Duberstein, Y. Conwell et al. 2001. Psychological vulnerability to completed suicide: A review of empirical studies. *Suicide Life-threat. Behav.* **31**: 367–385.

Coyne, J.C. 1976. Depression and the response of others. *J. Abnorm. Psychol.* **85**: 186–193.

deCatanzaro, D. 1981. Suicide and Self-damaging Behavior: A Sociobiological Perspective. New York: Academic.

Fessler, D.M.T. 2003. The male flash of anger: Violent response to transgression as an example of the intersection of evolved psychology and culture. In: Missing the Revolution: Darwinism for Social Scientists, ed. J. Barkow. Oxford: Oxford Univ. Press, in press.

Figueredo, A.J., V. Corral-Verdugo, M. Frias-Armenta et al. 2001. Blood, solidarity, status, and honor: The sexual balance of power and spousal abuse in Sonora, Mexico. *Evol. Hum. Behav.* **22**:295–328.

Finch, J.F., M.A. Okun, G.J. Pool, and L.S. Ruehlman. 1999. A comparison of the influence of conflictual and supportive social interactions on psychological distress. *J. Pers.* **67**:581–621.

Giddens, A. 1964. Suicide, attempted suicide, and the suicide threat. *Man* **64**:115–116.

Goldberg, D. 2001. Vulnerability factors for common mental illnesses. *Brit. J. Psychiatry* **178 (40)**:s69–s71.

Goldney, R.D., E. Dal Grande, L.J. Fisher, and D. Wilson. 2003. Population attributable risk of major depression for suicidal ideation in a random and representative community sample. *J. Affect. Disord.*, in press.

Hagen, E.H. 1996. Postpartum depression as an adaptation to paternal and kin exploitation. Human Behavior and Evolution Society 8th Ann. Conf., Northwestern Univ. Abstract. http://www.hbes.com/HBES/abst96.htm.

Hagen, E.H. 1999. The functions of postpartum depression. *Evol. Hum. Behav.* **20**: 325–359.

Hagen, E.H. 2002. Depression as bargaining: The case postpartum. *Evol. Hum. Behav.* **23**:323–336.

Harris, T. 2001. Recent developments in understanding the psychosocial aspects of depression. *Brit. Med. Bull.* **57**:17–32.

Hawton, K., D. Cole, J. O'Grady, and M. Osborn. 1982. Motivational aspects of deliberate self-poisoning in adolescents. *Brit. J. Psychiatry* **141**:286–291.

Henderson, S. 1974. Care-eliciting behavior in man. *J. Nerv. Ment. Disord.* **159**: 172–181.

Hirschfeld, R.M. 2000. History and evolution of the monoamine hypothesis of depression. *J. Clin. Psychiatry* **61(6)**:4–6.

Holsboer, F. 2000. The corticosteroid receptor hypothesis of depression. *Neuropsychopharmacology* **23**:477–501.

James, D., and K. Hawton. 1985. Overdoses: Explanations and attitudes in self-poisoners and significant others. *Brit. J. Psychiatry* **146**:481–485.

Kendler, K.S., C.O. Gardner, and C.A. Prescott. 2002. Toward a comprehensive developmental model for major depression in women. *Am. J. Psychiatry* **159**: 1133–1145.

Kendler, K.S., R.C. Kessler, M.C. Neale et al. 1993. The prediction of major depression in women: Toward an integrated etiologic model. *Am. J. Psychiatry* **150**:1139–1148.

Kendler, K.S., R.C. Kessler, E.E. Walters et al. 1995. Stressful life events, genetic liability, and onset of an episode of major depression in women. *Am. J. Psychiatry* **152**:833–842.

Kendler, K.S., L.M. Thornton, and C.O. Gardner. 2000. Stressful life events and previous episodes in the etiology of major depression in women: An evaluation of the "kindling" hypothesis. *Am. J. Psychiatry* **157**:1243–1251.

Kendler, K.S., L.M. Thornton, and C.O. Gardner. 2001. Genetic risk, number of previous depressive episodes, and stressful life events in predicting onset of major depression. *Am. J. Psychiatry* **158**:582–586.

Kennan, J., and R. Wilson. 1993. Bargaining with private information. *J. Econ. Lit.* **31**: 45–104.

Kreitman, N. 1977. Parasuicide. London: Wiley.

Lewis, A.J. 1934. Melancholia: A clinical survey of depressive states. *J. Mental Sci.* **80**:1–43.

MacKey, W.C., and R.S. Immerman. 2000. Depression as a counter for women against men who renege on the sex contract. *Psychol. Evol. Gender* **2**:47–71.

Maier, W., M. Gansicke, R. Gater et al. 1999. Gender differences in the prevalence of depression: A survey in primary care. *J. Affect. Disord.* **53**:241–252.

Mazure, C.M., M.L. Bruce, P.K. Maciejewski, and S.C. Jacobs. 2000. Adverse life events and cognitive-personality characteristics in the prediction of major depression and antidepressant response. *Am. J. Psychiatry* **157**:896–903.

McElreath, R. 2003. Reputation and the evolution of conflict. *J. Theor. Biol.* **220**: 345–357.

Nemeroff, C.B. 1996. The corticotropin-releasing factor (CRF) hypothesis of depression: New findings and new directions. *Molec. Psychiatry* **1**:336–342.

Nesse, R. 1991. What good is feeling bad? The evolutionary benefits of psychic pain. *Sciences* **31**:30–37.

Nesse, R. 1999. What Darwinian medicine offers psychiatry. In: Evolutionary Medicine, ed. W.R. Trevathan, E.O. Smith, and J.J. McKenna, pp. 351–373. Oxford: Oxford Univ. Press.

Nesse, R. 2000. Is depression an adaptation? *Arch. Gen. Psychiatry* **57**:14–20.

NIMH (National Institute of Mental Health). 2000. Frequently Asked Questions about Suicide. http://www.nimh.nih.gov/research/suicidefaq.cfm.

Pariante, C.M., and A.H. Miller. 2001. Glucocorticoid receptors in major depression: Relevance to pathophysiology and treatment. *Biol. Psychiatry* **49**:391–404.

Patel, V. 2001. Cultural factors and international epidemiology. *Brit. Med. Bull.* **57**: 33–45.

Platt, S., U. Bille-Brahe, A. Kerkhof et al. 1992. Parasuicide in Europe: The WHO/EURO multicentre study on parasuicide. I. Introduction and preliminary analysis for 1989. *Acta Psychiat. Scand.* **85**:97–104.

Presson, P.K., and V.A. Benassi. 1996. Locus of control orientation and depressive symptomatology: A meta-analysis. *J. Soc. Behav. Pers.* **11**:201–212.

Price, J., L. Sloman, R. Gardner et al. 1994. The social competition hypothesis of depression. *Brit. J. Psychiatry* **164**:309–315.

Rodseth, L., R.W. Wrangham, A.M. Harrigan, and B.B. Smuts. 1991. The human community as a primate society. *Curr. Anthro.* **32**:221–254.

Rotter, J.B. 1966. Generalized expectancies for internal versus external control of reinforcement. *Psychol. Mono.: Gen. Applied* **80**:1–28.

Rubinstein, A. 1982. Perfect equilibrium in a bargaining model. *Econometrica* **50**: 97–109.

Schieffelin, E.L. 1985. The cultural analysis of depressive affect: An example from New Guinea. In: Culture and Depression, ed. A.M. Kleinman and B. Good, pp. 101–133. Berkeley: Univ. of California Press.

Segrin, C., and J.P. Dillard. 1992. The interactional theory of depression: A meta-analysis of the research literature. *J. Soc. Clin. Psychol.* **11**:43–70.

Sheeber, L., H. Hops, and B. Davis. 2001. Family processes in adolescent depression. *Clin. Child Fam. Psych. Rev.* **4**:19–35.

Shimbo, D., J. Child, K. Davidson et al. 2002. Exaggerated serotonin-mediated platelet reactivity as a possible link in depression and acute coronary syndromes. *Am. J. Cardiol.* **89**:331–333.

Shostak, M. 1981. Nisa: The Life and Words of a !Kung Woman. New York: Vintage.

Smith, J. 1999. Where the Roots Reach for Water: A Personal and Natural History of Melancholia. New York: North Point.

Spence, M. 1974. Market Signaling. Cambridge, MA: Harvard Univ. Press.

Stengel, E. 1974. Suicide and Attempted Suicide. New York: Penguin.

Suarez, S.D., and G.G. Gallup. 1985. Depression as a response to reproductive failure. *J. Soc. Biol. Struct.* **8**:279–287.

Thornhill, R., and N.W. Thornhill. 1989. The evolution of psychological pain. In: Sociobiology and the Social Sciences, ed. R.W. Bell and N.J. Bell, pp. 73–103. Lubbock: Texas Tech Univ. Press.

Tooby, J., and L. Cosmides. 1990. The past explains the present: Emotional adaptations and the structure of ancestral environments. *Ethol. Sociobiol.* **11**:375–424.

Tooby, J., and L. Cosmides. 2000. Evolutionary psychology and the emotions. In: Handbook of Emotions, ed. M. Lewis and J.M. Haviland-Jones, 2d ed., pp. 91–115. New York: Guilford.

Tousignant, M., and M. Moldonado. 1989. Sadness, depression and social reciprocity in highland Ecuador. *Soc. Sci. Med.* **28**:899–904.

Trivers, R.L. 1972. Parental investment and sexual selection. In: Sexual Selection and the Descent of Man, 1871–1971, ed. B. Campbell, pp. 136–179. Chicago: Aldine.

U.S. Dept. of Health and Human Services. 1993. Depression in Primary Care, vol. 2. Treatment of Major Depression. AHCPR Publication No 93–0551. Rockville, MD: U.S. Dept. of Health and Human Services.

Ustun,T.B., and N. Sartorius. 1995. Mental Illness in General Health Care: An International Study. Chichester: Wiley and Geneva: World Health Organization.

Vinokur, A.D., and M. van Ryn. 1993. Social support and undermining in close relationships: Their independent effects on the mental health of unemployed persons. *J. Pers. Soc. Psych.* **65**:350–359.

Watson, P.J., and P.W. Andrews. 2002. Toward a revised evolutionary adaptationist analysis of depression: The social navigation hypothesis. *J. Affect. Disord.* **72**:1–14.

Weissman, M.M., R.C. Bland, G.J. Canino et al. 1999. Prevalence of suicide ideation and suicide attempts in nine countries. *Psychol. Med.* **29**:9–17.

Wells, K.B., A. Stewart, R.D Hays et al. 1989. The functioning and well-being of depressed patients: Results from the Medical Outcomes Study. *J. Am. Med. Assn.* **262**:914–919.

Wenegrat, B. 1995. Illness and Power: Women's Mental Disorders and the Battle between the Sexes. New York: New York Univ. Press.

Whiffen, V.E., and I.H. Gotlib. 1993. Comparison of postpartum and nonpostpartum depression: Clinical presentation, psychiatric history, and psychosocial functioning. *J. Consult. Clin. Psychol.* **61**:485–494.

WHO (World Health Organization). 2001. The World Health Report 2001. Health Systems: Improving Performance. Geneva: WHO.

Wolman, B.B. 1990. Preface. In: Depressive Disorders: Facts, Theories, and Treatment Methods, ed. B.B. Wolman and G. Stricker. New York: Wiley.

Wulsin, L.R., G.E. Vaillant, and V.E. Wells. 1999. A systematic review of the mortality of depression. *Psychosom. Med.* **61**:6–17.

Zahavi, A. 1975. Mate selection: A selection for a handicap. *J. Theor. Biol.* **53**:205–214.

Standing, left to right: Ed Hagen, Ernst Fehr, John Tooby, Manfred Milinski, Peter Hammerstein, Joan Silk, Tim Clutton-Brock, Richard McElreath
Seated, left to right: Michael Kosfeld, Dan Fessler, Margo Wilson

7

Group Report: The Role of Cognition and Emotion in Cooperation

Richard McElreath, Rapporteur

Timothy H. Clutton-Brock, Ernst Fehr, Daniel M. T. Fessler, Edward H. Hagen, Peter Hammerstein, Michael Kosfeld, Manfred Milinski, Joan B. Silk, John Tooby, and Margo I. Wilson

INTRODUCTION

Altruism, behavior which reduces the individual fitness of the actor while increasing the fitness of another organism, has attracted much attention from both biologists and economists because it seems to defy the logic of both natural selection and standard preferences. In biology, kin selection (Hamilton 1964) is the best-established explanation of the evolution and maintenance of altruistic behavior. However, many examples of apparent altruism defy explanation by kin selection, since they occur among unrelated individuals. The second best-established theory, reciprocal altruism (Trivers 1971), offers to explain substantial portions of this remainder. However, outside of humans, little good evidence exists, so its status is still undetermined. In addition, many examples of putative altruism in humans, particularly those of greatest interest to economists, defy explanation by reciprocal altruism, either because they occur within very large groups of individuals or occur without the possibility of reciprocation. Thus the challenge before us is to understand better the range of mechanisms that support cooperation, particularly outside kin selection.

In this chapter, we summarize our discussions of mechanisms that support altruism outside of kin selection. We felt it was important to focus our discussion on mechanisms. One of the strengths of Darwin's account of adaptations is that it not only explains why animals are often well-adapted to their environments, but also why they are often poorly adapted. If all Darwinism did was to predict that animals should be well-adapted, its predictions would be indistinguishable from Creationism. Instead, the theory of natural selection provides a mechanism by which adaptations as well as maladaptations are constructed. It is in this way that attention to mechanisms in the study of cooperation is scientifically productive. A model of cooperation that focuses only on outcome cannot easily predict when cooperation does not emerge. Simultaneously, without attention to errors in the functioning of cognitive machinery or flaws in specific algorithms, we may not be able to understand the design of the machinery we do find. Although the distinction between mechanism (proximate explanation) and function (ultimate explanation) is useful, it obscures the modern understanding that mechanisms have strong impacts on function.

Economists, like biologists, have been interested in the emergence and stability of cooperative behavior. They also have good reason to turn to mechanisms as assets in designing both models and experiments. A substantial body of experimental evidence now confirms that human behavior substantially deviates from the predictions made by standard models of selfish rationality. However, this confirms only that people do not have standard preferences, that their utilities do not emerge in a simple way from the explicit payoffs. Behavioral economics has emerged as a way of uniting traditional tools with a concern for dissecting the components of the utility functions behind economic theory, as well as exploring alternatives to optimizing strategies. These debates must focus on the details of how individuals, for example, infer intention and compute concepts such as "fairness." The specific form which rationality takes, the nature of algorithms in an individual's head, and the cues which individuals attend to and how they use them, all influence behavior in potentially cooperative settings.

This report is organized as follows. First, we discuss evidence for reciprocal altruism in animal societies, as well as specific mechanisms for the bookkeeping of past interactions. Next, we explore the role of reputation and strong reciprocity in dyadic cooperation. After these two sections on dyads, we discuss the role of reciprocal altruism, strong reciprocity, and reputation for cooperation in sizable groups of individuals, not just pairs. Finally, emotions may as well implement strategies in both dyadic and large-scale cooperation, and the nature of emotion mechanisms may powerfully affect our behavioral predictions in any of these contexts.

BOOKKEEPING

"Do unto others as they do unto you" is not quite the Golden Rule, but it is in the theory of reciprocal altruism. Trivers (1971) brought biologists' attention to the possibility of altruism contingent upon the altruism of other individuals. Axelrod's (1984) tournaments and Axelrod and Hamilton's (1981) model of reciprocal altruism went a long way toward popularizing the prediction that cooperation in pairs of unrelated individuals could be sustained if individuals (a) recognize one another, (b) individuals keep track of past interactions, and (c) contingently help those who helped in the past. Consequently, the "keeping track," or bookkeeping, of past interactions has been the focus of much work on reciprocal altruism, much as kin recognition has been in kin selection. We begin by reviewing the empirical evidence that bookkeeping allows unrelated animals to sustain cooperation. We then present theory and observations about the nature of bookkeeping strategies in dyads which suggest that, in some contexts, careful bookkeeping may not always be such a clear prediction after all.

Evidence of Bookkeeping in Nature

Outside of humans, good evidence of reciprocal altruism is quite limited. Hammerstein (Chapter 5, this volume) discusses significant flaws with several of the most widely cited studies of reciprocal altruism in nonhuman animals (see also Enquist and Leimar 1993). A number of studies do not explicitly examine contingency of aid. Instead, many studies, including those on nonhuman primates, simply provide correlations between help given and received for particular pairs of individuals (Silk, this volume). The main problem that arises in correlational studies of reciprocal altruism (as in all correlational studies) is that it is difficult to be certain that the association between two forms of behavior is not the product of some third variable that has not been measured. Thus, some researchers have reported a positive correlation between the amount of grooming within dyads and the amount of social support within dyads (Silk, this volume). It is possible that this correlation reflects contingent behavior: "I will continue to groom you as long as you respond to my solicitations for support." However, it is also possible that this correlation reflects a noncontingent preference for certain partners, such as close kin or age mates or familiar associates ("friends"). Correlational data are also problematic because they hide variation across dyads. If noncontingent cooperation among kin is common, then small, but selectively important amounts of reciprocal altruism among nonkin might be difficult to detect in group-level analyses. This would occur if, for example, all but one dyad in a social group were comprised of related individuals who cooperated without need for reciprocal altruism, since kin selection maintains cooperation in these dyads. However, the lone unrelated dyad might be maintained by reciprocal exchanges but vanish in a group-level analysis. Thus aspects of both the positive and negative evidence are still in question.

Experimental studies, in which contingencies are explicitly examined, provide more convincing evidence that individuals keep track of past exchanges and use that information to direct aid selectively, at least in nonhuman primates (reviewed in Silk, this volume). However, even when a study explicitly examines contingency, the evidence can remain unclear. This is because, in naturalistic settings, it is very difficult to detect contingencies in behavior. In vervets and macaques, grooming is linked to subsequent support (or apparent willingness to provide support) in experimental settings (Hemelrijk 1994; Seyfarth and

Cheney 1984); however, grooming is not consistently correlated with support among nonrelatives in naturalistic settings (Schino 2001). Among captive chimpanzees, possessors of food are more likely to share with former groomers than with others and are less likely to behave aggressively to attempts to share by former grooming partners than with others (de Waal 1997). However, the absolute magnitude of the effect of grooming on subsequent grooming is very small; and in dyads that groom often, the contingency disappears. The relevant time interval for judging contingent behavior is still unclear. Reciprocity may be more delayed in the more stable pairings but still maintain cooperation.

The entire literature is, however, not so ambiguous. Ungulates (e.g., impala; Hart and Hart 1992), some rodents (Stopka and Graciasova 2001), and some monkeys (Barrett and Henzi 2001; Cords 2002) exchange grooming reciprocally, taking turns grooming one another. Thus, A grooms B for a short period; then B grooms A; then A grooms B again, etc. In some cases, changes in the length of each grooming sequence within the bout are matched by the other partner. In baboons, however, time matching does not occur in all bouts; roughly 40% of all grooming bouts involve unilateral interactions (A groomed B, but was not groomed by B).

Henzi and Barrett (2002) have presented evidence which suggests that female baboons "trade" grooming for access to other females' newborn infants. In nearly all primate species (including humans), infants are extremely attractive to females other than their mothers. In macaques and baboons, females are quite eager to inspect, greet, and touch other females' infants, but do not hold, carry, or nurse them. Many researchers have noticed that females often use grooming to gain access to infants, but Henzi and Barrett were the first to show that the "price" (grooming time) females pay for access to infants depends on the relative rank of the mother and the handler. Mothers are groomed longer by lower-ranking than by higher-ranking females who want to handle their infants.

Additional evidence from shoaling fish suggests the importance of reciprocal altruism in maintaining cooperative dyads, through both evidence of immediate bookkeeping and the nature of cooperating groups. In the wild, when groups of sticklebacks (*Gasterosteus aculeatus*) have detected a predator, such as a pike, they do not normally flee or hide. Instead, single fish or small groups leave the school and approach the predator very closely, waiting a few moments within striking distance of the predator. One fish moves forward a bit, and if the other one follows, the first proceeds a bit more, perhaps monitoring the partner's continued cooperation. It has been shown experimentally that this behavior is contingent (Milinski 1987). The fish inspect repeatedly with the same partner in a way consistent with a contingent reciprocal strategy. (For a discussion of the controversy surrounding this evidence, cf. Dugatkin 1997.) Usually pairs, but not larger groups, of sticklebacks participate in these so-called predator inspection visits. This may seem puzzling, since the cost of predator inspection would be smaller in larger groups, due to risk dilution. However, theoretical work by

Boyd and Richerson (1988) suggests that reciprocal altruism is unlikely to evolve in large and even moderately sized groups. (This result is explained in a later section.) Among the sticklebacks, even the rarer, larger inspection groups have been shown to consist of several well-synchronized pairs (Milinski et al. 1990), not a large well-synchronized whole. These experiments and observations thus constitute indirect evidence of direct reciprocity, since altruism driven by reciprocity should be confined to small groups of individuals.

Similar evidence from social carnivores makes the same suggestion. Coalitions consisting of two to nine male lions take over groups of females and defend them against male rivals who persistently attempt to overthrow them. These coalitions can hold a group for two years on average, and during this time they father offspring. Defending the group against other lions is a risky altruistic behavior, since males who may defend less benefit from others' defense. Boyd and Richerson's (1988) prediction is fulfilled here as well. Packer et al. (1991) found that while successful coalitions of two or three male lions often consisted of unrelated individuals, larger groups consisted of close kin. One interpretation of these results is that, in the small coalitions, reciprocal altruism could successfully maintain cooperation. In larger coalitions, kinship was instead the only viable option. (Packer has another interpretation of these observations, invoking sharing paternity within the pride.)

In the preceding examples, the actual costs and benefits of the behaviors in question are very unclear. Part of the debate about bookkeeping in nature is about whether each example is indeed an example of altruism. It is very difficult to measure, or even estimate, the costs and benefits of alternative behaviors. Milinski et al.'s (1997) elegant and painstaking experiments with sticklebacks illustrate this point. Only after two years of investment in experimental design were they able to measure the risks associated with inspection behavior precisely. Fish who lag behind (and therefore "defect") are indeed less likely to be taken by the predator, although with a significantly nonzero probability. The probabilities of capture provide estimates of cost parameters and suggest that inspection really is costly to individuals, that closer inspection entails greater costs, and that "defection" reduces these costs. Furthermore, fish do not seem to be engaging in costly signaling of their own quality, as fish which advance further than their partners and then return to the same position are no better at escaping attacks, which casts doubt on one important alternative explanation. Another two years were needed to estimate the benefits of inspection behavior, which seem to be some function of the advantage of feeding in safety when the fish has information suggesting that the predator is not hungry and will not strike.

After all this careful experimental work, we still do not know how well these costs and benefits generalize to the wild, and perhaps because of this, predator inspection remains controversial (Dugatkin 1997). Milinski et al.'s studies illustrate that the lack of convincing evidence for reciprocal altruism in nature is partly due to the difficulty of measuring the relevant costs and benefits, as well as performing the correct contingency tests. Thus we should not yet conclude that the absence of evidence suggests the absence of contingent reciprocal strategies which maintain cooperation in pairs. Further, we think that this situation provides an appealing opportunity for thoughtful and careful empirical studies to make a big impact, whatever the results.

Cooperation without Bookkeeping

There is a conspicuous discontinuity between humans and other animals in the prevalence of reciprocal altruism. It requires no special methodology to demonstrate that human life relies on a series of exchanges among nonrelatives. Every time we pay for our groceries or revise our colleagues' manuscript, we are practicing some kind of reciprocal strategy. However, it is not entirely clear whether the same contingency mechanisms shape all kinds of cooperative dyadic relationships in human societies. Silk (this volume) reviews evidence that friendship in humans violates the contingency and bookkeeping predictions of reciprocal altruism theory. Reviewing a number of studies from social psychology, she argues that the evidence on human friendship suggests that friends do not keep careful accounts. In fact, the apparent or actual absence of bookkeeping is often taken as one of the best signals of friendship. Most of the evidence comes from Western subjects, and so these results may not generalize to most human societies. If they do, evolutionary theorists face the challenge of explaining either how some of the most significant cooperative relationships in humans might function without detailed bookkeeping or why individuals present the image that they are not keeping track.

Most people recall some proportion of interactions in friendships and other reciprocal relationships. We all have intuitions that people recall instances of aid or defection from the distant past, perhaps reciting such lists in angry moments. However, experimental evidence exists which suggests that people may be forgetting or not even bothering to store much more. Milinski and Wedekind (1998) performed an experiment designed to investigate the use of two different bookkeeping strategies in an iterated Prisoner's Dilemma (PD) setting. The first, Pavlov (Nowak and Sigmund 1993), attends to both its own and its partner's previous round payoffs, in deciding how to behave in the present. The second, Generous Tit-for-Tat (GTFT; Nowak and Sigmund 1992), simply copies what its partner did in the last round but sometimes cooperates when its partner defected. Since these two strategies differ in the amount of memory they require (Pavlov needs more), Milinski and Wedekind introduced a memory constraint into the game by requiring subjects to play a game of memory, in which they had to match symbols on the backs of a field of cards. After each round of the PD with a fixed partner, each subject was allowed to turn over two cards. If they did not match, the cards were turned back over. Subjects were told they would be paid the *product* of their scores in the iterated PD and the memory game, meaning a subject could not afford to ignore either game.

The results showed that subjects' behavior was more consistent with a GTFT strategy when under memory constraints, but with a Pavlovian strategy in the absence of those constraints. These results suggest that memory space is really a finite resource and that strategies which keep simple tidy books can therefore outperform those with detailed books, under the right conditions. This calls into question whether it is always practical for people to keep detailed accounts of interactions in long-term cooperative relationships. Instead, they may be tracking only recent interactions, or only interactions with substantial costs and benefits. Currently, we know of no evidence sufficient to answer these questions, since high-quality data on the life histories of human friendships are sorely lacking.

Theoretical work also suggests that strategies which keep more detailed accounts may not be more adaptive, in some environments. Bendor et al. (1991) conducted a computer tournament using a continuous variant of the repeated PD which casts some doubt on the intuition that Tit-for-Tat, like bookkeeping, is a good strategy in all reciprocal interactions. Bendor solicited computer strategies much like Axelrod (Axelrod and Hamilton 1981; Axelrod 1984) did during his tournaments. Strategies were paired at random and played a repeated game. During each round of the game, each player picks a number between zero and one. Larger numbers cost the player more and benefited its partner more. Individuals observed the other player's number, but with normal random error added. Strategies which kept running accounts, and attempted to return as much on average as they received, did badly. Tit-for-Tat also did badly. The strategies that did best were ones that chose a number that was some modest percentage larger than the number they observed their opponent use during the previous period. Bendor argues that account-keeping rules did badly because errors in perception caused them to walk randomly through the space between zero and one. Such strategies over-fit their observations, taking every deviation far too seriously. In contrast, strategies that were a little nicer than their opponent tended to bump up toward the maximum payoff without too much risk of exploitation and were robust in the face of perception errors. Of course, the nature of successful strategies does depend upon the mix of strategies in the population, and thus these results may not be robust. They do, however, suggest that we should be careful about the intuition that only account-keeping strategies can be successful and avoid exploitation.

To understand more fully the mechanisms that sustain dyadic cooperation in humans, we need both more theoretical work investigating the range of environments in which strategies that keep short and (as above) optimistic accounts do well, and more theoretically grounded empirical work investigating the nature of friendship and the ontogeny of cooperative relationships. The experimental and theoretical results above suggest that the optimal amount of bookkeeping may be low, given memory requirements and perception errors. In addition, which interactions one should regard as important for reciprocal altruism remains an open question. If interactions vary in the magnitude of benefits and costs, then attending only to substantial instances in which perception errors will have smaller effects, may be a better strategy than regarding all interactions as equally informative.

REPUTATION IN DYADIC COOPERATION

Although the issues in the preceding section concern dyads keeping track of past behavior, potential cooperators might also be interested in the past behavior of individuals with whom they have not yet themselves cooperated. Most people have a strong intuition that reputation, some index constructed from past social behavior, is important in human cooperation. Alexander (1987) suggested that *indirect reciprocity*, in which third parties either observe or hear about the behavior of members of their social groups, might support cooperation. About the same time, Sugden (1986) developed a small family of models of such a process. Similar ideas about the power of third-party knowledge have also arisen in noncooperative and nonhuman contexts, such as the formation of linear dominance hierarchies (Chase 1982; Chase et al. 2002; also Tomasello and Call 1997) and in animal conflict (Johnstone 2001).

Indirect reciprocity, if it works, must rely upon some distributed bookkeeping system, in which information about past behavior travels through social networks and regulates ongoing cooperative behavior. Boyd and Richerson (1989) modeled one version of Alexander's idea of indirect reciprocity, involving a circular chain of benefits. However, this mechanism supported cooperation under only small and very long-lived associations, much like reciprocal altruism. Although Sugden (1986) worked on the problem earlier and developed a plausible mechanism, it was not until Nowak and Sigmund's (1998a, b) models of indirect reciprocity that much interest in reputation mechanisms reemerged.

In this section, we review the theoretical work on reputation in dyadic cooperation as well as the experimental evidence. It is important to note that reputation in these models does not solve problems of cooperation in large groups. All of the cooperation here happens within dyads. We discuss reputation and other mechanisms which may maintain cooperation in larger groups in a later section.

Image Scoring and Standing

There are two components to any indirectly reciprocal strategy: (a) how the accounts are kept and (b) how the accounts are used to make decisions. Nowak and Sigmund (1998a, b) modeled indirect reciprocity with a system of bookkeeping they call *image scoring*. Image scoring works in the following way. Each individual in a social group is characterized by an image score, which is a positive or negative integer. Whenever an individual has the opportunity to aid another individual, this image score increases by one if he donates aid (cooperates) and decreases by one if he does not donate aid (defects). It is assumed that image scores are completely accurate and common knowledge: every individual knows (or has access to) the image score of every other individual, as well as his own, without error. Nowak and Sigmund then proposed a strategy which discriminates based upon image scores. If a discriminating cooperator is paired with an individual with an image score above a given threshold, the discriminator provides aid (cooperates). Otherwise, the discriminator refuses aid (defects). It is important to note that this strategy is insensitive to the effects of its behavior on its *own* image score. A discriminator of this kind will defect with an individual of low image score, even though that defection reduces her own image score by one unit. In this regard, the image scoring and discriminating strategy is providing altruistic punishment.

Some work demonstrates that image scoring can sustain cooperation. Nowak and Sigmund (1998b) modeled a world of 100 individuals in a single social group. Each generation, individuals were paired at random with one other individual to whom they had the option of providing aid, which was an altruistic act. After behavior, image scores were updated, and each individual was matched with another random individual. There were no fixed cooperating dyads. Nowak and Sigmund found that the discriminator strategy, although it never went to fixation against a pure defection strategy, sustained about a 40% frequency in the group over the long run.

Later simulation work challenges these results, however. Leimar and Hammerstein (2001) became interested in how well the image scoring results would generalize in a more realistic model. Theory always contains an antagonism between realism and tractability. We want theories which capture only the important details, but no more, lest the model become just as incomprehensible as reality. However, Nowak and Sigmund's model contained an assumption that does not fit the problem under study. In their simulations there existed only one social group, of only 100 individuals. Such a population structure is known to result in large amounts of drift, overwhelming selective forces. Furthermore, if we are thinking of a genetic model of human populations, even in the distant past, effective population sizes (N_e) were probably on the order of tens or hundreds of thousands (the low-bound estimate is around 10,000 over the last 1–2 million years; Relethford 1998). There has been some debate about these estimates, but the debates have focused on the probability that current simulations *under*estimate N_e, not that they *over*estimate it (Hey 1997; Wolfpoff 1998).

To see if this assumption of a small lone group made a difference, Leimar and Hammerstein simulated Nowak and Sigmund's image scoring model with a population of 100 groups of 100 individuals each (a maximum N_e of 10,000). Groups were linked by migration, such that when migration was reduced to zero, they could reproduce the Nowak and Sigmund results; with increasing amounts of migration, however, the results differed substantially. With even modest amounts of mixing among groups, image scoring and discrimination began to perform quite badly. The reason is that a complex interaction of powerful drift and selection were driving the cycles of evolution of the image scoring strategy, but in the larger effective population, drift was much weaker and these interactions did not arise.

In a genetic model, image scoring has some serious problems. It should not be overlooked that a model assuming cultural rather than genetic transmission is much less constrained in its assumptions about effective population size. For cultural transmission, Nowak and Sigmund's model might be a reasonable approximation of the dynamics.

A more serious problem with the image scoring strategy, which both genetic and cultural models face, is that it is easily invaded by strategies which Nowak and Sigmund did not consider. Leimar and Hammerstein introduced a strategy which attends only to its *own* image score, ignoring the image score of its partner. If such an individual's image score is above the discriminator strategy's threshold for providing aid, it defects. If its image score is below the threshold or equal to it, it cooperates. Introduced into Nowak and Sigmund's model, this strategy quickly replaces the image scoring and discriminating strategy. The reason is that discriminators help such image score seekers, and the image score seekers take advantage of discriminators.

To solve this problem of invadibility, Leimar and Hammerstein introduced a strategy invented by Sugden (1986) which instead keeps track of *standing*. An individual's standing can be either *good* or *bad*. An individual gains or retains good standing by providing aid to another individual. An individual loses good standing and attains bad standing by failing to aid another individual in good standing. Failing to aid an individual in bad standing, however, does not result in a loss of good standing. These are justified defections. They then considered a strategy, called the standing strategy, which provides aid to individuals with good standing but refuses to aid individuals in bad standing. They found that the standing strategy outperformed the image scoring strategy, even in the presence of execution and perception errors. Nowak and Sigmund (1998a) suggested that standing strategies would be more vulnerable to errors in perception than image scoring strategies. According to Leimar and Hammerstein's simulations, this is probably not true: although errors hurt the standing strategy, it still out-competed image scoring.

Image scoring suffers from two serious deficits: (a) it is exploitable by image-seeking strategies which defect after achieving high image scores and (b) it provides a form of altruistic punishment every time it defects on an individual with a low image score. The results above were produced in the absence of errors in knowledge of reputations. If reputations (i.e., image scores and standings) are known with some error, then image scoring might perform better, since accumulated scores would be less sensitive to random errors than binary standings. However, both strategies must be very sensitive to errors in knowledge (Nowak and Sigmund 1998a), so we await future work to address this question.

Experimental Evidence on Reputation Mechanisms

Theoretical work thus far suggests that standing strategies are more likely candidates for implementations of indirect reciprocity in human societies than are image-scoring strategies. Some of the most recent experimental work disagrees, however. Wedekind and Milinski (2000) showed that groups of eight subjects could sustain cooperation through indirect reciprocity, but these experiments were not designed to distinguish between image scoring and standing strategies. To investigate the specific mechanisms supporting indirect reciprocity, Milinski and colleagues (2001) conducted a series of experiments designed to tease apart image scoring and standing in a simplified indirect reciprocity situation. They set up groups of seven subjects in which one subject was actually a confederate instructed to always refuse to give aid, the "NO" player. Individuals with the opportunity to aid the NO player should refuse to do so whether they are using an image scoring or standing strategy. These strategies should respond differently to refusals to aid the NO player, if given the opportunity to aid players who just had the chance to aid the NO player. Image scorers should refuse to aid the individual who refused to aid the NO player. Individuals using a standing strategy should, however, provide aid to the same individual. The experimenters found that subjects' behavior was better explained by an image scoring than a standing strategy. Furthermore, individuals who refused aid to the NO player seemed to compensate for the damage to their image scores by being more generous to other individuals. Such compensation is hard to explain as a standing strategy, since justified defections would eliminate the need for compensating a defection. This result also hints at a strategy more complicated than the image-scoring strategies explained in the previous section.

Evidence from the Wason selection task (Wason 1968) provides less specific evidence about mechanism, but again suggests that people regulate their behavior toward others contingent upon reputation. (Cosmides [1989] relates the task to reciprocal altruism.) The human brain must serve as the input circuit for reputational memory. To examine the relationship between cheater detection in the Wason task and reputation, John Tooby and colleagues (pers. comm.) conducted experiments in which subjects read descriptions about persons who have the opportunity to cheat, and then either take advantage of the opportunity, or do not. The Wason task measures cheater detection through the proportion of logically correct card selections. If positive reputation information about a person deregulates cheater detection, then we should expect fewer correct card selections in social contract treatments. If negative reputation sharpens cheater detection, we should expect improved performance with the same instrument. The results indicate that prior acts of cheating by a person do not increase cheater detection. However, four refusals to cheat relax cheater detection, but only for that person, suggesting that reputation about specific individuals regulates attention to rule violations on an individual basis.

STRONG RECIPROCITY IN DYADIC COOPERATION

Fehr and Gächter (1998a, b, 2000), Gintis (2000), Henrich and Boyd (2001), Bowles and Gintis (2001), and Fehr et al. (2002) have focused attention on a strategy that differs fundamentally from reciprocal altruism and reputation mechanisms. They have called this strategy *strong reciprocity.* Strong reciprocity applies to two-person interactions as well as to n-person interactions with $n > 2$. A person is a strong reciprocator if she is willing (a) to sacrifice resources to be kind to those who are being kind (= strong positive reciprocity) and (b) to sacrifice resources to punish those who are being unkind (= strong negative reciprocity). The essential feature of strong reciprocity is a willingness to sacrifice resources for rewarding fair behavior and punishing unfair behavior *even if this is costly and provides neither present nor future material rewards for the reciprocator.* Whether an action is perceived as fair or unfair depends on the distributional consequences of the action relative to a neutral reference action (Falk and Fischbacher 1999). Fehr and Gächter (1998a, b) and Fehr et al. (2002) provide experimental evidence indicating that there exist many people who exhibit strong reciprocity and whose existence greatly improves the prospects for cooperation in dyadic as well as in n-person cooperation.

Despite the similarity of terms, it is important to distinguish strong reciprocity from "reciprocal altruism." In one economic (but not necessarily biological[1]) conception of the strategy, a reciprocally altruistic actor is only willing to help another actor if she expects long-term net benefits from the act of helping – call this a forward-thinking reciprocal altruist. In contrast, a strong reciprocator is willing to incur the costs of helping in response to kind acts of the other party even if there are long-term net costs from the act of helping. The distinction between strong reciprocity and forward-thinking reciprocal altruism can most easily be illustrated in the context of a *sequential* Prisoner's Dilemma (PD) that is played *only once*. In a sequential PD, player A first decides whether to defect or to cooperate. Then player B observes player A's action after which she decides to defect or to cooperate. To be specific, let the economic payoffs for (A, B) be (5, 5) if both cooperate, (2, 2) if both defect, (0, 7) if A cooperates and B defects, and (7, 0) if A defects and B cooperates. If player B is a strong reciprocator, she defects if A defected and cooperates if A cooperated because she is willing to sacrifice resources to reward a behavior that is perceived as kind. A cooperative act by player A, despite the economic incentive to cheat, is a prime example of such kindness. The kindness of a strong reciprocator is thus *conditional* on the perceived kindness of the other player. In contrast, a forward-thinking reciprocal altruist only cooperates if there are future returns from cooperation. Thus a

---

[1] Many people have modeled reciprocal altruism with simple rules that lack all consideration of long-term benefits, e.g., Axelrod and Hamilton (1981) and nearly all other evolutionary models of reciprocal altruism. In this case, telling strong reciprocity from reciprocal altruism becomes harder, but is clearly not impossible.

forward-thinking reciprocally altruistic player B will always defect in a sequential *one-shot* PD.

The structure of a sequential PD neatly captures the problem of economic and social exchanges under circumstances in which the quality of the goods exchanged is not enforced by third parties, like an impartial police and impartial courts. Fehr and colleagues (Fehr and Gächter 1998b; Fehr et al. 1993) describe the results of many generalized sequential PDs (often called gift exchange experiments or trust experiments) in which the parties are not constrained to pure "cooperate" or "defect" choices but can also choose several different intermediate cooperation levels. The upshot of these experiments is that there is a strong positive correlation between the level of cooperation of player A and the level of cooperation of player B. Depending on the details of the parameters, between 40–60% of the B-players typically respond in a strongly reciprocal manner to the choice of player A: Their cooperation reflects player A's cooperation level. If player A chooses zero cooperation, then strongly reciprocal B-players also choose zero cooperation. However, there are also typically between 40–60% of second movers who *always* choose zero cooperation irrespective of what player A does. These players thus exhibit purely selfish behavior.

It is important to emphasize that in all of these experiments, real money (sometimes up to three months' income) was at stake and players remained anonymous before, during, and after the experiment. There was no repeated interaction and the experimental subjects had no chance to build a reputation. Despite the absence of repeated interactions and reputation building opportunities, subjects in the role of player B reciprocated to cooperative actions of player A. Moreover, Gächter and Falk (2002) have shown that if subjects are given the chance to interact repeatedly in the generalized sequential PD, subjects in the role of player B strongly increase their cooperation rate. This was reasonable because in the condition with repeated interactions, player A could punish player B in the next period by ceasing to cooperate with B. The strong increase in the cooperation of player B in the repeated interaction condition suggests that human subjects are well aware of the difference between a one-shot interaction and a repeated interaction and that their choices are conditional on this difference.

There is an interesting extension of the generalized sequential PD if player A is given the additional option to punish or reward player B after observing the action of player B. In Fehr and Gächter (1998b) player A could invest money to reward or punish player B in this way. Every dollar invested into rewarding increased player B's earnings by $2.50, and every dollar invested into punishment of B reduced player B's earnings by $2.50. Since after the reward and punishment stage the game is over, a selfish player A will never reward or sanction in this experiment. In fact, many A-players rewarded player B for high cooperation and punished low cooperation. Moreover, subjects in the role of player B expected to be rewarded for high and punished for low levels of cooperation and, therefore, the cooperation rate of player B was much higher in the presence of a reward and punishment opportunity. Thus, it is not only the case that many B-players exhibit strongly reciprocal responses in the sequential PD. In the extended version of the sequential PD, in which A can punish or reward, B-players also expect A-players to exhibit strongly reciprocal behavior. This expectation, in turn, causes a large rise in the cooperation of the B-players relative to the situation in which A-players have no opportunity to reward or punish.

MECHANISMS IN *N*-PERSON COOPERATION

Boyd and Richerson (1988) have shown that reciprocal altruism should be confined to small groups of individuals. The theory is complicated in the details, but the intuition behind it is simple. Reciprocal altruists only do well when they are paired with other reciprocators. In all other cases, nasty strategies do better. This is because the only evolutionarily stable strategy in such a game is the one which cooperates only if everyone else cooperates as well. Otherwise, a few defectors will free ride on the efforts of the reciprocators and out-reproduce them. Furthermore, when groups of individuals are large, the chance of getting a group of all reciprocal altruists is very small. Consider, for example, a case in which individuals are grouped together in fives. Even if half of the population consists of reciprocal altruists, the chance of getting five reciprocators in a randomly formed group is 0.5^5, or 0.03. If groups are around twenty individuals or reciprocators are rare, the situation is truly hopeless. The standard solution to this problem is a small amount of assortative group formation, such as kinship. However, assortment will not help in the case of large groups, since the probability of getting a group consisting only of reciprocal altruists falls geometrically with group size. Even if groups are comprised entirely of full siblings ($r = 0.5$), and assuming again that half of the population is cooperators, a group of ten cooperators has a less than 5% chance of forming.[2]

Thus, cooperation that is contingent on cooperation of all other group members is unlikely to be an effective mechanism for cooperation in large groups. This poses a puzzle since humans often cooperate in large groups of unrelated individuals, groups in which benefits cannot be directed to specific individuals but must be disbursed to the entire group. Furthermore, indirect bookkeeping mechanisms discussed earlier do not apply here: indirect reciprocity as described involves pair-wise cooperation, not cooperation in sizeable groups.

In this section, we discuss mechanisms which may support cooperation in larger groups of unrelated individuals, which is sometimes called n-person cooperation. We discuss strong reciprocity as well as the role of reputation in the n-person setting.

---

2 Let p be the frequency of cooperators in the population as a whole. Let groups be comprised of n individuals with an average coefficient of relatedness r. Then the probability of sampling a group of all cooperators is $p \times \{r + (1 - r)p\}^{n-1}$.

Strong Reciprocity in *n*-Person Groups

Cooperation in *n*-person groups is best viewed as a problem of public goods provision. The crucial feature of a public good is that it is difficult or impossible to exclude other group members from the consumption of the good. Hence, those who do not contribute to the production of the good can also consume the good. In the public goods context, strong positive reciprocity means that individuals increase their own contribution to the good if they expect the other group members also to increase their contributions. Strong reciprocators thus condition their choices on the other group members' choices even in one-shot situations. Strong negative reciprocity means that individuals who cooperate are willing to punish those who defected, if given a chance to do so, even if punishment is costly for the punisher and yields no economic benefits whatsoever.

Strong Positive Reciprocity

Fischbacher et al. (2001) examined to what extent strong positive reciprocity is present in one-shot *n*-person public goods situations. In their experiment, a self-interested subject is predicted to defect fully, irrespective of how much the other group members contribute to the public good. However, only a minority of subjects behave in this way. About 50% of the subjects are willing to contribute to the public good if the other group members contribute as well. Moreover, these subjects contribute more to the public good the more they expect others to contribute, indicating a strongly reciprocal cooperation pattern. Only 10% of these subjects are willing to match the average contribution of the other group members, whereas 40% of the strongly reciprocal types contribute less than the average contribution of the other group members. Roughly 30% of the subjects behave in a fully selfish manner, always defecting irrespective of how much they expect others to contribute. The rest of the subjects exhibits either a quite erratic contribution pattern (6%) or a hump-shaped pattern (14%).

In Fehr and Gächter (2000, 2002), subjects repeat the public goods experiment over many periods. In each period the subjects choose simultaneously a contribution level. At the end of the period they are informed about the other group members' individual contributions, and then they proceed to the next period to choose again (simultaneously) the contribution level. This is repeated for six periods in total. In each period new groups are formed such that no subject meets another subject twice. This setting ensures that subjects can learn, over time, how to play the game without allowing for repeated interactions. It turns out that the contributions to the public good strongly decline over time, and toward the final period the vast majority of the subjects contribute little or nothing to the public good. This decline in cooperation can be neatly explained by the dynamics of the interaction between strongly reciprocal types and selfish types, as revealed by the results of Fischbacher et al. (2001): For any given expected average contribution of the other group members in period t, the strong reciprocators either match this average contribution or contribute somewhat less than the expected average contribution. Moreover, the selfish types contribute nothing. Thus, the actual average contribution in period t clearly falls short of the expected average contribution in period t, inducing the subjects to reduce their expectations about the other members' contributions in period $t + 1$. Due to the presence of reciprocal types, however, the lower expected average contributions in period $t + 1$ cause a further decrease in the actual contributions in $t + 1$. This process repeats itself over time until very low contribution levels are reached. Simulations conducted by Fischbacher et al. indicate that the described process captures the actual behavior of the subjects quite well. It is worth emphasizing that a similar decline in cooperation rates is observed in finitely repeated public goods experiments when the group composition remains stable over time. Thus, even if one allows (finitely) repeated interactions between the same people, cooperation cannot be sustained. Despite this decline, cooperation under stable group composition is, in general, higher than when groups are randomly rebuilt every period (see Fehr et al. 2002). This again indicates that subjects understand the difference between one-shot and repeated interactions and behave accordingly.

Note that the Boyd and Richersen (1988) account—why reciprocal altruism cannot explain cooperation in large groups—and the above account—why cooperation in one-shot public goods games cannot be sustained—rely on similar intuitions. Reciprocal altruism cannot flourish in large groups because even a small number of defectors induce a breakdown of cooperation. Likewise, strong positive reciprocity cannot sustain cooperation in one-shot public goods situations because the expectation of even a small number of selfish actors will induce the strongly reciprocal actors to cease to cooperate.

Strong Negative Reciprocity

The previously described public goods experiment is characterized by the absence of targeted punishment opportunities. In this situation subjects can only punish other group members for noncooperation by withdrawing their own cooperation. The withholding of cooperation always punishes all other group members irrespective of whether they contributed or defected. This is the deeper reason for why cooperation cannot be sustained in this setting. The situation changes, however, dramatically if targeted punishment opportunities are made available. This has been done by Fehr and Gächter (2000, 2002) by adding an additional stage at the end of every period. After subjects had made their simultaneous contribution decisions, and after they had been informed about the other group members' individual contributions, each subject in the group had the option of punishing each of the other subjects in the group. Each dollar invested in the punishment of one other group member reduced the income of the punished member by three dollars. When all subjects had made their punishment decisions, they moved to the next period in which they again first chose their contribution levels. The groups were again randomly rebuilt every period so that nobody met anybody else twice.

Selfish subjects will never punish in this situation because punishment is costly and in the future periods they meet only new members. This means that if there are only selfish subjects, the option to target the punishment to specific other individuals in the group is worthless. Since nobody punishes, and since in the absence of targeted punishment nobody has an incentive to cooperate, a group consisting of only selfish subjects will exhibit no cooperation. Strong reciprocators will, however, be willing to punish despite the costs because they view little or no cooperation as an unkind act that deserves to be punished. In fact, a majority of the subjects punished the defectors, and those who were punished increased their contributions in the next period. The existence of targeted punishment led to dramatic changes in overall contribution behavior. Already in the first period of the treatment with targeted punishment, cooperation rates were much higher than in the absence of targeted punishment. Moreover, whereas cooperation unraveled in the absence of targeted punishment, cooperation increased over time when targeted punishment was possible. This indicates that strong negative reciprocity can be a powerful mechanism for obtaining and maintaining cooperation in *n*-person groups.

Fehr and Gächter (2000) also conducted experiments with targeted punishment when the group composition remained stable over (finitely) many periods. Under these conditions it was possible to reach almost 100% cooperation, although in the presence of only self-interested actors the prediction is zero cooperation. Note that in the presence of a stable group composition, the punishment of other group members constitutes a second-order public good because the punished member will in general increase cooperation in the next period and all group members benefit from this increase. It is, therefore, important to distinguish this kind of punishment from punishment in which there is no public goods dilemma. This is the case in two-party interactions (see Clutton-Brock and Parker 1995), where the second-order dilemma is absent.

In view of the powerful effects of strong reciprocity on human cooperation, it is important to develop evolutionary models explaining this phenomenon. Gintis (2000) and Henrich and Boyd (2001) have developed models showing that strong reciprocators persist in evolutionary equilibrium. The challenge for these models is that in the presence of a mix of selfish and cooperative (but nonpunishing) players, those who cooperate and do not punish will do better than those who cooperate and punish because the latter bears the costs of punishing the defectors. However, these evolutionary scenarios remain controversial because they rely on group selection arguments. Chapters 19–23 (this volume) explore in more detail the theory of the evolution of punishment in large groups. An important question for future work is to examine the empirical plausibility of these group selection accounts. Another important yet unsolved question is whether the heterogeneity of behaviors observed in laboratory experiments concerns stable personality differences. Is there such a thing as a strong reciprocator and a selfish type, or do the same subjects sometimes exhibit strongly reciprocal behavior and sometimes purely selfish behavior? How stable are the propensities to reciprocate across time, different games, and different contexts? We are unaware of any good data which address these questions, providing an opportunity for interesting future work.

Reputation and *n*-Person Cooperation

Milinski et al. (2002) studied whether the insertion of reputation in public goods games through interaction with indirect reciprocity games can maintain *n*-person cooperation. They tested this idea with groups of six subjects each. By alternating rounds of a public goods game and an indirect reciprocity game, they found that contributions in the public goods game were maintained at a high level. The results suggest that the need to maintain reputation for the indirect reciprocity game maintained contributions to the public good. However, if subjects no longer expected rounds of indirect reciprocation, contributions to the public good quickly dropped to typically low levels. Thus reputation can maintain cooperation in a public goods game at a level similar as in the punishment experiments of Fehr and Gächter (2000, 2002). Reputation has been shown to raise cooperation levels in subsequent direct reciprocity games also, probably because it builds up trust (Wedekind and Braithwaite 2002).

EMOTIONS

One view of emotions popular in the social and biological sciences is that emotions should be invoked to explain deviations from the norms of rationality. Loewenstein's (1996) work on hot and cold cognition, for example, provides compelling evidence that emotional states affect cognition, although the discussion and experimental design are framed in ways that emphasize the maladaptive consequences of their effects. One gets the impression from much work in these traditions that we would all be better off without emotions. Another view, held in different forms by psychologists in the tradition of Herbert Simon's bounded rationality, evolutionary psychologists, and many others, is that emotions are inseparable and adaptive parts of human decision-making, not forces which necessarily lead us astray. These views suggest ways in which emotion mechanisms process information, together with the more traditionally "cognitive" parts of cognition, to produce adaptive decisions in the real world or environments relevant to the design of human cognition.

We use "emotions" here to refer to a wide category of things people commonly call "feelings." Emotions may prune decision trees, direct attention to specific aspects of the environment, and even prevent our more conscious cognitive apparatus from causing us harm. For example, territorial spiders locked in combat are much easier to approach than those not locked in combat. Attention is a finite resource for any organism, and it is easy to see how focusing on one's opponent, in a situation in which one can die in a few seconds, is an adaptation, not purely a cognitive constraint. Fear in humans probably serves a similar function by directing attention to specific threats. Similarly, Bechara, Damasio, and colleagues (Bechara et al. 1994, 1997, 2000; Damasio 1994) have shown how emotions may be eminently cognitive, weighing probabilities in so-called "multi-arm bandit" tasks. They had normal and brain-damaged subjects participate in a card-stack task. In such tasks, the subject has between two and four stacks of cards, face down, in front of him. He may turn over the card on the top of any stack. In doing so, he receives the payoff printed on the face of the card. Card stacks vary in their expected payoffs, as well as their variances. This task continues for many rounds. During this time, individuals slowly converge on the stack with the highest expected payoff, although this choice behavior seems driven more by impression of "good" and "bad" stacks than conscious understanding of payoff differences. However, some brain-damaged subjects who exhibit low affect never converge on the highest payoff stack nor do they display anticipatory skin reactions of risky choices (as do normal subjects). Even in cases in which brain-damaged subjects developed accurate feelings of "good" and "bad" stacks, they failed to make choices accordingly. These results suggest that emotions play an important information processing role.

Another key feature of emotions is that they are sometimes not penetrable by other parts of cognition. Rozin et al. (1986) performed experiments in which an experimenter gives a subject fudge and then asks the subject (in a between subjects design) if they would be willing to eat more of the same fudge in (a) the shape of a disc or (b) in the shape of feces. Even though the subject knows consciously that the substance is the same fudge they have already eaten, most subjects refuse to eat the fudge in the shape of feces. One interpretation of this and similar experiments (there are many; e.g., Rozin et al. 1986) is that the cues which prime disgust—one of the emotions that regulate consumption—operate independently of other cues. Thus disgust's power over behavior is strong enough such that propositional knowledge that the "dog feces" is really fudge cannot penetrate, leading subjects to forgo a benefit. Although this example might be interpreted as maladaptive behavior on the part of the subjects, it is easy to see how it illustrates adaptive design: in a broad range of environments, objects which resemble feces are not good to eat. Since information about the exceptions is likely difficult to acquire, relying upon a simple set of cues (color, shape) may be more adaptive on average than bothering to learn about each possible food, when the costs of a mistake are likely quite high. Contrived experiments can always make subjects and their cognitive mechanisms look foolish, and we think there is little harm and much more promise in searching for cogent adaptive explanations to be refined and tested.

In this final section, we report on several avenues for exploring emotions as mechanisms that support cooperation in humans. We limit discussion to humans not because of any species prejudice about emotion or its importance in cognition and behavior but rather because data on emotions in nonhuman animals is quite sketchy. We think, however, that the issues explored here suggest ways to investigate the impact of the analogs of human emotions in other animals.

Emotion Mechanisms for Supporting Cooperation

Fessler (1999; Fessler and Haley, this volume) discusses the roles of human emotions in supporting cooperative institutions. One key emotion implicated in cooperative strategies seems to be anger. Cooperative individuals respond with anger to the noncooperative behavior of others, and this appears to motivate them to inflict costs on these defectors. Experiments also find that potential defectors typically anticipate these angry responses (Fehr and Gächter 2002). Thus anger may instantiate part of the mechanisms which lead to strong reciprocity. Also of interest are the eminently normative emotions of shame and pride. Unlike guilt, shame appears to be a human universal and may motivate compliance to norms, including norms which regulate prosocial behavior. Pride is the positive pole of this experience and may function to provide subjective rewards for norm adherence, just as shame provides subjective punishment. Fessler (1999) lays out an evolutionary argument for the function of these emotions in cooperation. Barr (2001) has found that shame can motivate cooperation in experimental games. Bowles and Gintis (this volume) also discuss the role of emotions in regulating cooperative behavior.

Recent evidence using the Wason selection task also suggests that the emotional state is a key part of the instantiation of cooperative strategies. Chang (2002) had subjects complete a mood induction exercise for a specific emotion before completing the social contract version of the Wason selection task (Cosmides 1989). Subjects who successfully completed negative mood induction exercises were significantly better at cheater detection than those who completed positive mood induction exercises (63% vs. 34% correct card selections, respectively). The performance in the negative mood case is similar to usual social contract conditions. However, the positive mood situation led to significantly lower performance than is the norm. This effect of emotional state provides additional evidence that emotions can either direct or deregulate an individual's attention to specific kinds of information or disengage information processing related to cooperative strategies. These behavioral results echo the suggestions of other work by Fehr and Gächter (2002), who found that punishment in a public goods game was motivated by anger, as indicated by subjects' self-reports.

Emotions and Honest Signals

Economists, political scientists, and biologists have long been interested in commitment problems. In many game theoretic situations with sequential play, in which one player moves before the other, the first player has the advantage and gets her way, since the first move restricts the payoffs available to the second player. The second player, however, can grab the strategic advantage if she can "burn her bridges" such that she is constrained to choose an option that is unattractive to the first player. This can be accomplished by really burning one's bridges or by providing credible signals that one is committed to an option. For example, in animal contests, the costs of escalated fights often exceed the value of the resource under dispute. By attacking, a first mover can therefore force a second into retreating from a resource, since it is would be more costly for the second to fight than to flee. If, however, the second animal can commit itself to retaliate any aggression, the first no longer gets a higher average payoff by attacking. Similarly, in situations in which individuals are willing to cooperate if they can be assured that the second player will also cooperate, commitment on the part of the second player can be adaptive.

Signals of intent from the second player are one solution. The trouble, however, is in keeping such signals honest. One puzzling fact about human emotions, unlike the emotions of other animals, is that many are linked to species-typical, fixed, and involuntary facial expressions. Although chimpanzees have some seeming analogs of fixed expressions which correspond to probable emotions, the human repertoire is vast in comparison. Some explanation of this fact is required. It is possible that other animals have similar signals which are olfactory. Whether this is the case or not, some explanation of what exactly these emotions and their expressions are signaling is needed.

Frank (1988), among others, has suggested that involuntary emotional states can help cooperators coordinate by providing solutions to the commitment problem. However, why would natural selection not favor individuals who could fake emotional displays and therefore exploit cooperators? One possibility is that the production of emotional displays is physiologically costly. However, no careful and accepted argument exists as to why this might be the case. Also, for a costly signaling argument, what is important is that the signal be *more* costly for the liar than the honest signaler. Cost alone will not suffice to evolve an honest signal. A careful argument along these lines may be possible, but to our knowledge has not yet emerged in the literature on emotion.

One requirement that all such theories must face is: if there is supposedly a simple and easy-to-evolve signaling mechanism supporting cooperation, then we are left with the mystery of why other animals, and especially other primates who have rich social lives and highly analogous and probably homologous emotions, have not evolved it. One possibility is that smaller-scale primate societies have less opportunity to benefit from cooperation; thus they may have evolved similar mechanisms, but on a smaller scale. However, other primates (e.g., hamadryas baboons) sometimes live in quite large social groups, as large or larger than many human foraging groups. In additional, the size of cooperating groups is partly a result of the evolution of cooperation mechanisms and therefore cannot be easily regarded as an inert exogenous variable.

Given the existence of individuals such as actors and actresses who can convincingly manipulate the overt expression of their emotions, it is worth considering the possibility that natural selection could lead to the ability to fake emotions but that there is some other reason that such lying would not be advantageous in the long run. A problem with our intuitions about signaling equilibria is that almost all models of signaling in animals involve one-shot games. Many people are convinced that honest signals in situations in which animals have at least partly conflicting interests require costly displays or are otherwise simply revealing or unfakeable due to constraints. Silk et al. (2000) have recently provided a simple and intuitive model which explains how honest cheap signals can evolve among unrelated individuals even when interests conflict. The key is to allow repeated interaction and reputation formation. In species as diverse as sparrows and baboons, interactions with the same individuals are often repeated. Silk et al. were inspired by the existence of apparent low-cost and honest signals of intent in a variety of nonhuman species that live in stable social groups. The appropriate contrast, of course, is not between one-shot and repeat interactions but between low and high probability of continuing interacting. Their model shows that high probabilities of continued interaction may drastically change our intuitions about what sorts of signals we should expect to find in nature.

Maynard Smith (1991, 1994) has shown that honest low-cost signals can evolve when interests of individuals are at least partly aligned; they must order the payoffs in the same way. However, these and similar results arise from models which assume that individuals interact only once. Introducing repeat interaction and a memory for events of deception (a signaling reputation of a sort) changes the conclusions. Honest cheap signals can evolve in repeat interactions where they would not be stable in finite relationships. Human emotion displays may have a similar character. Additionally, Farrell and Rabin (1996) have demonstrated that honest cheap signals can be stable when there are substantial conflicts of interest, even in a one-shot game, provided that parties have sufficient incentive to coordinate with one another. An appreciation of these two results, the effects of repetition and coordination, should lead to new ideas about the nature of emotional signals.

Depression as a Bargaining Strategy

Future models of human sociality need to incorporate strategies beyond reciprocity and signaling. In particular, when a cooperative strategy ceases to provide fitness benefits for one of the participants in a cooperative venture, she may find it advantageous to attempt to renegotiate the terms of the venture. Hagen (this volume) proposes that the symptoms of clinical depression — such as loss of interest in virtually all activities — might be elements of a bargaining strategy: an individual who has suffered a serious social loss withholds the benefits she is providing to other group members until they agree to improve the terms of her "social contract." This theory, based on a review of the empirical evidence on clinical depression in Western and non-Western cultures, explicitly links emotions, signals, and bargaining theory to challenge the prevailing view of unipolar depression as a pathology.

Error Management and the Design of Emotion

In reviewing Bendor's (1991) results about the evolution of reciprocity in a stochastic environment, we saw that errors can affect the adaptive design of mechanisms, at least in principle. At the broadest level, emotions, being the product of natural selection, can be expected to reflect the same principal of error management that is to be biased or weighted in such a fashion that, if errors are to occur, they are more likely to be of the sort that, under ancestral conditions, were less rather than more costly (Buss 2001; "error management," Haselton and Buss 2000; Nesse 2001; "smoke-detector principle," Williams and Nesse 1991). The design of disgust, the emotion which guards against contamination (Rozin et al. 1986), may an be an example of error management, because it appears to be elicited when merely superficial cues suggest that contamination is possible. For example, people refuse to eat fudge shaped like feces. Note that error management is operating primarily in the initial interpretation-of-the-stimulus phase of the emotion process (i.e., "Is this fudge or feces?").

By the same token, it is reasonable to expect that error management may affect subjects' interpretation of the tasks they are asked to perform in experimental situations. The interpretation of the "meaning" of cues from the environment is part and parcel of the experience of an emotion ("Is that a shadow in the woods or a jaguar?"). Because the costs of mistaking an iterated game for a one-shot game may have been greater than the costs of the reciprocal error, it is possible that players in one-shot games (particularly when cues are ambiguous) experience emotions appropriate to iterated games and behave accordingly. Except when the format of an experimental game closely matches a familiar cultural practice (Henrich et al. 2001), subjects may experience the game context as somewhat alien, hence calling for interpretation. This interpretation is likely to be subject to the influence of error management effects that stem from both the evolved predispositions and the repertoire of experience. Thus, it is possible that subjects react with anger to perceived transgressions (e.g., inequitable divisions in one-shot ultimatum games) and with shame to perceived disapproval (as with verbal punishment in commons games; Barr 2001) despite the fact that both anger and shame have utility primarily in long-standing interactions.

There is, however, also a competing interpretation of these emotions which stresses that interactions with low probabilities of future encounters have been quite frequent in evolutionary history (see Fehr and Henrich, this volume, and Gintis 2000). In addition, the costs of mistakenly treating an encounter with a low or zero probability of future interactions as an event with a high probability of future interactions may have been quite dangerous so that individuals who were able to distinguish cognitively and emotionally between low- and high-frequency interactions had better survival chances. For instance, treating a stranger like a friend may have been quite costly because it enabled the stranger to exploit the situation and cheat, whereas being cheated by a friend is constrained by the implicit threat of withholding future cooperation. In fact, most modern humans well understand that the probability of being cheated in one-shot interactions in a foreign town or country is higher than in interactions with colleagues and friends. This capacity to distinguish low- from high-frequency encounters, and to behave accordingly, is also documented in the experiments of Gächter and Falk (2002) and Fehr et al. (2002). The competing view is also more optimistic about the human capacity to have emotions that are fine-tuned to low- and high-frequency interactions. Most people probably experience more anger when cheated by a close friend than when cheated by a stranger because the feelings of betrayal tend to be stronger when cheated by a friend.

It is a well-established fact that a substantial fraction of humans cooperate with unrelated strangers even if the shadow of the future or the possibility to build a reputation is absent. Whether the emotions that help sustain cooperation in these low-frequency encounters are ill- or well-adapted to the low-frequency situation is an important topic for future research. We need to know more about subjects' actual default assumptions when they are in one-shot encounters in the laboratory and about the cues that affect the default assumptions. We also need to know more about the details of our evolutionary history, about the likelihood of low-frequency interactions, and about the costs of mistakenly treating one-shot encounters as repeated encounters. By experience, subjects can be persuaded that their default assumptions are in error; however, it remains an empirical question as to how much, and under what conditions, such defaults continue to influence decisions. Interpreting the design of emotion mechanisms in this light suggests both new experiments to tease apart the cues involved as well as new theory exploring the evolution of strategies in an environment with stochastically varying group sizes.

Emotions as Mechanisms That Manipulate Time Horizons

Aggression and punishment as strategies which change the behavior of other individuals rely upon a fundamental logic: Reactions to current transgressions must be sufficiently costly to the target to deter future transgressions. However, deterrence is costly. It is costly for one individual to inflict harm on another, and these costs must be paid in the present even though the benefits will be reaped in the future. This leads to a puzzle because humans, like virtually all other animals studied, steeply discount the future. Anger may effectively solve this problem, motivating people to respond to transgressions and overriding the tendency to discount the future (Daly and Wilson 1988; Lerner and Keltner 2001; Fessler and Haley, this volume). In fact, anger sometimes seems to be disproportionate to the magnitude of the transgression, perhaps because the anger system sums the costs of prospective future transgressions and then substitutes this sum for the actual cost of the present transgression (Frank 1988). Reputational effects may magnify emotional responses because the payoffs of deterrence are multiplied when third parties observe the response or hear others gossip about the response. Thus, anger may be expressed even in one-shot interactions if reputational effects are important (Nisbett and Cohen 1996).

CONCLUSION

A number of problems remain unsolved for understanding cooperation outside kin selection. In this report, we have summarized the group discussion of cognitive and emotional mechanisms which instantiate possible solutions. This discussion has certainly not been exhaustive. Several important topics remain unexplored. Many mechanisms which were selected by inclusive fitness may have been later exapted (i.e., put to a new purpose) to serve roles in nonkin cooperation, and we have neglected phylogeny in almost every aspect of the discussion. Theory of mind and the attribution of intentions is a large and important topic in cognition and cooperation, which we have only touched upon here. Our discussion of justified defections in indirect reciprocity invokes intentionality and suggests that individuals use attributed intentions in guiding their cooperative behavior, and strategies in the iterated PD such as Contrite Tit-for-Tat (Boyd 1989) necessarily invoke the communication of intentions.

Many of the experiments and studies we have discussed, especially with respect to human friendship, are inadequate to address many of the newer questions. With respect to human friendship, this is because the studies in social psychology were conducted with different questions in mind. Thus a number of new experiments and observations will be needed to address the concerns raised in this report. We have tried to suggest such empirical investigations where obvious, but we think that inventive experimenters and fieldworkers will see many more, just as ingenious theoreticians will no doubt see many promising modeling possibilities that we have missed.

REFERENCES

Alexander, R.D. 1987. The Biology of Moral Systems. New York: Aldine de Gruyter.

Axelrod, R. 1984. The Evolution of Cooperation. New York: Basic.

Axelrod, R., and W.D. Hamilton. 1981. The evolution of cooperation in biological systems. *Science* **211**:1390–1396.

Barr, A. 2001. Social dilemmas, shame-based sanctions, and shamelessness: Experimental results from Rural Zimbabwe. In: CSAE Working Paper WPS/2001.11. Oxford: Centre for the Study of African Economics.

Barrett, L., and S.P. Henzi. 2001. The utility of grooming in baboon groups. In: Economics in Nature, ed. R. Noë, J.A.R.A.M. van Hooff, and P. Hammerstein, pp. 119–145. Cambridge: Cambridge Univ. Press.

Bechara, A., A.R. Damasio, H. Damasio, and S. Anderson. 1994. Insensitivity to future consequences following damage to human prefrontal cortex. *Cognition* **50**:7–15.

Bechara, A., H. Damasio, and A. Damasio. 2000. Emotion, decision making, and the orbitofrontal cortex. *Cerebral Cortex* **10**:295–307.

Bechara, A., H. Damasio, D. Tranel, and A.R. Damasio. 1997. Deciding advantageously before knowing the advantageous strategy. *Science* **275**:1293–1295.

Bendor, J., R.M. Kramer, and S. Stout. 1991. When in doubt ...: Cooperation in a noisy prisoner's dilemma. *J. Conflict Resol.* **35**:691–719.

Bowles, S., and H. Gintis. 2001. The evolution of strong reciprocity. Discussion Paper, Univ. of Massachusetts at Amherst.

Boyd, R. 1989. Mistakes allow evolutionary stability in the repeated prisoner's dilemma game. *J. Theor. Biol.* **136**:47–56.

Boyd, R., and P.J. Richerson. 1988. The evolution of reciprocity in sizeable groups. *J. Theor. Biol.* **132**:337–356.

Boyd, R., and P.J. Richerson. 1989. The evolution of indirect reciprocity. *Soc. Netw.* **11**:213–236.

Buss, D. 2001. Cognitive biases and emotional wisdom in the evolution of conflict between the sexes. *Curr. Dir. Psychol. Sci.* **10**:219–223.

Chang, A. 2002. The Relationship between Recalling a Relevant Past Experience and Vigilance for Cheaters and Altruists. B.Sc., Dept. of Psychology, McMaster Univ., Hamilton, Ontario, Canada.

Chase, I.D. 1982. Dynamics of hierarchy formation: The sequential development of dominance relationships. *Behaviour* **80**:218–240.

Chase, I.D., C. Tovey, D. Spangler-Martin, and M. Manfredonia. 2002. Individual differences versus social dynamics in the formation of animal dominance hierarchies. *Proc. Natl. Acad. Sci. USA* **99**:5744–5749.

Clutton-Brock, T.H., and G.A. Parker. 1995. Punishment in animal societies. *Nature* **373**:209–216.

Cords, M. 2002. Friendship among adult female blue monkeys. *Behaviour* **139**:291–314.

Cosmides, L. 1989. The logic of social exchange: Has natural selection shaped how humans reason? Studies with the Wason selection task. *Cognition* **31**:187–276.

Daly, M., and M. Wilson. 1988. Homicide. New York: Aldine de Gruyter.

Damasio, A.R. 1994. Descartes' Error: Emotion, Reason, and the Human Brain. New York: Grosset and Putnam.

de Waal, F.B.M. 1997. The chimpanzee's service economy: Food for grooming. *Evol. Hum. Behav.* **18**:375–386.

Dugatkin, L.A. 1997. Cooperation among Animals: An Evolutionary Perspective. Oxford: Oxford Univ. Press.

Enquist, M., and O. Leimar. 1993. The evolution of cooperation in mobile organisms. *Anim. Behav.* **45**:747–757.

Falk, A., and U. Fischbacher. 1999. A theory of reciprocity. Working Paper No. 6. Zurich: Institute for Empirical Research in Economics, Univ. of Zurich.

Farrell, J., and M. Rabin. 1996. Cheap talk. *J. Econ. Persp.* **10**:103–118.

Fehr, E., U. Fischbacher, and S. Gächter. 2002. Strong reciprocity, human cooperation, and the enforcement of social norms. *Hum. Nat.* **13**:1–25.

Fehr, E., and S. Gächter. 1998a. How effective are trust- and reciprocity-based incentives? In: Economics, Values and Organization, ed. A. Ben-Ner and L. Putterman, pp. 337–363. Cambridge: Cambridge Univ. Press.

Fehr, E., and S. Gächter. 1998b. Reciprocity and economics: The economic implications of *Homo reciprocans*. *Eur. Econ. Rev.* **42**:845–859.

Fehr, E., and S. Gächter. 2000. Cooperation and punishment in public goods experiments. *Am. Econ. Rev.* **90**:980–994.

Fehr, E., and S. Gächter. 2002. Altruistic punishment in humans. *Nature* **415**:137–140.

Fehr, E., G. Kirchsteiger, and A. Riedl. 1993. Does fairness prevent market clearing?: An experimental investigation. *Qtly. J. Econ.* **108**:437–460.

Fessler, D.M.T. 1999. Toward an understanding of the universality of second-order emotions. In: Beyond Nature or Nurture: Biocultural Approaches to the Emotions, ed. A. Hinton, pp. 75–116. New York: Cambridge Univ. Press.

Fischbacher, U., S. Gächter, and E. Fehr. 2001. Are people conditionally cooperative? Evidence from a public goods experiment. *Econ. Lett.* **71**:297–404.

Frank, R. 1988. Passions within Reason: The Strategic Role of the Emotions. New York: Norton.

Gächter, S., and A. Falk. 2002. Reputation and reciprocity: Consequences for the labor relation. *Scand. J. Econ.* **104**:1–25.

Gintis, H. 2000. Strong reciprocity and human sociality. *J. Theor. Biol.* **206**:169–179.

Hamilton, W.D. 1964. The genetical evolution of social behavior. II. *J. Theor. Biol.* **7**:17–52.

Hart, B.L., and L.A. Hart. 1992. Reciprocal allogrooming in impala, *Aepyceros melampus*. *Anim. Behav.* **44**:1073–1083.

Haselton, M.G., and D.M. Buss. 2000. Error management theory: A new perspective on biases in cross-sex mind reading. *J. Pers. Soc. Psych.* **78**:81–91.

Hemelrijk, C.K. 1994. Support for being groomed in long-tailed macaques, *Macaca fasicularis*. *Anim. Behav.* **48**:479–481.

Henrich, J., and R. Boyd. 2001. Why people punish defectors: Weak conformist transmission can stabilize costly enforcement of norms in cooperative dilemmas. *J. Theor. Biol.* **208**:79–89.

Henrich, J., R. Boyd, S. Bowles et al. 2001. In search of *Homo economicus*: Behavioral experiments in 15 small-scale societies. *Am. Econ. Rev.* **91**:73–78.

Henzi, S.P., and L. Barrett. 2002. Infants as a commodity in a baboon market. *Anim. Behav.* **64**:915–921.

Hey, J. 1997. Mitochondrial and nuclear genes present conflicting portraits of human origins. *Mol. Biol. Evol.* **14**:166–172.

Johnstone, R.A. 2001. Eavesdropping and animal conflict. *Proc. Natl. Acad. Sci. USA* **98**:9177–9180.

Leimar, O., and P. Hammerstein. 2001. Evolution of cooperation through indirect reciprocity. *Proc. Roy. Soc. Lond. B* **268**:745–753.

Lerner, J., and D. Keltner. 2001. Fear, anger, and risk. *J. Pers. Soc. Psych.* **81**:146–159.

Loewenstein, G. 1996. Out of control: Visceral influences on behavior. *Org. Behav. Hum. Dec. Proc.* **65**:272–292.

Maynard Smith, J. 1991. Honest signalling: The Philip Sidney game. *Anim. Behav.* **42**:1034–1035.

Maynard Smith, J. 1994. Must reliable signals always be costly? *Anim. Behav.* **47**:1115–1120.

Milinski, M. 1987. TIT FOR TAT in sticklebacks and the evolution of cooperation. *Nature* **325**:433–435.

Milinski, M., J.H. Lühti, R. Eggler, and G.A. Parker. 1997. Cooperation under predation risk: Experiments on costs and benefits. *Proc. Roy. Soc. Lond. B* **264**:1239–1247.

Milinski, M., D. Pfluger, D. Külling, and R. Kettler. 1990. Do sticklebacks cooperate repeatedly in reciprocal pairs? *Behav. Ecol. Sociobiol.* **27**:17–21.

Milinski, M., D. Semmann, T.C.M. Bakker, and H.-J. Krambeck. 2001. Cooperation through indirect reciprocity: Image scoring or standing strategy? *Proc. Roy. Soc. Lond. B* **268**:2495–2501.

Milinski, M., D. Semmann, and H.-J. Krambeck. 2002. Reputation helps solve the "tragedy of the commons." *Nature* **415**:424–426.

Milinski, M., and C. Wedekind. 1998. Working memory constrains human cooperation in the prisoner's dilemma. *Proc. Natl. Acad. Sci. USA* **95**:13,755–13,758.

Nesse, R.M. 2001. The smoke detector principle: Natural selection and the regulation of defenses. In: Unity of Knowledge: The Convergence of Natural and Human Science, ed. A.R. Damasio, A. Harrington, J. Kagan et al., vol. 935, pp. 75–85. New York: New York Academy of Sciences.

Nisbett, R.E., and D. Cohen. 1996. Culture of Honor: The Psychology of Violence in the South. Boulder, CO: Westview Press.

Nowak, M.A., and K. Sigmund. 1992. Tit for tat in heterogenous populations. *Nature* **355**:250–253.

Nowak, M.A., and K. Sigmund. 1993. A strategy of win-stay, lose-shift outperforms tit for tat. *Nature* **364**:56–58.

Nowak, M.A., and K. Sigmund. 1998a. The dynamics of indirect reciprocity. *J. Theor. Biol.* **194**:561–574.

Nowak, M.A., and K. Sigmund. 1998b. Evolution of indirect reciprocity by image scoring. *Nature* **393**:573–577.

Packer, C., D.A. Gilbert, A.E. Pusey, and S.J. O'Brien. 1991. A molecular genetic analysis of kinship and cooperation in African lions. *Nature* **351**:562–565.

Relethford, J.H. 1998. Genetics of modern human origins and diversity. *Ann. Rev. Anthro.* **27**:1–23.

Rozin, P., L. Millman, and C. Nemeroff. 1986. Operation of the laws of sympathetic magic in disgust and other domains. *J. Pers. Soc. Psych.* **50**:703–712.

Schino, G. 2001. Grooming, competition, and social rank among female primates: A meta-analysis. *Anim. Behav.* **62**:265–271.

Seyfarth, R.M., and D.L. Cheney. 1984. Grooming, alliances, and reciprocal altruism in vervet monkeys. *Nature* **308**:541–543.

Silk, J.B., E. Kaldor, and R. Boyd. 2000. Cheap talk when interests conflict. *Anim. Behav.* **59**:423–432.

Stopka, P., and R. Graciasova. 2001. Conditional allogrooming in the herb-field mouse. *Behav. Ecol.* **12**:584–589.

Sugden, R. 1986. The Economics of Rights, Co-operation, and Welfare. New York: Blackwell.

Tomasello, M., and J. Call. 1997. Primate Cognition. Oxford: Oxford Univ. Press.

Trivers, R. 1971. The evolution of reciprocal altruism. *Qtly. Rev. Biol.* **46**:35–57.

Wason, P.C. 1968. Reasoning about a rule. *Qtly. J. Exp. Psych.* **20**:273–289.

Wedekind, C., and V.A. Braithwaite. 2002. The long-term benefits of human generosity in indirect reciprocity. *Curr. Biol.* **12**:1012–1015.

Wedekind, C., and M. Milinski. 2000. Cooperation through image scoring in humans. *Science* **288**:850–852.

Williams, G.C., and R.M. Nesse. 1991. The dawn of Darwinian medicine. *Qtly. Rev. Biol.* **66**:1–22.

Wolfpoff, M. 1998. Concocting a divisive theory. *Evol. Anthro.* **7**:1–3.

8

Does Market Theory Apply to Biology?

Samuel Bowles[1] and Peter Hammerstein[2]

[1]Santa Fe Institute, Santa Fe, NM 87501, U.S.A.
[2] Institute for Theoretical Biology, Humboldt University, 10115 Berlin, Germany

ABSTRACT

Traditionally, market models in economics describe interactions in which the commodities traded are subject to complete contracts that are enforceable at no cost. Such contracts do not exist among other animals. In conventional economic models, there is also no account of who meets whom, what the traders know, and how they settle on a transaction, whereas these aspects play a major role in biological market models. From this point of view, the scope for applying market theory to biology appears very limited. Recent developments in economics, however, may allow for fruitful interdisciplinary cooperation. These developments include what one leading economist termed "the abrogation of the law of supply and demand" accomplished by the introduction of principal-agent models, based on the incomplete nature of contracts and the traders' limited information. There is an important convergence of thought in both disciplines, and biologists have recently identified a variety of interesting examples beyond the basic mating market. Some of these examples resemble labor markets and may be illuminated by principal-agent models. A look at the mating market shows that adopting an economist's perspective provides a comprehensive model of the market, the components of which are now well understood by biologists. Finally, there are striking parallels between the signaling games studied in biology and economics, the value of education and the peacock's tail having much in common.

INTRODUCTION: WHY BOTHER WITH MARKETS?

When we buy a basket of apples, the interaction with the farmer is mutually beneficial: we receive a commodity while the farmer gets money in return. Mutual benefits alone, however, are not sufficient to explain cooperation. We tend to refuse a particular trade if we know that a better deal can be obtained elsewhere and is worth the effort of moving and searching. Similar market phenomena seem to exist in the nonhuman animal world. When a female mates with a male, she receives sperm while the male "cashes in" on the eggs that his sperm fertilize. The mutual benefits, however, do not imply that they are worth the trade. Females often refuse a particular mate if superior partners are available. The preferred partners might offer "nuptial gifts," more valuable sperm, lower risk of picking up sexually transmitted diseases, or better abilities to care for the offspring (if this can be expected at all).

Ever since 1838 when Charles Darwin read the classical economist Thomas Malthus, the emergent properties of competitive interactions have been prominent in biological thinking. The analogy between animal mating and human trade led much later to the metaphor of *mating markets* in behavioral ecology. Recently, a more general field of research on *biological markets* has emerged (Noë and Hammerstein 1994, 1995; Schwartz and Hoeksema 1998; Noë et al. 2001; Simms and Taylor 2002). The reason behind this broadening in scope is that *partner choice* plays an important role in social interactions other than mating (Hammerstein, Chapter 5, this volume) and that many cooperative *exchanges* take place within and between species.

In songbirds called lazuli buntings (*Passerina amoena*), for example, the following exchange seems to take place: territorial males give juvenile-looking males access to their high-quality territories[1] and are compensated through offspring benefits that result from copulations with the juvenile's mate (Greene et al. 2000). The juvenile-looking males are yearlings. In general, yearlings differ markedly in their plumage color, ranging from very dull to bright (adult looking). Color plays a crucial role in social partner choice. Adult males behave very aggressively toward brightly colored yearlings. In contrast, they sometimes show extreme tolerance toward dull-looking males, who then use this opportunity to settle near the adult in its high-quality habitat. Greene and his collaborators interpret this as a cooperative relationship, whereby the dull yearling benefits from the habitat quality as it allows him to attract a female and produce offspring with her. Rather than posing a threat to the adult, the presence of the dull male makes it possible for the adult to obtain extra-pair fertilizations — a mutually beneficial trade (revealed by DNA fingerprinting). Young birds with bright plumage coloration probably compete for territories as if they were adults. Green et al. report that both the dullest and the brightest yearlings generally obtained high-quality sites. It would be difficult to understand this empirical result without considering the *trade* among males in the *social partner market.* Noë and Hammerstein (1994) analyzed a similar scenario for purple martins.

Experimentalists have conducted several other market studies in which social partnerships are observed. For example, Bshary investigated the relationship between cleaner fish and their "customers" (i.e., other fish from which they remove ectoparasites). Cleaners live in coral reefs and have customers from the immediate neighborhood as well as from the open sea. Local customers for which long-distance moves are costly are cleaned less well than long-range travelers, who can easily switch between cleaning stations and thereby exert partner

[1] We refer to territory in a broad sense as the habitat controlled by the adult male.

choice (see Bshary and Noë, this volume). This is exactly what one would expect from the economic theory of *monopolistic competition*: buyers with few alternative sources of supply will have less advantageous transactions than those who can shop around.

Biology does not lack market examples, and it is obvious that many important insights can be gained if the market is properly reflected in studies of cooperation. Biologists have only begun, however, to develop a general market theory, and thus it seems important at this stage to ask what might be learned from economics. At first glance, the scope for simple interdisciplinary "trade" may seem rather limited. Traditional concepts of economic markets appear to be particularly unsuitable for biologists. A cursory look at biological mating markets confirms this view. Some outstanding puzzles in biological market theory, however, demonstrate that there is some convergence of theory development in biology and economics.

MARKETS IN BIOLOGY AND ECONOMICS ARE NOT THE SAME

The recent success of the market analogy in biology comes somewhat as a surprise to economists, for standard market models in economics appear to be a poor template for studying interactions among nonhuman animals. There are three reasons for this:

- First, canonical economic agents deploy extraordinary cognitive capacities unique to humans in pursuit of their self interest. By contrast, biological market traders at most perform an "as if" fitness maximization, and this is the product of population-wide dynamics, not of intentional behavior.
- Second, conventional economic models determine prices and other equilibrium outcomes in markets without representing the actual interactions among traders. In contrast to biological market models, there is no account of who meets whom, what the traders know, and how they settle on a transaction. In this sense, there is not even an economic theory of the price-setting process.
- Third, most market models describe interactions in which the goods and services traded are subject to complete contracts that are enforceable at no cost to the exchanging parties. This means that the explicit terms of the exchange cover all aspects of the trade of interest to the trader, and, once decided upon, these terms are not subject to cheating. Human contracts of this type are unique in nature.

As a result, conventional market models are silent on issues of considerable interest to biologists, including the determinants of bargaining power, how cheating is controlled, how the terms of a trade are determined in a biological exchange, and how power can be exercised in a highly competitive environment in which all traders have many alternative transactions (Bowles 2003). Henzi and Barrett (2002) conclude from their study of grooming among chacma baboons that "if biological markets are to be fully applicable to primate groups (and those of other social animals), then the potentially distorting effect of dominance needs to be incorporated into the framework." Economic models taking account of the importance of power (Coase 1937; Simon 1951) and social and genetic affinity in the exchange process (Sahlins 1974) have long existed, as have approaches that eschew the conventional but unrealistic assumptions concerning the cognitive capacities of economic agents (Becker 1962; Alchian 1950; Simon 1955). However, these contributions have made little impact on economic theory until recent years.

MATING MARKETS AND THE ABROGATION OF THE LAW OF SUPPLY AND DEMAND

Let us now return to the oldest market paradigm in biology. Mating markets are *implicitly* involved in most evolutionary studies of partner choice, reproduction, and sex differences. They also set the stage for conflict among and between the sexes (for a review, see Hammerstein and Parker 1987). It is, therefore, interesting to give an *explicit* picture of these markets.

Driven by a strong inclination to take facts into account, biologists have collected numerous pieces of evidence suggesting that the *supply* of sperm exceeds *demand* in many animal species. Let us take Bateman's (1948) famous mating experiment as an example. He demonstrated that male fruit flies (*Drosophila melanogaster*) can strongly increase their reproductive success by copulating with several partners, whereas the reverse is not true for females. Combined with the *sex ratio* argument that males and females are produced in roughly equal numbers, this indicates the following: The aggregated fertilization services offered by males substantially exceed female demand in the fruit fly population. It would seem that females should require commodities other than sperm as the appropriate "price for their eggs." But in fruit flies, *sperm is all they get*.[2] The law of supply and demand does not apply.

This "law" states that, in a market economy, the forces of supply and demand push the price toward the level at which the quantity supplied and the quantity demanded are equal, a result termed "market clearing." Biologists have many reasons to be critical of such a simplistic view of the world, and recently the same holds true for economists. Joseph Stiglitz, recipient of the Nobel Prize in economics, wrote of the "abrogation of the law of supply and demand" accomplished by recent development in microeconomics. The conventional market model, termed *Walrasian* after Leon Walras (1834–1910), one of the founders of

[2] Note, however, that female insects can receive other material with the seminal fluid. The advantages to the female are often marginal, and in the case of *Drosophila* seminal fluid actually shortens the life span of the female (Fowler and Partridge 1989).

neoclassical economics, has for the most part been superseded. The new market theory is quite different from the old and takes as its foundational assumptions the incomplete nature of contracts (biologically speaking, the possibility of cheating, exploitation, etc.) and the traders' limited information about the trades being offered and accepted by other traders. The new *post-Walrasian* microeconomics provides models of markets — labor markets, credit markets, markets for goods of variable quality — in which market clearing does not occur, even in a competitive equilibrium (Bowles 2003). We will see that this new approach may help resolve some outstanding puzzles in the theory of biological markets.

Returning to the mating scenario, let us now look at the *market entry problem.* Given the excessive supply of sperm, why is half the population entering the "male side of the market" instead of producing eggs? In other words, why are males and females often produced in roughly equal numbers? The initial attempt to explain *sex ratio evolution* was made by Darwin in the first edition of his monograph on *The Descent of Man* where he implicitly resorted to group selection reasoning. But he abandoned this in the second edition. Almost in a state of despair he had to admit that "the whole problem is so intricate that it is safer to leave its solution for the future" (Darwin 1874, p. 399). The evolutionary explanation of sex ratios is not quite as difficult as it appeared to Darwin. Sex ratio theory developed soon after he raised the problem but only reached the attention of later generations via Ronald Fisher (1930).

Fisher's presentation of the theory can be rephrased in economic terms. Assuming that mothers determine the sex of their offspring, a female acts like an *investor,* allocating resources to sons and daughters to obtain as many grandchildren as possible. As soon as the population sex ratio is biased, it pays to invest in the rarer sex. This is so because, looking at the entire population, members of the less abundant sex produce collectively as many offspring as those of the more abundant sex (in diploid organisms, every grandchild has exactly one genetic father and one genetic mother so the only way there could be fewer fathers in the population is that they would on average have more offspring). On average, therefore, individuals of the rarer sex have more children. Thus selection acts in favor of the unbiased sex ratio. Of course, this is only the basic idea; it has since been elaborated (e.g., Charnov 1982). In particular, one can allow the organism to choose being a male or female independent of the mother. Under many circumstances, the result is the same.

Even if, for these reasons, the *supply of males* cannot adjust we still have to ask why males do not adjust the *supply of sperm* to a "sperm-saturated market." The answer lies again in Bateman's fruit fly experiment and in the assumption that sperm production is not very costly. If males can increase their reproductive success by having more than one mate, they should produce enough sperm for fertilizing the eggs of two or more females. In addition, the more sperm competition there is, the more sperm is required (Parker 1970; Parker and Ball 2001). Sperm competition results from successive inseminations by different males, whose sperm compete for access to eggs. Comparing different primates species, Harcourt et al. (1981) showed that the size of testes correlates with the degree of promiscuity typically found. Even in animal populations with social monogamy, a somewhat excessive production of sperm is to be expected, since extra-pair copulations are not unheard of in humans and have been demonstrated for a number of socially monogamous animal species.

In contrast to males whose reproductive potential is enhanced by the low cost of sperm, females are severely limited in their reproductive potential by the high cost of egg production (or in mammals by viviparity). This sex difference in reproductive potential gave rise to the term *asymmetric mating market*. Biologists think that the asymmetry in reproductive potential is perhaps the main key to understanding the morphological, physiological, and behavioral *differences* between the sexes.

The asymmetric mating market explains nicely why males *compete* for access to females and why females are in a strong position to exert precopulatory or postcopulatory *mate choice*. The peacock's tail has probably evolved in response to female choice. But why do females not use this advantage and sell their eggs at a higher price instead of contenting themselves with a beautiful tail? In the presence of excess supply of sperm, what prevents price adjustment from clearing the mating market as the Walrasian market model would predict? It would seem that females should prefer male partners who offer "commodities," such as nutrients or parental care, in addition to sperm.

The same puzzle arises in the theory of human labor markets. If labor is chronically in excess supply, what prevents the unemployed workers from offering employers a more attractive package, promising to work harder for the same wage? Or given that markets do not clear, so that jobs are typically scarce and workers abundant, why do employers not sell jobs, charging a fee to the prospective worker as a condition of employment?

The problem in both human and other markets is that the relevant contracts are not enforceable and this appears to be a serious impediment to the "package deal." The workers' promise to work harder is not enforceable, nor is the employer's promise not to fire the worker once the fee has been paid. Among other animals, it is easy to "promise" paternal care and forgo the effort when it is due.

Occasionally, a package deal has evolved. In sea horses and giant water bugs, for example, males make a major parental effort and care intensively for their offspring, whereas females "only" provide the eggs. How does evolution force a male to carry the burden of parental care? To address this, we describe an abstract scenario inspired by R.L. Smith's work (1997) on the giant water bug.

Suppose we look at a stage in the evolution of an abstract animal species where parental care is absent. Females deposit their eggs on plants that line their freshwater habitats. Female foraging, however, takes place at other locations. Since males compete for access to females, they defend territories that contain the plants required for egg laying. Females deposit their eggs in male territories.

At this stage in our evolutionary story, there is no reason for a female to remain with her eggs. To the contrary, she will pursue foraging activities to produce more offspring.

Next, a change in the environment occurs that calls for parental care to ensure egg survival. In principle, both males and females have an interest in the survival of their joint offspring. However, if she leaves after depositing the eggs, he is caught in a situation where he has the last move in the interaction sequence. If he does not care, the brood is gone. Sometimes it is bad to have the last move. Since he would not benefit from deserting his territory, he cannot benefit from ending the spatial association with the eggs. Strategically he is thus in a weak position to "pass the buck" in the parental investment game. What does he do? It is easy to imagine evolution imposing the burden on his broad shoulder. (The game theoretic logic behind mate desertion is discussed by Hammerstein [2001].)

Our scenario shows that sex role reversal is possible in evolution. The effects of the basic market asymmetry are indeed more subtle and less supportive of our cultural stereotypes than popular presentations of sociobiology would make us believe (for further discussion of reversals in the relative strength of sexual selection on males and females, see Lorch [2002]).

To conclude this section on mating markets, it would appear that the advancements in this impressive field of research have mostly been made by biologists. Yet, as we just saw, there are many bridges to economics, and we maintain that looking at mating markets through the economist's eyes is extremely useful if one aims to "assemble" the various pieces of sexual selection theory to study the whole picture that emerges.

PRINCIPALS, AGENTS, AND POWER IN BIOLOGICAL EXCHANGES

In the animal world, egalitarian societies are the exception rather than the rule. Often, a fraction of the male population controls access to high-quality habitats, leaving the rest of the males to contend with what is left. The weak receives permission to settle within the otherwise defended territory, but a service to the strong must be rendered in return. What kind of service would this be, and why can the strong rely on this service when opening the door to his "estate"? The answer provided by post-Walrasian microeconomics is that the *power* of the strong males to keep weak males off their territories enables them to act as *principals* in trades with *agents*. A *principal* benefits from the actions of an *agent* but cannot use an enforceable contract to bind the agent to do the actions that are optimal from the principal's standpoint. The principal must therefore exercise power to induce the agent to act in accordance with the principal's interests. What follows is a biological example (worked out mathematically by Bowles and Hammerstein, unpublished manuscript).

Among lazuli buntings, the adult territory owner is the principal. He allows a dull-colored yearling—his agent—to settle in the habitat area that he controls. The yearling, of course, is interested in attracting a mate and producing offspring with her, and the habitat of the territory owner increases his chances. Due to the imbalance of power between the males, the adult can "steal" copulations from the yearling by mating with the yearling's partner. Copulation is one service the territory holder gets, and he can count on it as long as the yearling's mate agrees. It is not, however, in the interest of the territory holder to push the "adultery" to an extreme. Monopolizing all copulations could be countered by the yearling withdrawing his investment in parental care, another service that he provides. If this logic drives the adult male's behavior (via natural selection), the yearling is better off with the trade than without it, just as the employed worker is better off with the job than without it.

In an evolutionary equilibrium that reflects this logic, the trade can take place and the birds do not have to "worry" about commitment and enforceable contracts. The exercise of power is thus essential to the way the market works.

Noë and Hammerstein (1994) created a tale between a fictitious boa "constructor" and "shadowbirds" to make a similar point. The female boa "constructs" a nest mound in an open desert environment upon which she lays her eggs. The snake has all it takes to guard her nest successfully against egg predators, but her eggs are still at risk from solar radiation. To protect the eggs from thermal stress, the snake benefits from cooperation with a shadowbird. If permitted by the boa (the principal), the female shadowbird adds her eggs to the boa's nest and subsequently shades the nest with a fan-like tail. The trade is mutually advantageous because the boa protects her social partner from predation. The amount of shade is determined by a morphological characteristic, namely tail length. Cheating by the bird is, therefore, not an issue. Conversely, it is assumed that the bird serves the boa better for shade than for a meal, so that the boa has no incentive for "breaking the social contract."

The larger the bird's tail, the larger the fitness of the boa. However, it is the bird that pays the price for the tail in this tale, as it could shade its own eggs with relatively shorter feathers. If there are typically fewer birds than boas, evolution will tune the boa's mind to accept birds with short tails. This resembles the situation of an employer seeking employees when few are available; almost anyone

who meets minimum needs has to be accepted. As we all know, when there are many candidates, the job market looks different. For the boa constructor this means that it can exert choice and thereby create a selection pressure on shadow birds to evolve elongated tails.

This shows, at least theoretically, that exaggerated or understated morphological traits may result from social partner choice, not just from sexual selection. Let us interpret the lazuli bunting example in this spirit. Among the males that are able to reproduce, some are dull and resemble juveniles whereas others are brightly colored and thus look like adults. What incentive does a male capable of reproduction have to delay plumage maturation? Dull plumage seems to signal the denial of territorial claims in the lazuli bunting. The signal comes with the moult and cannot be changed during the season. At first glance, this appears to be a self-imposed obstacle, but social choice exerted by territorial adults generates the advantage of being "dressed as a juvenile." The market determines the extent to which this dress code holds.

The lazuli bunting example and the boa–shadowbird tale share a strong resemblance with human labor markets. Both adult bunting and boa act as principals hiring a helper, whose job it is to increase the principal's breeding success. We think that it is, therefore, possible to draw on recent developments in economics and model the biologist's boas and buntings in the spirit of modern market models, in which cheating is a theme and workers may be lazy if they wish.

EDUCATION AND THE PEACOCK'S TAIL

As shown above, signaling can play a crucial role in biological markets. The peacock's tail demonstrates this even more impressively than the coloration phenomena observed in lazuli buntings. Although it is tempting to compare the peacock's signal with advertising observed in human economic activities, there are important differences. Human advertising can easily manipulate mental mechanisms because they operate in the modern world and not in the environment of evolutionary adaptation. For animals, however, this situation is much simpler: we expect countermeasures to work. Empirical attempts to understand male advertising and female choice have kept an industry of biological research busy for at least two decades and it remains difficult to understand all the details (e.g., Bradbury and Vehrencamp 1998, 2000).

At the theoretical level there has been a long-lasting discussion about the so-called *handicap principle* which goes back to Zahavi (1975). When Zahavi first expressed his idea, that animals acquire costly handicaps just to impress others, he failed to convince the community of theoretical biologists. However, subsequently Pomiankowski (1987), Grafen (1990), and Gintis, Smith and Bowles (2002) showed that the handicap principle can be expressed in coherent mathematical models. The basic idea is that a signal that is costly to send—and more costly for some than others—will not be easily faked, so other animals can infer that those sending the signals are those with lower costs.

Let us have a quick look at the easiest way to approach this issue (following Siller and Hammerstein, in prep.) and compare it with modeling in economics.

Consider a theoretical bird population in which males have elongated tail feathers and females base their mating decisions on tail length. Assume that males have the opportunity to adjust the intensity of their signal s (tail length) to their own physical condition (i.e., health state, vigor, etc.). We call this condition the sender's type t and allow the signal to be conditional on t. The signal s is received by the female who rewards the sender an amount $b(s)$ in terms of offspring, where b is increasing in s. Tail length is not for free and the male has to pay an amount c for its signal, where c is increasing in s. Now, to create an appropriate model for Zahavi's handicap principle, we have to assume that this cost depends not only on the level of the signal s but also on the male's type t (i.e., on its physical condition).

In this signaling game, the payoff w to a male is the reward from the female minus the cost of the tail, i.e., $w = b - c$. Let us assume that the population is at an evolutionary equilibrium and that females prefer males with longer tails. Consider two males with the following characteristics:

Male 1: signal s_1, type t_1
Male 2: signal s_2, type t_2.

Assume that male 1 has the longer tail. By just looking at tails in the equilibrium population what can we (and the females) conclude about the underlying types and costs?

Consider the difference in tail length, $\Delta s = s_1 - s_2$, with $\Delta s > 0$. For the two birds, what is the cost of having the longer tail as compared to the shorter tail? That is, given the type of each bird, how expensive would it be to increase tail length? We can express this by:

$$\Delta c_{male1} = c(s_1, t_1) - c(s_2, t_1) \text{ and } \Delta c_{male2} = c(s_1, t_2) - c(s_2, t_2).$$

In economics one would call Δc_{male1} and Δc_{male2} the comparative costs of signal 1 for male 1, male 2, respectively. At evolutionary equilibrium, where both males play best responses to the females' and other males' behavior, the long tails must be a best reponse for birds of type 1 and short tails a best response for birds of type 2, or $w(s_1, t_1) \geq w(s_2, t_1)$ and $w(s_2, t_2) \geq w(s_1, t_2)$. These inequalities imply the following comparative fitness advantage for type 1:

$$w(s_1, t_1) - w(s_2, t_1) \geq w(s_1, t_2) - w(s_2, t_2). \tag{8.1}$$

If the benefit function b depends only on the signal and not on the bird's type, then this inequality implies that

$$\Delta c_{male2} \geq \Delta c_{male1}. \tag{8.2}$$

Thus, a necessary condition for the signaling equilibrium is that if either of the two males has a greater additional cost for growing the longer tail, this will be the one that we observe with the shorter tail. The equilibrium does not permit males to cheat with their tails.

This is the essence of "handicap mathematics" and it gives the reader a good foretaste of how to formalize Zahavi's idea. Of course one wants to complete the argument and inquire about sufficient conditions for an equilibrium that separates the types. Depending on the appropriate assumptions, evolutionarily stable states exist in which the sender's type can be inferred from the signal observed by the receiver.

It must be emphasized that biologists have struggled with the handicap principle without paying much attention to the existence of similar models in economics, dating from about the same time as Zahavi's initial paper. Anyone who knows signaling theory will recognize the striking similarity between what we just discussed in relation to Zahavi's thoughts and signaling games in economics.

Ultimately, the worlds of biology and economics are perhaps not so different. Let us, therefore, end the chapter with celebrating this proximity. We move on to human affairs and present an *economic version of the peacock's tail*.

In our school days, when we had doubts about the value of learning "exotic" things such as Latin, mathematics, or the capital cities of Europe, our teachers tried to console us by explaining that education serves to prepare us for our future lives and is not intended to just get us through school. It appears that Nobel Laureate Michael Spence (1973) was not quite convinced by his teachers' advice because he posed the following theoretical question: Does the acquisition of *higher levels of education* lead to *higher wages* even if education fails to improve a person's productivity? Spence's model of the job market can be formulated as follows. A person's type (health, talent, productive ability) is randomly determined. The person can then choose a level of education conditional upon talent, it being less costly for the more talented to continue in school. Following completion of schooling, two firms observe the person's education (but not the person's talent) and make simultaneous wage offers. The person accepts the higher offer or flips a coin in case of a tie.

Spence found that an equilibrium can exist in which education signals talent and higher education implies higher wage. The reason is that only the talented persist in long years of schooling, so employers use years of schooling as a signal of the unobservable trait, talent. This result is remarkable because education is costly in the model and does not increase a person's productivity — quite like a peacock's tail.

CONCLUDING REMARKS

Biologists have discovered fascinating examples of market-like interactions and have made considerable progress in understanding these markets. In light of this success it seems unfortunate that traditional models from economics do not easily apply to biology. We have argued, however, that with the development of post-Walrasian microeconomics, the interdisciplinary gap is shrinking, and we have indicated how some bridges can be built. One bridge, however, remains to be mentioned. Inspired by the general equilibrium concept from economics, biologists should perhaps dare to step beyond the analysis of dyadic and other small numbers interactions and consider the population-level interactions among more than a single market. Such investigations would be essential to understand, for example, why the disadvantages of the locally based cleaner fish persist in equilibrium. Conversely, economists should take more seriously the idea that humans are animals after all and not quite as distinct from the rest of nature as traditional modeling approaches might make us believe.

ACKNOWLEDGMENTS

We would like to thank the Santa Fe Institute and the *Deutsche Forschungsgemeinschaft* (SFB 618 *Theoretische Biologie*) for support of this research. Ronald Noë provided valuable comments and help with drawing the endangered species of shadowbirds.

REFERENCES

Alchian, A. 1950. Uncertainty, evolution, and economic theory. *J. Pol. Econ.* **58**:211–221.

Bateman, A.J. 1948. Intra-sexual selection in *Drosophila*. *Heredity* **2**:349–368.

Becker, G.S. 1962. Irrational behavior and economic theory. *J. Pol. Econ.* **70**:1–13.

Bowles, S. 2003. Microeconomics: Behavior, Institutions, and Evolution. Princeton, NJ: Princeton Univ. Press.

Bradbury, J.W., and S.L. Vehrencamp. 1998. Principles of Animal Communication. Sunderland, MA: Sinauer.

Bradbury, J.W., and S.L.Vehrencamp. 2000. Economic models of animal communication. *Anim. Behav.* **59**:259–268.

Charnov, E.L. 1982. The Theory of Sex Allocation. Princeton, NJ: Princeton Univ. Press.

Coase, R.H. 1937. The nature of the firm. *Economica* **4**:386–405.

Darwin, C. 1874. The Descent of Man in Relation to Sex. 2nd ed. London: J. Murray.

Fisher, R. 1930. The Genetical Theory of Natural Selection. Oxford: Clarendon.

Fowler, K., and L. Partridge. 1989. A cost of mating in female fruit flies. *Nature* **338**: 760–761.

Gintis, H., E.A. Smith, and S. Bowles. 2002. Costly signalling and cooperation. *J. Theor. Biol*. **213**:103–119.

Grafen, A. 1990. Biological signals as handicaps. *J. Theor. Biol.* **144**:517–546.

Greene, E., B.E. Lyon, V.R. Muehter, et al. 2000. Disruptive sexual selection for plumage coloration in a passerine bird. *Nature* **407**:1000–1003.

Hammerstein, P. 2001. Games and markets: Economic behaviour in humans and other animals. In: Economics in Nature: Social Dilemmas, Mate Choice and Biological Markets, ed. R. Noë, L.A.R.A.M. van Hooff, and P. Hammerstein, pp. 1–19. Cambridge: Cambridge Univ. Press.

Hammerstein, P., and G.A. Parker. 1987. Sexual selection: Games between the sexes. In: Sexual Selection: Testing the Alternatives, eds. J.W. Bradbury and M.B. Andersson, pp. 119–142. Dahlem Workshop Report LS39. Chichester: Wiley.

Harcourt, A.H., P.H. Harvey, S.G. Larson, and R.V. Short. 1981. Testis weight, body weight and breeding system in primates. *Nature* **293**:55–57.

Henzi, S.P., and L. Barrett. 2002. Infants as a commodity in a baboon market. *Anim. Behav.* **63**:1–7.

Lorch, P.D. 2002. Understanding reversals in relative strength of sexual selection on males and females: A role for sperm competition? *Am. Nat.* **159**:645–657.

Noë, R. and P. Hammerstein. 1994. Biological markets: Supply and demand determine the effect of partner choice in cooperation, mutualism and mating. *Behav. Ecol. Sociobiol*. **35**:1–11.

Noë, R. and P. Hammerstein. 1995. Biological markets. *Trends Ecol. Evol.* **10**:336–339.

Noë, R., J.A.R.A.M. van Hooff, and P. Hammerstein, eds. 2001. Economics in Nature: Social Dilemmas, Mate Choice and Biological Markets. Cambridge: Cambridge Univ. Press.

Parker, G.A. 1970. Sperm competition and its evolutionary consequences in insects. *Biol. Rev.* **45**:525–567.

Parker, G.A., and M.A. Ball. 2001. Information about sperm competition and the economics of sperm allocation. In: Economics in Nature: Social Dilemmas, Mate Choice and Biological Markets, ed. R. Noë, J.A.R.A.M. van Hooff, and P. Hammerstein, pp. 221–244. Cambridge: Cambridge Univ. Press.

Pomiankowski, A. 1987. The "handicap principle" does work — sometimes. *Proc. Roy. Soc. Lond. B* **127**:123–145.

Sahlins, M. 1974. Stone Age Economics. Chicago: Aldine.

Schwartz, M.W., and J.D. Hoeksema. 1998. Specialization and resource trade: Biological markets as a model of mutualisms. *Ecology* **79**:1029–1038.

Simms, E.L., and D.L. Taylor. 2002. Partner choice in nitrogen-fixation mutualisms of legumes and rhizobia. *Integ. Comp. Biol.* (formerly: *Am. Zool.*) **42**:369–380.

Simon, H. 1951. A formal theory of the employment relation. *Econometrica* **19**:293–305.

Simon, H. 1955. A behavioral model of rational choice. *Qtly. J. Econ.* **69**:99–118.

Smith, R.L. 1997. Evolution of paternal care in the giant water bugs (*Heteroptera: Belostomatidae*). In: Social Competition and Cooperation among Insects and Arachnids, ed. J. Choe and B. Crespi, vol. II. Evolution of Sociality. Princeton, NJ: Princeton Univ. Press

Spence, M. 1973. Job market signalling *Qtly. J. Econ.* **87**:355–374.

Zahavi, A. 1975. Mate selection: A selection for a handicap. *J. Theor. Biol.* **53**:205–214.

9

Biological Markets

The Ubiquitous Influence of Partner Choice on the Dynamics of Cleaner Fish – Client Reef Fish Interactions

Redouan Bshary[1] and Ronald Noë[2]

[1]Department of Zoology, University of Cambridge, Cambridge CB2 3EJ, U.K.

[2]Ethologie et Ecologie comportementale des Primates, Université Louis Pasteur, 67000 Strasbourg, France

ABSTRACT

The applicability of biological market theory with its emphasis on partner choice is explored using the interactions between the cleaner wrasse *Labroides dimidiatus* and its "client" reef fish as a model system of mutualism. Cleaners have small territories, which the majority of reef fish species actively visit to invite inspection of their surface, gills, and mouth. Clients benefit from the removal of parasites while cleaners benefit from the access to a food source. Some client species (choosy clients) have large home ranges that cover several cleaning stations, whereas other clients have small ranges and have access to one cleaning station only (resident clients). Field observations, field manipulations, and laboratory experiments revealed that whether or not a client has choice options influences several aspects of both cleaner and client behavior. Cleaners give choosy clients priority of access. Choosy clients switch partners if cheated by a cleaner (= cleaner feeds on mucus/scales), whereas resident clients punish cheats. Cleaners and resident clients, but not choosy clients, build up relationships before normal cleaning interactions take place. Cleaners are particularly cooperative if choosy clients are bystanders of an interaction but less so when resident clients are bystanders. When it comes to the frequency of cheating by cleaner fish, however, partner choice options are overrun by client control mechanisms: predatory clients are far less often cheated than nonpredatory clients, irrespective of choice options. Future research needs to focus more on empirical testing of game theory so that this new information can be used to formulate deductive models.

INTRODUCTION

On human markets, goods or services are traded against money or other goods. It is well established that changes in the ratio between the supply of a good/service and the demand cause changes in its price. This is because when the supply is higher than the demand, potential buyers are able to compare prices and choose the cheapest offer. This choosiness of buyers is crucial, as it exerts pressure on sellers to outcompete each other with lower prices in order sell their goods. These principles have recently been applied to cooperative (within species) and mutualistic (between species) interactions in animals (Noë et al. 1991; Noë and Hammerstein 1994, 1995; Noë 2001). The biological market approach has two major goals:

1. Explain adaptations that are the result of "market selection." Market selection explains the evolution of mechanisms enabling partner choice as well as adaptations that increase the chances of being chosen in cooperative and mutualistic systems. There is an obvious overlap with the evolution of mate preference and the selection for secondary sexual characters driven by mate choice, but we propose to use the label "market selection" only in relation to cooperation outside the mating context. A crucial difference is that covariance between choice mechanisms and chosen traits typical for sexual selection (Fisher 1930) is unlikely to occur under market selection, notably when trading partners belong to different species.
2. Predict changes in exchange rates of commodities due to changes in the supply/demand ratio between these commodities.

Mechanisms, such as partner choice and outbidding competitors, are relevant in three major fields within behavioral ecology: sexual selection, intraspecific cooperation, and interspecific mutualism (see below). To date, surprisingly little attention has been paid to the selective force of partner choice outside the context of sexual selection. Similarly exchange ratios have been studied in the context of mating markets but not in the context of cooperation and mutualism. This is probably due to an obsession with partner control in the theoretical approaches to cooperation, as a result of a historically determined fascination with the evolution of altruism. Classical models of cooperation are usually based on an iterated version of the so-called Prisoner's Dilemma (PD) game (Axelrod and Hamilton 1981). In short, a PD exists if two players, who can either cooperate or defect, receive a higher payoff from defection independently of the partner's behavior; if both cooperate, however, they receive a higher payoff than if both defect. Because of the payoff structure, cooperation is highly unlikely in single-round PD games, but in iterated versions several strategies have been proposed that lead to evolutionarily stable cooperation (reviewed by Dugatkin 1997). In other words, once all individuals of a given population play certain cooperative strategies, the population cannot be invaded by individuals playing any defecting strategy. The cooperative strategists, on average, gain higher payoffs than the individuals playing a defective strategy.

At first glance, it appears that partner control models and biological market models address different problems invoked in cooperative interactions. "Traders" on biological markets may react to partners that do not yield enough "profit" by terminating the cooperative relationship. The negative effects of this decision to the actor can be limited if he can switch to an alternative partner. The biological market paradigm explicitly excludes the use of force to extort commodities from partners in the same sense that mainstream economical models also ignore the possibility of theft and robbery. In partner control models, which usually focus on the interactions of two players only, all forms of defections and reactions to defections are considered, from simply terminating the cooperation to the use of force and other forms of coercion. Partner control models thus cover the whole spectrum of cooperative to exploitative interactions. In reality, there are obvious connections between partner choice and partner control issues. Ultimately, the two approaches have to be brought together. To keep things simple at first, however, it may be desirable to study partner choice and partner control separately.

In this chapter we use data on cleaner fish – client reef fish interactions to explore how market theory may explain (a) payoff asymmetries between cooperating partners, (b) differences in strategic options that are available to partners, (c) how partner choice interacts with partner control mechanisms, and (d) how incomplete control over the resources traded may lead to a shift from mutualism to parasitism.

BIOLOGICAL MARKETS IN NATURE: A SHORT OVERVIEW OF THE EVIDENCE

Noë and Hammerstein (1994, 1995) concentrated on cooperative and mutualistic systems in which two classes of traders can be distinguished that control different commodities. In such systems, "bartering" can take place through the mechanism of partner switching. Consistent partner choice over many generations can lead to the selection of specific adaptations comparable to secondary sexual characters, which evolve under selection through mate choice. Noë and Hammerstein (1994) used the example of delayed plumage maturation in purple martins based on data presented by Morton et al. (1990). Male purple martins may control several nest cavities, most of which are "martin houses" provided by humans in this day and age. The male can only use one cavity for breeding, since the species has obligatory biparental care. However, he can allow a pair consisting of a yearling male and his mate to breed in another cavity under his control. By copulating with the female of the subordinate pair, the dominant male gains some extra offspring that are cared for by another male. The subordinate males also gain, since they usually sire some chicks produced by their mate. Without access to this nesting opportunity, they would not have any offspring at all. Noë and Hammerstein (1994) proposed that the delayed plumage maturation in the purple martins is selected for through the choice made by the dominant males among the yearlings. By accepting only males that carry obvious signs of subordinance as subletters they would lower the risk of challenges to their own position. Females probably prefer males with a mature plumage and males with delayed plumage would not be able to hold a nest cavity for very long, since they would attract immediate attacks from other adult males.

The weak point in this account is that the situation is rather unnatural: the evolution of delayed plumage maturation over the period that humans provide grouped nest cavities in the form of martin houses is unlikely. However, Greene et al. (2000) recently provided a very similar example in another colorful North American bird, the lazuli bunting, observed in a natural setting. In this species, dominant males do not accept yearling males with delayed plumage maturation in adjacent cavities but in adjacent territories. Greene et al. carefully checked that the conditions for market selection were met.

Under market selection, the evolution of mechanisms for implementing choice can be expected in the choosing class. Bull and Rice (1991) hypothesized such a mechanism for the fig/fig-wasp system: the selective abortion of overexploited figs. Although this is probably not the mechanism that figs use to control overexploitation (Herre and West 1997), the abortion mechanism has been described for similar pollination mutualisms (e.g., Pellmyr and Huth 1994; Fleming and Holland 1998). Both sides in a market may evolve mechanisms that ensure that the partner (a) cannot rob a commodity and (b) has to pay a price as high as the law of supply and demand allows. Flowers may, for example, have structures that force pollinators to stay longer and take up more pollen, instead of offering more nectar.

Biological markets can be found throughout the living world. As long as there is a possibility to exchange commodities and at least one of the trading partners can exert choice, the market mechanism may influence the outcome of the interaction. One should not forget, however, that commodities from abiotic sources may play a role as well. Such resources can sometimes be in competition with commodities provided by one of the trader classes. This is, for example, the case in the biological markets involving mycorrhiza (Schwartz and Hoeksema 1998; Wilkinson 2001) and rhizobia (Simms and Taylor 2002), i.e., mutualisms in which plants exchange nutrients with fungi and bacteria, respectively. These markets can be strongly influenced by abiotic sources of these nutrients, including artificial fertilizers (West et al. 2002).

Exact quantitative predictions of exchange rates on the basis of supply/demand ratios are hard to make, both for human economic systems and for biological systems. It is easy, however, to predict in which direction the exchange rate will change when the supply/demand ratio shifts. The most straightforward and simple example published to date is probably by Henzi and Barrett (2002): Baboon females like to inspect and handle the infants of other females, but they are not allowed to do so without "paying" for it by grooming the mother first. Henzi and Barrett predicted that females would have to groom longer when there were fewer infants available in the group, a prediction that held both when mothers were dominant over the would-be handlers of their infant and when the mothers were subordinate. There was a quantitative difference between the two cases: the use of force can alter or even overrule market effects (see below).

Together with further examples of biological markets given in Noë et al. (1991), Noë and Hammerstein (1994, 1995), and Noë (2001), these examples show that market theory provides a general framework to study payoff distributions among partners in sexual selection, cooperation, and mutualism. We will now explore the applicability of market theory and its current limits by providing a detailed overview on a field and experimental study on the mutualism between the cleaner wrasse *Labroides dimidiatus* and other reef fish species.

A CASE STUDY: MUTUALISM BETWEEN THE CLEANER WRASSE *L. DIMIDIATUS* AND CLIENT REEF FISH

In coral reefs from the Red Sea through the Indo-pacific to the Great Barrier Reef in Northeastern Australia, the best way to see a wide variety of fish species within a short time period is to locate a cleaner wrasse, *L. dimidiatus*. These fish are visited by the majority of other reef fish species who often "pose" for the cleaner by spreading their pectoral fins and stopping coordinated swimming, leading to "head up" or "head down" postures, depending on the species. Posing is a signal to the cleaner fish that the visitor seeks its service, which comprises the removal of ectoparasites and dead or infected tissue from the body surface, the gills, and sometimes even inside the mouth. The deal seems to be straightforward: the cleaner fish gets access to an easy meal and the so-called "client" gets cleaned. *L. dimidiatus* is just one of a large variety of cleaning organisms but it is probably the best studied (for reviews, see Losey et al. 1999; Côté 2000). The interaction between *L. dimidiatus* and its clients has attracted considerable attention, one reason being that it is particularly suited to test game theoretical models of cooperation among unrelated organisms. Individual *L. dimidiatus* may have more than 2000 interactions per day, eat about 1200 parasites per day, and may reduce the parasite density on clients by a factor of 4–5 (Grutter 1999). Most recently, it was found that clients without access to cleaner fish in their natural environment showed a higher stress response to capture than clients with access to cleaner fish (R. Bshary, R. Oliveira, and A. Canario, unpubl. data). The difference can be seen as a net difference between costs and benefits of cleaning interactions. There is thus little doubt that interactions between cleaner fish and clients are overall mutualistic, i.e., to the benefit of both partners.

There are several forms of conflict in these interactions. One of the most well known involves the possibility of predatory clients defecting by eating the cleaner rather than letting it inspect — a problem that Trivers (1971) used to explain his concept of reciprocal altruism. He proposed that predators refrain from eating cleaners despite the energy gain being higher than the gain from the removal of parasites in one single interaction. Refraining from eating cleaners only becomes advantageous for the predator when the repeated removal of parasites by a particular cleaner leads to a larger benefit than eating that cleaner. Trivers's idea has not yet been tested but there is some evidence that predators may eat cleaners under some circumstances (Côté 2000). In this chapter, we do not limit ourselves to the game between cleaners and predatory clients but rather focus on the strategies cleaner fish use when confronted with different types of clients and how these clients deal with them. It is important to note that the majority of client species feed on plankton or graze and are hence not potential predators of cleaners.

Recent experimental evidence suggests that cleaners regularly cheat clients by consuming client mucus and scales, as the consumption of these food items is not linked to the removal of ectoparasites (Bshary and Grutter 2002a). In addition, there might be conflicts between cleaners and clients over the timing and duration of interactions. This conflict is most obvious when two or more clients seek inspection from the same cleaner simultaneously so that the cleaner has to choose between them. As we shall see, the solutions to these conflicts are often, but not always, dependent on the ratio between supply and demand.

Methods

We refer to a single species of cleaner wrasse: *L. dimidiatus*. All reported field observations and field experiments were conducted at Ras Mohammed National Park, Egypt. Data were either collected while sitting 2–3 m in front of a cleaning station or by following individual clients. Interactions between cleaners and clients were first observed and then the key information was noted on a Plexiglas plate. Laboratory experiments were conducted at the Lizard Island Research Station, Great Barrier Reef, Australia. For methodological details, we refer to the respective publications of the original data.

Game Structure

For a thorough appreciation of what happens during interactions, it is important to outline the strategic options of cleaners and clients. Note that the game structure applies to *L. dimidiatus* but may be different for other cleaner fish species.

1. *Repeated game structure:* Individual clients seek inspection about 5–30 times per day, in extreme cases more than 100 times a day (Grutter 1995). Even clients with territories or home ranges that cover large reef areas may interact repeatedly with the same individual cleaner fish. Cleaners may have more than 2000 cleaning interactions per day (Grutter 1995). Cleaners are territorial and usually move within very confined areas of a few cubic meters, which led to the term "cleaning station." Clients can tell individual cleaners apart through site recognition (unless there is a pair of cleaners at one station) and/or individual recognition. Experimental evidence indicates that cleaners can recognize individual clients (Tebbich et al. 2002), although it is hard to quantify how many they actually recognize under natural conditions.

2. *Choice options:* Due to the cleaners' site fidelity, a client decides whether or not to visit a cleaner and to seek an interaction; cleaners only have the option of ignoring a visiting client. Members of some client species (also called "residents") have small territories or home ranges, which allow them access to only one cleaning station. Cleaners, thus, have veto power (Noë 1990) in that they have exclusive access to these clients without competition from other cleaners. Cleaners are, therefore, expected to be in a strong position with respect to the service quality: they can cheat residents more often and let them wait longer than nonresident clients. Individuals of nonresident client species have home ranges that cover several cleaning stations (further called "choosy clients"). In these cases, cleaners compete among each other to attract these clients. As each cleaner remains at its respective cleaning station, competition can only take place through outbidding each other with better service quality, not through aggressive interference or actively approaching clients. Clients with the option to choose between cleaners are therefore expected to show preference for the cleaners that offer the best service.
3. *Asymmetry in possible sanctions:* About 15% of client species are predatory in the sense that they hunt fish that are the size of cleaners. In interactions between cleaners and predatory clients, both have symmetrical strategic options, i.e., each party can either cooperate or cheat. However, the effects of cheating on the partner are highly asymmetric: an exploited predator only loses a little bit of mucus, whereas an exploited cleaner loses its life. In interactions between cleaners and the 85% nonpredatory client species (grazers or plankton feeders), the strategic options are asymmetric: the cleaner can cooperate as well as cheat whereas the client has no option to exploit the cleaner to its own advantage. These clients can only control the duration of an interaction, i.e., by swimming off, they can terminate an interaction.

A PURE MARKET GAME DETERMINED BY SUPPLY AND DEMAND: CLIENT–CLIENT COMPETITION OVER PRIORITY OF ACCESS TO CLEANER FISH

Sometimes a client arrives at a cleaning station while the cleaner is inspecting another client; sometimes two or more clients seek the inspection of a cleaner fish simultaneously. In such cases, individual clients compete directly over access to the cleaner fish. This competition takes place only through inviting the cleaner for inspection, not through aggressive displacement (Bshary 2001). Market theory makes the following predictions for these situations:

1. Cleaners should give priority to choosy clients over resident clients. The reason is that choosy clients can visit another cleaning station rather than queuing for service. A cleaner that ignores a choosy client in favor of a resident client could lose access to a food source, and selection is

expected to work against that decision rule. Resident clients have no strategy available to push a cleaner into giving them priority of access. They have to come back at a time that is more convenient for the cleaner if they want to be inspected at all. A cleaner's access to this food source is thus just delayed if it ignores the resident's invitation for inspection in favor of another client.

2. The preference for choosy clients should be independent of the clients' quality as a food patch. Optimal foraging theory based on the marginal value theorem (Charnov 1976) would lead to an alternative prediction: as client size strongly correlates with parasite load (Grutter 1995), the decisions of the cleaner fish should be determined by the relative body sizes of clients that present themselves simultaneously.

Field observations were in line with predictions based on market theory. Cleaners switched from resident clients to choosy clients 51 times but only once the other way round; when a choosy client and a resident client invited inspection simultaneously, the cleaners inspected the choosy client in 65 out of 66 observations (Bshary 2001). These results were independent of the clients' relative body sizes, as the choosy clients were smaller than the residents in 12 (switching) and in 21 (queuing) occasions. Following individual long-nosed parrotfish, *Hipposcarus harid*, for up to 120 minutes, Bshary and Schäffer (2002) confirmed that choosy clients are indeed active players in the game. The probability of a client returning to the same cleaning station for its next inspection was high (median of 13 individuals: 60%) if it was inspected, but low (median: 0%, $n = 13$) if it had been ignored in favor of another (choosy) client.

It could have been that choosy clients are better food patches than residents, independent of size, because they might visit cleaning stations less frequently than residents to optimize foraging, or that choosy clients are generally more infected than residents because they traverse larger areas, which might make them more vulnerable to infestation. Therefore, Bshary and Grutter (2002b) experimentally removed food patch quality as a confounding variable. Cleaner fish were kept in aquaria that had a compartment inaccessible to the cleaner. A lever construction allowed two Plexiglas plates of similar shape and color, but of different sizes, to be moved in and out of the cleaners' compartment. On both plates, equal amounts of mashed shrimp meat were spread over an area of 4 cm^2. Cleaners could thus choose between two food patches of equal quality and, when the two plates were presented simultaneously, had to decide which one to inspect first. The lever construction was used to mimic differences in behaviors between resident clients and choosy clients. Two differences could potentially provide cleaners with clues about clients being either residents or choosy under natural conditions. First, residents are often willing to queue for inspection when the cleaner is busy with another client, but choosy clients swim off in such situations. Second, residents may revisit the cleaner shortly after being ignored, whereas choosy clients switch to another cleaning station. We mimicked these two cues in two experiments.

Experiment 1: The "resident plate" was left in the cleaners' compartment until the cleaner finished foraging, whereas the "choosy plate" was retrieved immediately if the cleaner started eating from the other plate.

Experiment 2: The plate that was first ignored was invariably pulled back, but the resident plate was pushed in again as soon as the cleaner had finished foraging on the choosy plate; the choosy plate was not returned.

Cleaners were repeatedly confronted with these situations, and market theory predicted that the cleaners should develop a preference for the choosy plate despite a lack of difference between the two plates regarding patch quality. In both experiments, a significant proportion of cleaners soon fed from the "choosy" plate first (median values for trials 16–20 in Experiment 1 and trials 11–15 in Experiment 2: 80%). In combination with the field observations, these results provide strong support for the market theory. The partner choice options (and the use of them) of choosy clients select for cleaner fish that give them priority of access over residents.

A Limitation of Market Theory: Predicting Cheating Frequencies by Cleaner Fish

From a client's perspective, a good cleaning service does not only include getting priority of access but also that the cleaner searches for parasites and refrains from eating healthy client tissue. From a cleaner's perspective, however, searching for parasites is time consuming whereas client mucus is readily available. Therefore, a conflict between cleaner and client exists over the cleaner's foraging behavior. Most client species do not win this conflict entirely (cf. below) and fortunately, defections by the cleaner become visible to the observer through short jolts performed by clients in response to cleaner fish mouth contact. There is experimental evidence that client jolts are not related to the removal of parasites; to the contrary, clients jolt more frequently the less parasite infested they are (Bshary and Grutter 2002a). Jolt rates are therefore a good correlate of cleaner fish cheating rates. (This does not exclude the possibility that a jolt may sometimes occur in response to the removal of a parasite.) In an experiment in which anaesthetized parasite-free client surgeon fish *Ctenochaetus striatus* were added to a cleaner fish in an aquarium, most cleaners scraped the surface of their clients rather than feeding on prawns that were provided as an alternative food source (Bshary and Grutter 2002a). It thus appears that when clients cannot control cleaner fish behavior, cleaners are likely to cheat. So how do clients control cheating frequencies of cleaners, and what factors may explain variation among different client species with regard to cleaner fish defection frequencies?

Market theory predicts that choosy clients use their option to play cleaners off against each other to control cheating partners. It is therefore predicted that choosy clients swim off when the cleaner bites them and visit a different cleaner for their next inspection (similar to situations where cleaners ignore them as discussed above). Resident clients, however, lack this option of switching partners.

Market theory therefore predicts that residents have to accept more frequent cheating by the cleaner than choosy clients.

Field observations revealed that choosy clients do indeed use their choice option to control defecting cleaners. Long-nosed parrotfish came back to the very same cleaning station for their next inspection if the previous one had ended without a conflict due to cheating by the cleaner in 65% of observations; they, however, switched to another cleaning station if the cleaner had bitten them in about 90% of observations (Bshary and Schäffer 2002). Partner choice is not the only option for clients to keep the cleaner in check. As suggested by Trivers (1971), predatory clients can eat the cleaner. All clients have the option to attack the cleaner and chase it around, a strategy which may represent "punishment" sensu Clutton-Brock and Parker (1995). A strategy based on punishment includes three steps:

1. An individual A performs an act that increases its own fitness at the expense of the fitness of another individual B.
2. In response, individual B performs an act that is temporarily spiteful as it reduces the fitness of both individuals.
3. As a consequence, individual A will change its behavior in a way that is costly for itself but increases the fitness of individual B.

Note this strategy does not imply any causal understanding by the players. This form of punishment is exactly what happens during interactions between cleaners and resident clients. Cheating by cleaners was often followed by the clients chasing the cleaners (> 60% of responses). As a result of chasing, the interaction terminated in 95% of observations ($n = 195$). During the next interaction between the cleaner and the same individual, which is usually delayed by a few minutes, the cleaner refrained from cheating (median value of jolt frequencies in 36 client species is 0/100 s in interactions following chasing, compared to ~5/100 s on average). In summary, there are three ways in which different client species may control cleaner fish defection: kill and eat (predators), partner switching (choosy clients), and punishment (resident clients).

Predatory client species jolted less frequently than nonpredatory client species (Bshary 2001). Within the predator category, there was no significant difference between resident predators and choosy predators. Basically, predatory clients did not jolt (median values for both resident and choosy predatory species = 0/100 s interactions). Nonpredatory choosy clients, on average, jolted less frequently than nonpredatory resident clients. However, this result appeared to be confounded by size, in particular in resident species. There was a significant negative correlation between resident client size and jolt rates ($r_s = -0.51$, $n = 24$ species, $p < 0.01$). This correlation is not caused by larger resident clients being more likely to punish cleaners as there was no correlation between resident client size and probability of punishment (both small and large species punish > 60% of cleaner cheats). When only resident and choosy client species of similar size were considered, no significant difference in jolt rates between the two client categories were found ($p > 0.1$) (Bshary 2001). It thus appears that the predators' option to kill the cleaner leads the cleaners to engage in an unconditional cooperative strategy during interactions with predators. In conclusion, being or not being able to choose among cleaners does not predict cleaner fish cheating rates. This is because the choice option of choosy clients is just one of several potential mechanisms, of which being able to retaliate cheating seems to be the most efficient. Predatory clients are cheated least frequently probably because they could eat cleaners; the punishment of cheating cleaners by resident clients is about as effective as the switching strategy of choosy clients. Nevertheless, the market situation still determines who has to invest in partner coercion in the form of "punishment" and who has not.

INTEGRATING MARKET THEORY INTO PARTNER CONTROL THEORY

Partner control theory in the context of cooperation is heavily biased toward the iterated PD game (IPD) (review in Dugatkin 1997). Since Axelrod and Hamilton (1981) used computer tournaments in the IPD and found that "Tit-for-Tat" (start cooperative and consequently play the strategy your partner played in the previous round) emerged as the superior strategy, many extensions have been explored (Dugatkin 1997). Recently, Roberts and Sherratt (1998) explored variable payoffs, and Nowak and Sigmund (1998) explored the evolution of altruistic behavior based on indirect reciprocity (help to receive future help from the observers rather than the recipient). However, most examples of intraspecific cooperation and interspecific mutualism do not seem to fit the models. Bshary and Grutter (2002a) argued that this is because the participants in most empirical examples have different strategic options and thus should be analyzed with the help of asymmetrical games. One class of traders usually lacks the option to gain anything from exploiting the partner, whereas the other class has the option to gain from cheating. A typical example is the cleaner fish – client mutualism explained above but there are plenty of others (Bshary and Grutter 2002a). For a further introduction into these systems, see Noë et al. (1991) and Noë (2001).

Because of the asymmetric structure of many cooperative and most mutualistic games, the IPD is not the appropriate framework to analyze which conditions lead to cooperative outcomes. In contrast, partner choice options will be of major importance in understanding the structure of the underlying game and the payoff configuration. As we argued above, partner choice can be a mechanism of partner control in asymmetric games. However, partner choice options (or the lack of them) may influence partner control games in other ways as well, as we illustrate below with further data on the cleaner fish – client mutualism.

Interaction between Partner Choice and the Need for Building up Relationships

Consider a model of reciprocal altruism where the investment of both partners can be variable. Roberts and Sherratt (1998) proposed that a strategy called "raising the stakes" might prove to be an evolutionary stable solution to this game. Raising the stakes means that an individual will initially invest very little into its partner and will stepwise increase investment if the partner returns at least the same amount. If the investment is not met, the individual stops being altruistic. Thus, relationships are built up carefully and therefore no defecting strategy can yield large benefits, as only established relationships consist of partners trading large favors. Testing the predictions of the model within the asymmetric games between cleaners and clients by transferring cleaners, Bshary (2002b) found no support for a "raising the stakes" strategy. Introduction of a cleaner at a new locality affects its relationships with local residents and choosy clients differently. Interactions between transferred cleaners and residents were very different from established relationships. During observations that began two hours after transfer, cleaners refrained from cheating (median: 0/100 s interaction) and provided tactile stimulation to the dorsal area of the clients' body during the entire interaction in about 80% of encounters. Tactile stimulation is a special treat that cleaners offer to their clients, usually to make clients that are unwilling to interact slow down for inspection or as a reconciliatory gesture after a conflict due to cleaner fish cheating (Bshary and Würth 2001). Tactile stimulation is incompatible with foraging and thus costly for the cleaner. Whether tactile stimulation yields any benefits to clients is currently unknown but clients readily accept this treat. Resident clients frequently fled from approaching transferred cleaners (median: 16% of all interactions) and often chased cleaners without apparent reason, i.e., without the cleaners trying to approach them (median: 26% of all interactions). After 24 hr, these behaviors of cleaners and clients were still significantly elevated compared with a control group in which cleaners were caught and released at their original cleaning station. (Fleeing and unprovoked aggression hardly ever occur when relationships are established.) Thus, cleaners and residents clearly build up relationships with transaction specific investments but opposite to the predictions of raising the stakes. Initial heavy investment by both partners is necessary to gain eventually the benefits of cleaning interactions between established partners. A likely explanation for this pattern is that residents seem to "expect" heavy cheating by the cleaner and show off their ability to punish first. Note that the behavior of residents toward immigrated cleaners restricts advantages that cleaners might gain from a roving strategy and works in favor of stationary cleaners.

Interactions between experimentally transferred cleaners and choosy clients were "normal" from the very beginning with respect to the parameters measured, i.e., duration of interactions, client jolt rate, and cleaners providing tactile stimulation. Cleaners were typically transferred over distances of 200–400 m coastline, which virtually excludes the possibility that the choosy clients already knew them from their previous station. This experiment again shows that market theory is essential to understand the strategies played by both cleaners and clients. It explains why relationships first go through a phase of investment in trust-building behavior when clients have no option to choose another partner and have to rely on their ability to punish to control their partner. No such trust-building phase is necessary when the client can switch to another cleaner.

Interaction between Partner Choice and Indirect Reciprocity

As in reciprocal altruism, an individual may help another improve his chances to receive help. In "indirect reciprocity," however, the altruistic act improves an individual's chances of obtaining help in return, not only from the recipient, but from bystanders who witnessed the altruistic act. By behaving altruistically, the altruist gains something like an "image score" or "social prestige" (Alexander 1987; Zahavi 1995), and individuals with a high score are the most likely ones to receive help by others or to be chosen as cooperation partners. This idea has been modeled by Nowak and Sigmund (1998) and by Leimar and Hammerstein (2001); Wedekind and Milinski (2000) provided experimental evidence for its adaptive value in humans. Transferring the logic of indirect reciprocity to the cleaner fish – client mutualism, it might pay for cleaners to refrain from cheating current clients if onlookers base their decision to invite inspection on what they witness: "Stay if you see a cooperative cleaner, flee if you witness a cheating cleaner." (Remember that cheating is often followed by the client darting off or chasing the cleaner and that these behaviors probably provide an easily observable cue for bystanders as opposed to looking for client jolts.) The prediction based on image scoring is opposite to what is predicted by market theory. The market paradigm suggests that the cleaner can drop its service quality and hence cheat more frequently when there is a temporarily high demand for cleaning. Clients indeed seem to base their decision to seek inspection on what they witness: field observations revealed that if clients supposedly saw a positive interaction, they almost always invited inspection (median: 100%), and if they saw a negative interaction, they rarely invited inspection (median: 15%) (Bshary 2002a). Thus, cleaners have an image score that depends on how cooperatively they behave. In turn, cleaners cheated current clients less frequently in the presence of bystanders than in the absence of bystanders (A. D'Souza and R. Bshary, unpubl. data). Distinguishing between resident bystanders and choosy bystanders, the effect is significant only in the presence of choosy clients (negative correlation between current client jolt rate and number of choosy bystanders for 15 out of 16 species), not in the presence of resident clients (negative correlation between current client jolt rate and number of resident bystanders for 9 out of 14 species). This apparent differentiation by cleaners, which should be backed up with further data, makes sense in the light of market theory as only image scoring choosy clients may decide to visit another cleaner instead.

There is an interesting twist to the story: choosy client species that visit a cleaning station in large schools (*Abudefduf vaigiensis, Caesio lunaris*) jolt extremely frequently (about 20 jolts/100 s interactions), whereas jolt rates in interactions between cleaners and single-visiting individuals of the same species are "normal" compared to other client species (about 5 jolts/100 s interactions) (A.

D'Souza and R. Bshary, unpubl. data). A possible explanation is that individuals in schools are not moving independently of the other school members and therefore stay around at the station as long as the school does, even if the cleaner cheats. School cohesion might thus offset image scoring, and in its absence, the market effect of a temporarily high demand for cleaning service leads to low service quality by cleaners.

CONCLUSIONS AND OUTLOOK

We believe that market theory, with its emphasis on partner choice, provides a useful framework to study payoff distributions and strategy sets of collaborators in intraspecific cooperation and interspecific mutualism. In particular, market theory generates testable predictions about exchange rates that may result in payoff asymmetries between partners, whereas partner control models based on the IPD do not. As partner choice occurs to some extent in most natural systems, partner choice by the limiting class of traders can be a powerful control mechanism to control defection by members of the common class of traders. Therefore, future partner control models must have partner choice options incorporated into their assumptions to become more realistic. Until now, the incorporation of partner choice within the framework of the IPD has focused on defectors being able to rove (see references in Dugatkin 1997) and has concluded that partner switching may hinder the evolution of cooperation. Other recent models, however, suggest that partner choice may enhance cooperation (Ferriere and Michod 1995).

Pure free markets will probably turn out to be very rare in nature. The form and outcome of most cases of cooperation and mutualism will be determined by more than partner choice and outbidding alone. The market effect is but one of several sources of leverage cooperating individuals have over each other. Other sources are:

- The option of simply terminating the relationship.
- The option to switch from a cooperative to an exploitative strategy, described as "defection" in IPD models.
- The use of force to influence the outcome of current or future interactions. This is the sort of leverage described in models of dominance, in models of conflict, and in punishment models.
- The possibility to steal desired commodities, as in parasitic relationships.
- The possibility to influence interactions of the partner with third parties, as described in models that take "image scoring" into consideration.

Such sources of leverage contribute to what is known as the "power" one individual or group has over another in the economic, political, and sociological literature (Bowles and Hammerstein, this volume; Lewis 2002). Some or all of these sources of power may influence a single mutualistic relationship simultaneously. To keep things traceable, we need to understand fully the interactions with one or two sources of sole power. Ultimately, all six, and perhaps more, building blocks should be integrated in what can be called "power models" of cooperation. It seems to us that the most promising way forward is to use well-understood empirical examples to guide the building of more complex models, in order to avoid the sterile theorizing of the recent past. Our remarks below are intended to guide the further development of market models.

Modeling Market Theory

Earlier, we stated that market models might (a) help explain the evolution of traits beneficial to the partner but detrimental to its bearer as well as (b) provide quantitative predictions about exchange rates of commodities based on supply/demand ratios. Noë and Hammerstein (1994, 1995; see also Noë 2001) have taken initial steps toward solving the first problem, but the second issue has barely been addressed, apart from the obvious prediction that exchange rates should shift in favor of a trader class whose commodity becomes in short supply. The second problem has been totally ignored in cooperation theory until now. This has partly been due to historically determined blindness, but also because it may prove to be a more difficult issue to resolve. In sexual selection theory, this problem has not been ignored and several models have been proposed to explain, for example, the relationship between the operational sex ratio and parental investment. The same problem exists in cooperative and mutualistic interactions. Flowering plants, lycaenid butterfly larvae, and aphids have to decide how many units of sugar they offer to pollinators or ants. Similarly, helpers have to "decide" how much food they provide to the territory owners' offspring, and cleaners have to decide how often they cheat their clients.

Incomplete Control over Trading and a Shift from Mutualism to Parasitism

To explore the effects of partner choice, it is best to assume that each trading partner has complete control over the resource or commodity it offers. In reality, however, forceful exchange of goods or robbery may occur frequently. In particular, in situations where one partner cannot control whether an interaction takes place or when it ends, the other partner may be more likely to cheat. Based on cleaner fish – client data, Johnstone and Bshary (2002) developed a model with the following properties.

1. One class of traders in a potentially mutualistic interaction has the option to vary the degree to which it exploits its partner, assuming that exploitation yields a higher payoff than cooperation.
2. The potential victim has variable degree of control over the duration of an interaction.

It turned out that as long as the potential victim has sufficient control of the duration, interactions were mutualistic for both participants. With decreasing control and increasing temptation to defect, the outcome of interactions became more and more parasitic. The model yields three important implications: First, cooperation may be stable when only one partner has the option to cheat. Second, despite the temptation to cheat, cooperation may evolve in one-round interactions. Finally, two figurative screws, namely the temptation to cheat and the degree of control by the potential victim, can be turned to explain interactions shifting between mutualism and parasitism. As it stands, the model is very simple in its assumptions. Incorporating the possibility for the potential victim to switch partners would probably make the mutualism more stable. In general, we predict that combining the market effect with any other source of leverage may offer a powerful approach to explain transitions from mutualism to parasitism.

Statements and Open Questions

- Partner choice should be recognized as one of several sources of leverage cooperating organisms have over each other. Other such sources are the option to refuse interactions, to switch from cooperation to exploitation, the use of force, the possibility to steal commodities, and the effect on the "image score" of the partner.
- Partner choice options, influenced by the costs of choosing and by supply and demand are major predictors of payoff distributions among cooperators and mutualists. Future research must take the step from qualitative predictions in payoff shifts to quantitative predictions of "exchange rates."
- Participants in cooperation and mutualism typically have different options, which means that asymmetric games are the appropriate paradigms. Although these asymmetries do not affect market theory, it is clear that IPD models are useless to explain or predict the evolution or the strategies played in these systems. To understand the evolution of cooperation between unrelated individuals, we need asymmetric strategy sets.
- As it stands, the emphasis of future research should be on collecting quantitative data that allow the development of deductive models as a basis for new empirical research. A lack of empirical studies that confirm the predictions of available partner control models, most of which were developed unconstrained by facts, may reflect the lack of realism of these models.
- Restricted market games in which individuals may have incomplete control, either over the recourse they trade or over the course of interactions, are most likely to yield a framework that may help to understand under which circumstances symbioses may be commensalistic or parasitic rather than mutualistic.
- Cooperation and defection are often seen as a hallmark of Machiavellian intelligence, the idea that primates have their large neocortex to cope with a complex social environment (Byrne and Whiten 1988). Nowak and Sigmund (1998) proposed that indirect reciprocity based on image scoring may have been crucial for the evolution of human societies. As it stands, these phenomena may well occur in a wide variety of taxa. We therefore wonder how cognitive abilities or constraints influence game structures and how one could potentially distinguish "complex" cooperation from "brainless" cooperation.

ACKNOWLEDGMENTS

We thank the organizers for providing us this great opportunity to write a background paper. We also thank Samuel Bowles and other participants for valuable comments. R. Bshary is funded by a Marie Curie Fellowship of the European Union.

REFERENCES

Alexander, R.D. 1987. The Biology of Moral Systems. New York: Aldine de Gruyter.

Axelrod, R., and W.D. Hamilton. 1981. On the evolution of co-operation. *Science* **211**:1390–1396.

Bshary, R. 2001. The cleaner fish market. In: Economics in Nature, ed. R. Noë, J.A.R.A. van Hooff, and P. Hammerstein, pp. 146–172. Cambridge: Cambridge Univ. Press.

Bshary, R. 2002a. Biting cleaner fish use altruism to deceive image scoring clients. *Proc. Roy. Soc. Lond. B* **269**:2087–2093.

Bshary, R. 2002b. Building up relationships in asymmetric cooperative interaction between cleaner fish and client reef fish. *Behav. Ecol. Sociobiol.* **5**:365–371.

Bshary, R., and A.S. Grutter. 2002a. Asymmetric cheating opportunities and partner control in a cleaner fish mutualism. *Anim. Behav.* **63**:547–555.

Bshary, R., and A.S. Grutter. 2002b. Experimental evidence that partner choice is the driving force in the payoff distribution among cooperators or mutualists: The cleaner fish case. *Ecol. Lett.* **5**:130–136.

Bshary, R., and D. Schäffer. 2002. Choosy reef fish select cleaner fish that provide high service quality. *Anim. Behav.* **63**:557–564.

Bshary, R., and M. Würth. 2001. Cleaner fish *Labroides dimidiatus* manipulate client reef fish by providing tactile stimulation. *Proc. Roy. Soc. Lond. B* **268**:1495–1501.

Bull, J.J., and W.R. Rice. 1991. Distinguishing mechanisms for the evolution of cooperation. *J. Theor. Biol.* **149**:63–74.

Byrne, R.W., and A. Whiten, eds. 1988. Machiavellian Intelligence. Oxford: Clarendon Press.

Charnov, E.L. 1976. Optimal foraging, the marginal value theorem. *Theor. Popul. Biol.* **9**:129 136.

Clutton-Brock, T.H., and G.A. Parker. 1995. Punishment in animal societies. *Nature* **373**:209–215.

Côté, I.M. 2000. Evolution and ecology of cleaning symbioses in the sea. *Oceanogr. Mar. Biol. Ann. Rev.* **38**:311–355.

Dugatkin, L.A. 1997. Cooperation among Animals: An Evolutionary Perspective. Oxford: Oxford Univ. Press.

Ferriere, R., and R.E. Michod. 1995. Invading wave of cooperation in a spatially iterated prisoner's dilemma. *Proc. Roy. Soc. Lond. B* **259**:77–83.

Fisher, R.A. 1930. The Genetical Theory of Selection. Oxford: Clarendon.

Fleming, T.H., and J.N. Holland. 1998. The evolution of obligate pollination mutualisms: Senita cactus and senita moth. *Oecologia* **114**:368–375.

Greene, E., B.E. Lyon, V.R. Muehter et al. 2000. Disruptive sexual selection for plumage coloration in a passerine bird. *Nature* **407**:1000–1003.

Grutter, A.S. 1995. Relationship between cleaning rates and ectoparasite loads in coral reef fishes. *Mar. Ecol. Prog. Ser.* **118**:51–58.

Grutter, A.S. 1999. Cleaner fish really do clean. *Nature* **398**:672–673.

Henzi, S.P., and L. Barrett. 2002. Infants as a commodity in a baboon market. *Anim. Behav.* **63**:915–921.

Herre, E.A., and S.A. West. 1997. Conflict of interest in a mutualism: Documenting the elusive fig wasp–seed trade-off. *Proc. Roy. Soc. Lond. B* **264**:1501–1507.

Johnstone, R.A., and R. Bshary. 2002. From parasitism to mutalism: Partner control in asymmetric interactions. *Ecol. Lett.* **5**:634–639.

Leimar, O., and P. Hammerstein. 2001. Evolution of cooperation through indirect reciprocity. *Proc. Roy. Soc. Lond. B* **268**:745–753.

Lewis, R.J. 2002. Beyond dominance: The importance of leverage. *Qtly. Rev. Biol.* **77**:149–164.

Losey, G.C., A.S. Grutter, G. Rosenquist et al. 1999. Cleaning symbiosis: A review. In: Behaviour and Conservation of Littoral Fishes, ed. V.C. Almada, R.F. Oliveira, and E.J. Goncalves, pp. 379–395. Lisbon: Instituto Superior de Psichologia Aplicada.

Morton, E.S., L. Forman, and M. Braun. 1990. Extrapair fertilizations and the evolution of colonial breeding in purple martins. *Auk* **107**:275–283.

Noë, R. 1990. A veto game played by baboons: A challenge to the use of the Prisoner's Dilemma as a paradigm for reciprocity and cooperation. *Anim. Behav.* **39**:78–90.

Noë, R. 2001. Biological markets: Partner choice as the driving force behind the evolution of cooperation. In: Economics in Nature: Social Dilemmas, Mate Choice and Biological Markets, ed. R. Noë, J.A.R.A.M. van Hooff, and P. Hammerstein, pp. 92–118. Cambridge: Cambridge Univ. Press.

Noë, R., and P. Hammerstein. 1994. Biological markets: Supply and demand determine the effect of partner choice in cooperation, mutualism and mating. *Behav. Ecol. Sociobiol.* **35**:1–11.

Noë, R., and P. Hammerstein. 1995. Biological markets. *Trends Ecol. Evol.* **10**:336–339.

Noë, R., C.P. van Schaik, and J.A.R.A.M. van Hooff. 1991. The market effect: An explanation for pay-off asymmetries among collaborating animals. *Ethology* **87**:97–118.

Nowak, M.A., and K. Sigmund. 1998. Evolution of indirect reciprocity by image scoring. *Nature* **393**:573–577.

Pellmyr, O., and C.J. Huth. 1994. Evolutionary stability of mutualism between yuccas and yucca moths. *Nature* **372**:257–260.

Roberts, G., and T.N. Sherratt. 1998. Development of cooperative relationships through increasing investment. *Nature* **394**:175–179.

Schwartz, M.W., and J.D. Hoeksema. 1998. Specialization and resource trade: Biological markets as a model of mutualisms. *Ecology* **79**:1029–1038.

Simms, E.L., and D.L. Taylor. 2002. Partner choice in nitrogen-fixation mutualisms of legumes and rhizobia. *Integ. Comp. Biol.* (formerly: *Am. Zool.*) **42**:369–380.

Tebbich, S., R. Bshary, and A.S. Grutter. 2002. Cleaner fish, *Labroides dimidiatus*, recognise familiar clients. *Anim. Cogn.* **5**:139–145.

Trivers, R.L. 1971. The evolution of reciprocal altruism. *Qtly. Rev. Biol.* **46**:35–57.

Wedekind, C., and M. Milinski. 2000. Cooperation through image scoring in humans. *Science* **288**:850–852.

West, S.A., E.T. Kiers, E.L. Simms, and R.F. Denison. 2002. Sanctions and mutualism stability: Why do rhizobia fix nitrogen? *Proc. Roy. Soc. Lond. B* **269**:685–694.

Wilkinson, D.M. 2001. Mycorrhizal evolution. *Trends. Ecol. Evol.* **16**:64–65.

Zahavi, A. 1995. Altruism as a handicap: The limitations of kin selection and reciprocity. *J. Avian Biol.* **26**:1–3.

10

The Scope for Exploitation within Mutualistic Interactions

Judith L. Bronstein

Department of Ecology and Evolutionary Biology, University of Arizona,
Tucson, AZ 85721, U.S.A.

ABSTRACT

Cooperative interactions between species (mutualisms) are exploited by individuals and species that obtain benefits directed at mutualists, but that provide nothing in return. The persistence of mutualism in the face of exploitation has been considered an evolutionary problem akin to the persistence of cooperation in intraspecific interactions; as in the latter case, it is generally assumed that mutualism cannot persist over evolutionary time unless exploitation is kept under fairly tight control. A review of empirical studies indicates, however, that exploitation of mutualism is in fact a diverse suite of phenomena differing both in evolutionary origins and evolutionary significance. A few forms of exploitation pose a real evolutionary threat to mutualism, but many others do not. Furthermore, mechanisms commonly believed to have evolved to police exploitation may in fact have a range of evolutionary origins. For example, selective delivery of rewards to the best partners may function first and foremost as a resource conservation strategy and only secondarily as a means to punish exploiters. A pluralistic approach is clearly required to understand how mutualism functions in the presence of exploiters. On the empirical side, it is necessary to move beyond inferences relevant to a few, rather atypical mutualist–exploiter interactions, and toward experimental studies of carefully chosen model systems. On the theoretical side, new approaches are needed to reflect the reality of who exploiters are and the impacts they can inflict upon mutualisms.

INTRODUCTION

Humans seem to be instinctively drawn to examples of cooperative behaviors in other species. For example, Herodotus discussed how plovers remove leeches from crocodiles' mouths ("The crocodile enjoys this, and never, in consequence, hurts the bird"). Aristotle, Cicero, Pliny, and others repeated this story and added others, drawing moral lessons that showed the importance of "friendships" in maintaining nature's balance (Boucher 1985). This idyllic view of harmonious coexistence among cooperating species persisted well into the twentieth century. With a few notable exceptions, a fact rather obvious to those interested in within-species cooperation (at least among humans) escaped attention for many years: many individuals within supposedly cooperating species fail to cooperate some, or even all, of the time. Furthermore, most interspecific cooperative interactions are exploited by other species that are able to purloin and profit from the benefits that cooperators exchange (Bronstein 2001b).

A substantial body of theory has accumulated around predicting the conditions that favor persistence of cooperation *within* species, but the question of how cooperation *between* species (mutualism) can persist in the face of exploitation has been relatively ignored (cf. Bull and Rice 1991; Noë and Hammerstein 1994; Yu 2001). In general, these have been assumed to be kindred evolutionary problems: in both cases, it is assumed that exploitation, if unchecked, has the ability to extinguish cooperation.

This view of exploitation of mutualisms has developed in a relative void of accurate information on the nature of mutualism, the forms of exploitation to which it is subject, the mechanisms that might function to limit such exploitation, and evidence that such mechanisms actually exist in nature. My goal in this chapter is to provide a thorough and up-to-date empirical review of these phenomena. Hopefully, this will provide a useful background for further theoretical discussions of old issues but will also challenge theoreticians with some new issues. I develop the idea that exploitation of mutualism is far from a unitary phenomenon, although theorists have at times been tempted to treat it that way. Hence, somewhat different suites of ecological and evolutionary concepts will be most applicable for explaining the origins, impacts, and potential responses to different kinds of exploitation. Most of these concepts remain to be developed by theoreticians or examined by empiricists in the field.

MUTUALISM, SYMBIOSIS, AND COOPERATION

Diverse terms are currently in use to define interspecific cooperative interactions. Some of these terms have well-accepted alternative meanings, however. Given such inconsistencies, the terminology we choose to employ is more than strictly a semantic matter: it can determine whether we are all trying to explain the same phenomena. Here, I comment specifically on the terms *mutualism*, *symbiosis*, and *cooperation*.

I use the term *mutualism* to refer to all mutually beneficial, interspecific interactions, regardless of their specificity, intimacy, or evolutionary history. The term was first used in a biological context by Pierre van Beneden, a Belgian zoologist, in 1873 ("There is mutual aid in many species, with services being repaid with good behaviour or in kind, and *mutualism* can well take its place beside *commensalism*"). Albert Bernhard Frank and Anton de Bary independently coined the term *symbiosis* a few years later in an attempt to group physiologically intimate interactions independent of their parasitic, commensal, or mutualistic outcome. A tendency began soon thereafter to use the terms symbiosis and mutualism interchangeably. Confusion has continued to the present day. Here, I retain the original usage of these terms (Boucher 1985): some, but not all, mutualisms are symbiotic and some, but not all, symbioses are mutualistic.

Whereas mutualism denotes two-species beneficial interactions, *cooperation* has usually been used somewhat more vaguely to denote benefits in a within-species context. Some researchers have used the terms mutualism and cooperation interchangeably, however, probably following the lead of Axelrod and Hamilton (1981) in their seminal work on the evolution of cooperation. Cooperation has also been used to refer to the subset of mutualisms that are not obligate or, more casually, to denote the attribute of mutualism that contrasts with "conflict" (e.g., Herre et al. 1999). Adding further confusion, various subsets of interactions *within* species have been referred to as mutualisms. In particular, intraspecific interactions in which cooperation is an incidental result of individually selfish behaviors have been dubbed "by-product mutualisms." We have recently come full circle, however: that particular term has recently been co-opted back into the study of mutualism (e.g., Leimar and Connor, this volume).

This working group focused on *mutualism*. In contrast, the Dahlem Workshop as a whole examined a much broader set of phenomena (including those that take place within mutualism) that we can think of as *cooperation.*

THE EXPLOITATION OF MUTUALISM

Mutualisms have been considered to be biological markets in which species offer their partners commodities that are relatively inexpensive for them to produce, in exchange for commodities that are more expensive or impossible for them to produce (Noë and Hammerstein 1994; Schwartz and Hoeksema 1998). The existence of such commodities provides a powerful enticement for exploitation by individuals that can obtain them while providing nothing in return.

A Definition of Exploitation

An exploiter of a mutualism is an individual that obtains a benefit offered to mutualists but that does not reciprocate (Bronstein 2001b).

Ever since exploitative phenomena were first recognized within mutualism, there have been persistent difficulties in finding the appropriate words to describe them. Exploitation itself is a less than ideal term because its usage can be taken to imply that the mutualists themselves are *not* fundamentally exploitative. In fact they are: the net effect of mutualism to each participant is highest when it is able to maximize the benefit it receives from its partner at the least possible cost. However, exploitation of mutualists by mutualists is at least to some degree reciprocal, since it generates the potential for a net return of benefits to both partners. Exploitation by nonmutualists, in contrast, is unilateral.

Many of the terms used in place of exploitation, including defection, robbery, thievery, slavery, and (most commonly) cheating, are problematic as well. All of these terms hold implications of intent and subterfuge inappropriate outside of the human context. Furthermore, by using such language, we risk being drawn into assumptions about organisms' abilities to recognize and to respond behaviorally to these phenomena. Exploitation of mutualisms has also commonly been termed "parasitism" (e.g., Yu 2001), but this usage sets up as-yet untested expectations that parasites of interactions function, ecologically and evolutionarily, like parasites of individuals. In any case, it is arguably misleading to use a single term for a group of phenomena that, as I will argue below, have distinctly different evolutionary origins and implications.

Three related phenomena have not typically been considered to be exploitation of mutualism per se. Predators of mutualists are not usually treated as exploiters, even though some exploiters do kill mutualists in the process of obtaining commodities from them. Nor are species that feed on structures used to advertise rewards to mutualists (e.g., flowers) usually considered exploiters, even though they certainly can disrupt and lower the benefits of mutualism. Finally, species that are in fact mutualists, but relatively ineffective or inefficient ones, are not usually treated as exploiters. It should be recognized, however, that a fine line exists between all of these antagonistic phenomena and certain forms of exploitation discussed below. Their ecological and evolutionary parallels will be worth exploring in the future.

Exploiters Are Diverse and Ubiquitous

Empirical examples of exploitation are commonly cited in the theoretical literature on mutualism, but the same few interactions seem to be invoked each time (see sections below on CASE STUDIES OF EXPLOITATION and Model Systems for Studying Exploitation). Here, I provide a broader survey of the ecological distribution of exploitation, to point out both the ubiquity and the diversity of exploitative phenomena in nature (see also Bronstein 2001b).

Exploitation can be found within all of the commonly recognized classes of mutualism. In *transportation mutualisms,* the benefit to one partner is movement, either of itself or its gametes, to a location more favorable for growth or reproduction. The best-known examples are biotic pollination and seed dispersal. The species receiving the service of transportation provides some reward in exchange, usually food. Exploitation on the part of both partners has been abundantly documented in these mutualisms. Organisms requiring transport may advertise rewards but deliver none. Certain plants, for instance, trick pollinators into visiting nectarless flowers that closely mimic those of nectar-rich species (Little 1983). The other partner in transportation interactions may be exploitative as well, by collecting rewards but not transporting or even destroying the associate or its gametes (e.g., Maloof and Inouye 2000).

In *protection mutualisms*, one partner protects its associate from negative influences of its biotic or abiotic environment, in exchange for a food reward or for protection against its own antagonists or the physical environment. Exploiters exist on both sides of these interactions. Cleaners, for instance, feed on hosts' ectoparasites but most also regularly feed on host tissue itself, sometimes inflicting great damage to their hosts. Hosts can also cheat, by eating cleaners (Bshary and Noë, this volume). Certain ants collect rewards produced by their plant or insect partners, then either fail to attack their partner's enemies or consume the partner itself (e.g., Yu and Pierce 1998). Batesian mimics gain the benefit of protection enjoyed by Müllerian mutualists without paying the price of producing distasteful compounds (Gilbert 1983), functioning as exploiters in these systems.

Nutrition mutualisms are those in which one species obtains one or more essential nutrients from its partner. Many interactions mentioned above are nutrition mutualisms from the perspective of one partner (e.g., pollination and seed dispersal from the perspective of animals, protection from the perspective of the defender, and cleaning from the perspective of the cleaner). As noted, all of these mutualisms are subject to exploitation. Exploitation has also been identified in nutritional symbioses. For example, certain mycorrhizal fungi take plant carbon but transfer few nutrients back to their hosts (Smith and Smith 1996). Some strains of *Rhizobium* bacteria either do not benefit or actively harm their leguminous hosts (Denison 2000). Lichens are attacked by diverse parasitic fungi that "enslave" the algae and confer no benefit in return (Richardson 1999).

Exploiters can be found taking advantage of both the rewards and services exchanged by mutualists. The examples of exploitation mentioned above break down into four classes: cases in which rewards are obtained but not provided (e.g., parasitic cleaner fish), services are obtained but not provided (Batesian mimics), rewards are obtained but services are not provided (nectar robbers), and services are obtained but rewards are not provided (nectarless floral mimics). Furthermore, exploiters are found in both specialized and generalized mutualisms. For example, plants are exploited by nonpollinating flower visitors whether they are associated with a wide array of mutualistic pollinators (Maloof and Inouye 2000) or with a single one, as in the case of yuccas (see below section on Yuccas and Yucca Moths).

Three Forms of Exploitation

Exploiters can be divided into three distinct types: exploiter species, individuals within mutualistic species that act purely as exploiters, and individuals that switch between mutualistic and exploitative strategies (Bronstein 2001b). The distinction is an important one: I will subsequently argue that these modes of exploitation have different evolutionary origins and consequences, and that different forms of control will be effective against them.

Exploitation as a Pure Species-level Behavior

Probably the most ubiquitous exploiters are species in which all individuals exhibit traits that permit them to obtain the benefits of mutualisms but never to deliver benefits in return. Exploiter species exhibit a wide array of evolutionary origins and degrees of specialization. Many belong to lineages having no evolutionary history of participation in the mutualism being exploited; they are likely to have become associated with the mutualism long after its establishment. These species are often fairly generalized foragers able to make use of relatively undefended rewards intended for mutualists. Since exploiters like these are often only transiently associated with mutualisms, they may be of little interest in the context of how mutualisms arise and persist over evolutionary time. However, mechanisms to control them may nevertheless be strongly favored by natural selection. For instance, ants lacking any evolutionary history of pollination behaviors commonly steal floral nectar. Their activities in flowers can lead both to pollen loss and reduced seed set. Diverse traits have been recognized that help plants successfully deter floral visits from ants (e.g., Ghazoul 2001).

Other exploiter species belong to clades in which mutualism is extensively represented. Species like these are relatively specialized for exploitation of one or a very few mutualisms. In some cases, novel traits apparently have arisen and/or traits involved in performing mutualistic actions have been lost. For example, many floral mimics have converged on complex visual cues offered by distantly related model species, while at the same time losing the ability to produce nectar (Little 1983).

Exploitation as a Pure Behavior within Mutualistic Species

Exploitation can alternatively be a behavior exhibited by certain individuals within species that also include individuals that act mutualistically. This phenomenon has received very little study to date; it requires observing individuals over substantial periods, ideally their entire lifetimes. Automimicry, in which some individuals in a population mimic others that provide a mutualistic reward or service, falls into this category. For example, in some plant species, certain individuals fail to produce nectar but attract sufficient "mistake" visits to reproduce successfully (e.g., Golubov et al. 1999).

Exploitation and Mutualism as Alternative Behaviors of the Same Individual

Finally, exploitation can be a conditional behavior exhibited by an individual that also acts mutualistically. Examples include nectar-robbing insects that also pollinate flowers of the same plant, most cleaners (which inflict occasional or regular damage to host tissue), and most ant defenders (which occasionally attack their mutualists rather than their mutualists' enemies).

Conditional exploitation appears common in cases where very slight differences in a partner's behavior or in the ecological context itself is sufficient to shift an interaction from reciprocal exploitation (i.e., mutualism) to unilateral exploitation. For example, feeding behaviors of cleaner fish may only benefit their hosts when hosts' parasite loads are exceptionally high. Other conditional exploiters are clearly making choices between acting mutualistically or nonmutualistically towards their partners, possibly in response to their partners' behaviors. It is specifically for this class of exploitation that the Prisoner's Dilemma (Axelrod and Hamilton 1981) may serve as a useful model.

THE CONTROL OF EXPLOITATION

It is widely assumed that the presence of exploiters within mutualisms necessitates a response, if the mutualism is to persist over evolutionary time. Later in this chapter I will question the logic underlying this assumption; here, as background, I review four classes of control mechanisms that might exist. Several others have been proposed recently by Yu (2001).

Mechanisms That Promote Partner Fidelity

When two individuals (or lineages) live in intimate association and interact repeatedly during their lifetimes, beneficial or harmful actions of one partner feed back directly on its own success. In this way, exploitation can cease to be in the individual interests of the exploiter (Herre et al. 1999). Such *partner fidelity* (Bull and Rice 1991) has been investigated in particular depth as an explanation for how slowly reproducing, vertically transmitted symbionts can persist evolutionarily in the presence of lineages exhibiting higher reproductive rates and higher virulence. Frank (1996) has argued that the maintenance of single strains of vertically transmitted symbionts favors cooperation and thus primarily benefits hosts, whereas escape from vertical transmission and the mixing of lineages are in the interests of the symbionts themselves. This conflict of interest lends some credence to the idea that host mechanisms that reduce the mixing of symbiont lineages may have had their evolutionary origins in exploiter control.

Mechanisms That Permit Partner Choice

Partner fidelity mechanisms cannot explain how mutualism can persist when symbionts are transmitted across generations horizontally rather than vertically. In these cases, multiple strains of symbionts differing in cooperativeness may in fact coexist within a host; this is the case, for instance, in legume–*Rhizobium* interactions (Denison 2000). Nor are these mechanisms useful in most nonsymbiotic relationships, in which individuals may interact only once and thus do not experience negative feedback from their own failure to cooperate. It has been widely argued that mutualisms like these can only persist if there are mechanisms for *partner choice*, wherein individuals can assess the quality of potential partners, then choose with whom to associate and/or how long the association will last (Bull and Rice 1991; Yu 2001). Cases in which better mutualists appear to be differentially rewarded have been interpreted as evidence for traits fostering partner choice, as have interactions in which exploiters and the worst mutualists are differentially punished (e.g., Denison 2000); these two mechanisms are somewhat different, although they are commonly treated as a single phenomenon. Cases in which reward investment increases under ecological conditions in which mutualists are especially beneficial (e.g., Leimar and Axén 1993) have also been treated as evidence for partner-choice adaptations.

Mechanisms That Exclude Exploiters

To date, field research has revealed limited evidence for partner-choice mechanisms (Bronstein 2001b). Among empiricists, a more widely accepted explanation is that mutualists have evolved traits that effectively keep commodities away from exploiters altogether. For instance, in certain plant species, floral corollas are very long, very thick, and/or tightly clustered, traits that supposedly prevent nonpollinators from gaining access to nectar. Although many of these traits may function in this context, in many cases we can identify common exploiters able to circumvent putative defenses against them. It is possible, in fact, that coevolutionary races of adaptation and counteradaptation may be taking place between mutualists and exploiters. Such a process has recently been postulated within an interaction between a virulent fungal pathogen and a three-way mutualism involving leafcutter ants, their mutualistic fungi, and a bacterium that suppresses pathogen growth (Currie 2001).

Mechanisms That Increase Tolerance to Exploitation

Finally, it is possible that mutualists have evolved mechanisms not to exclude or to punish exploiters, but rather to tolerate their effects. That is, natural selection may have acted to reduce the cost, rather than the frequency, of exploitation. Selection favoring plant tolerance to herbivory rather than exclusion of herbivores has received extensive attention in recent years (Stowe et al. 2000). A parallel mechanism may explain why exploiters seem ubiquitous in mutualisms but inflict surprisingly low fitness costs (see section below on How Costly are Exploiters?).

CASE STUDIES OF EXPLOITATION

In this section, I briefly summarize two of the most thoroughly studied and heavily cited examples of exploitation within mutualism. These case studies highlight the diversity of exploitative phenomena within even a single form of mutualism, as well as the range of mechanisms that might keep different forms of exploitation in check. A third case study, on interactions between cleaner fish and their hosts, is provided in this volume by Bshary and Noë.

Yuccas and Yucca Moths

All of the 35–50 species in the genus *Yucca* (Agavaceae) are obligately pollinated by yucca moths (*Tegeticula* and *Parategeticula*, Prodoxidae). Female moths, newly emerged from the soil where they overwinter, copulate and then actively collect pollen from a fresh flower. They then fly to another inflorescence, lay eggs directly into the pistil, then actively transfer pollen to the receptive stigma of the same flower. As the seeds begin developing, yucca moth larvae emerge and begin consuming some of them. Hence, these mutualists are, fundamentally, plant antagonists: they pollinate to provision their young, who proceed to destroy many of the seeds initiated by their mothers' actions. The interaction is mutualistic only to the extent that any intact seeds are left when the larvae have finished feeding and have left the fruit. Most yuccas are also associated with nonpollinating yucca moth species closely related to the pollinators, commonly called "cheaters." Cheaters have retained larval seed consumption behaviors, but adults have lost the modified mouthparts used to collect pollen, as well as the behaviors associated with the act of pollination (Pellmyr and Leebens-Mack 2000). Thus, yuccas must cope with two kinds of seed predators: ones that provide benefits of pollen delivery and removal, and ones that do not.

About 5–20% of all seeds that yuccas initiate are consumed by pollinator and cheater larvae (Bronstein 2001a). In light of this substantial cost, it has been expected that yuccas should be under strong selection to reduce seed consumption (while at the same time retaining seed initiation). They do in fact have traits that function in this context. The best known of these traits is selective fruit abortion: newly initiated fruits that bear particularly high egg loads are aborted, apparently selecting for pollinator females that lay few eggs per flower (or, as more commonly described in the literature, punishing females that overexploit their plants). This mechanism has become the most heavily cited sanction against exploiters within a mutualism.

In fact, the problem of exploitation in yuccas and how it can be policed is considerably more complicated (and more interesting) than commonly believed:

- All of the closest relatives of yuccas abort insect-damaged fruits, although in all of them except yuccas, the insects in question are not the offspring of mutualists. Thus, the yucca mutualism appears to have evolved against a background of selective abortion. Abortion may have been secondarily co-opted as a sanction against exploiters, but it was not selected for in this context.
- Although the selective abortion mechanism is often treated as common to all yuccas, it is not. In fact, alternative mechanisms exist

that also control exploitation by pollinators. *Yucca baccata*, for example, appears to kill eggs and young larvae within fruits, then to retain those fruits to maturity (Addicott and Bao 1999).

- Selective fruit abortion (where it exists) may effectively limit exploitation by the pollinators, but it cannot control the damaging cheater species. Cheaters sidestep this sanction by emerging after the brief abortion period has concluded, then laying their eggs into fruits rather than into flowers. In fact, it is likely that cheaters have evolved (or retained) a late phenology as a means of avoiding this sanction. Cheaters deposit far more eggs per fruit than pollinators and have a substantial impact on seed production. How, if at all, do yuccas limit their effects?
- Other forms of exploitation also exist in this mutualism. For example, conditional exploitation is well documented among pollinator yucca moths: some individuals apparently skip the time-consuming pollen-deposition phase when ovipositing into a yucca flower that has previously been pollinated by another individual (Addicott and Tyre 1995). Again, how (if at all) do yuccas control these behaviors?
- Finally, the existence of cheater moths is sometimes used as evidence that mutualistic species have the capacity to evolve into pure exploiters of their partner species. Cheater yucca moths are in fact derived from mutualists. However, it currently appears that cheater and mutualist clades radiated independently, relatively early in the evolution of the association with yuccas. The mutualist and cheater species associated with an individual host plant species are therefore not generally each other's sister species but represent different radiations that independently colonized or co-speciated with that host (Pellmyr and Leebens-Mack 2000). The more distant evolutionary origins of now-speciose cheater clades might well have been linked to defection from mutualism, but individual cheater species associated with individual yucca species do not themselves represent lapsed mutualists.

Thus, there are multiple conflicts and modes of conflict resolution in the yucca–yucca moth mutualism. Current understanding of the evolutionary origins of both exploitation and sanctions against it in this system do not fit easily into simple scenarios invoking defection and policing.

Lycaenids and Ants

Most of the 6000 species in the Lepidopteran family Lycaenidae (which makes up about one-third of all butterfly species) associate with ants, usually during the larval stage, in interactions that range from parasitic to mutualistic (Pierce et al. 2002). The mutualistic relationship is based on an exchange of protection for nutrients. Ants collect droplets of nutrients that lycaenid larvae secrete from a specialized gland; ants protect their food source, and thus the larvae themselves from their natural enemies. Secretions are expensive for lycaenids to produce, at least in the subset of species obligately associated with ants: larvae that produce more secretions grow more slowly and/or to a smaller size (Pierce et al. 2002). Not surprisingly, lycaenids exhibit behavioral strategies to reduce these costs without sacrificing the protection that they require, permitting interpretations of this mutualism in terms of laws of supply and demand (Noë 2001). For example, in different lycaenid species, the number of droplets secreted *per capita* has been shown to be higher when the interaction with ants is first established, when more ants are present, when lycaenids perceive a higher risk of enemy attack, and when lycaenids occur singly rather than in aggregations (Leimar and Axén 1993; Axén et al. 1996; Axén and Pierce 1998).

These strategies have also been discussed in a second context, social control of exploitation (Leimar and Axén 1993; Yu 2001). By this argument, if investment into cooperation can be varied gradually, then an individual that responds to variation in its partner's contribution not only can minimize the cost of its own investment in cooperation, but can also avoid being exploited by its partner. Without a means to control the degree to which the partner is rewarded, it is argued, the partner would be expected to provide nothing at all. Strategic regulation of rewards by lycaenids might thus have evolved as a means to control exploitation. Alternatively, Noë (2001) has argued that exploiter control might be a by-product of a resource conservation strategy. In other words, lycaenids may have evolved to dole out energetically expensive rewards only when absolutely necessary. This might secondarily serve to reward better mutualists differentially.

As in the case of yucca–yucca moth mutualisms, the natural history and potential consequences of exploitation in lycaenid–ant mutualisms are actually fairly diverse and complex. Among the many twists in the story that bear further investigation are these:

- Costs of secretions appear to be negligible among lycaenid species that have relatively loose, generalized interactions with ants; these represent by far the majority of ant-tended lycaenids but are currently much more poorly studied. Presumably, the benefit of reward regulation in these species should not be based on resource conservation, even though that might be true in the case of the more ant-dependent lycaenids. Thus, further studies of strategic behaviors exhibited within facultative ant–lycaenid interactions would seem particularly valuable if we wish to interpret these behaviors in a context of exploiter control.
- Ant species that confer poorer protection may in some cases actually be rewarded *more* by lycaenids, not less (Axén 2000). Presumably, it takes more unaggressive than aggressive ants to achieve a given level of protection. The logic of such a reward strategy seems quite reasonable; it is important to recognize, however, that it represents the exact *opposite* of a sanctioning strategy (since the poorest partners are rewarded more,

not less). In what other kinds of mutualisms might this strategy be found?

- The efficacy of fine-tuned strategic responses by lycaenids may well be minimal at times and places where lycaenid secretions are not highly valued by ants. For example, ants may ignore lycaenids altogether at times of year when ants are protein- rather than carbohydrate-starved, and/or when superior carbohydrate sources are available (Weeks and Bronstein, submitted). More generally, a deeper understanding of the strategic features of these interactions will require consideration of the ant, not only the lycaenid, perspective.

CHALLENGES FOR UNDERSTANDING MUTUALIST–EXPLOITER INTERACTIONS

The persistence of mutualism in the face of exploitation has been considered an evolutionary problem akin to the persistence of cooperation in intraspecific interactions; as in the latter case, it has generally been assumed that mutualism cannot persist unless exploitation is kept under fairly tight control. In this section I reconsider the empirical data on exploitation with the aim of evaluating this assumption. I first examine how much of a cost exploiters actually inflict upon mutualisms. Then, I clarify the subset of exploiters that may really pose an evolutionary threat. I conclude by reconsidering the two model systems of exploitation discussed above. I will argue that they may in fact constitute special cases, cautioning against using them to represent "the norm" of mutualist–exploiter interactions.

How Costly Are Exploiters?

A common assumption underlying ideas about the evolutionary effects of exploitation is that individuals that do not reciprocate within mutualisms will inflict significant fitness costs on their partners. However, empirical research has detected a continuum of effects of exploiters. In particular, surprisingly few exploiters have yet been demonstrated to inflict a measurable fitness cost to the mutualists with whom they associate (Bronstein 2001b). For example, nonmutualistic ants associated with a few ant-plants "castrate" their plants by pruning buds and flowers, increasing vegetative plant growth but reducing sexual reproduction to near zero (Yu and Pierce 1998). The costs of many or even most kinds of exploitation are likely to range from fairly low to negligible, however. For instance, floral nectar is commonly consumed by organisms that confer no benefits to plants but that also inflict no measurable harm (Maloof and Inouye 2000). At least some of these rewards appear to be energetically cheap to produce (Bronstein 2001b). Losing a small proportion of a small energy investment is unlikely to be costly, especially since reward volumes rarely scale closely with the benefits obtained from mutualists. Similarly, the negative fitness effects of visiting a partner that advertises food rewards but delivers none (such as a nectarless floral mimic) are elusive. An unrewarded visit entails a net loss of energy as well as some missed opportunity costs; however, to my knowledge, no attempts have been made to quantify these effects.

There appears to be no pattern as to which kinds of exploiters, i.e., pure exploiter species, pure exploiter individuals within mutualist species, or conditional exploiters, are most costly to mutualisms. Elsewhere I have offered three predictions about the costliness of exploitation (Bronstein 2001b).

1. The cost of exploitation should increase with the amount invested into the stolen commodity and the ease with which that commodity can be replaced.
2. For mutualisms in which partner success increases with the amount of attention received from mutualists, the cost of exploitation should increase with the degree to which organisms are avoided by their mutualists after they have been exploited.
3. The costs and benefits of exploitation should vary with the ecological context of the interaction, much like the costs and benefits of mutualism itself. For example, where plant growth is water- or resource-limited, the costs of replacing robbed food rewards should be higher. Similarly, the cost to pollinators of making an unrewarded visit will depend on the degree to which they are limited energetically by rewards available elsewhere in the floral neighborhood. The cost of losing rewards to species that do not confer protective services is probably negligible when enemies are rare, but potentially very high when enemies are abundant.

Ultimately, the (unanswered) question is this: in cases where the cost of being exploited is low, nonexistent, or highly context-dependent, how essential is it to sanction exploiters if mutualism is to persist over evolutionary time?

What Kind of Threat Does Exploitation Pose?

The question I just posed was phrased to reflect the usual assumption about the nature of exploitation of mutualisms. Exploitation is seen as a temptation to mutualists that, if accepted and not policed, will threaten the very persistence of mutualism.

A review of the empirical literature suggests that the problem is a more multifaceted one, reflecting the fact that exploitation of mutualism itself is not a unitary phenomenon. First, consider exploiter species (see section on *Exploitation as a Pure Species-level Behavior*), probably the most common exploiters of mutualism. If they increase in numbers at the expense of mutualists, they may threaten the *ecological* persistence of the mutualism they exploit, but not its *evolutionary* persistence (in the sense that mutualist numbers may dwindle in their presence, but mutualists will still be mutualists). Furthermore, evidence is growing that their effect on ecological persistence of mutualism may not be very great either. New models strongly suggest that ecological coexistence of mutualists and exploiters is rather easy to obtain, especially if mutualists and exploiters are allowed to exhibit slightly different dispersal strategies (Yu et al. 2001; Morris et al. 2003; Wilson et al. 2003). In cases where specialized exploiters are particularly damaging, their effects on the mutualism will drive themselves to extinction as well. Models like these can help explain the ubiquity of exploiter species without any recourse whatsoever to evolutionary mechanisms of control.

Second, consider species in which purely exploitative and purely mutualistic individuals coexist (section on *Exploitation as a Pure Behavior within Mutualistic Species*). If exploitation threatens mutualism, why has the mutualistic strategy not been extinguished? Conversely, if sanctions against exploitation have evolved in these interactions, why does the exploitative strategy still persist? In such cases, it is possible that mild partner discrimination against exploiters is sufficient to reduce exploiter advantage but not to extinguish it completely. For example, half of all honey mesquite, *Prosopis glandulosa*, individuals produce no nectar. Current evidence suggests that this is a stable polymorphism. Nectarful individuals attract 21 times more pollinators, whereas nectarless individuals produce more pollen (possibly using resources saved on nectar), leading to roughly equal reproductive success between nectarful and nectarless morphs (Golubov et al. 1999). Theories of model-mimic coexistence should prove helpful for understanding the ecological and evolutionary dynamics of mutualist–exploiter interactions like these.

It is most clearly in interactions involving conditional exploiters (section on *Exploitation and Mutualism as Alternative Behaviors of the Same Individual*) in which an uncontrolled advantage to exploitation might conceivably eliminate mutualism altogether. These are the interactions to which both Prisoner's Dilemma models of cooperation (Axelrod and Hamilton 1981; Yu 2001) and biological markets models (Noë and Hammerstein 1994; Noë 2001; Bshary and Noë, this volume) most easily apply. Even in these cases, sanctions may not always be essential for mutualism to persist. It is widely assumed that for individuals that have a choice, exploitation is a superior strategy to cooperation. However, this is not *necessarily* the case. It is also possible that exploitation itself is a fairly ineffective strategy to which mutualists rarely resort, or a strategy that is only profitable under a very limited set of ecological conditions. If this is the case, then conditional exploiters might quickly revert back to mutualism when conditions change. This self-regulatory mechanism may explain, for instance, why many pollinators have the ability to pursue nectar-robbing strategies as well, but use pollination as their "default" strategy in the apparent absence of any plant sanctions enforcing this behavior (Bronstein, unpublished data). Clearly, empirical studies are needed that compare the costs and benefits of mutualistic versus exploitative strategies in conditional exploiter species.

Model Systems for Studying Exploitation

I mentioned earlier that theoretical treatments of mutualist–exploiter dynamics have invoked a very narrow set of such interactions. Are these well-studied examples broadly typical of exploitation, or are they actually special cases fairly far from "the norm"?

Consider first yucca–yucca moth interactions (see above). These are highly atypical pollination mutualisms, from the perspective both of exploitative options available to partners and of responses that might control exploitation. First, the reward that yucca-associated moths exploit (developing seeds, i.e., a portion of the partner's reproductive output) is far more costly to their partners than are the vast majority of mutualistic rewards. Second, these interactions are obligate and frequently species specific, characteristics that are highly unusual among plant–pollinator relationships. Extreme specialization means that any traits disfavoring exploitative genotypes will likely affect the genetic composition of future generations; in effect, exploiters have nowhere to run. In contrast, in more generalized interactions, disfavored exploiters can and probably do switch to more tolerant host species. Finally, in the yucca system, sanctions against exploitative behaviors rely crucially on a time delay: it is not exploiters themselves (i.e., females that lay "too many" eggs) that are punished, but their offspring instead. In the large majority of pollination mutualisms, the exploiter itself obtains an immediate benefit (such as a food reward) and then departs. In these cases, any sanction would have to be imposed directly on the exploiter. However, rewards such as nectar must be invested in and made available before the interaction takes place, making such a sanction virtually impossible.

If there is any system that would be expected to exhibit the same kinds of exploiter control as exhibited by yuccas, it should be the fig–fig-wasp pollination mutualism. Like the yucca–yucca moth mutualism, the interaction between figs and fig wasps is obligate, usually species specific, involves seed consumption by offspring of the pollinators, and is exploited by diverse specialized taxa related to the pollinator clade (Bronstein 2001a). Yet, it has proven very difficult to find any sanctions by which figs control potential overexploitation by their associates: selective fruit abortion, in particular, is entirely absent. (The fundamental difference between yuccas and figs in this regard seems related to a cost of exploiter control experienced by figs but not yuccas. Seed-consuming fig wasps disperse the pollen of their natal fig when they have finished feeding, so that punishment of exploiters would inevitably lower reproductive success of the same individual plant [Bronstein 2001a].) If yucca control of exploiters is not helpful in predicting sanctions in the ecologically very similar fig mutualism, it is not surprising that it is a very poor model for more generalized pollination mutualisms based on substantially different systems of reward.

Are lycaenid–ant interactions typical protection mutualisms, particularly with regard to the ability to police exploitation? Certainly, they are more typical of protection mutualisms than the yucca–moth interaction is typical of pollination mutualisms. Most animal–animal protection interactions would appear to offer scope for assessment of partners' willingness to invest in relationships, and for responses to those assessments. Such strategies are in fact currently being studied in some of these associations, including cleaning symbioses (Bshary and Noë, this volume). On the other hand, possibilities for immediate assessment and response to exploitation are undoubtedly more constrained in the ecologically similar protection mutualisms between animals and plants. The rewards that plants produce for ants that defend them against herbivores cannot be selectively doled out or withdrawn based on the quality of protection plants are receiving at a given moment. Furthermore, many of these rewards (although not all of them) are extremely cheap for plants to produce. In these two senses, plants' rewards for defensive ants resemble plants' rewards for pollinators more than they resemble insects' rewards for defensive ants. This does not mean that plants lack the means to conserve rewards for protectors. For example, rewards may only be invested in once plants have established long-term contact with a mutualist population (Letourneau 1990), or the quantity and/or quality of rewards may increase upon attack by herbivores, leading to the recruitment of more defenders (Heil et al. 2001). It is an open question whether reward conservation strategies like these also serve to reward differentially the best mutualists or to punish exploiters.

Thus, both of these frequently cited model systems, as well as the one discussed by Bshary and Noë in this volume, share two features. First, the rewards provided to mutualists are quite costly relative to other, ecologically similar mutualisms; thus, exploitation is costly too. Second, control of investment in mutualism takes place more or less simultaneously with the act of exploitation; this paves the way for responses to exploitation that impact the exploiter individual (or her progeny) directly. A more pluralistic view of exploiter control ultimately needs to consider the vast array of mutualist–exploiter interactions that do not exhibit these features.

In this volume, Bergstrom et al. have issued a call for the identification and development of experimental model systems for studying mutualism. The review I have provided here has hopefully clarified that exploitative phenomena are central features of virtually any mutualism of interest, including most of the potential model systems highlighted by Bergstrom et al. Thus, the model-systems approach has great potential for shedding light on the most general features of exploitation in mutualism. There is, however, another advantage to this approach. Studying exploitation within mutualisms that are simultaneously being studied for other features, including patterns of specificity, evolutionary history, and population regulation, is likely to reveal new and fundamental properties of these fascinating interactions that have eluded biologists' attention until now.

ACKNOWLEDGMENTS

I thank all the participants of this Dahlem Workshop, especially the members of Group 2, for helpful feedback and stimulating discussion. Laurent Keller and Olle Leimar provided especially useful comments for revising the manuscript. Some of the work reported here was funded by a National Science Foundation grant (DEB 99-73521).

REFERENCES

Addicott, J.F., and T. Bao. 1999. Limiting the costs of mutualism: Multiple modes of interaction between yuccas and yucca moths. *Proc. Roy. Soc. Lond. B* **266**:197–202.

Addicott, J.F., and A.J. Tyre. 1995. Cheating in an obligate mutualism: How often do yucca moths benefit yuccas? *Oikos* **72**:382–394.

Axelrod, R., and W.D. Hamilton. 1981. The evolution of cooperation. *Science* **211**:1390–1396.

Axén, A.H. 2000. Variation in behavior of lycaenid larvae when attended by different ant species. *Evol. Ecol.* **14**:611–625.

Axén, A.H., O. Leimar, and V. Hoffman. 1996. Signalling in a mutualistic interaction. *Anim. Behav.* **52**:321–333.

Axén, A.H., and N.E. Pierce. 1998. Aggregation as a cost-reducing strategy for lycaenid larvae. *Behav. Ecol.* **9**:109–115.

Boucher, D.H. 1985. The idea of mutualism, past and future. In: The Biology of Mutualism, ed. D.H. Boucher, pp. 1–28. New York: Oxford Univ. Press.

Bronstein, J.L. 2001a. The costs of mutualism. *Am. Zool.* **41**:127–141.

Bronstein, J.L. 2001b. The exploitation of mutualisms. *Ecol. Lett.* **4**:277–287.

Bull, J.J., and W.R. Rice. 1991. Distinguishing mechanisms for the evolution of cooperation. *J. Theor. Biol.* **149**:63–74.

Currie, C.R. 2001. A community of ants, fungi, and bacteria: A multilateral approach to studying symbiosis. *Ann. Rev. Microbiol.* **55**:357–380.

Denison, R.F. 2000. Legume sanctions and the evolution of symbiotic cooperation by rhizobia. *Am. Nat.* **156**:567–576.

Frank, S.A. 1996. Host-symbiont conflict over the mixing of symbiotic lineages. *Proc. Roy. Soc. Lond. B* **263**:339–344.

Ghazoul, J. 2001. Can floral repellents pre-empt potential ant-plant conflicts? *Ecol. Lett.* **4**:295–299.

Gilbert, L.E. 1983. Coevolution and mimicry. In: Coevolution, ed. D. Futuyma and M. Slatkin, pp. 263–281. Sunderland, MA: Sinauer.

Golubov, J., L.E. Eguiarte, M.C. Mandujano et al. 1999. Why be a honeyless honey mesquite? Reproduction and mating system of nectarful and nectarless individuals. *Am. J. Bot.* **86**:955–963.

Heil, M., T. Koch, A. Hilpert et al. 2001. Extrafloral nectar production of the ant-associated plant, *Macaranga tanarius*, is an induced, indirect, defensive response elicited by jasmonic acid. *Proc. Natl. Acad. Sci. USA* **98**:1083–1088.

Herre, E.A., N. Knowlton, U.G. Mueller, and S.A. Rehner. 1999. The evolution of mutualisms: Exploring the paths between conflict and cooperation. *Trends Ecol. Evol.* **14**:49–53.

Leimar, O., and A.H. Axén. 1993. Strategic behavior in an interspecific mutualism: Interactions between lycaenid larvae and ants. *Anim. Behav.* **46**:1177–1182.

Letourneau, D.K. 1990. Code of ant-plant mutualism broken by parasite. *Science* **248**:215–217.

Little, R.J. 1983. A review of floral food deception mimicries with comments on floral mutualism. In: Handbook of Experimental Pollination Biology, ed. C.E. Jones and R.J. Little, pp. 294–309. New York: Van Nostrand Reinhold.

Maloof, J.E., and D.W. Inouye. 2000. Are nectar robbers cheaters or mutualists? *Ecology* **81**:2651–2661.

Morris, W.F., J.L. Bronstein, and W.G. Wilson. 2003. Three-way coexistence in obligate mutualist-exploiter interactions: The potential role of competition. *Am. Nat.,* in press.

Noë, R. 2001. Biological markets: Partner choice as the driving force behind the evolution of mutualisms. In: Economics in Nature, ed. R. Noë, J.A.R.A.M. van Hooff, and P. Hammerstein, pp. 93–118. Cambridge: Cambridge Univ. Press.

Noë, R., and P. Hammerstein. 1994. Biological markets: Supply and demand determine the effect of partner choice in cooperation, mutualism and mating. *Behav. Ecol. Sociobiol.* **35**:1–11.

Pellmyr, O., and J. Leebens-Mack. 2000. Reversal of mutualism as a mechanism for adaptive radiation in yucca moths. *Am. Nat.* **156**:S62–S76.

Pierce, N.E., M.F. Braby, A. Heath et al. 2002. The ecology and evolution of ant association in the Lycaenidae (Lepidoptera). *Ann. Rev. Entomol.* **47**:733–771.

Richardson, D.H.S. 1999. War in the world of lichens: Parasitism and symbiosis as exemplified by lichens and lichenicolous fungi. *Mycol. Res.* **103**:641–650.

Schwartz, M.W., and J.D. Hoeksema. 1998. Specialization and resource trade: Biological markets as a model of mutualisms. *Ecology* **79**:1029–1038.

Smith, F.A., and S.E. Smith. 1996. Mutualism and parasitism: Diversity in function and structure in the "arbuscular" (VA) mycorrhizal symbiosis. *Adv. Bot. Res.* **22**:1–43.

Stowe, K.A., R.J. Marquis, C.G. Hochwender, and E.L. Simms. 2000. The evolutionary ecology of tolerance to consumer damage. *Ann. Rev. Ecol. Syst.* **31**:565–595.

Wilson, W.G., J.L. Bronstein, and W.F. Morris. 2003. Coexistence of mutualists and exploiters on spatial landscapes. *Ecology*, in press.

Yu, D.W. 2001. Parasites of mutualisms. *Biol. J. Linn. Soc.* **72**:529–546.

Yu, D.W., and N.E. Pierce. 1998. A castration parasite of an ant-plant mutualism. *Proc. Roy. Soc. Lond. B* **265**:375–382.

Yu, D.W., H.W. Wilson, and N.E. Pierce. 2001. An empirical model of species coexistence in a spatially structured environment. *Ecology* **82**:1761–1771.

11

By-product Benefits, Reciprocity, and Pseudoreciprocity in Mutualism

Olof Leimar[1] and Richard C. Connor[2]

[1]Department of Zoology, Stockholm University, 106 91 Stockholm, Sweden
[2]Department of Biology, University of Massachusetts at Dartmouth, North Dartmouth, MA 02747, U.S.A.

ABSTRACT

The concepts of by-product mutualism and reciprocal altruism have played important roles for theories of mutualistic interactions between unrelated organisms. By-product mutualism could explain the evolution of traits that primarily benefit their bearer and benefit other individuals as a side effect, whereas reciprocity could explain the evolution of traits that entail costly investments in other individuals. The concept of pseudoreciprocity — in which an individual invests in another to acquire or enhance by-product benefits obtained from that individual — is an alternative theory of the evolution of investment in unrelated individuals. Such pseudoreciprocity represents an important category of cooperation and may well be the main explanation for existing examples of investment in unrelated organisms. A reason for the prevalence of pseudoreciprocity, particularly in comparison with reciprocity, could be that there are many ways in which investments can yield by-product benefits, so that there will be many opportunities for the evolution of such investment from a noninvesting state.

INTRODUCTION

For mutualistic interactions between unrelated organisms, including members of different species, a useful starting point for evolutionary analysis is that traits influencing the interaction can evolve only if they are advantageous for their bearers. There is a multitude of ways in which an advantage can be gained — some direct and others operating indirectly through an influence on other organisms — and there will be a corresponding richness in the types of interaction. An evolutionary theory of such interactions should focus on traits that play a role for the interaction; the net fitness effects, which form the basis for classification into mutualism or parasitism, are less important for theoretical understanding. Nevertheless, there is scope for an evolutionary theory of traits that sometimes promote mutual benefit. A central issue for such a theory must be the evolution of traits that can benefit others.

There are several ways in which an individual could derive benefit from interacting with an unrelated organism (Table 11.1) (cf. Connor 1995). The transfer of benefit to the individual could primarily be the result of traits of the other organism. These characteristics of the other organism might be adaptations for generally benefiting the individual, provided that there is an element of common interest. For instance, if partners tend to succeed or fail together, there can be an incentive for investing in a partner's success, at least to some extent. In other cases it may be more appropriate to regard the transfer of benefit either as a means by the other organism to influence the behavior of the individual — perhaps attracting the individual with a food reward — or as a side effect of characteristics that are adaptive for the organism for other reasons. The transfer of benefit could also be a consequence of adaptations in the benefiting individual. In the latter case, the individual can be thought of as exploiting some characteristic of the other organism.

Our aim here is to examine and exemplify a number of concepts that are important for the evolutionary analysis of traits that play a role in providing benefit for others. In much previous work, two ideas stand out as having been particularly influential: that benefits to others may be by-products of traits or behaviors that directly benefit the individual itself (West-Eberhard 1975; Brown 1983) and that mechanisms of reciprocity could support the evolutionary stability of investing in a partner (Trivers 1971; Axelrod and Hamilton 1981). An important

Table 11.1 Classification of benefits in mutualistic interactions.*

Category	Description
invested benefits	An individual benefits as a consequence of traits in another, which traits have evolved for the purpose of influencing the individual in a way that has been advantageous for the other.
by-product benefits	An individual benefits as a consequence of traits in another, which traits have *not* evolved for the purpose of influencing the individual, but may instead have been advantageous for other reasons.
purloined benefits	An individual obtains a benefit from another as a consequence of its own traits, which traits have evolved for the purpose of obtaining this benefit from the other.

*The classification emphasizes the selective background of traits and is highly idealized in ascribing simple functions to traits that may in reality have a more complex selective background, for instance because benefits appear as a consequence of the interaction of traits in different individuals.

distinction is that by-product mutualism appears not to explain traits that primarily benefit others, whereas such traits might be consistent with reciprocity. A tempting but incorrect conclusion is then that examples of investments in unrelated individuals ought to be examples of reciprocity. Instead, there is also the possibility that an individual invests in another because this would enhance the by-product benefits obtained from the other. This type of interaction has been called pseudoreciprocity (Connor 1986, 1995) and could be one of the main explanations for the occurrence of investments in unrelated individuals.

An entirely hypothetical illustration of pseudoreciprocity is as follows (Connor 1986): a bird whose nest is in a hot, sunlit area spends some time and effort "fertilizing" a nearby bush, causing the bush to grow taller and shade the bird's nest. The bird has then performed a beneficent act for the bush, but the bush has not performed a beneficent act for the bird in return. Rather, the bird's return benefit is an incidental effect or by-product of a self-promoting act (growth) on the part of the bush.

The suggestion that pseudoreciprocity is a major category of cooperation has been strongly criticized (Mesterton-Gibbons and Dugatkin 1992, 1997) on the grounds that it would be a redundant category and an unnecessary complication of the fundamental division into by-product mutualism and reciprocity. Here, we counter this criticism and argue that pseudoreciprocity and, more generally, the concept of investment in by-product benefits play natural and important roles when applying basic ideas of adaptation to the phenomenon of mutualism. In addition to providing an understanding of the evolution of investments in interspecific mutualisms, pseudoreciprocity may also be operating in intraspecific interactions, for instance in cooperatively breeding animals where helpers may gain by recruiting additional group members, and in some mating systems where sperm competition provides an incentive for males to transfer nuptial gifts to females at mating.

THE EVOLUTION OF INVESTMENT IN UNRELATED INDIVIDUALS

The general procedure for determining whether some characteristic is an adaptation, and for what function it then has been adapted, is to identify the circumstances, either in the past or at the present time, that selected for the appearance or maintenance of the characteristic (Futuyma 1998). Let us apply this to traits entailing investment in unrelated individuals and entailing direct costs for the investing individual. If an individual would benefit in some way from the presence of another organism, it is clear that an investment in increasing the organism's survival could be adaptive. Another general and important circumstance favoring investment would be the presence of behavioral mechanisms or plastic phenotypes in other organisms that make them respond to investment in a way that becomes favorable for the investing individual. The hypothetical example of the fertilization–growth response in the bush is one such mechanism. There are many others, for example, the propensity of potentially beneficial visitors, like pollinators, to be attracted by rewards.

Reciprocity as an Adaptation

Although there is no universally agreed upon definition of reciprocity, an unambiguous case would be unilateral investment by one individual in another on one occasion, followed by a reverse investment on a clearly separate occasion, and so on in an alternating fashion (Trivers 1971). Usually, one would also include situations showing similarity to the play of Tit-for-Tat in a two-person repeated Prisoner's Dilemma game (PD) (Axelrod and Hamilton 1981). If the investments are concurrent, ongoing activities that are regulated by some sort of bookkeeping mechanism, i.e., a mechanism serving to regulate current investment based on benefits received and previous investments, it may be preferable to speak of trading rather than reciprocity, since it may not be possible to identify any discrete "chunks" of investment that are reciprocated. In either case, whether there is ideal reciprocity or just some form of trading, investing in the partner could be viewed as an adaptation to a behavioral mechanism in the partner, namely the partner's tendency to deliver returns in proportion to the investment. Note that there may also be other situations where two organisms invest in each other, and these investments are adaptations to behavioral mechanisms in the partner, but where the mechanisms are different from bookkeeping. For instance, each organism could invest in by-product benefits from the other (Connor 1995). For mobile organisms, behaviors such as abandoning an unproductive partner but staying with a sufficiently productive one, which is not really bookkeeping, can also promote investment, provided there is some cost of partner switching (e.g., Friedman and Hammerstein 1991; Connor 1992; Enquist and Leimar 1993).

Strictly from the point of view of evolutionary game theory, the perspective above would seem incomplete, because the requirement that investment is an adaptation to a mechanism in the partner is only part of a demonstration of evolutionary stability. Thus, one might also require that each individual's behavioral mechanism is an adaptation to the circumstances of the interaction and, even further, that it is an optimal adaptation, in order to say, for instance, that there is adaptive reciprocity. The Tit-for-Tat strategy for the repeated PD game cannot be regarded as an adaptation in this sense (Selten and Hammerstein 1984). However, with random errors in the execution of investments, strategies based on "good standing" can be optimal adaptations (Boyd 1989); similar kinds of strategies can be optimal adaptations when there is random variation in the ability to perform investment or in the partner's need for the investment (Leimar 1997). Nevertheless, when thinking about the characteristics of real organisms, it seems wise to be somewhat more lenient in one's requirements to avoid falling into the trap of extreme adaptationism.

A General Definition of Pseudoreciprocity

The term pseudoreciprocity was originally applied to situations where one individual invests to acquire or enhance by-product benefits from the other, but where there is no similar investment in the other direction (Connor 1986, 1995). It seems rather natural to broaden the definition of the term to also include cases where each party invests in by-product benefits from the other (such relationships were classified as by-product investment – by-product investment in Connor 1995). Thus, the term pseudoreciprocity could cover any mutualistic interaction where there are investments in by-product benefits. This usage has the advantage of making pseudoreciprocity a main competing alternative to reciprocity and trading as an explanation for mutualistic relationships with costly investments in unrelated individuals.

Pseudoreciprocity and Evolutionary Stability

When an individual invests in by-product benefits from another organism, the traits of the other organism that produce the by-product benefits need not be adaptations for the particular circumstances of the interaction. Thus, in the bird-bush example, the growth response of the bush need not be an adaptation for interacting with fertilizing birds but might just be a plastic response to variation in the availability of nutrients that has evolved independently of any interactions with birds. The example illustrates that pseudoreciprocity could evolve through adaptation of only one of the parties of the interaction. In such a case, one would not expect complete evolutionary stability. Although the organism producing the by-product effect would tend to benefit from its own response, the response would not be adaptively fine-tuned to the particular situation. The fact that pseudoreciprocity identifies only the minimal circumstances for investment in others to evolve should be seen as an advantage of the concept because this gives it a potentially broad range of application.

Pseudoreciprocity can, of course, also be an evolutionarily stable strategy (ESS). The analysis is particularly simple if an interaction can be modeled as a sequence of two unambiguous steps: an initial investment by one individual, followed by the response of the other. In such a case, backward induction can be used to determine evolutionary stability. For an ongoing or repeated interaction, the issue of evolutionary stability becomes more delicate if one allows for the possibility that the organism receiving investment might adjust its response in a way that leads to increased investment. Thus, the bush in our example might do better by growing more laterally before increasing its height, in this way forcing the bird to perform greater fertilization to achieve the desired shade. Just as in any other interaction between unrelated individuals with little common interest, there is always the possibility that various means of exploitation evolve.

Is Pseudoreciprocity a Case of By-product Mutualism?

According to Mesterton-Gibbons and Dugatkin (1992, p. 278; 1997, p. 556), "pseudoreciprocity, in which one individual manipulates another for the benefit of both, ... is simply asymmetric by-product mutualism (with asymmetry caused by sequential action)." It is, of course, true that the individual's investment in another is ultimately self-serving, so that the fitness effect on the other individual can in this sense be regarded as a by-product. Similar things can be said about any adaptation with fitness effects on others, including reciprocity, which thus has no particular status in this regard. However, when discussing the transfer of benefits between unrelated individuals, it is of basic interest to identify the circumstances under which investing in another organism can be an adaptation. Investments in by-product benefits, which pseudoreciprocity entails, correspond to a broad set of such circumstances and may apply to many existing mutualistic interactions, making it an important concept.

EXAMPLES OF INVESTMENT IN BY-PRODUCT BENEFITS

Activities such as seeking nourishment and shelter are ubiquitous among animals. This may be a reason why cases of investment in by-product benefits often follow the pattern of "hosts" offering food or shelter to "visitors," in this way gaining by-product benefits from the visits (Cushman and Beattie 1991). The benefits can, for instance, be that visitors act to defend their food source or shelter against enemies of the host. In the following, we use the mutualism between lycaenid larvae and ants to illustrate this kind of interaction. For interactions within species, the host–visitor situation is less common; however, there are other possibilities for investment in by-product benefits. We discuss helping at the nest and nuptial gift giving as examples that may be interpreted in this way.

Interactions between Lycaenid Larvae and Ants

The Lycaenidae is a large family of butterflies that contains the blues, coppers, and hairstreaks. A considerable proportion of the over six thousand lycaenid species associate with ants during parts of the life cycle, most commonly during the larval stage (reviewed in Pierce 1987; Fiedler 1991; Pierce et al. 2002). A small minority of the associations are clearly parasitic, with lycaenid larvae either being predators on ant brood or acting as cuckoos inside ant nests. The majority of the associations are usually considered mutualistic and are based on nutritional rewards delivered by lycaenid larvae in exchange for protective benefits of ant attendance. The interactions typically take place on the larval host plant. Lycaenid larvae possess—in varying degrees—a suite of adaptations for influencing ant behavior. The most important of these adaptations may be the so-called dorsal nectar organ, which is a gland situated dorsally on the seventh abdominal segment of a larva and which is capable of secreting a nutritious liquid, packaged into discrete droplets. The delivery of droplets is stimulated by ant attendance and causes the ants to treat a larva as a food source to be defended against possibly competing intruders. As a result, a larva's mortality rate from attacks by invertebrate predators and parasitoids is reduced. For some lycaenid species, larvae in the field have only a small or no chance of surviving into adulthood without ant protection, making ant association an obligate relationship, whereas other species have a looser, facultative association with ants.

The lycaenid–ant relationship seems to be a case of pseudoreciprocity, in which lycaenid larvae invest in food rewards and receive the by-product benefits of ant defence. The ant defence should be regarded as a by-product because it is typical of ant behavior vis-à-vis any food source. Some enemies of a larva, such as predatory insects, would also be enemies of the ants, since these predators will remove the food source. The most important enemies of lycaenid larvae are likely to be parasitoids, which lay eggs that hatch into parasitic larvae that grow inside and eventually kill the butterfly larva, often around the time it pupates. The parasitoids are not really enemies of the ants, since they seem not to interfere with a lycaenid larva's ability to deliver food rewards. The reason the ants still protect lycaenid larvae from parasitoids is probably that ants attack more or less any type of seeming intruder.

Behavioral Mechanisms of Lycaenid Larvae and Attending Ants

Looking more closely at the behavioral mechanisms regulating lycaenid–ant interactions, the overall impression is that lycaenid adaptations serve to control the number of attending ants to a level that corresponds to a larva's need for protection. For instance, lycaenid larvae tend to deliver more droplets as a response to higher ant attendance; however, for species that are not completely dependent on ants for survival, this relationship holds only up to a moderate degree of attendance, after which the delivery rate of rewards either levels off or decreases with increasing attendance (Figure 11.1) (Leimar and Axén 1993; Fiedler and Hageman 1995; Fiedler and Hummel 1995; Axén et al. 1996; Axén and Pierce 1998). As a result, more ants will attend larvae with greater need for ant protection. For ant allocation of workers to a food source, the general principle is that more workers will be allocated to a richer source (Figure 11.2) (Crawford and Rissing 1983). Joint operation of the lycaenid and ant behavioral mechanisms (Figures 11.1, 11.2) will then determine the delivery of reward and level of ant attendance in a given situation. This outcome should be regarded as a dynamic balance that adjusts itself over a period of a few minutes (Axén et al. 1996).

This picture is complicated by a situation that leads us to believe that ants often allocate more workers to lycaenid larvae than would in principle be needed

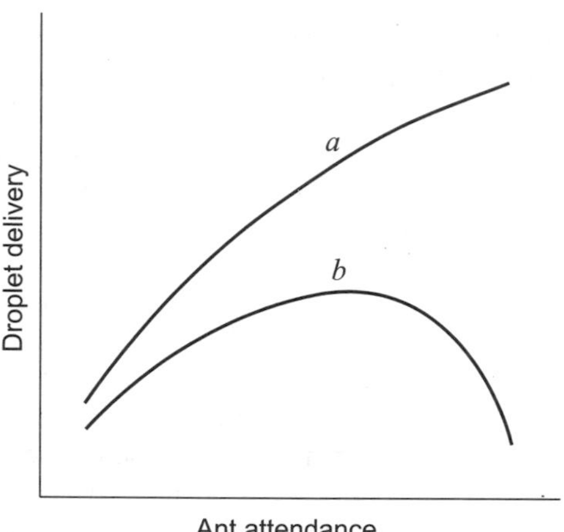

Figure 11.1 Schema of a lycaenid larva's rate of reward delivery as a function of ant attendance. Curve (a) encourages a high level of attendance and could be a lycaenid species with strong dependence on ant protection. For curve (b), which could correspond to a species less dependent on ants, the ants ought not to allocate very many workers to the larva, since this decreases the rewards. Note that the curves depict a hypothetical steady state situation under some given circumstances. Other factors influence droplet delivery in addition to the average level of ant attendance, e.g., a larva's age and the quality of its diet.

to collect the nutritious reward. When several ants tend a larva, most of them are not retrieving droplets but are standing or walking either on the larva or next to it, while sometimes palpating it with their antennae. Assuming — for the sake of the argument — that ant behavior has been adapted for efficient handling of lycaenid larvae, these "extra" workers could play the role of an investment by the ant colony, serving to stimulate a larva to increase its rate of reward delivery. At least for certain ranges of investment, the combination of ant and larval behavioral mechanisms would then show some similarity to bookkeeping (see Figures 11.1 and 11.2; cf. Leimar 1997). Another explanation for the "extra" workers is that they are there to defend the colony's food source. Since tending ant species are not known to allocate such extra workers to other types of food sources, this latter explanation is perhaps unlikely. In either case, the benefit to lycaenid larvae of ant defence would be a by-product of normal ant behavior.

The assumption that ant behavior is well adapted to the handling of lycaenid larvae is, however, rather doubtful. It may well be that lycaenid larvae — through skillful interference with the chemical communication system of the ants — manage to manipulate the ants into overestimating their value as food sources. There are also several ways in which ants could increase the average rate of reward delivery of a larva, but which they seldom use. It has been found that a larva will sharply increase its delivery of droplets when it perceives itself to be under attack from enemies or when ants return to it after an interruption in attendance (Leimar and Axén 1993; Axén et al. 1996; Agrawal and Fordyce

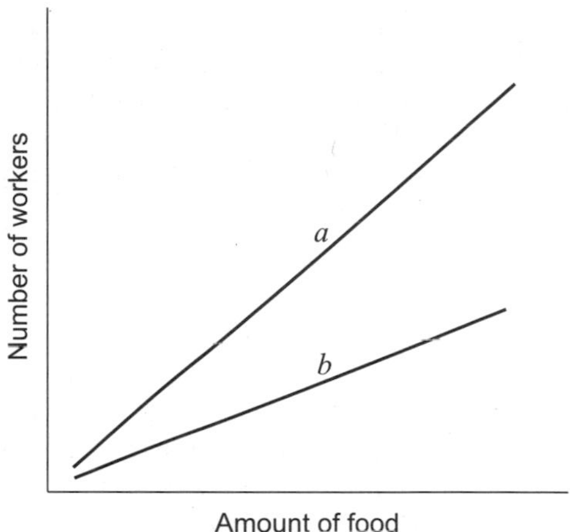

Figure 11.2 An ant colony's allocation of workers as a function of the rate of delivery of a food source. For efficient (ideal free) foraging, one would expect the allocation of workers to be directly proportional to the rate of delivery. This will hold over small additional food sources that are simultaneously available at equivalent positions near the colony. The allocation of workers per amount of resource will also depend on factors like the colony state (e.g., the level of "hunger"), the overall availability of food sources, the distance from the food source to the colony, and the time of day. For instance, line (a) could correspond to a situation where the colony has a strong need for the particular resource (e.g., carbohydrates vs. protein) and line (b) to a situation with less intense need. A number of aspects of ant behavior will jointly determine the allocation of workers illustrated in the figure, for instance the individual worker's tendency to deposit trail pheromones when returning from the food source — which is influenced by how readily she was able to obtain her load — and the degree of "enthusiasm" with which she is met when she hands over her load to nest mates inside the colony.

2000). Thus, if ants "attack" a larva occasionally or leave it unattended for a few minutes and then return, they would substantially increase their benefits.

If allocating too many workers to lycaenid larvae in fact hurts a colony, the interaction would be a case of parasitism. However, if the ants still benefit from the interaction and the larvae actually need the protection, the larval investment and ant protection would be pseudoreciprocity. Since ecological conditions like the food abundance for ants and the population densities of enemies of lycaenid larvae may fluctuate in time and space, the nature of the lycaenid–ant interaction may also vary from mutualism to parasitism (cf. Bronstein 1994).

Markets and Investment

A situation where hosts invest to attract visitors can be likened to a market where hosts supply a commodity for which there is demand among the visitors. One might then expect the evolutionarily stable level of host investment to be influenced by "market forces" (Noë and Hammerstein 1994, 1995). Examples of such market forces are the relative population sizes of hosts and visitors as well as the availability of alternative commodities for the visitors. The host–visitor association could still be described as pseudoreciprocity, but the relationship between supply and demand would influence the evolution of investment. One might even argue that the market situation provides a very basic explanation for the presence and the magnitude of investment.

Several issues make the matter more complex. Visitors could be attracted by traits that do not directly benefit them but only signal the presence of reward. Competition for visitors might then just increase investment in advertisement, perhaps even at the expense of investment in reward, as might be the case in flowers lacking nectar. The by-product benefits of the visits could also be shared among a group of hosts, for instance among neighbors, resulting in decreased investment by the hosts ("tragedy of the commons").

Lycaenid–ant interactions provide an example. The larvae of some lycaenid species occur in aggregations, and these species invariably associate with ants. Pierce et al. (1987) have suggested that group living could be a way for the larvae to increase the protection from ant attendance and/or to decrease the cost of the association. In a series of experiments, Axén and Pierce (1998) demonstrated that larvae of the group-living lycaenid *Jalmenus evagoras* modify the rate of reward delivery as a function of group size. A solitary larva secretes considerably more droplets than a group member (when controlling for the number of ants directly attending the larva; see Figure 11.3). A likely explanation is that a larva will, to some extent, be protected by the presence of ants on other group members. The reason why the rate of secretion does not drop to zero may be that a larva needs some directly attending ants, which can alert other nearby ants in

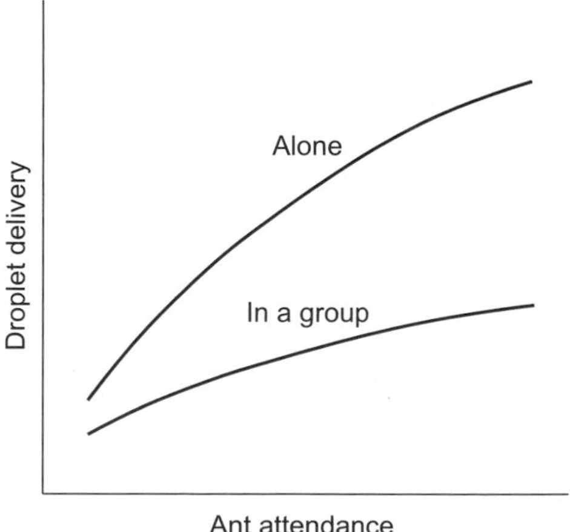

Figure 11.3 Illustration of a normally group-living lycaenid larva's rate of reward delivery as a function of per capita ant attendance, in two different situations. The rate of delivery is considerably higher when the larva happens to be alone compared to when it is in a larval group.

case of danger. Another reason could be that the larvae in a group are related; however, since several females oviposit on the same host plant and the larvae move about on the plant, the within-group relatedness is probably not so high (Pierce et al. 1987). From the point of view of the ants, the larval aggregations seem disadvantageous. Axén and Pierce (1998) estimated that ants would nearly triple their rewards if they were to break up naturally occurring aggregations into singletons (but the ants do not attempt this). The study indicates that larvae do not compete for ant attendance within a group. Nevertheless, larvae have to compete with other food sources of the ants, which could be other larval groups. For instance, the reason a singleton larva delivers more droplets should be to attract enough ants away from foraging elsewhere. The idea of a market (Noë and Hammerstein 1994, 1995), at least within the home range of a single ant colony, thus seems valid.

Intraspecific Helping at the Nest

Several authors have suggested that subordinates in communally breeding birds and mammals might "help at the nest" because of benefits received later when the subordinate becomes a breeder and the young it helped have survived to engage in anti-predator behavior, territorial defence, or feeding the former helper's offspring (Woolfenden 1975; Brown 1978, 1983; Rood 1978; Clutton-Brock 2002). As noted by Connor (1986), this would make helping at the nest a candidate for intraspecific pseudoreciprocity. Connor (1986) also pointed out that selection would not favor helping to promote the survival of young that help in return unless helping is already favored by kin selection, or there are return benefits that derive from nonhelping behaviors (e.g., territorial defence). In either case, there would be by-product benefits from the presence of additional group members acting in their own interest. A recent formal model of cooperative breeding and group augmentation by Kokko et al. (2001) substantiates this. They found that helping was readily favored with "passive group augmentation," i.e., return benefits from territorial defence, etc. When there was only "active group augmentation," i.e., return benefits only from active helping, Kokko et al. (2001) concluded that helping must occur initially for other reasons (e.g., helping kin) before selection would favor helping to augment group size.

Pseudoreciprocity in Insect Mating Systems

Mating effort can sometimes represent an investment in by-product benefits (Connor 1986). An example could be the nutritious "nuptial gifts" that are part of male mating effort in certain insects (Boggs 1990). The gift is transferred to the female at mating and may increase her reproductive output. Nuptial gifts could function to increase a male's chances of acquiring matings, but they could also function as investments in a male's own offspring or devices to increase a male's success in sperm competition. In insect groups where nuptial gifts are part of the male ejaculate, there is no effect on the probability of mating; however, the relative importance of investment in the male's own offspring versus sperm competition has been debated. For butterflies, where the issue has been studied experimentally (Wiklund et al. 1993), it appears that nuptial gifts partly represent an investment in eggs that the male will fertilize and partly serve to increase the male's sperm competition success. Butterfly sperm competition is characterized by last male advantage, meaning that the male that most recently copulated with a female will fertilize most eggs laid by her, up to the time she mates again. The importance of nutritious ejaculates for sperm competition is that a female will wait longer until remating if she receives a larger gift (Kaitala and Wiklund 1995), and thus will lay more eggs fathered by the male. In this way, a larger investment serves to increase the male's success in sperm competition. For a female, the delay in remating after receiving a larger nuptial gift is directly in her own interest. Females in species with nuptial gifts can be said to forage for matings (Kaitala and Wiklund 1994) to increase their own reproductive output. Since there will be a limit on the rate at which a female processes the resources in gifts, it will be in the female's interest to vary the intervals between matings in proportion to the resources received. Only part of the resources in a male's ejaculate will find their way into eggs fertilized by the male; the remainder will provision eggs that are fertilized by other males (Wiklund et al. 1993). The overall conclusion is then that males invest partly in their own offspring and partly in by-product benefits from female reproductive behavior. The relative importance of these two factors for the evolution of male investment is, however, not known. Nevertheless, it seems likely that pseudoreciprocity plays an important role in certain insect mating systems.

MUTUAL INVESTMENTS

For cases where there are clear investments from both sides of a relationship, and only a limited degree of common interest, it is of interest to determine for what function the investments are adaptations. Some form of bookkeeping is, of course, a possibility but may be unusual. The return benefits of an investment could come about in many other ways, more resembling by-product benefits than bookkeeping.

Mycorrhizal Symbiosis and Investment in By-product Benefits

The majority of the species of vascular plants form mycorrhiza with fungal mycelium in the soil (Smith and Read 1997). It is thought that this association was instrumental for the original colonization of land by plants more than 400 million years ago (Simon et al. 1993). The association is based on transport of organic carbon from plant photosynthesis to the fungal partner as well as a transport of soil mineral nutrients, such as like phosphorus, from fungus to plant.

The physiology and molecular biology of the interaction are quite complex and not yet completely understood for any of the various types of mycorrhiza (Smith and Read 1997). However, barring some examples of parasitism, it seems likely that mycorrhizal symbiosis is a case of mutual investment of plant and fungus in each other. The transport of molecules — of organic carbon by the plant and of mineral nutrients by the fungus — to the site of contact between the organisms, across a plasma membrane to a so-called interfacial apoplast, where the partner can extract the molecules, is a process that involves the expression of several genes of both plant and fungus (Harrison 1999).

Many mycorrhizal fungi depend obligately on plants, and most plants benefit from mycorrhizal interactions, although they can grow without them. If a single plant interacted with a single fungus, their mutual investments could be explained by common interest in a "team project" of joint growth. In reality, plants often interact with several mycelial networks, corresponding to different fungi, and each of these networks may be connected to several plants. Thus, a plant can perhaps avoid investing in a fungus and still draw benefit from it if the fungus is maintained by other plants, and vice versa for the fungus. This might severely limit the evolution of investment on both sides, in a way analogous to the "tragedy of the commons."

One could speculate that some kind of bookkeeping, localized to a single mycorrhizal site of interaction, would maintain the investments. There is, however, no evidence for any such trading mechanism. On the contrary, there are reasons to doubt that the interaction works in this way. A common observation is that a young seedling plant can benefit from being surrounded by mature plants, because the seedling can draw nutrients from a hyphal network that is being maintained by the mature plants (Smith and Read 1997). Thus, it seems that it is quite possible for a plant to receive nutrients without "paying for them."

It is perhaps more likely that by-product benefits play a role in maintaining the investments. One potentially important factor is that the plant–fungus interaction takes place in a dynamical and spatial setting, where roots grow and senesce and where different fungal mycelia compete with each other for access to plants. For instance, a fungus could invest in a plant to stimulate additional root growth, which the fungus will be in a good position to colonize (Figure 11.4). It is known that root growth can be plastic, resembling foraging for elevated nutrient concentrations in the soil (Robinson 1994). This kind of response is particularly strong for non-mycorrhizal plants, where roots absorb the nutrients directly, and it is less pronounced when fungal hyphae of mycorrhizal symbionts instead explore the soil. Nevertheless, the roots of mycorrhizal plants also respond to nutrients (e.g., Jackson et al. 1990), so that local root growth as a by-product benefit of fungal investment is a reasonable hypothesis.

Competition between mycorrhizal fungi could also be responsible for by-product effects of plant investment. In an experiment with the Scots pine, Saikkonen et al. (1999) found that defoliation of a tree led to a changed composition of mycorrhizal associates. Since defoliation reduces a plant's ability to

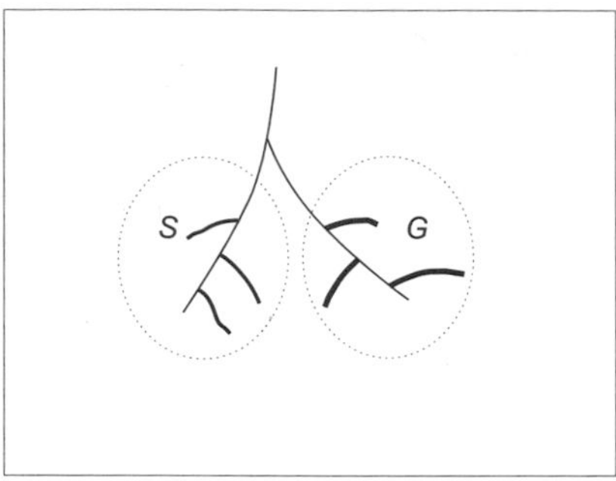

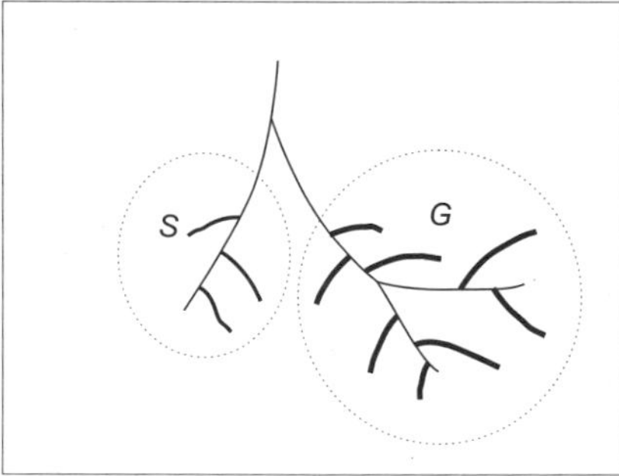

Figure 11.4 Local root growth is a possible by-product benefit of fungal investment in a plant. The top panel shows a hypothetical situation where two fungi have colonized different parts of a growing root. One fungus (S) is "stingy" and contributes few nutrients to the plant, whereas the other fungus (G) is more "generous." Assuming that the plant will allocate root growth primarily to those parts of the root that are successful in collecting nutrients, the investing fungus may have a competitive advantage (because of spatial proximity) in attempting to colonize new parts of the root. As shown in the bottom panel, the investing fungus might then form mycorrhiza with a greater proportion of the root, and thus enjoy the benefit of a greater proportion of the flow of carbon from the plant.

photosynthesize, the response was interpreted as an outcome of competition between fungal types with different carbon demands. This interpretation by Saikkonen et al. (1999) implies that a plant's level of photosynthate investment in associated fungi has the effect of a choice of fungal partner. Provided that a more "expensive" partner gives better returns to a plant that can "afford" this partner, there may thus be an incentive for the plant to invest in giving such a partner a competitive advantage.

BY-PRODUCT MUTUALISM AND COORDINATION

In by-product mutualism, there is little or no investment in other organisms, and the transfer of benefit is a side effect of traits that are present for other reasons, for instance, for providing direct benefit to their bearer. This concept of by-product mutualism can be elaborated by noting that there may be traits that benefit their bearer directly, but which have been selectively modified because of feedback of fitness effects from an influence on the behavioral mechanisms of other organisms. One example could be aphid secretion of "honeydew," which may be collected by attending ants (Way 1963). Since the aphid honeydew is a waste product that needs to be excreted regardless of the presence of ants, it cannot be regarded as an investment in the ants. However, aphids seem to benefit from the presence of ants, protecting them from enemies and keeping their local environment clean. Consequently, aphids time their secretions to the presence of ants and use special body postures to display and deliver droplets. The ants also communicate with the aphids, stroking them with the antennae to encourage the release of honeydew. Both ants and aphids act essentially in their own immediate interest and cannot be said to invest in each other; however, they show a considerable degree of coordination of their actions (Douglas and Sudd 1980). Such coordinated by-product mutualism, where the partners are adapted both to pursue their own immediate interests and to deal with each other's behavioral mechanisms, represents an important category of cooperation.

A common form of coordinated by-product mutualism is when individuals join together for the purpose of some project, such as cooperative hunting (e.g., Dugatkin 1997). In addition to coordinating their activities, individuals may also modify the degree to which they invest in the activities of the project (Leimar and Tuomi 1998). If benefits of the project are shared among group members, there may typically be a decrease in each individual's investment in the activity with increasing group size, following the logic of the "tragedy of the commons." In bird parental care, a single female shows a higher rate of brood provisioning than one working with a mate (Houston and Davies 1985).

Joint Tasks and Division of Labor

Recently, Anderson and Franks (2001) defined teams as exhibiting a division of labor and team tasks as comprised of different subtasks that must be performed concurrently. They defined group tasks as requiring concurrent action by multiple individuals, but where everybody performs the same task, without any division of labor. A group task might well be regarded as a special case of a team task. An example of a group task is coordinated fishing by cormorants (Bartholomew 1942). Stander (1992) described a team task performed by lions in which individuals take different strategic positions, "wing" or "center," in a group hunt. In these examples, both group and team tasks represent coordinated by-product mutualism.

Anderson and Franks (2001) extended the team concept to putative cases of reciprocal altruism. For example, they recognized three subtasks performed by supposedly reciprocating vampire bats (Wilkinson 1984): foraging, regurgitation (which is reciprocated), and receiving blood. However, it seems unhelpful to consider the acts of giving and receiving altruism to be subtasks in a team task, because the receiving individual does not invest in the execution of any project but rather just receives a benefit.

Team tasks can include reciprocity if it is demonstrated that two individuals perform subtasks concurrently to acquire a benefit that only one of them enjoys, and that individuals take turns receiving the benefit in a series of tasks. In the same sense, reciprocity can be a part of group behavior with no division of labor, provided there is concurrent action and turn taking in receiving benefits. A hypothetical example would be a pair of individuals jointly hunting a single prey that only one of them will eat, and where this role of beneficiary alternates. A second example discussed by Anderson and Franks (2001) — coalition formation in male olive baboons (Packer 1977) — could in principle correspond to a team task with reciprocity. The suggestion has been that a male solicits the assistance of his coalition partner to engage a high-ranking male, giving the soliciting male an opportunity to mate with a female that is guarded by the high ranking male (Packer 1977). However, further studies have shown that male baboons do not exhibit a division of labor, nor do they take turns consorting females taken from high-ranking males (Bercovitch 1988; Noë 1992). Rather, their coalitions are by-product mutualisms in which, by coordinating their actions, each male enjoys a better chance of obtaining a female than he would have by acting alone.

A striking example of teams in interspecific mutualism is the cooperative feeding association between humans and dolphins, of which the most detailed accounts come from Mauritania and Brazil (Busnel 1973; Pryor et al. 1990). In the team task, humans and dolphins perform concurrent subtasks to reach the common goal of capturing fish. The humans place or throw nets into the water, which serve as a barrier against which the dolphins can catch fish. Since the dolphins drive fish into the nets, they also produce benefit for their human partners. This is a by-product mutualism with a division of labor, but with some investment by humans in the Mauritania case as they signal to the dolphins upon sighting a school of fish. Presumably, the association is a result of cultural evolution on both the human and the dolphin side. The signals produced by humans to initiate cooperative foraging with dolphins can be seen as (rather small) investments needed to initiate the execution of coordinated by-product mutualism, so that these interactions contain elements of pseudoreciprocity.

DISCUSSION

Distinguishing in which way benefits acquired in interactions are consequences of adaptations of either of the interacting parties is a useful method of analysis of mutualism. This may seem like stating the obvious, but in our opinion the study of mutualism is in need of a more thorough analysis, both of the selective background of traits playing a role in interactions and of the side effects of these traits. Important traits may also have evolved in other contexts than the mutualistic interaction, or otherwise not be perfectly fine-tuned to the circumstances of the interaction.

Our presentation has primarily dealt with benefits deriving from traits of other organisms, either invested benefits or by-product benefits. Benefits can also be adaptive consequences of an individual's own traits, and these benefits have been referred to as purloined by Connor (1995; see Table 11.1). In principle, there could be interspecific mutualisms with purloined benefits on both sides, resulting in net benefits for the parties involved, although interactions where one organism gains by-product benefits from the exploitative behavior of another seem more likely. An example would be a seed predator that benefits a plant through the dispersal of some of its seeds.

By exploitation, one often means the presence of traits for which the advantage is directly linked to a disadvantage for another organism, although one might also regard other traits, including investments, as ultimately a form of exploitation (Bronstein 2001 and this volume). Adaptations for exploitation may lead from mutualism to parasitism, either through reduced investments by one party or through increased benefits purloined by one party, as well as the appearance of novel ways to purloin benefits. Concerning by-product benefits, they could be lost or reduced if there is a change in the original adaptation that produced the by-product benefits. Such a change may well shift the interaction from mutualistic to parasitic, but saying that there is increased exploitation seems inappropriate as an explanation of the change. Although the interaction may have shifted from mutualistic to parasitic, there has not been selection on the former mutualist to reduce benefits; the reduction simply occurs as a side effect. An example might be a pollinator becoming more efficient at extracting nectar rewards and, at the same time, becoming less efficient in transporting pollen (cf. Bronstein 2001).

The perspective we have used that benefits to an individual are either primarily due to its own traits or to the traits of another (Table 11.1) is clearly an idealization. For certain cases of mutualism, the basic principle is that benefits are obtained only through the interaction of traits of different individuals: warning coloration and Müllerian mimicry would be examples. These can be regarded as special forms of by-product mutualism, where coordination — in behavior or in other traits — is crucial for any benefit to be obtained. The concept of synergistic selection (Maynard Smith 1982) can be applied to such phenomena. More generally, coordination of directly beneficial behaviors or traits to acquire or enhance by-product benefits, which we have referred to as coordinated by-product mutualism, is a very widespread type of mutualistic adaptation.

Nevertheless, the evolution of investment in unrelated organisms is conceptually perhaps the most basic mutualistic adaptation, and reciprocity and pseudoreciprocity are the main competing explanations for such investments. Our arguments in favor of pseudoreciprocity have been that reciprocity is a quite special adaptation, whereas pseudoreciprocity could come about in many ways. Game theory analyses show that reciprocity in principle may evolve; however, these arguments say rather little about the likelihood of this happening for real organisms, in comparison with the likelihood of the evolution of investment in by-product benefits. There are several ways in which investment can yield by-product benefits, and some of these are associated with ubiquitous activities, such as foraging and growth. This could be the reason for the scarcity of reciprocity and the much wider distribution of pseudoreciprocity.

ACKNOWLEDGMENTS

We thank the participants of the Dahlem Workshop for helpful suggestions in improving the presentation. We also thank Minna-Maarit Kytöviita and Juha Tuomi for valuable comments. This work was supported by grants from the Swedish Research Council.

REFERENCES

Agrawal, A.A., and J.A. Fordyce. 2000. Induced indirect defence in a lycaenid-ant association: The regulation of a resource in a mutualism. *Proc. Roy. Soc. Lond. B* **267**: 1857–1861.

Anderson, C., and N.R. Franks. 2001. Teams in animal societies. *Behav. Ecol.* **12**:534–540.

Axelrod, R., and W.D. Hamilton. 1981. The evolution of cooperation. *Science* **211**: 1390–1396.

Axén, A.H., O. Leimar, and V. Hoffman. 1996. Signalling in a mutualistic interaction. *Anim. Behav.* **52**:321–333.

Axén, A. H., and N. Pierce. 1998. Aggregation as a cost-reducing strategy for lycaenid larvae. *Behav. Ecol.* **9**:109–115.

Bartholomew, G.A. 1942. The fishing activities of double-crested cormorants on San Francisco Bay. *Condor* **44**:13–21.

Bercovitch, F.B. 1988. Coalitions, cooperation, and reproductive tactics among adult male baboons. *Anim Behav.* **36**:1198–1209.

Boggs, C.L. 1990. A general model of the role of male-donated nutrients in female insects' reproduction. *Am. Nat.* **136**:598–617.

Boyd, R. 1989. Mistakes allow evolutionary stability in the repeated Prisoner's Dilemma game. *J. Theor. Biol.* **136**:47–56.

Bronstein, J.L. 1994. Conditional outcomes in mutualistic interactions. *Trends Ecol. Evol.* **9**:214–217.

Bronstein, J.L. 2001. The exploitation of mutualisms. *Ecol. Lett.* **4**:277–287.

Brown, J.L. 1978. Avian communal breeding systems. *Ann. Rev. Ecol. Syst.* **9**:123–156.

Brown, J.L. 1983. Cooperation: A biologist's dilemma. *Adv. Study Behav.* **13**:1–37.

Busnel, R.G. 1973. Symbiotic relationship between man and dolphins. *NY Acad. Sci. Trans.* **35**:112–131.

Clutton-Brock, T. 2002. Breeding together: Kin selection and mutualism in cooperative vertebrates. *Science* **296**:69–72.

Connor, R.C. 1986. Pseudo-reciprocity: Investing in mutualism. *Anim. Behav.* **34**: 1562–1584.

Connor, R.C. 1992. Egg-trading in simultaneous hermaphrodites: An alternative to Tit-for-Tat. *J. Evol. Biol.* **5**:523–528.

Connor, R.C. 1995. The benefits of mutualism: A conceptual framework. *Biol. Rev.* **70**: 427–457.

Crawford, D.L., and S.W. Rissing. 1983. Regulation of recruitment by individual scouts in *Formica oreas* Wheeler (Hymenoptera, Formicidae). *Insectes Soc.* **30**: 177–183.

Cushman, J.H., and A.J. Beattie. 1991. Mutualism: Assessing the benefits to hosts and visitors. *Trends Ecol. Evol.* **6**:193–195.

Douglas, J.M., and J.H. Sudd. 1980. Behavioural coordination between an aphis (*Symydobius oblongus* von Heyden; Hemiptera, Callaphidae) and the ant that attends it (*Formica lugubris* Zetterstedt; Hymenoptera, Formicidae): An ethological analysis. *Anim. Behav.* **28**:1127–1139.

Dugatkin, L.A. 1997. Cooperation among Animals. Oxford: Oxford Univ. Press.

Enquist, M., and O. Leimar. 1993. The evolution of cooperation in mobile organisms. *Anim. Behav.* **45**:747–757.

Fiedler, K. 1991. Systematic, Evolutionary, and Ecological Implications of Myrmecophily within the Lycaenidae (Insecta: Lepidoptera: Papilionoidea). Bonner Zool. Monog. 31. Bonn: Zoologisches Forschungsinstitut und Museum Alexander Koenig.

Fiedler, K., and D. Hageman. 1995. The influence of larval age and ant number on myrmecophilous interactions of the African grass blue butterfly, *Zizeeria knysna* (Lepidoptera: Lycaenidae). *J. Res. Lepid.* **31**:213–232.

Fiedler, K., and V. Hummel. 1995. Myrmecophily in the brown argus butterfly, *Aricia agestis* (Lepidoptera: Lycaenidae): Effects of larval age, ant number, and persistence of contact with ants. *Zoology* **99**:128–137.

Friedman, J.W., and P. Hammerstein. 1991. To trade or not to trade: That is the question. In: Game Equilibrium Models. I. Evolutionary and Game Dynamics, ed. R. Selten, pp. 257–275. Berlin: Springer.

Futuyma, D.J. 1998. Evolutionary Biology. 3d ed. Sunderland, MA: Sinauer.

Harrison, M.J. 1999. Molecular and cellular aspects of the arbuscular mycorrhizal symbiosis. *Ann. Rev. Plant Physiol.* **50**:361–389.

Houston, A.I., and N.B. Davies. 1985. The evolution of cooperation and life history of the dunnock, *Prunella modularis*. In: Behavioural Ecology: Ecological Consequences of Adaptive Behaviour, ed. R.M. Sibly and R.H. Smith, pp. 471–487. Oxford: Blackwell.

Jackson, R.B., J.H. Manwaring, and M.M. Caldwell. 1990. Rapid physiological adjustment of roots to localized soil enrichment. *Nature* **344**:58–60.

Kaitala, A., and C. Wiklund. 1994. Polyandrous female butterflies forage for matings. *Behav. Ecol. Sociobiol.* **35**:385–388.

Kaitala, A., and C. Wiklund. 1995. Female mate choice and mating costs in the polyandrous butterfly *Pieris napi* (Lepidoptera: Pieridae). *J. Insect Behav.* **8**:355–363.

Kokko, H., R.A. Johnstone, and T.H. Clutton-Brock. 2001. The evolution of cooperative breeding through group augmentation. *Proc. Roy. Soc. Lond. B* **268**:187–196.

Leimar, O. 1997. Reciprocity and communication of partner quality. *Proc. Roy. Soc Lond. B* **264**:1209–1215.

Leimar, O., and A.H. Axén. 1993. Strategic behaviour in an interspecific mutualism: Interactions between lycaenid larvae and ants. *Anim. Behav.* **46**:1177–1182.

Leimar, O., and J. Tuomi. 1998. Synergistic selection and graded traits. *Evol. Ecol.* **12**: 59–71.

Maynard Smith, J. 1982. The evolution of social behaviour: A classification of models. In: Current Problems in Sociobiology, ed. Kings College Sociobiology Group, pp. 22–44. Cambridge: Cambridge Univ. Press.

Mesterton-Gibbons, M., and L.A. Dugatkin. 1992. Cooperation among unrelated individuals: Evolutionary factors. *Qtly. Rev. Biol.* **67**:267–281.

Mesterton-Gibbons, M., and L.A. Dugatkin. 1997. Cooperation and the Prisoner's Dilemma: Towards testable models of mutualism versus reciprocity. *Anim. Behav.* **54**:551–557.
Noë, R. 1992. Alliance formation among male baboons: Shopping for profitable partners. In: Coalitions and Alliances in Humans and Other Animals, ed. A.H. Harcourt and F.B.M. de Waal, pp. 285–321. Oxford: Oxford Univ. Press.
Noë, R., and P. Hammerstein. 1994. Biological markets: Supply and demand determine the effect of partner choice in cooperation, mutualism and mating. *Behav. Ecol. Sociobiol.* **35**:1–11.
Noë, R., and P. Hammerstein. 1995. Biological markets. *Trends. Ecol. Evol.* **10**:336–339.
Packer, C. 1977. Reciprocal altruism in *Papio anubis*. *Nature* **265**:441–443.
Pierce, N.E. 1987. The evolution and biogeography of associations between lycaenid butterflies and ants. In: Oxford Surveys in Evolutionary Biology, ed. P.H. Harvey and L. Partridge, vol. 4, pp. 89–116. Oxford: Oxford Univ. Press.
Pierce, N.E., M.F. Braby, A. Heath et al. 2002. The ecology and evolution of ant association in the Lycaenidae (Lepidoptera). *Ann. Rev. Entomol.* **47**:733–771.
Pierce, N.E., R.L. Kitching, R.C. Buckley et al. 1987. The costs and benefits of cooperation between the Australian lycaenid butterfly, *Jalmenus evagoras*, and its attendant ants. *Behav. Ecol. Sociobiol.* **21**:237–248.
Pryor, K., J. Lindbergh, S. Lindbergh, and R. Milano. 1990. A human-dolphin fishing cooperative in Brazil. *Mar. Mamm. Sci.* **6**:77–82.
Robinson, D. 1994. The response of plants and their roots to non-uniform supplies of nutrients. *New Phytol.* **127**:635–674.
Rood, J.P. 1978. Dwarf mongoose helpers at the den. *Zeitschrift Tierpsychol.* **48**:277–278.
Saikkonen, K., U. Ahonen-Jonnarth, A.M. Markkola et al. 1999. Defoliation and mycorrhizal symbiosis: A functional balance between carbon sources and below-ground sinks. *Ecol. Lett.* **2**:19–26.
Selten, R., and P. Hammerstein. 1984. Gaps in Harley's argument on the evolutionary stability of learning rules and in the logic of "tit for tat." *Behav. Brain Sci.* **7**:115–116.
Simon, L., J. Bousquet, R.C. Levesque, and M. Lalonde. 1993. Origin and diversification of endomycorrhizal fungi and coincidence with vascular land plants. *Nature* **363**:67–69.
Smith, S.E., and D.J. Read. 1997. Mycorrhizal Symbiosis. San Diego: Academic Press.
Stander, P.E. 1992. Cooperative hunting in lions: The role of the individual. *Behav. Ecol. Sociobiol.* **29**:445–454.
Trivers, R.L. 1971. The evolution of reciprocal altruism. *Qtly. Rev. Biol.* **46**:33–57.
Way, M.J. 1963. Mutualism between ants and honeydew-producing Homoptera. *Ann. Rev. Entomol.* **8**:307–341.
West-Eberhard, M.J. 1975. The evolution of social behavior by kin selection. *Qtly. Rev. Biol.* **50**:1–33.
Wiklund, C., A. Kaitala, V. Lindfors, and J. Abenuis. 1993. Polyandry and its effect on female reproduction in the green-veined white butterfly (*Pieris napi* L.). *Behav. Ecol. Sociobiol.* **33**:25–33.
Wilkinson, G.S. 1984. Reciprocal food sharing in the vampire bat. *Nature* **308**:181–184.
Woolfenden, G.E. 1975. Florida scrub jay helpers at the nest. *Auk* **92**:1–15.

12

The Red King Effect

Evolutionary Rates and the Division of Surpluses in Mutualisms

Carl T. Bergstrom[1] and Michael Lachmann[2]

[1]Department of Zoology, University of Washington, Seattle, WA 98195–1800, U.S.A.
[2]Max Planck Institute for Mathematics in the Sciences, 04103 Leipzig, Germany

ABSTRACT

Mutualisms generate surpluses. Although much of the theoretical literature to date has focused on mechanisms by which cooperation is stabilized so that these surpluses can continue to be produced and enjoyed, we address a second question: how will these surpluses be distributed among the participants? We approach this question from an evolutionary game theory perspective, exploring how the coevolutionary process "selects" an equilibrium division of the surplus from among the many possibilities.

We place particular emphasis on the importance of the relative rates of evolution of the two species. Contrary to the Red Queen hypothesis, which suggests that fast evolution is favored in coevolutionary interactions, we find that slowly evolving species are likely to gain a disproportionate fraction of the surplus generated through mutualism. This occurs because on an evolutionary timescale, slow evolution effectively ties the hands of a species, allowing it to "commit" to threats and thus "bargain" more effectively with its partner over the course of the coevolutionary process.

INTRODUCTION

Mutualist partners benefit mutually, by definition. That is to say, when individuals engage in an interspecific mutualism, they enjoy benefits above and beyond what they would have enjoyed in the absence of the interaction. (Using the terminology from economics, we call these benefits the surplus generated by the mutualism.) Despite the bilaterally advantageous nature of such interactions, the participants in a mutualism rarely have entirely coincident interests. Each would benefit from altering the arrangement so as to increase its own share of the surplus at the expense of its partner. How are mutualisms established and maintained despite these conflicts? This question can be subdivided: What prevents a mutualism from breaking down as individuals find ways to exploit their partners over evolutionary time? If mutualism does not break down, what determines the allocation of the surplus among partners?

What Prevents the Exploitation of Mutualism?

To date, the majority of the empirical and theoretical studies of mutualism evolution have focused on this question. Structurally, the theoretical issue — how cooperation is maintained despite incentives to defect — is very similar to that addressed in the extensive literature on the evolution of intraspecific cooperation. Since partners are not conspecifics, however, the kin selection explanations commonly employed to explain intraspecific generosity cannot be invoked to explain the interspecific analog. Instead, investigators have typically searched for mechanisms that deter cheating (or at least ameliorate the cost of being cheated) by more direct means. Such mechanisms include reciprocal altruism, partner choice, sanctioning, and by-product mutualism or pseudoreciprocity. Bergstrom et al. (this volume) provide an overview of these alternatives. Thus, we will not consider this question in detail here.

How Will the Benefits of Mutualism Be Divided?

Far less attention has been given to the matter of what happens once the mutualistic association is somehow stabilized. (Welcome exceptions include Bowles and Hammerstein [this volume] and some of the "biological markets" literature, including Bshary and Noë [this volume].) In particular, how will the benefits from the interaction be allocated among the participants? Though a mutualistic interaction offers benefits to both species, the two species will obviously have different interests with respect to the actual division of the surplus: each would benefit from gaining a larger share.

In some cases, the goods being "traded" are provided in very different currencies and the "exchange rates" between them are essentially set by mechanistic constraints. In such cases, division of the surplus is straightforward. Cleaning mutualisms, such as those described by Bshary and Noë (this volume), provide one of the best examples. In these interactions, the cleaner gets the benefit of a ready food source, and the "client" gets the benefit of having its parasite load reduced. The potential for cheating — cleaners feeding on live tissue or clients preying on cheaters, for example — adds a degree of extra complexity, as does competition among clients for cleaners. Nonetheless, if market forces or other mechanisms do ensure cooperation between a cleaner and a client, the division of the benefits is relatively straightforward (Bshary and Noë, this volume).

The allocation of benefits, however, is not always so clear, as we can see by observing the mutualistic association between ants and lycaenid butterfly caterpillars (Pierce 1987, 2001). These caterpillars, largely protected by the ants from parasitoids (a huge contributor to mortality), enjoy enormous increases in survivorship to and during pupation (Pierce and Mead 1981; Pierce and Easteal 1986). Consequently, they can afford an extended developmental period, during which they are able to generate a sugar- and protein-rich exocrine secretion with which to purchase continued protection at the expense of a reduced rate of growth (Hill and Pierce 1989; Baylis and Pierce 1992; Pierce et al. 1987). In this situation, there is no single obvious division of the surplus. In general, then, at what rate should the lycaenids provision their ant attendants? In addition, how much should the ants "demand" in return for tending to the caterpillars?

Evolutionary Rate and the Coevolutionary Process

Here, we describe the way in which dynamic evolutionary game theory can be used to explore how surpluses will be divided among mutualist partners. We will pay particular attention to the role of evolutionary rate in determining the properties (in particular, the allocation of benefits) of mutualisms. Partners in coevolutionary interactions may evolve at different rates for a number of reasons, including differences in generation time, differences in the importance of the interaction, differences in population size, and differences in the amount of segregating genetic variation (Dawkins and Krebs 1979).

Theoretical and empirical studies of coevolution have explored the consequences of evolutionary rates and coevolutionary races in substantial detail; however, the present approach represents something of a departure from these earlier studies in its emphasis on mutualistic interactions. Most previous analyses have dealt with antagonistic coevolution, such as that between predators and prey or hosts and parasites. In these situations, species pairs become locked into "rat races" (Rosenzweig 1973) or "arms races" (Dawkins and Krebs 1979) with each rushing to evolve the upper hand in the interaction. The end result is a Red Queen process (Van Valen 1973), in which the two species each have to evolve rapidly just to keep up with one another. As Lewis Carroll wrote, "it takes all the running you can do, to keep in the same place."

Do mutualisms evolve by similar dynamical processes, with species racing to keep ahead of their partners (Herre et al. 1999)? Is a rapidly evolving species likely to fare better than a slowly evolving one? Here we describe how these questions can be addressed using an alternative approach to modeling the evolution of mutualism (Bergstrom and Lachmann 2003) and summarize new results which suggest that, in contrast to the Red Queen theory, slower rates of evolution may lead to favorable outcomes in the evolution of mutualism.

METHODS FOR MODELING MUTUALISM

Game theory is the study of decision making in a social context. As such, game theory provides a set of tools for analyzing the decision problem that an individual faces when her fate depends both on her own choices and on the choices of others. Traditionally, game theory has focused on identifying Nash (or related) equilibria: combinations of strategies for each participant such that no participant can gain from a unilateral change in strategy. Although this approach has proven to be extremely valuable in biology, many strategic situations or "games" turn out to have multiple equilibria, and the basic theory does little to distinguish among them (Samuelson 1997).

Resolving this *equilibrium selection* problem requires some sort of extension to the basic Nash equilibrium framework. One of the most successful extensions derives from the work of Maynard Smith and Price (1973). These authors studied how evolutionary processes (e.g., evolution by natural selection) would lead to the selection of certain strategies in populations of game-playing individuals. In general, this evolutionary game theory approach assumes that a population of agents play a given game against one another repeatedly.[1] The agents change their strategies at some rate, based on their own past experiences or those of others. Strategy change is assumed to be myopic, toward immediate improvement with no consideration of the long-term consequences. Agents may occasionally mutate or experiment, trying new strategies at random. Examples of such processes include evolution by natural selection in asexual or sexual populations, cultural transmission systems in which individuals copy successful neighbors, and learning processes in which individuals alter their strategies in accord with their previous payoffs.

Among these processes, the replicator dynamics plays a central role, in that (a) it corresponds to simple deterministic biological model of asexual reproduction with fitnesses proportional to expected payoffs, (b) it is relatively simple to analyze, and (c) many other processes can be shown to share with it the same equilibrium points and stability properties (Samuelson and Zhang 1992; Cressman 1997), and in some cases, even the same dynamics (Binmore et al. 1995; Schlag 1998). Throughout this chapter, we use the replicator dynamics as model of evolution by natural selection. However, the aforementioned convergence properties imply that our findings will also pertain to systems in which strategies change by other processes (e.g., learning) as well.

In many simple coevolutionary interactions, players come from two separate populations to engage in pairwise interactions. Such circumstances can be modeled using bimatrix games (Weibull 1995; Hofbauer 1996; Hofbauer and Sigmund 1998), also known as role asymmetric games (Maynard Smith 1982), in which the two populations have distinct payoff matrices and strategy frequencies. Here, we restrict ourselves to consideration of two-player bimatrix games. The simplest of these are 2×2 games, which can be represented by the following payoff matrix:

[1] Though the game is assumed to be played repeatedly, most evolutionary game theory models do not endow players with the sort of individual recognition and memory of past events that are necessary to play "responsive" strategies, e.g., Tit-for-Tat. Instead, models typically focus on how strategies for playing the 1-shot game evolve over time.

	L	R
U	a, e	b, f
D	c, g	d, h

With a bit of arithmetic manipulation, we can derive the replicator dynamics for these simple bimatrix games, where the players come from two separate populations with evolutionary rates n and m respectively (Bergstrom and Lachmann 2003). Here, x is the frequency of L players in population 1, y is the frequency of U players in population 2, and $\pi(D, z)$ is the payoff to choosing strategy D when a fraction z of the other population plays strategy L:

$$\begin{aligned} \dot{x} &= mx\left(\pi\,(L,y) - [\pi\,(L,y)x + \pi\,(R,y)(1-x)]\right) \\ \dot{y} &= ny\left(\pi\,(U,x) - [\pi\,(U,x)y + \pi\,(D,x)(1-y)]\right) \end{aligned} \tag{12.1}$$

Qualitatively, these 2 × 2 games allow only a limited range of dynamic behaviors. We can see this by examining a strategically equivalent game; equivalent replicator dynamics can always be constructed by renormalizing matrix (1) so that the off-diagonal elements are zero (Hofbauer and Sigmund 1998):

	L	R
U	α, β	$0, 0$
D	$0, 0$	γ, δ

Setting $\alpha = a - c$, $\beta = e - f$, $\gamma = d - b$, and $\delta = h - g$, the evolutionary dynamics are preserved.[2] Qualitatively, (generic) 2 × 2 games afford four different types of evolutionary dynamics, characterized by what happens along each edge (Hofbauer and Sigmund 1998; see Figure 12.1).

Types I and II have only one stable equilibrium to which the dynamics always converge, and thus these games are of little interest so far as equilibrium selection is concerned. Type III has two stable equilibria, one at the upper right corner and one at the lower left corner. Type IV has no stable external equilibria, but only the mixed strategy equilibrium in the interior of the strategy frequency

---

[2] Although this renormalization does not alter the evolutionary dynamics for a population playing this game, it is important to note that by renormalizing in this way we do not necessarily preserve the relative value of (U,L) and (D,R) outcomes to each player. Player 1 may receiver a higher payoff from the (U,L) combination when the game is written in form (1), and a higher payoff from the (D,R) combination when the game is written in form (3). This renormalization can actually transform a coordination team game (in which both players prefer the same equilibrium) into a Battle-of-the-Sexes game (in which each player prefers a different Nash equilibrium). Here we are interested in more than just evolutionary dynamics: we wish to compare payoffs across equilibria (cf. section on LOCAL DYNAMICS OF MUTUALISM) and in structured population models for which equilibrium payoffs determine carrying capacities (see section on HIGHER-LEVEL POPULATION STRUCTURE). Since the renormalized form (3) of the game does not preserve these comparisons, we will break from common convention and work with games in their unnormalized forms.

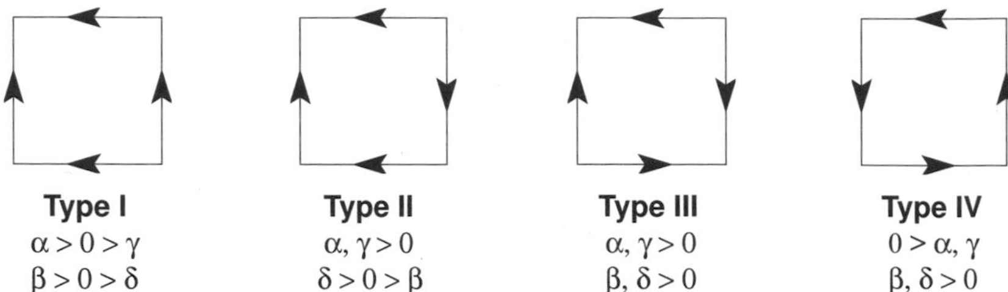

Figure 12.1 The four basic types of evolutionary dynamics for 2 × 2 games. Adapted from Hofbauer and Sigmund (1998).

space. Because we are interested in how the evolutionary process chooses among a set of possible equilibrium divisions of the mutualistic surplus, the games we examine here are Type III games with two equilibria.

As in previous studies, we consider interactions in which both players stand to gain from the interaction if they can find a way to cooperate. However, the approach described here differs in two respects. Rather than looking at what prevents breakdown, we examine how the gains from the interaction will be distributed between the two players, *in the absence of incentive to defect on an established cooperative arrangement.* Thus, instead of examining a Type I Prisoner's Dilemma interaction with one Nash equilibrium, we examine a Type III coordination-type game with two Nash equilibria. This provides us with a simple model that shares a common feature of many game-theoretic interactions: multiple Nash equilibria exist, but different players have different "preferences" over the set of equilibria. We would like to understand which equilibrium will be selected in an evolutionary system. Second, we go "back to basics," in the sense that we will examine only the simple one-shot 2 × 2 game dynamics. The basic rationale for doing so is simple. Regardless of the complex strategies of reward and punishment, regardless of partner choice and market function, regardless of the series of moves and countermoves involved, successful mutualistic interaction will ultimately generate a surplus, and this surplus will ultimately have to be divided. We defer the issue of how the mutualism is enforced and how bargaining proceeds, so as to concentrate on the role of the evolutionary dynamics in shaping the division of the surplus. By doing so and by choosing a simple 2 × 2 game with its small strategy space as in our model, we can examine the question of surplus division in the simplest possible context. Once we understand the workings of this system, we can extend the model in any number of ways. In the final section, we speculate on the likely outcomes of such extensions.

LOCAL DYNAMICS OF MUTUALISM

We are interested in how organisms split the surplus from a nascent mutualism. This problem is closely related to bargaining problems treated in economics (Nash 1950, 1953; Rubinstein 1982): two or more individuals seek to establish a mutually beneficial agreement (e.g., how to divide a surplus) by common consensus, but their interests conflict regarding the precise terms of the agreement (Osborne and Rubinstein 1990). Frequently in these games, many possible divisions of the surplus are stable in the Nash equilibrium sense. Given this multitude of equilibria, what sort of division should we expect to observe in practice? One can imagine a host of models to explain how such a division could take place, and indeed the study of such models is a major component of bargaining theory (Osborne and Rubinstein 1990). Although an axiomatic approach (Nash 1950) or rationality considerations (Rubinstein 1982) can resolve the many possible equilibria, it might be more appropriate to employ population-based evolutionary models to the study of mutualisms.

Then how, precisely, should we model this situation? For example, how can we model a scenario in which two individuals have to split a surplus of three units? Unfortunately, dynamic evolutionary models can be difficult to apply to full-blown bargaining scenarios because of the infinite strategy spaces of these games. Fortunately, one can learn a great deal by looking at the evolutionary dynamics of populations playing simpler one-stage games.

One of the classic one-stage games used is known as the Nash bargaining game (Nash 1953; Osborne and Rubinstein 1990). Two players have to divide a surplus of 3 units. Each player simultaneously "demands" an amount of the surplus. If the two demands sum to 3 units or less, each player gets the amount that she demanded. If the total of the two demands exceeds 3 units, each player gets 0. Because any demand from 0 to 3 is a legitimate strategy in the Nash bargaining game, even this game has an infinite space. To study the evolutionary dynamics, we will make yet another simplification and look at a "discrete" or "mini-game" form (Skyrms 1996; Sigmund et al. 2001):

	Generous	Selfish
Selfish	2, 1	0, 0
Generous	1, 1	1, 2

In this mini-game form of the Nash bargaining game, each player can demand either 1 or 2 units of the surplus; the players receive their demands so long as the two demands are compatible with a total surplus of 3 units. Let us now extend this model slightly by replacing the (1,1) payoffs to mutual generous offers with a payoff (k, k):

	Generous	Selfish
Selfish	2, 1	0, 0
Generous	k, k	1, 2

When $k = 1$, we have the Nash bargaining mini-game, as shown above. When $k = 1.5$, the entire surplus is retained and split evenly; the game becomes a Hawk–Dove game with resource benefit 1 and cost 3 of fighting. When $k = 0$, two generous offers lead to a coordination failure as severe as that resulting from two selfish ones: players suffer a complete loss of mutualistic surplus and a standard battle-of-the-sexes game results. Thus, parameter k plays an important role in determining the effect of evolutionary rate on equilibrium selection.

We begin by looking at the dynamics of this game with $k = 1$. Figure 12.2 shows a set of evolutionary trajectories for the space of strategy frequencies for species 1 on the y axis and species 2 on the x axis, under the replicator dynamics (2) with the two populations evolving at equal rates. Almost every trajectory ends at one of two resting points: the upper left corner in which species 1 enjoys a favorable division of the surplus, or the lower right corner in which species 2 enjoys a favorable division. The eventual end point is determined by the initial frequencies; the set of all points from which the dynamics lead to a given equilibrium is called the domain of attraction of that equilibrium. The diagonal line running from lower left to upper right corners represents the separatrix between the two domains of attraction. All points on the same side of this separatrix go to the same equilibrium. The horizontal line running through the middle of the strategy space separates the points at which species 2 evolves to be more generous (above this line) from those at which it evolves to be more selfish (below this line). The vertical line strikes a similar division for species 1. These two lines together partition the strategy space into four quadrants, discussed further below.

Clearly, the ultimate division of the mutualistic surplus will depend on the starting strategy frequencies in each species. Thus we cannot answer the question, "How will the surplus be split?" without knowing where the system started. Nonetheless, one reasonable measure of the likelihood of various outcomes is simply the relative size of the various domains of attraction. All else being equal, we might expect that equilibria with large domains of attraction will be reached more often than equilibria with small domains of attraction.

What determines, however, the sizes of the domains of attraction? Bergstrom and Lachmann (2002) show that both the game payoffs and the relatively evolutionary rates matter. In particular, the relatively evolutionary rates of the two species determine the way that the separatrix curves across the strategy space.

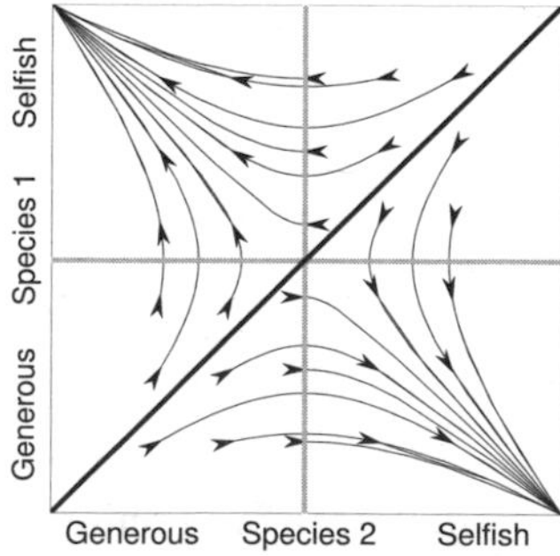

Figure 12.2 Evolutionary trajectories when $k = 1$. Horizontal and vertical lines show the places at which the change in strategy frequency switches direction for players 1 and 2, respectively. Diagonal is the separatrix between the two domains of attraction.

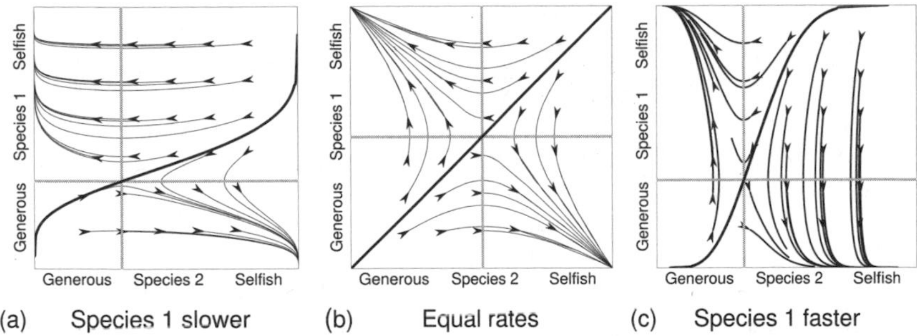

Figure 12.3 The effect of evolutionary rate on domains of attraction when $k = 1.5$: (a) species 1 evolves eightfold slower than species 2; (b) equal rates of evolution; (c) species 2 evolves eightfold slower. The slower that species 1 evolves, the larger the domain of attraction around its favored equilibrium at the upper left-hand corner.

Depending on the payoffs of the game, this can either increase or decrease the domain of attraction for the slower player. Figure 12.3 shows the strategy space and evolutionary dynamics for $k = 1.5$. Here the domain of attraction of player 1's favored equilibrium (the upper left corner) increases as player 1's relative rate of evolution decreases. This is the first manifestation of what we call the "Red King effect."

Note that as species 1 evolves at an increasingly slower rate, intense movement across the strategy space occurs along the horizontal axis. This strategy change occurs as the result of evolutionary change by species 2. This increases the fraction of the upper right-hand quadrant that goes to species 1's favored equilibrium, while decreasing the fraction of the lower left-hand quadrant. Relative evolutionary rates do not matter in the upper left- and lower right-hand quadrants; any point in either of these quadrants goes to the equilibrium in the same quadrant regardless of evolutionary rates.

Thus, the effect of evolutionary rate on the size of domains of attraction depends on the chance that the starting point is in the lower left quadrant versus the upper right quadrant. As summarized by Figure 12.4, the fast-evolving species "gets" the lower left quadrant and "loses" the upper right one.[3]

What determines, however, the quadrant in which the coevolutionary process is likely to begin? One important factor will be the size of each quadrant. As k increases, the area of the upper right quadrant—where slow evolution is favored—also increases. Indeed, the slowly evolving species will have a larger domain of attraction around its favored equilibrium whenever $k > 1$, whereas the

---

[3] Our results may explain a curious observation reported by Doebeli and Knowlton (1998; see also Figure 3C therein). In their simulations of mutualism evolution, based on an iterated Prisoner's Dilemma model, they found that the more slowly evolving species received higher payoffs.

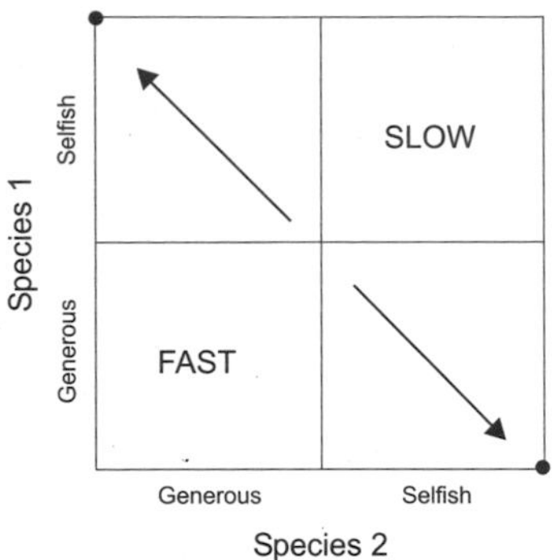

Figure 12.4 Summary of local dynamics. In the upper left and lower right quadrants, all evolutionary trajectories reach the upper left and lower right equilibria (black dots) respectively. In the upper right quadrant, the slow species reaches its favored equilibrium; in the lower left quadrant, the fast species reaches its favored equilibrium.

faster-evolving species will have a larger domain of attraction when $k < 1$ (Bergstrom and Lachmann 2003).

This result makes intuitive sense in light of bargaining theory. In bargaining games, it is well known that there may be a strategic advantage to "having one's hands tied" during bargaining. This is valuable because threats of a constrained player become more credible, while threats against this player are rendered ineffective. Since susceptibility to threats often acts as a major determinant of the strength of one's bargaining position, this is a significant advantage.

The Red King effect can be seen as simply this: a slowly evolving species has its hands tied in the coevolutionary interaction by which division of the surplus is "negotiated." Here the bargaining process does not take place within a single play of the game, but rather occurs over the course of the coevolutionary interaction between the players. In other words, the coevolutionary process can be viewed as a bargaining process through which the two species arrive at an equilibrium to the Nash bargaining game through a series of evolutionary moves and counter moves. In this bargaining process, fast evolution does not allow a species to outrun a partner — it simply causes this species to yield to whatever threats are made. This is captured by the local dynamics described earlier.

Of course, the initial proposals brought to the table by the negotiating parties will also have a major impact on the outcome of a negotiation. In the mutualism example considered here, if both species initially ask for more than their share of the proverbial pie, susceptibility to threat will be important. What will be the initial proposals that the species bring to the table? We explore this question below.

HIGHER-LEVEL POPULATION STRUCTURE

Evolutionary game theory typically assumes that the populations of players materialize fully formed and out of thin air at the beginning of the evolutionary process under consideration. Obviously, the real situation is somewhat more complicated. When populations are formed anew, their members must have come from *somewhere*, and this somewhere may have had a significant influence on the strategies that they bring with them to the new population.

Thus, when potential mutualists come together in a given location, what should we imagine about their past histories, their distribution of strategy choices, and so forth? One straightforward approach is to assume that upon founding a new patch, individuals use the same initial strategies that they had employed in their natal patches. We can model this by looking at a structured population of players, in which the dynamic process of strategy change treated above occurs in parallel in a set of distinct local patches. Each local patch then sends out migrants to join existing patches or to found new patches. There is an extensive literature on the workings of such structured-population models (Bergstrom 2002). Here we have selected to work with one of the simplest of these models, the haystack-type model (Maynard Smith 1964; Cohen and Eshel 1976). We expect that other structured population models will yield qualitatively similar results in most cases.

Our haystack model works as follows. The environment is divided into a set of local patches. Every "season," a small number of founder individuals of each the two species colonize each patch. Once colonization has occurred, within each patch during the course of a single season the strategy frequencies change according to the local dynamics characterized in the previous section.

Note that these local dynamics characterize changes in strategy frequencies but not in population size. In the structured population model, we are also interested in how population sizes change according to the strategies played. For simplicity, we will assume that within the course of a single season, each species grows to a carrying capacity in each patch. The exact magnitude of the carrying capacity for each species reflects the "favored" or "disfavored" nature of the equilibrium reached in the patch. That is, a species will have a higher carrying capacity in a given patch if it reaches its favored equilibrium than if it reaches its disfavored equilibrium. We will assume that each season is sufficiently long that every local subpopulation reaches an equilibrium with respect to strategy frequencies, so that we only need to specify carrying capacities for the two equilibria and not for any out-of-equilibrium combinations of strategy frequencies.

At the end of the season, patch boundaries are erased. Individuals disperse, and subpopulations are formed of individuals chosen at random from the global population. A new season then begins and the process starts anew.

Figure 12.5 shows how the domain of attraction around each equilibrium shifts as we take into account the higher-level population structure. Under local dynamics, domains of attraction are equal in size for $k = 1$. However, global dynamics favor slowly evolving species. This species (species 2) has a larger domain of attraction around its favored equilibrium at the lower right-hand corner.

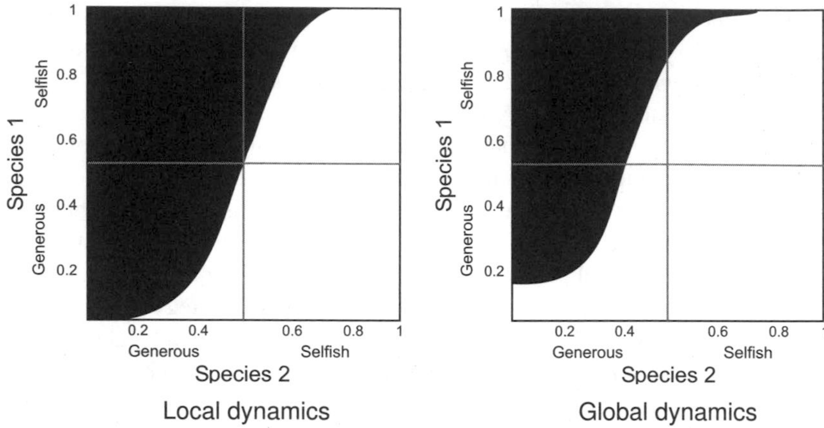

Figure 12.5 Domains of attraction in the local dynamics and for the higher-level population structure with $k = 1$, when species 1 evolves 8-fold faster than species 2. For the global dynamics, each new patch is founded by 9 individuals and each species' carrying capacity at its favored equilibrium is 4 times that at the disfavored equilibrium.

Why does this happen? Consider the process by which a new subpopulation is formed. Members arrive from other subpopulations. Subpopulations at the equilibrium where species 1 is favored, i.e., where species 1 is playing selfishly, have a higher carrying capacity for species 1 and thus contribute more species 1 individuals than do subpopulations where species 2 is favored. Therefore, the odds are that a majority of incoming species 1 individuals will have arrived from a subpopulation in which they were playing selfishly. Similarly, a majority of incoming species 2 individuals will most likely have come from a subpopulation in which *they* were playing selfishly. Consequently, when a new subpopulation is first established, the majority of players therein are likely to be playing selfishly: the newly formed population is likely to begin with a set of strategy frequencies belonging to the shaded quadrant in Figure 12.6. We know that local dynamics favor the slow evolver under these circumstances. Thus in each newly formed subpopulation, slowly evolving species will have a relative advantage.

We can visualize this argument as follows: if a proportion s of the patches reaches an equilibrium that favors species 1, and $(1 - s)$ reach one that favors species 2, and the relative size of the carrying capacities for species 1 and 2 are α and β, then at the end of a season the proportion of individuals of species 1 playing the selfish strategy in the global pool will be $s/(s + (1 - s)\alpha)$, and the proportion of species 2 playing the selfish strategy will be $(1 - s)/(1 - s + s\beta)$.

Thus the relative frequencies of the strategy types in the global pool will lie somewhere along the dark curve depicted in Figure 12.6. This curve passes through the upper-right quadrant, where slow evolution is favored, and not through the lower-left one, where fast evolvers are favored.

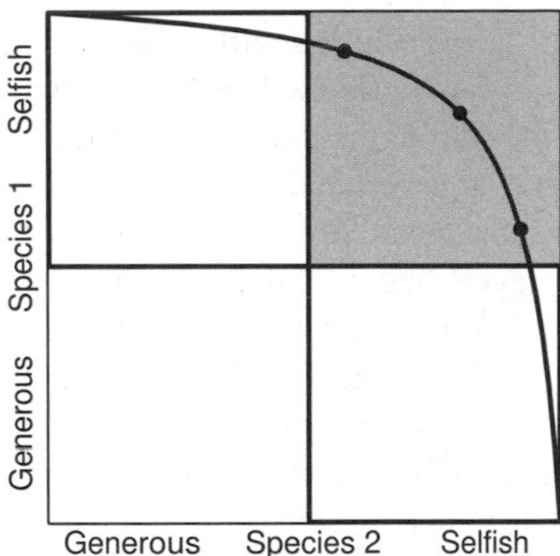

Figure 12.6 Summary of global dynamics. Each subpopulation ends with either species 1 playing generous and species 2 playing selfish, or visa versa. The fraction of subpopulations in each state determines the expected frequencies in each new subpopulation at the start of the next season. Recall that local dynamics favor slowly evolving species in the shaded upper right quadrant. Carrying capacity ratios are $\alpha = 4$ and $\beta = 4$.

Once again we can see these results in light of bargaining theory. As mentioned above, the initial proposals brought to the table by the negotiating parties will also have a major impact on the outcome of a negotiation. In the mutualism example considered here, if both species initially ask for more than their share of the proverbial pie, susceptibility to threat will be important. But what will be the initial proposals that the species bring to the table? We have argued that in coevolutionary interactions, population structure bears critically upon this question. If new patches are formed by immigrants from other patches, individuals will come together prepared by evolution to pursue a division similar to that which they were receiving in their previous patches. When carrying capacity of a patch is affected by the division of mutualistic surplus, most players entering a new patch will arrive "demanding" more than half of the surplus. This situation (when parties do not initially agree on the division because both expect a majority share) is precisely when it pays to have one's hands tied in the negotiations.

DISCUSSION

Beyond Mini-games

Thus far, we have discussed the evolutionary dynamics associated with populations playing simple 2×2 games. What happens when the interactions in question are broader in scope? What happens, for example, when individuals of the two species are playing the full Nash bargaining game in which each can demand any amount from 0 to 3?

For this game, a full analysis of the local dynamics, which take place on an infinite-dimensional simplex and which will depend on many particulars of the model, would be very difficult. Nonetheless, analogy to the 2×2 game provides us with a good sense of how evolution will proceed. In the full game, just as in the 2 × 2 counterpart, slow evolution will be favored when players are demanding too much, so that the sum of the demands exceeds 3 and bargaining breaks down. By starting with a large demand and evolving slowly, one species forces the other to "yield" and to demand less than half of the total. Fast evolution will be favored when the players begin by demanding less than 3, because the fast evolver will be able to adjust its demand to claim the remainder of the 3 units.

Moreover, as in the 2 × 2 version of the game, the higher-level population dynamics will ensure that players come together with demands in the region where slow evolution is favored: these dynamics will act to bring together players who are demanding too much, rather than too little. Imagine a global population composed of subpopulations which have reached a range of different equilibrium arrangements: $(\delta, 3-\delta), (2\,\delta, 3-2\,\delta), \ldots, (3-\delta, \delta)$. At the beginning of a new season, the majority of species 1 players will come from populations where species 1 had a high carrying capacity, i.e., populations where species 1 was receiving a relatively large fraction of the total benefits. Similarly, species 2 players will come from populations in which species 2 had a high carrying capacity, i.e., populations where species 2 was receiving a relatively large fraction of the total benefits. Thus the large majority of the newly founded subpopulations will be composed of players who together demand a total exceeding 3. This is the region in which slow evolution is favored; consequently we expect the higher-level population structure to favor slow evolution in the full Nash bargaining game as well.

Interactions among Humans

Here we have focused primarily on mutualistic associations among nonhuman agents evolving by natural selection. Will similar processes apply to human interactions? We argue, they may. Various processes of strategy change, including learning and copying behaviors, can yield qualitatively similar outcomes to those observed in systems that change according to replicator dynamics.

This is all good and well, but human beings (or even real-world mice) do not live in haystacks with the sort of structure modeled above. Can analogous higher-level selection processes nonetheless operate? We stress that these sorts of structured-population dynamics require neither small founding populations (see Bergstrom and Lachmann 2003) nor that some sort of life-or-death group selection take place. Such a process only requires that *the majority of players in each new round come from places where they did well in the previous round.* This could occur for many reasons. For example, in human interactions, players may decide whether to continue participating in some two-sided interaction based on their past experience. Players who have done well may continue to engage in the interaction, whereas those who have done poorly may choose to opt out and do something else instead. Under certain circumstances, players who reached the favored outcome will return to play again, players who reached the disfavored outcome will take the outside option, and the conditions will be met for the Red King effect to operate on the higher level of population structure.

This phenomenon may be rather general to human interactions, such that the individuals who choose to participate at any given time are either new to the interaction in question or have a past record of success. Thus individuals choosing to play may have a higher-than-average expected return from the game. In bargaining games of this sort, this means that the individuals choosing to participate will enter each new situation asking for more than an even share.

SUMMARY

The study of interspecific mutuality allows biologists an unparalleled opportunity to explore the mechanisms beyond kin selection by which coordination and cooperation can evolve. Although much of the theoretical literature to date has focused on mechanisms by which cooperation is stabilized, we addressed the issue of how benefits that arise as a consequence of mutualism are distributed among the participants. We have given particular attention to the role of evolutionary rate in determining coevolutionary outcomes. Most notably, recent results suggest that, contrary to the Red Queen hypothesis, slow evolution may actually lead to favorable outcomes in some coevolutionary interactions.

ACKNOWLEDGMENTS

We thank Naomi Pierce and Larry Samuelson for extensive and valuable discussion, and gratefully acknowledge the many useful suggestions and comments provided by Eric Alden Smith, Sam Bowles, and the other participants of this Dahlem Workshop. This research was supported in part by the Santa Fe Institute.

REFERENCES

Baylis, M., and N.E. Pierce. 1992. Lack of compensation by final instar larvae of the myrmecophilous lycaenid butterfly, *Jalmenus evagoras* for the loss of nutrients to ants. *Physiol. Entomol.* **17**:107–114.

Bergstrom, C.T., and M. Lachmann. 2003. The Red King effect: When the slowest runner wins the coevolutionary race. *Proc. Natl. Acad. Sci. USA* **100**:593–598.

Bergstrom, T.C. 2002. Evolution of social behavior: Individual and group selection. *J. Econ. Persp.* **16**:67–88.

Binmore, K., L. Samuelson, and R. Vaughan. 1995. Musical chairs: Modelling noisy evolution. *Games Econ. Behav.* **11**:1–35.

Cohen, D., and I. Eshel. 1976. On the founder effect and the evolution of altruistic traits. *Theoret. Pop. Biol*. **10**:276–302.

Cressman, R. 1997. Local stability of smooth selection dynamics for normal form games. *Math. Soc. Sci.* **34**:1–19.

Dawkins, R., and J.R. Krebs. 1979. Arms races between and within species. *Proc. Roy. Soc. Lond. B* **205**:489–511.

Doebeli, M., and N. Knowlton. 1998. The evolution of interspecific mutualisms. *Proc. Natl. Acad. Sci. USA* **95**:8676–8680.

Herre, E.A., N. Knowlton, U.G. Mueller, and S.A. Rehner. 1999. The evolution of mutualisms: Exploring the paths between conflict and cooperation. *Trends Ecol. Evol.* **14**:49–53.

Hill, C.J., and N.E. Pierce. 1989. The effect of adult diet on the biology of butterflies. 1. The common imperial blue, *Jalmenus evagoras*. *Oecologia* **81**:249–257.

Hofbauer, J. 1996. Evolutionary dynamics for bimatrix games: A Hamiltonian system? *J. Math. Biol.* **34**:675–688.

Hofbauer, J., and K. Sigmund. 1998. Evolutionary Games and Population Dynamics. Cambridge: Cambridge Univ. Press.

Maynard Smith, J. 1964. Group selection and kin selection. *Nature* **201**:1145–1147.

Maynard Smith, J. 1982. Evolution and the Theory of Games. Cambridge: Cambridge Univ. Press.

Maynard Smith, J., and G.R. Price. 1973. The logic of animal conflict. *Nature* **246**:15–18.

Nash, J.F. 1950. The bargaining problem. *Econometrica* **18**:155–162.

Nash, J.F. 1953. Two-person cooperative games. *Econometrica* **21**:128–140.

Osborne, M.J., and A. Rubinstein. 1990. Bargaining and Markets. New York: Academic.

Pierce, N.E. 1987. The evolution and biogeography of associations between lycaenid butterflies and ants. In: Oxford Surveys in Evolutionary Biology, ed. P.H. Harvey and L. Partridge, vol. 4, pp. 89–116. Oxford: Oxford Univ. Press.

Pierce, N.E. 2001. Peeling the onion: Symbioses between ants and blue butterflies. In: Model Systems in Behavioral Ecology, ed. L.A. Dugatkin, pp. 41–56. Princeton, NJ: Princeton Univ. Press.

Pierce, N.E., and S. Easteal. 1986. The selective advantage of attendant ants for the larvae of a lycaenid butterfly, *Glaucopsyche lygdamus*. *J. Anim. Ecol.* **55**:451–462.

Pierce, N.E., R.L. Kitching, R.C. Buckley et al. 1987. Costs and benefits of cooperation between the Australian lycaenid butterfly, *Jalmenus evagoras* and its attendant ants. *Behav. Ecol. Sociobiol.* **21**:237–248.

Pierce, N.E., and P.S. Mead. 1981. Parasitoids as selective agents in the symbiosis between lycaenid butterfly larvae and ants. *Science* **211**:1185–1187.

Rosenzweig, M.L. 1973. Evolution of the predator isocline. *Evolution* **27**:84–94.

Rubinstein, A. 1982. Perfect equilibrium in a bargaining model. *Econometrica* **50**:97–109.

Samuelson, L. 1997. Evolutionary Games and Equilibrium Selection. Cambridge, MA: MIT Press.

Samuelson, L., and J. Zhang. 1992. Evolutionary stability in asymmetric games. *J. Econ. Theory* **57**:363–391.

Schlag, K. 1998. Why imitate, and if so, how? A bounded rational approach to multi-armed bandits. *J. Econ. Theory* **78**:130–156.

Sigmund, K., C. Hauert, and M.A. Nowak. 2001. Reward and punishment. *Proc. Natl. Acad. Sci. USA* **98**:10,757–10,762.

Skyrms, B. 1996. Evolution of the Social Contract. Cambridge: Cambridge Univ. Press.

Van Valen, L. 1973. A new evolutionary law. *Evol. Theory* **1**:1–30.

Weibull, J.W. 1995. Evolutionary Game Theory. Cambridge, MA: MIT Press.

Standing, left to right: Carl Bergstrom, Laurent Keller, Herb Gintis, Olof Leimar, Martin Daly, Ronald Noë
Seated, left to right: David Queller, Redouan Bshary, Richard Connor, Steve Frank, Judie Bronstein

13

Group Report: Interspecific Mutualism

Puzzles and Predictions

Carl T. Bergstrom, Rapporteur

Judith L. Bronstein, Redouan Bshary, Richard C. Connor, Martin Daly, Steven A. Frank, Herbert Gintis, Laurent Keller, Olof Leimar, Ronald Noë, and David C. Queller

INTRODUCTION

In 1859, Charles Darwin wrote:

> *If it could be proved that any part of the structure of any one species had been formed for the exclusive good of another species, it would annihilate my theory, for such could not have been produced through natural selection.*
>
> —*The Origin of Species*, Chapter 6

This was a bold prediction indeed! Many species were known to provide one another with benefits, and a good fraction of these appeared to have evolved elaborate mechanisms by which to do so. The evolutionary study of *interspecific mutualism*—defined as mutually positive interactions between species—aims to explain such observations by identifying the direct or indirect benefits that accrue to an actor from the actions that benefit its partner.

A hundred and forty years after Darwin stepped out onto this proverbial limb, where do we stand? Has the branch given way under the weight of his bold prediction, or has the branch held firm as theory and observation remain in close accord? Have we addressed Darwin's challenge to our full satisfaction, or do we need to seek new principles by which to explain the full range of observed interactions? More specifically, what observations do we need to explain, and what explanations do we have to offer? In this report, we describe the major conceptual foundations that are applied to the study of mutualism, and we ask what questions remain unanswered.

MAJOR QUESTIONS

To provide a thorough and satisfying evolutionary explanation of a given phenomenon, we typically need to identify mechanisms responsible for two separate processes: First, how did the phenomenon of interest initially arise? Second, what prevents its dissolution once established? These two questions stand at the center of the study of mutualism and comprise our first two major questions in the field of mutualism:

- By what processes do mutualisms form?
- How do mutualisms persist over time, despite the ever-present threat of exploitation from within or by third parties?

Beyond obtaining a basic answer to each of these problems, we would like to be able to say at least something about how the ecological and evolutionary context affects the biological outcomes observed. In essence, we seek to understand how the properties of mutualisms change in response to changes in underlying parameters. In general, we would like to know:

- What factors promote or inhibit the formation and persistence of mutualisms?
- What factors influence the partitioning of benefits between the mutualist partners?

All of these questions can be addressed on both ecological and evolutionary timescales. For example, if we ask how mutualisms are formed, we can explore how partners find one another in real time and how the ecological dynamics of growth and dispersal act to structure the patterns of mutualistic association in time and space. Alternatively, we could explore the processes by which novel mutualisms arise over evolutionary time among previously unassociated species pairs or out of other forms of interspecies interaction (e.g., parasitism; see Table 13.4). In this report we focus primarily on what happens over evolutionary time; however, we recognize the crucial importance (and ultimate interdependence) of both scales.

THE RANGE OF MUTUALISMS

Biological mutualisms span such a broad range of natural histories and evolutionary origins that mutualism as a concept cannot easily be shoehorned into any simple, single situational template. For this reason, it would be a tremendous stretch to argue that any particular system is itself the "archetypal" mutualism. That said, would it even be useful to consider mutualisms as a class of interactions delimited by a set of common features and subject to some single conceptual framework? A pessimist might answer "no" and argue that mutualism is a catch-all category, an unrelated grab bag of interactions with no hope for conceptual unification.

We, the authors (or at least some subset thereof), prefer a more optimistic view: Mutualism represents an ecologically, evolutionarily, and taxonomically rich spectrum of biological phenomena that remain in large part unexplored or even undescribed. By identifying the salient structural features of mutualistic interactions — both those features relevant to the nature of the interaction in real time and those relevant to the ontogeny of the mutualism on an evolutionary scale — we should ultimately be able to highlight relevant connections among these diverse systems. Moreover, we expect the diversity of mutualistic systems to manifest these various structural features in a multiplicity of combinations, creating a network of relationships whereby any pair of different mutualisms share some characters in common and differ in others. In Table 13.1, we list some of the key features that we consider to be important in determining the nature, outcome, and evolutionary history of mutualistic systems. Any one mutualism is characterized by a number of features from this list, and different pairs have different features in common (an ideal situation in which to take a comparative approach to understanding the consequences of game structure!)

In Table 13.2, we briefly summarize a few of the better-studied examples of mutualism and describe them in terms of combinations of the properties listed in Table 13.1. Among the various types of mutualism listed in Table 13.2, only in the shared-benefit mutualisms do all individuals, regardless of species, contribute the same good. In these mutualisms, there is not really a trading market (Bowles and Hammerstein, this volume; Bshary and Noë, this volume) at all, nor can a "price" or "exchange rate" be computed. Instead, these shared benefit mutualisms typically involve exchange (or production) of *information*. The laws governing information-sharing work differently from those governing the exchange of ordinary physical commodities, for information can be transferred to another without reducing that enjoyed by the donor (Lachmann et al. 2001).

CONCEPTUAL TOOLS AND FOUNDATIONS

In the field of intraspecific cooperation, theoretical foundations (e.g., Hamilton 1964; Trivers 1971; Axelrod and Hamilton 1981) developed alongside empirical observation and experimentation. The study of mutualism biology has had quite a different history; this history may in part explain the relation (or lack thereof!) between theory and empirical work in this area. The study of mutualism largely arose out of efforts to understand the elaborate natural histories of species pairs interacting to their mutual benefit (e.g., Buchner 1965; Janzen 1966). When theorists first began to take note of the "mutualism problem," they brought with them a body of conceptual and mathematical models largely developed within other disciplines. The theory of reciprocal altruism, based largely on models of the iterated Prisoner's Dilemma (PD) game, was an early colonist, having been imported from the study of intraspecific cooperation. More recently, a second wave of theoretical concepts — market theory and

Table 13.1 Key variables that differ across mutualisms.

Properties of the Game Payoffs

1. Magnitude of benefits: The benefit that members of a given species reap from their participation in a mutualism can range from marginally greater than zero to an opposite extreme in which all fitness comes through the mutualism, as is the case for obligate mutualists.
2. Magnitude of investment: For each partner, investment can range from nothing to enormous; investments can be fixed at the onset of the interaction or variable across its course; investments can be symmetric or asymmetric between partners; investments can be concealed, revealed, or even extravagantly signaled.
3. Cost of being cheated: The consequences of a partner's defection range from negligible to fatal.
4. Potential for sanction: Some species may be able to impose substantial costs on their mutualist partners; others will have little opportunity to negatively influence their associates.

Availability of "Outside Options"

5. Obligacy: Mutualistic association can increase an individual's fitness from a baseline of zero (in obligate mutualisms) or from something greater than zero (in facultative mutualisms).
6. Specificity: Some mutualisms feature only one partner species for each species (species-specific mutualism); in others, partner species are substitutable such that any of a number of species may be able to step into a partner role (nonspecific mutualism).
7. Opportunity for choice: Partner choice can range from highly important to nonexistent.

Ecological Structure and Evolutionary Dynamics

8. Population structure: Partners may be clumped or spread evenly across the habitat; there may or may not be significant genetic structure across space.
9. Symmetry: Both obligacy and specialization can be symmetrical between partners or asymmetrical (e.g., obligate on one side, facultative on the other).
10. Duration of association: The durations of mutualistic associations can range from single-shot and fleeting encounters to life-long partnerships with highly iterated interactions.
11. Influence of third parties: Though typically modeled as dyadic interactions, mutualisms commonly involve third species that influence the outcome or magnitude of the mutualism. The third species may be responsible for raising (or even creating) the benefits of the mutualism. Alternatively, it may lower the benefits and/or stability of the mutualism by exploitation.
12. Evolutionary rate: Mutualist partners may have similar or widely divergent generation times and evolutionary rates.

a suite of related ideas—has been imported from the field of economics. In this section, we consider these and other conceptual foundations for understanding mutualistic interactions.

To understand the relations among these core concepts, we find it helpful to distinguish clearly between the structure of the pair formation, on one hand, and

Table 13.2 Well-studied empirical systems and some phenomena they exemplify.

Protection Mutualisms

General features: Mutualistic only in the presence of third species (antagonists of one partner); protection traded for food

Examples:
- Cleaning: (Bshary and Noë, this volume): Generalized; facultative; extensive and reciprocal partner choice and partner-recognition mechanisms; high cognitive abilities; minimal investment
- Ant-tending of lycaenids (Bronstein, this volume; Leimar and Connor, this volume): Range from generalized to specialized, from obligate to facultative, from mutualistic to parasitic; partner recognition at least by ants; adaptively plastic reward production ranging from cheap to expensive

Similar mutualisms: Ant-tending of aphids, ant-tending of plants

Transportation Mutualisms

General features: Food traded for transport of self or gametes

Examples:
- Obligate pollination of yuccas (Bronstein, this volume): Symmetrically obligate and species-specific; high reward investment, fixed before onset of interaction; costly exploitation by mutualists and other species
- Generalized pollination: Varying symmetry of obligacy and specificity; reward investment fixed before outset of interaction; exploitation by mutualists and other species that varies widely in costs.

Similar mutualisms: Obligate pollination of figs; generalized seed dispersal

Nutrition Mutualisms

General features: Food traded for food/protection; often but not always symbiotic

Examples:
- Plant–mycorrhizal symbiosis: Varying symmetry of obligacy and specificity; can range from mutualistic to parasitic across species and across gradients of resource availability for individual species pairs; at least one-sided partner choice

Similar mutualisms: Plant–rhizobium symbiosis; light organ symbioses; gut symbioses

Shared-benefit Mutualisms

General features: Multispecies aggregations that benefit participants via shared vigilance or defense

Examples:
- Mixed-species foraging: Highly facultative and generalized; negligible investment; may involve more than two species simultaneously

Similar mutualisms: Müllerian mimicry complexes

the population-genetic structure in which mutualistic phenotypes and behaviors evolve, on the other. The former essentially concerns the number of separate classes or categories from which partnerships are assembled. Are mutualist

Table 13.3 Types of cooperative interaction.

Partner Classes	*Gene Pools*	*Example*
1	1	Coalition formation among baboons (Noë 1994)
2	1	Lazuli bunting"tenant" system (Bowles and Hammerstein, this volume)
		Biparental care (Clutton-Brock 1991)
1	2	Mixed flock aggregations (Lima 1995)
2	2	Cleaning mutualisms (Bshary and Noë, this volume)
		Protection mutualisms (Bronstein, this volume)
		Pollination mutualisms (Bronstein, this volume)

pairs composed of individuals paired from each of two mutually exclusive groups (e.g., mating pairs of one male and one female; cleaning mutualistic pairs composed of one cleaner species and one client species), or are they drawn from one homogeneous class (e.g., coalition partners taken from the set of all individuals in a population)? The structure of pair formation can play a significant role in determining the nature of the interaction as well as the structure and stability of the pairing (see e.g., the two-sided matching literature: Roth and Sotomayer 1990; Bergstrom and Real 2000).

We contrast this to the structure of the *gene pool* in which the cooperative strategies evolve. Do both partners belong to a common gene pool (as is the case in examples of intraspecific cooperation), or is each pair composed of one member from each of two separate populations (as is the case in interspecific mutualism)? This distinction can be crucial in determining the evolutionary dynamics by which strategies are ultimately selected. In Table 13.3, we summarize the possible combinations of pairing structure and gene pool structure, and provide biological examples of each combination. Four basic theoretical frameworks used to understand mutualism evolution are described below.

Reciprocity

As mentioned above, reciprocal altruism was an early — and largely unsuccessful — invader from intraspecific cooperation theory. In his treatment of reciprocity, Trivers (1971) gave a detailed account of cleaning symbiosis and argued that the phenomenon is likely to be an example of interspecific reciprocal altruism. Axelrod and Hamilton (1981) extended this theoretical stance, by applying the repeated PD game to interspecific interactions. In addition to cleaning symbiosis, Axelrod and Hamilton suggested a range of applications covering most of the general categories of mutualism listed in Table 13.2. Both Trivers (1971) and Axelrod and Hamilton (1981) stressed the importance of detecting and punishing cheaters: an individual must not be able to get away with defection without others being able to retaliate effectively.

The papers by Trivers (1971) and by Axelrod and Hamilton (1981) came to be regarded as providing a general conceptual foundation for the evolution of cooperation between unrelated individuals, between as well as within species. However, this position has in turn resulted in a growing discontent among biologists interested in the evolution of mutualism. The perceived weakness of the theory of reciprocal altruism is not that the logic of the arguments supporting it appear faulty, but rather that there appear to be few examples of reciprocal altruism that have held up to closer scrutiny (e.g., Bronstein, this volume; Bshary and Noë, this volume; Hammerstein, Chapter 5, this volume; Leimar and Connor, this volume). Thus, it would seem that Tit-for-Tat reciprocity is a logically feasible, but in practice marginal, form of interspecific cooperation.

By-product Effects and Pseudoreciprocity

Certain traits or behaviors that have evolved to benefit an individual directly might also benefit others, as a side effect of their primary function. For example, Müllerian mimics receive by-product benefits from one another, as members of one species "train" predators to avoid other similarly colored and similarly dangerous species as well. Such by-product benefits may have been instrumental for the evolutionary origin of many existing mutualisms (Connor 1995).

Nevertheless, all benefits of mutualism cannot be regarded as by-products of other activities. Many costly traits, such as nectar production, must be interpreted as investments that primarily benefit other organisms and, as suggested in the introduction, these are the traits that most desperately need to be explained by any successful evolutionary theory of mutualism.

Reciprocal altruism provides a candidate explanatory framework, but as we have noted it appears to have limited applicability. *Pseuodoreciprocity*— that idea that investments in unrelated individuals have evolved to enhance by-product benefits obtained from these individuals—may be a more common explanation for both the origin and the maintenance of mutualistic associations. Investment in by-product benefits could have played a role for the origin of certain mutualisms. For example, Tilman (1978) discovered that ants that are attracted to extrafloral nectaries on black cherry trees reduced herbivore damage from caterpillars. Apparently, this reduction in herbivory was the by-product benefit that favored investment in the ants by the trees, in the form of nectar production. Such investment is likely to be of even greater importance for the further adaptive modification of mutualistic interactions, whether derived from by-product benefits or from initially parasitic interactions (Table 13.4). Leimar and Connor (this volume) argue that pseudoreciprocity should replace reciprocity as the dominant explanatory framework for the evolution of investments in unrelated individuals.

Markets and Partner Choice

The term "biological market" was introduced to highlight the commonalities among human economic markets, mating markets, and cooperation markets

Table 13.4 Mutualism may arise in different ways depending upon the nature of the benefits exchanged. Traits that have evolved to benefit an individual directly may produce incidental or *by-product* benefits for others; individuals may extract benefits at a cost to others (*purloined*), or individuals may *invest* in others at a cost to themselves. All mutualisms may originate with one or both parties receiving by-product benefits, yielding three different routes to mutualism (see text for discussion of the examples). If it were discovered that the origin of any mutualism fell into one of the three categories that does not include by-product benefits, including reciprocal altruism, **Prediction 3** (see text) would be falsified. Adapted from Connor (1995).

		Mutualist 2		
		By-product	*Purloined*	*Invested*
Mutualist 1	*By-product*	Müllerian mimicry	Origin of insect pollination mutualisms	Ant–black cherry tree
	Purloined		?	?
	Invested			? (Reciprocal altrusim)

(Noë and Hammerstein 1994, 1995). In human markets, buyers are "choosy": they seek out sellers who offer the best prices. This choosiness pressures sellers to compete with one another to offer lower prices, thus forming a crucial link between supply, demand, and the exchange ratio of commodities. As Bshary and Noë (this volume) illustrate for cleaner fish – client interactions, the same process and basic principles apply to nonhuman systems as well; market theory can be used to understand the flow of resources among any organisms that exchange commodities that they cannot take from one another by force.

The biological market approach has two major goals:

1. To explain adaptations in organisms involved in cooperation or mutualism that are due to "market selection," i.e., that have evolved under the pressure of partner choice. The obvious parallel is the evolution of secondary sexual characters driven by mate choice.
2. To predict changes in exchange rates of commodities due to shifts in the supply and demand curves. Again, these dynamics have been well studied in the context of economics and sexual selection, but have received far less attention in the context of cooperation and mutualism.

Sanctions, Power, and Partner Control

Although the biological market analogy can be extremely useful, biological markets and their economic counterparts — at least as typically abstracted — differ from one another in important ways. In standard neoclassical economic theory, agents (e.g., buyers and sellers, employers and employees) are assumed to be able to establish complete contracts. That is, individuals are able to make fully binding commitments regarding the terms of any exchange, and these commitments are enforceable at zero cost. By contrast, biological markets typically offer no analogous way of establishing binding and freely enforceable contracts.

In the absence of complete contracts, the participants in market exchanges have to bring about their desired outcomes by alternative means, for example, by the strategic use of rewards for fulfilling an agreement or punishment for failing to do so. The ability to make effective use of such strategic incentives is termed *power* in recent economic models designed to address situations such as labor agreements, in which contracts are not in practice complete (see Bowles and Hammerstein, this volume). In short, we can describe power as follows:

> We say that agent A has *power* over agent B if A can gain advantage over B by threatening B with punishment, and B has no analogous counter-response.

With an example drawn from lazuli bunting mating systems, Bowles and Hammerstein (this volume) illustrate the way in which power can be exerted in biological systems. They find that in this system, models accounting for power relations better explain the division of benefits among participants than do models based upon biological markets with complete contracts (see also Bergstrom and Lachmann, this volume).

Models based on power can also account for observed inefficiencies in social equilibria among animals that cannot be explained in a simple biological markets framework. Bowles and Hammerstein (unpublished manuscript) stress that in biological markets *or* economic markets, when power is employed in lieu of complete contracts and neither party has absolute power, equilibrium outcomes will often be Pareto inefficient, because Pareto efficiency can often be obtained only through trade with enforceable contracts. Although, in principle, these ideas will apply to interspecific mutualisms, we stress that many current examples are drawn from intraspecific interactions: Bowles and Hammerstein examine the landlord-tenant system among lazuli bunting, Reuter and Keller (2001) predict this sort of inefficiency arising from the exercise of power in hymenopteran sex-ratio conflict, and of course the original economic theories were derived to explain human intraspecific behavior.

PREDICTIONS

In the previous section, we briefly surveyed the current suite of conceptual tools available to address the issues surrounding the evolution of mutualism. Each is appealing in its own way, but how do we know which of these tools are right for the job? How do we avoid driving nails with a screwdriver and turning bolts with a hammer? Conceptual constructs can prove their utility by helping us organize the facts that we already have collected, but often a stronger challenge can be brought to bear upon our conceptual foundations through direct contact between theory and empirical data (Hilborn and Mangel 1997).

To bring about such an encounter, one requires that the theory generate testable predictions. In this spirit, we offer (and briefly motivate) a set of such predictions here. Rather than hedging our bets in mortal terror of possibly being proven wrong at some future date, we have deliberately stated these predictions in strong forms that are more likely than their timorous counterparts both to generate debate and to collapse ultimately under the weight of an accumulated body of empirical evidence.

Prediction 1: Reciprocal altruism will never be observed in interspecific mutualism. Reciprocal altruism requires both the presence of adequate cognitive complexity to handle accounting and individual recognition as well as the absence of alternative mechanisms sufficient to enforce mutualistic behavior. We conjecture that this combination of circumstances will rarely, if ever, be present in the interspecies associations that spawn interspecies mutualisms.

Prediction 2: Among organisms with relatively well-developed cognitive systems, many by-product mutualisms are the results of learning rather than the results of adaptation for that particular interaction. Therefore, if one partner is replaced with some phylogenetically related and/or physiologically similar but typically nonsympatric species, the individuals involved will be able to establish mutually beneficial interactions despite the novelty of the partnership, and they will be able to do so on the timescale of individual lifetimes.

Prediction 3: All interspecific mutualisms began, evolutionarily, from an association with by-product benefits to at least one party. Some by-product benefits are necessary on at least one side in order to select for further development of the interspecific association.

Prediction 4: Most mutualisms neither require nor exhibit sanctioning behavior on the part of either partner. Partner choice and individually beneficial response to undercontributing partners will be sufficient to motivate and enforce cooperative behavior. Where mechanisms for imposing sanctions do exist, they will be co-opted rather than evolved directly as sanctions.

PUZZLES

Thus far in our understanding of mutualism, a number of observations remain baffling. Although, at this point, we cannot lay out a set of Hilbertian problems for the study of mutualism, we would like to suggest the following puzzles as possible areas of focus for future empirical and theoretical development.

Puzzle 1: Many mutualist partners appear to be quite poorly coadapted and poorly fine-tuned to profit maximally from the mutualistic interaction. This is surprising, given that opportunities for exploitation of mutualist partners appear to be plentiful (Leimar and Connor, this volume; Bronstein, this volume). How can we explain the limited success of mutualist partners in finding adaptive solutions to the problem of extracting maximal resources from the interspecific interaction?

Puzzle 2: Why do adaptations for imposing sanctions on an interspecific mutualist partner appear to be so rare? Individuals commonly engage in self-interested behavior (e.g., switching partners when paired with a noncooperator), which has the side effect of imposing costs on an uncooperative partner. (Could we call this by-product punishment, in analogy to by-product mutualism?) However, we rarely find examples in which an individual regulates or manipulates the behavior of an interspecific partner behavior by actively imposing costs on the partner at a direct cost to itself. Are there basic theoretical reasons why such behaviors are unlikely to evolve, much as the evolution of intraspecific punishment presents a free-rider problem (Frank 1995)? Is partner choice and/or "by-product punishment" sufficiently effective such that selection is simply too weak to generate active sanctioning? Are we failing to observe active sanctioning behavior even though it *is* present in a considerable number of systems? Is sanctioning simply unnecessary, because the evolutionary process rarely generates the mutational combinations to produce variants capable of taking advantage of their partners?

Puzzle 3: Why are mutualisms so commonly exploited by third parties, and why do they so seldom have built-in mechanisms for the prevention of such exploitation? Is the answer to this question more or less the same as the answer to Puzzle 2, or is dealing with "exploitation from outside" a fundamentally different sort of evolutionary problem?

Puzzle 4: Mutualist partners often use signals to coordinate their actions and contributions. What mechanisms (if any) prevent the evolution or persistence of deceptive signaling strategies and the ultimate dissolution of the communication system? Should we expect these mechanisms to be similar to those involved in the maintenance of intraspecific honest signaling?

By way of closing this section, we should note that each of these puzzles challenges the reader to explain a claim or pattern derived from our current assessment of the body of empirical evidence. In any or all of these cases, the resolution may lie in further theoretical or conceptual development; alternatively, it is possible that our current assessment of the data is incorrect and that the patterns around which the puzzles are founded could prove to be unsupported. In either case, explicit theoretical models can serve to create the logical framework in which to organize these data and address these puzzles. We hope that these puzzles will serve to stimulate development of such models.

Model Systems and Experimental Tests

The field of mutualism biology has been blessed with a spectacular diversity of field systems. However, the flip side of this lucky coin is that the particular systems that have received thorough attention thus far may be neither the most representative of mutualistic interactions in general (see Bronstein, this volume), nor the most tractable for observational and experimental study. Although no single system can capture the range of processes and phenomena observed across all mutualisms, the development of a small number of model systems could conceivably serve to accelerate progress in the field.

What makes a good model system? Clearly, this will depend on which questions one wishes to ask. In the study of nonobligate interspecific mutualism, we are often particularly interested in the consequences of the parameters listed in Table 13.1. Unfortunately, for most systems, many of these parameters either cannot be, or have not been, quantified. As a result, quantitative prediction will at best be difficult and at worst be a futile exercise in curve-fitting. We propose an alternative approach: Within a particular model system, one can manipulate these basic parameters and then observe the qualitative consequences. If one were interested in the role of partner choice on the frequency of punishment behavior, one could manipulate partner availability on each side of a facultative mutualism and measure the resulting changes in punishment frequency.

Another important consideration is the ability to perform experimental manipulations; potential model systems differ substantially in this respect. Some are more easily brought into the laboratory than others. Obligate mutualisms may be more difficult to manipulate broadly than facultative ones. Generation times, and thus the potential for experimental evolution, vary by orders of magnitude. So which systems would make good models? We propose two sets, corresponding to two timescales. Although this list is unavoidably biased toward the inclusion of the research systems that we know the best, we hope that it can nonetheless serve as a useful starting point.

Single-generation Experimental Manipulation

By observing short-term responses of species to changes in the behavior, condition, availability, and other characteristics of their mutualist partners, one can explore possible answers to some of the major questions with which this report began: How are mutualisms stabilized, and to what degree do species exhibit adaptive plasticity in their responses to the specific circumstances of the mutualism? However, a caveat is in order for studies of this kind. It will be crucial to work with a system that exhibits natural variation in the parameters that will be manipulated. Otherwise, one would not expect an evolved plasticity of response to be manifested by members of the population.

Possible model systems:

1. *Cleaner fish – client mutualism.* In this volume, Bshary and Noë describe the cleaner fish – client mutualism, a strong candidate for study on this timescale. The system has the notable advantage of being amenable to both field and laboratory study. Moreover, the relatively advanced

cognitive capabilities of the partners and evidence of extensive learning suggests that this system may be a particularly good place to study the role of cognitive function and learning mechanisms in the establishment and maintenance of mutualism.

2. *Ant–lycaenid systems.* Evolutionary associations between ants and lycaenid butterflies are some of the most extensive in terms of the number of species involved, with an estimated 4500 related species on the lepidopteran side spanning the range of symbiotic associations from parasitism through mutualism (Pierce 1987, 2001; Pierce et al. 2002; Bronstein, this volume) As such, this ensemble appears to be a particularly promising system for comparative research. Because of the huge number of species involved, this system also offers extraordinary potential for investigation of the underexplored relationships between phylogeny and mutualistic association; ongoing phylogenetic work by Pierce and colleagues will provide the necessary background for such investigations.

Experimental Evolution

Over the past decade, evolutionary biology has been deeply enriched by the development of procedures and model systems allowing the experimental study of evolution in real time in the laboratory. This approach would also seem to be highly promising for the study of mutualism. Mutualistic interactions, including biofilm formation, syntrophy, and various forms of environmental conditioning, appear to be common in bacterial communities. Moreover, recent evidence of interspecific bacterial signaling (Bassler 2002) strongly suggests coordinated mutualistic activity.

Possible model systems:

1. *Bacterial systems:* Several investigators have developed laboratory models of multispecies bacterial biofilms (see, e.g., Tolker-Nielsen and Molin 2000); these could serve as useful evolutionary models. Bacterial mutualisms — obligate or facultative — could also be constructed de novo. An artificial obligate mutualism could be created by knocking out complementary functions from two bacterial species so as to induce mutual dependence in certain selective environments. Such a protocol would allow the investigator to observe actually the initial steps in the ontology of mutualism and to explore the role of the gene transfer in the mutualism evolution. Moreover, biofilm systems are likely to provide useful insight into the population dynamics and regulation of mutualist partners.
2. *Bacterium–plasmid associations*: The interaction between bacteria and their semi-autonomous, horizontally transferred plasmid molecules may also merit consideration as a model system for the study of mutualism.

> Unlike bacteriophages and other parasitic replicons, plasmids most likely enjoy a mutually beneficial relationship with their hosts. Current theory and empirical work suggests that plasmids persist in bacterial populations only when they actively benefit their hosts under at least some environmental conditions (Bergstrom et al. 2000). Bacteria and novel plasmids are known to exhibit rapid coevolution in response to one another's genetic makeup (Levin and Lenski 1983) and may also provide interesting models of fitness compensation or even "addiction" in the absence of beneficial effects (Levin et al. 2000).

Other promising potential systems include plant–rhizobium interactions (Denison 2000), cnidarian–algae symbioses (Baghdasarian and Muscatine 2000), and bacterial–insect (Moran 2001) or bacterial–nematode associations (Burnell and Stock 2000).

CONCLUSION: WHY STUDY MUTUALISM?

To date, the study of mutualism has proceeded largely out of a desire to explore the extraordinary natural histories of mutualist species and mutualistic associations. This remains a fascinating area, a rich and still proportionally uncharted territory. In addition, there are further reasons why the study of mutualism will serve to address important, basic issues in ecology and evolutionary biology.

First, the study of interspecific mutualistic associations offers the opportunity to explore the mechanisms — from sanctioning to partner choice to pseudroreciprocity — that maintain cooperative behavior even in the absence of kin selection. These mechanisms are likely to be of fundamental importance as guarantors of prosocial behavior not only in interspecific interactions throughout the tree of life, but also in the human (intraspecific) interactions among nonkin that are ubiquitous in large modern societies (McElreath et al., this volume; Henrich et al., this volume).

Second, many of the major transitions in evolution (Maynard Smith and Szathmáry 1995) have involved the formation of mutually beneficial associations that ultimately became new levels of organization. A more detailed understanding of the origin and ontogeny of interspecific mutualisms can further help our efforts to understand the most important occurrences in the history of life.

Third, the study of mutualism to date has focused on the dyadic relationship between partner–species pairs. Of course, each of these pairwise species interactions is in fact embedded in a larger community-level context. To understand the function of mutualistic systems *fully*, we will have to understand the community-level context as well. As such, the study of mutualism dynamics should stimulate the development of additional connections between ecological and evolutionary processes and timescales. Ultimately, studies of multispecies phenomena that build upon mutualism should have significance for conserving and restoring species in a rapidly changing world.

REFERENCES

Axelrod, R., and W.D. Hamilton. 1981. The evolution of cooperation. *Science* **211**:1390–1396.

Baghdasarian, G., and L. Muscatine. 2000. Preferential expulsion of dividing algal cells as a mechanism for regulating algal-cnidarian symbiosis. *Biol. Bull.* **199**:278–286.

Bassler, B.L. 2002. Small talk: Cell-to-cell communication in bacteria. *Cell* **109**:421–424.

Bergstrom, C.T., M. Lipsitch, and B.R. Levin. 2000. Natural selection, infectious transfer, and the existence conditions for bacterial plasmids. *Genetics* **155**: 1505–1519.

Bergstrom, C.T., and L.A. Real. 2000. Toward a theory of mutual mate choice: Lessons from two-sided matching. *Evol. Ecol. Res.* **2**:493–508.

Buchner, P. 1965. Endosymbiosis of Animals with Plant Microorganisms. rev. Engl. ed. New York: Interscience Publ.

Burnell, A.M., and S.P. Stock. 2000. Heterorhabditis, Steinernema and their bacterial symbionts: Lethal pathogens of insects. *Nematology* **2**:31–41.

Clutton-Brock, T.H. 1991. The Evolution of Parental Care. Monographs in Behavior and Ecology. Princeton, NJ: Princeton Univ. Press.

Connor, R.C. 1995. The benefits of mutualism: A conceptual framework. *Biol. Rev.* **70**:427–457.

Denison, R.F. 2000. Legume sanctions and the evolution of symbiotic cooperation by rhizobia. *Am. Nat.* **156**:567–576.

Frank, S.A. 1995. Mutual policing and repression of competition in the evolution of cooperative groups. *Nature* **377**:520–522.

Hamilton, W.D. 1964. The genetical evolution of social behaviour. I and II. *J. Theor. Biol.* **7**:1–52.

Hilborn, R., and M. Mangel. 1997. The Ecological Detective: Confronting Models with Data. Princeton, NJ: Princeton Univ. Press.

Janzen, D.H. 1966. Coevolution of mutualism between ants and acacias in Central America. *Evolution* **20**:249–275.

Lachmann, M., S. Szamado, and C.T. Bergstrom. 2001. Cost and conflict in animal signals and human language. *Proc. Natl. Acad. Sci. USA* **98**:13,189–13,194.

Levin, B.R., and R.E. Lenski. 1983. Coevolution in bacteria and their viruses and plasmids. In: Coevolution, ed. J. Futuyama and M. Slatkin, chap. 5, pp. 99–127. Sunderland, MA: Sinauer Associates.

Levin, B.R., V. Perrot, and N.W. Walker. 2000. Compensatory mutations and the population genetics of adaptive evolution in asexual populations. *Genetics* **154**: 985–997.

Lima, S.L. 1995. Collective detection of predatory attack by social foragers: Fraught with ambiguity? *Anim. Behav.* **50**:1097–1108.

Maynard Smith, J., and E. Szathmáry. 1995. The Major Transitions in Evolution. Oxford: Oxford Univ. Press.

Moran, N.A. 2001. The coevolution of bacterial endosymbionts and phloem-feeding insects. *Ann. Missouri Bot. Gard.* **88**:35–44.

Noë, R. 1994. A model of coalition formation among male baboons with fighting ability as the crucial paramater. *Anim. Behav.* **47**:211–213.

Noë, R., and P. Hammerstein. 1994. Biological markets: Supply and demand determine the effect of partner choice in cooperation, mutualism and mating. *Behav. Ecol. Sociobiol.* **35**:1–11.

Noë, R., and P. Hammerstein. 1995. Biological markets. *Trends Ecol. Evol.* **10**:336–339.

Pierce, N.E. 1987. The evolution and biogeography of associations between lycaenid butterflies and ants. In: Oxford Surveys in Evolutionary Biology, ed. P.H. Harvey and L. Partridge, vol. 4, pp. 89–116. Oxford: Oxford Univ. Press.

Pierce, N.E. 2001. Peeling the onion: Symbioses between ants and blue butterflies. In: Model Systems in Behavioral Ecology, ed. L.A. Dugatkin, pp. 41–56. Princeton, NJ: Princeton Univ. Press.

Pierce, N.E., M.F. Braby, A. Heath et al. 2002. The ecology and evolution of ant association in the lycaenidae Lepidoptera. *Ann. Rev. Entomol.* **47**:733–771.

Reuter, M., and L. Keller. 2001. Sex ratio conflict and worker production in eusocial hymenoptera. *Am. Nat.* **158**:166–177.

Roth, A.E., and M.A.O. Sotomayer. 1990. Two-sided Matching. Cambridge: Cambridge Univ. Press.

Tilman, D. 1978. Cherries, ants and tent caterpillars: Timing of nectar production in relation to susceptibility of catepillars to ant predation. *Ecology* **59**:686–692.

Tolker-Nielsen, T., and S. Molin. 2000. Spatial organization of microbial biofilm communities. *Microb. Ecol.* **40**:75–84.

Trivers, R.L. 1971. The evolution of reciprocal altruism. *Qtly. Rev. Biol.* **46**:35–57.

14

Power in the Genome

Who Suppresses the Outlaw?

Rolf F. Hoekstra

Laboratory of Genetics, Wageningen University, 6703 BD Wageningen, The Netherlands

ABSTRACT

Genomic parasites are present in virtually every species and often in large numbers. As a rule they seem to have a long life, although negative effects on the individual host diminish over time. Genetic systems exhibit both features that promote and prevent genomic parasitism. The most important features promoting parasitism are sex and interindividual somatic fusion. Powerful mechanisms that prevent genomic parasitism are meiotic segregation and recombination, somatic incompatibility systems, and various molecular processes that suppress transposon activity. It is suggested that Leigh's "parliament of the genes" concept in a modified interpretation may contribute to the suppression of genomic parasites.

INTRODUCTION

For several decades, evolutionary biologists have been well aware that the genome is not a collection of genes working harmoniously toward a common goal: the successful development and functioning of the individual organism. Examples of "selfish" or "parasitic" genes — elements in the genome with harmful effects on the organisms that carry them but nevertheless capable of maintaining themselves in the genome over many generations (Werren et al. 1988) — are now well known. They are of considerable evolutionary interest since they represent agents that cause maladaptation as well as escape mechanisms that ought to prevent their occurrence or promote their quick removal. Their existence invites questions regarding the level of "genetic criminality" that can be tolerated by a species, how genetic systems are designed to minimize the threat they impose, and how possible countermeasures evolve. In this chapter I discuss a few examples of genomic "outlaws." I consider how common they are, for how long they manage to stay, and how we understand their ability to spread despite negative effects on their host's fitness. Thereafter I discuss features of genetic systems that promote or prevent the occurrence of genomic parasites and conclude with some general ideas about cooperation in the genome.

SOME EXAMPLES OF GENOMIC OUTLAWS

Because of the limited space and focus on general principles, I limit myself to a few of the better-known classes of genomic parasites in eukaryotes, with some regret totally omitting interesting phenomena of genomic parasitism in prokaryotes.

Transposable Elements

Transposable elements (TEs) are segments of mobile DNA that are able to move through the genome. Many of them contain the genes necessary for their own replication, others depend partly on enzymes provided by other TEs. They have been found in every species in which searches have been conducted and often occur in huge numbers, making up at least 10–20% of the genome. By their act of transposition (inserting a copy somewhere else in the genome), active transposons are capable of accumulating in the genome and inducing insertions, deletions, and chromosomal rearrangements. They are therefore likely to have played an important role in evolution. Their occurrence in extremely high copy numbers as well as their lack of genes coding for structural proteins that function in the host's phenotype, strongly suggest a status of parasitic DNA. However, at present it is not quite clear whether they all should solely be viewed as genomic parasites. Different regions of the genome may show widely different TE densities. For example, a 525 kb region on the human X chromosome has a TE density of 89%, whereas the four homeobox gene clusters contain regions of about 100 kb with less than 2% TEs (The Genome International Sequencing Consortium 2001). Remarkably, although in the human genome many TEs occur mainly in AT-rich (i.e., gene poor) DNA, consistent with the interpretation of TEs as genomic parasites, some class of TE called SINEs (Short INterspersed Elements) occur mainly in GC-rich DNA, which contains a high density of coding genes. There are indications that they are particularly frequent near actively described genes, suggestive of some useful function (The Genome International Sequencing Consortium 2001).

Segregation Distorters

An allele or chromosome is a segregation distorter (SD) if it is recovered in excess of 50% of the gametes from heterozygous parents. Another term for segregation distortion is meiotic drive, since it is the fair meiotic segregation that is corrupted, at least when segregation distortion occurs in the sexual cycle. The best-studied systems are SDs in *Drosophila melanogaster*, the *t*-haplotype in

Mus musculus and *Mus domesticus*, and Spore killers in *Neurospora intermedia*. When a driving gene is on a sex chromosome, the segregation distortion results in a skewed sex ratio, an easily recognized phenotype. Detection of segregation distortion on an autosome depends on the presence of an associated phenotypic effect. Many SDs may go unnoticed because of the lack of such a phenotypic effect. The exception is formed by fungi with ordered asci, where every instance of segregation distortion in principle has a phenotype, as explained below. Many genetic aspects of segregation distortion are reviewed in Lyttle (1991). All well-studied systems appear to consist of two components: a gene coding for a toxin and a second gene providing the antidote to the toxin. A meiotic drive complex consists of a tightly linked combination of a toxin production (or "killer") allele at the toxin locus and an antidote production allele at the immunity locus. A wild-type chromosome would have the linked combination of a nonkiller allele at the toxin locus and a sensitive allele at the immunity locus. A third type of chromosome—the nonkiller allele linked to the antidote allele—would represent a resistant chromosome type that does not cause segregation distortion but prevents distortion when heterozygous with a distorter chromosome. The fourth type of chromosome—the killer allele linked to the sensitive allele—is not observed because it is suicidal.

Simple population genetic arguments predict that whenever segregation distortion results in an increase in absolute number of progeny carrying the driving allele from a heterozygous parent, the distorter will increase in frequency, even when it is associated with detrimental effects on the individual organism, provided the fitness loss is not too severe (Hiraizumi et al. 1960). This justifies interpreting SDs as genomic parasites. Both SDs and the *t*-haplotypes are associated with clear fitness costs to individuals that are homozygous for the distorter allele. The population genetics of segregation distortion in fungi ("spore killing") is fundamentally different. Fungi have a haploid life cycle, in which the meiotic products are the offspring from a cross. Meiotic drive here implies basically killing up to 50% of the sibs. Therefore, Spore killer genes have only an extremely weak selective advantage when rare, and the occurrence of fungal Spore killers is therefore not yet understood.

Cytoplasmic Male Sterility

Cytoplasmic male sterility (CMS) is the maternally inherited inability to produce functional male gametes in individuals from an otherwise hermaphroditic species. The phenomenon has been described almost exclusively in plants and occurs in some 5% of the flowering plants (Laser and Lersten 1972). Theoretical analysis predicts that CMS may be maintained in natural populations if female (i.e., male sterile) plants have a higher reproductive success than hermaphrodites (Gouyon and Couvet 1987; Frank 1989). A CMS mutation is selected if it enhances female fitness, irrespective of its effects on male fitness because it is not transmitted via male gametes anyway. However, the individual fitness of male sterile plants is likely to be lower than that of hermaphrodites because they lack successful male gametes. Thus, a CMS can be qualified as parasitic or selfish because it may spread in a population despite negative effects on individual fitness. In almost all cases investigated, the male sterility mutation appeared to be located in the mitochondrial genome (reviewed in Schnable and Wise 1998).

Mitochondrial Plasmids

In general, fungal growth is indeterminate and can be extended "indefinitely" in the lab by means of successive inoculations onto fresh substrate. However, in several fungi belonging to the ascomycete genera *Podospora*, *Neurospora*, and *Aspergillus*, senescence has been described. When started from a single spore, a mycelium (the fungal soma consisting of a multinucleate hyphal network) expands at a constant rate during a characteristic period, which is followed by progressive slowing down of growth, a decline in fertility, and finally death. This process is associated with increasing titers of small DNA molecules in the mitochondria and increasing disruption of normal mitochondrial functioning (reviews by Griffiths 1992, 1995; Bertrand 2000). For reasons that are not yet fully understood, mitochondria that have been rendered dysfunctional by actions of these mitochondrial plasmids proliferate rapidly within the mycelium and gradually displace functional mitochondria containing wild-type mtDNA molecules. Thus we have a situation typical of genomic parasites: mitochondrial mutations at a selective advantage at the level of mitochondria but associated with a disadvantageous phenotype at the level of the fungal individual. The latter qualification, however, should perhaps be treated with some caution, since a clear fitness disadvantage of senescence under natural conditions has yet to be demonstrated.

INTRAGENOMIC CONFLICT MAY ALLOW THE SPREAD OF PARASITIC GENES

A general framework for understanding the occurrence of parasitic genes is based on the concept of *intragenomic conflict* (Cosmides and Tooby 1981; Hurst et al. 1996). Natural selection works whenever there is differential reproductive success among hereditary variants. Those variants with the highest reproductive success will be selected, i.e., are expected to increase in relative frequency. However, organisms are in some sense like Russian dolls: nested hierarchies of replication levels, at many of which natural selection can occur. Thus natural selection may operate between organisms, but also between cells within an organism and between genetic elements within its cells. For example, selection favors genotypes that contribute to greater individual reproductive success; however, within an individual, it may favor certain cell type variants that continue to replicate while their colleagues have stopped proliferating. The resulting tumor indicates the success of the mutant cell type, but not of the individual organism in which it occurs. Similarly, within cells, certain mitochondrial variants may outreplicate the resident type, not necessarily to the benefit of the cell or the multicellular individual. Examples are mitochondrial *petite* mutations in yeast and mitochondrial plasmids in fungi that are associated with senescence (Griffiths 1995). Whenever a trait is favored at one level, but selected against at a higher level, intragenomic conflict occurs. Transposable elements are favored because their intragenomic replication rate is higher than that of "normal" genes but disfavored because of the costs they impose on the organism. SDs are favored because of their enhanced segregation but disfavored because of their deleterious effects on the phenotype of the individuals carrying a distorter gene. Mitochondrial male sterility genes are selected because, despite their adverse effects on male function, they favor female fitness and therefore enhance their own fitness relative to wild-type mitochondria because they are maternally transmitted. The balance of selection forces on mitochondrial senescence plasmids is still poorly understood. Here, however, an advantage at some level of the genetic system must compensate for the disadvantage present at the organismal level.

A short discussion on terminology is perhaps relevant here. As is clear from the examples mentioned, in intragenomic conflicts we tend to view the highest level involved as the focal level, from which we judge the net result of natural selection working at several levels simultaneously. This is also apparent from the terminology used: a gene is parasitic or selfish if its effect on the individual is negative while being positive at some lower level, such as a mutation promoting the occurrence of a certain tumor. On the other hand, a mutation that would be more effective in suppressing tumor development would be selected at the individual level and selected against at the lower cellular level. Logically, this could also be viewed as an intragenomic conflict. Such a mutation, however, would not be called parasitic or selfish but rather would be viewed as adaptive. Clearly, adaptive means adaptive to the higher level, at least up to the level of individual. One may, of course, also consider conflicts of natural selection between still higher levels, say between individual and population, and there a mutation with positive effects on individual fitness, but negative to the population, would be a "parasitic" mutation. The converse, an "altruistic" mutation, would then be considered as "adaptive." I am not convinced that this terminology is completely acceptable; however, this issue is beyond the scope of this chapter.

GENOMIC CRIMINALITY FIGURES

To what extent are genomes burdened with parasitic genetic elements? Just as with criminality in human communities, the answer to this question is difficult to give because of an unknown amount of hidden, i.e., not officially registered, criminality. Unless identifiable on the basis of a typical DNA sequence, as is the case for many TEs, a parasitic element in the genome can only be recognized if it is associated with a phenotypic effect. Otherwise, it will remain unnoticed. How long can parasitic genes play an active role before (if ever) they are neutralized?

Transposons are very common and seem to be present in every species. The recent completion of the human genome sequencing project has allowed precise estimates of their occurrence in the human genome. About 45% of the genome is currently recognized as belonging to some class of TE. Much of the remaining "unique" DNA is probably also derived from ancient TE copies that have diverged too far to be recognized as such. A large majority of TEs is not active anymore, and their ancestry and approximate age can be inferred by making use of the fact that they derive via accumulation of random mutations from once active transposons. The sequence of the ancestral active elements can be deduced from phylogenetic analysis of the now existing sequences (The Genome International Sequencing Consortium 2001). Most TEs appear to have extremely long lives, some lineages being at least 150 million years old. In general, once having become nonfunctional as a consequence of accumulated mutations, TEs are cleared from the genome only very slowly, the process requiring tens of millions of years.

Frequency estimates of the occurrence of segregation distortion are very uncertain because they may often occur without a phenotypic effect. For this reason it is understandable that an appreciable number of known cases of meiotic drive involve genes distorting the sex ratio. However, fungi in which the haploid nuclei resulting from meiosis are linearly arranged within an ascus provide unique opportunities to analyze abnormal segregation for precisely the same reason that they have played such a big role in the classical experiments by Lindegren and others on fundamental aspects of genetic linkage, meiotic recombination, and gene conversion (see Whitehouse 1973; Perkins 1992). Any meiotic drive system in such fungi will be observed as *spore killing* in a cross between a driving and a sensitive strain: the degeneration and early abortion of half the spores in a certain proportion of the asci. Van der Gaag et al. (2000) reported that among 99 newly isolated *Podospora anserina* strains from Wageningen, 23% contain a meiotic drive element; altogether at least 7 different meiotic drive systems were identified in this population. Viewed in this way, segregation distortion can be concluded to be common. On the other hand, assuming that the number of coding genes per genome is on the order of 10^4, the probability per locus of a segregation-distorting allele will be on the order of 10^{-5}, implying that non-Mendelian segregation at nuclear loci is rare. Data on spore killing in other fungal populations, though less extensive, show roughly the same picture (Van der Gaag et al. 2000). Not much can be said about the age of the fungal meiotic drive systems; however, molecular data on the *t*-haplotype in mouse suggest that this distortion system arose at least 2–4 million years ago, predating the *domesticus–musculus* split (Hammer et al. 1989), and it is still active as a genomic parasite in 10–20% of the animals in these species.

As mentioned earlier, cytoplasmic male sterility is not uncommon among plants (e.g., on the order of a few percent among flowering plants). From what is known of the genetic details of the mitochondrial sterility mutations, CMS must have evolved many times independently. Moreover, it is not uncommon for a single species, even a single population, to contain several CMS systems (e.g., Van Damme 1983).

As to the frequency of occurrence of mitochondrial plasmids, they seem to be very common in fungi, although in many cases without a clear phenotypic effect (Griffiths 1995). If we limit our attention to the plasmids that are associated with fungal senescence, it is quite significant that in *Podospora anserina* every natural strain isolated so far contains mitochondrial plasmids and is subject to senescence. In this case one might say that the whole species seems to have been taken over by a genomic parasite. In *Neurospora*, strains isolated from some places, particularly from the Hawaiian islands, have been shown to contain senescence-associated plasmids, although these populations appear to be polymorphic for this factor. Some strains are infected, some are plasmid free.

GENERAL FEATURES OF GENETIC SYSTEMS THAT ALLOW OR PROMOTE GENOMIC PARASITISM

An essential aspect of the dynamics of parasitic mutations is the replication timescale difference that frequently exists between different replication levels within an individual organism. For example, cells may divide many times during a single individual generation, mitochondria divide many times during a single cellular generation, and mitochondrial plasmids may replicate many times during a single mitochondrial generation. This provides the scope for short-term increase in frequency of a parasitic mutation during a single generation of the next higher level. However, its ultimate fate is linked to that of the higher level, where selection may remove those units with lowered fitness as a result of the effects of the parasitic mutation. It follows that under a strictly clonal or vegetative regime of vertical transmission, the evolutionary prospects of parasitic mutations are quite limited. The population or species consists of a set of clones with no possibility of genetic exchange among them. In such a genetic system, natural selection has an easy job of weeding out those clones that carry the burden of parasitic genes. An example of a system that comes close to this scheme is the budding yeast *Saccharomyces cerevisiae*, when kept under conditions of exclusive asexual reproduction. Regularly *petite* mutations do occur, enjoy their short-term success at the mitochondrial level, but disapppear as a result of impaired growth of their carriers. Another example is the lack of LINE-like and gypsy-like retrotransposons in bdelloid rotifers, the famous "ancient asexuals" (Arkhipova and Meselson 2000).

However, the picture changes dramatically in genetic systems with the capacity for sex or somatic fusion. Both processes allow the transfer of genetic material from one line of descent to another, not unlike infectious transmission of microbes from a carrier to an uninfected individual. Therefore, genetic systems that involve sex and/or somatic fusion offer much better prospects for parasitic genes because, in principle, their evolutionary fate is no longer exclusively coupled to that of their carrier. They may "jump" to other genetic lineages and, under suitable conditions, spread through a population. Since almost all genetic systems known involve some form of sex or somatic fusion, let us now examine which features have evolved in (para)sexual systems to prevent or hinder the spread of parasitic mutations.

GENERAL FEATURES OF GENETIC SYSTEMS THAT PREVENT OR HINDER GENOMIC PARASITISM

Meiotic Segregation and Recombination

The science of genetics rests on the basis provided by Mendel, who discovered the regularities of genetic transmission of chromosomal genes in sexual crosses, the so-called Mendelian Laws. One of these states that each member of a gene pair in a heterozygote is included in 50% of the gametes. An important consequence of this "Mendelian Lottery" is that the transmission probability of an allele is independent of the quality of its phenotypic effect. A heterozygous carrier transmits a mutation with a severe deleterious effect just as likely as the wild-type allele. Still, the lottery hinders the spread of parasitic (or "selfish") alleles that would combine an advantage at segregation with a negative effect on individual fitness. Indeed, a fair segregation mechanism seems a prerequisite for adaptive evolution, since only then can each allele (be it the "wild-type" or a novel mutation) be put to the test by natural selection at a higher replication level. The fair segregation of chromosomal genes results from the mechanistic machinery of meiosis. Meiosis guarantees an orderly segregation of chromosomes so that every gamete receives the proper number and type of chromosomes. At the level of single loci, it leads in heterozygotes to the production of as many gametes with one allele as with the other. However, as mentioned above, segregation distortion or meiotic drive does occur, caused by selfish elements that are able to distort the segregation, both at the level of chromosomes and at the level of alleles. Genic SDs seem to involve postmeiotic killing of (sister) meiotic products; however, it should be noted that not all species are vulnerable to this type of meiotic drive. Consider a distorter mutation A that kills or inhibits other haploid cells of the same meiosis (i.e., cells not containing A): the number of functional gametes produced will be halved. This seems to be the basic mechanism of meiotic drive in the best-described systems (Lyttle 1991). It follows that in species with external fertilization, with little or no competition between sister gametes, there is no evolutionary advantage to the driving allele A. Indeed, if there is a cost to distortion, the mutation A would not spread. Only in species with strong competition between sister gametes would the mutation have a chance, provided an individual with only half the normal number of gametes is still fertile. This explains why segregation distortion in diploid animals only occurs typically in males.

Another aspect of meiosis that hinders the spread of parasitic genes is recombination. The two-component nature of a meiotic drive system (toxin and antidote) has the following consequences: First, a novel distorter will not easily arise as it requires two mutations. Second, recombination in a double heterozygote will tend to disrupt the killer + immunity combination on the driving chromosome. It is probably not coincidental that the well-known SDs in *Mus*, *Drosophila*, and *Neurospora* all are located on chromosomal segments with suppressed recombination close to the centromere.

In conclusion, the rigid machinery of fair meiotic segregation in conjunction with meiotic recombination is a severe hindrance for distorters to evolve. Whether meiotic recombination has evolved as a safeguard against distorters is difficult to say. This scenario does not seem very likely, since species exist that are not at risk for contracting genic distorters but still have recombination. Examples are marine species with broadcast fertilization, in which sperm are shed into water so that competition between sister gametes is absent or very weak.

Uniparental Transmission of Cytoplasmic DNA

Because mitochondria and chloroplasts cannot be synthesized de novo, these organelles must originate from preexisting ones and must be passed on to the next generation. A striking and fairly universal feature of cytoplasmic inheritance is uniparental transmission: in the great majority of eukaryotic organisms, the offspring from a cross inherit organelle genes from only one of the two parents, mostly the female (for reviews, see Birky 1996, 2001). There is a remarkable variation in mechanisms of uniparental transmission among different organisms, indicating many instances of independent evolution. The transmission of cytoplasmic genes from one parent can be blocked at any stage of sexual reproduction, at gametogenesis, at fertilization, or postfertilization. A strict control of intracellular replication and subsequent segregation of cytoplasmic (mitochondrial and chloroplast) genomes at gametogenesis is not known to exist in present-day organisms and indeed would presumably be difficult to establish for multicopy genomes whose replication is not synchronized with that of the cell. Without control but with biparental transmission, a mutant germline mitochondrial genome with replication advantage over the wild-type genome would outcompete the latter and gain transmission to the majority of the gametes. This segregation advantage, if great enough, could outweigh negative effects on the host's fitness and spread through the population. Natural selection is thus expected to favor variants that minimize the possibilities for spread of such deleterious cytoplasmic mutations. One possible mechanism is to restrict the role of transmitting cytoplasmic organelles to one of the parents. Although perhaps at first sight this would not seem to represent fair transmission, it is in fact unbiased because each individual has one father and one mother, and therefore any mutation has equal chance of getting transmitted. A deleterious mitochondrial mutation would not be expected to spread under uniparental transmission because it would remain confined to the segment of the population that is derived along the female line from the female in which the original mutation occurred. The scenario for evolution of uniparental inheritance has been modeled by Hoekstra (1990), Hurst and Hamilton (1992), and Law and Hutson (1992).

Somatic Incompatibility

Somatic fusions between individuals that are not genetically identical are possible in a number of taxa. This phenomenon occurs most notably in fungi, where hyphae may fuse (anastomosis), but also in sponges, tunicates and other marine colonial invertebrates. Below, I limit myself to fungi, for which factors that control anastomosis are best known.

When different conspecific mycelia grow close to each other, under certain conditions their hyphae may form anastomoses, enabling cytoplasmic continuity between the two mycelia. The resulting common mycelium contains a mixture of the nuclei of the fusion partners (hence the name *heterokaryon*) and of their cytoplasms (*heteroplasmon*). However, whether or not a heterokaryon results from a confrontation between two individuals depends on their genotype at a number of so-called *het*-loci. Only allelic identity (fungi generally have haploid nuclei) at all *het*-loci allows extensive fusion and heterokaryon formation. This situation is termed somatic compatibility. A single allelic difference at one of the *het*-loci is sufficient to trigger a somatic incompatibility response, resulting in the destruction of the hyphal cells in the contact zone. Somatic incompatibility prevents the formation of a heterokaryon. Since the number of segregating *het*-loci in a fungal population often appears to be on the order of 10, with mostly 2 alleles per locus, this criterion normally limits heterokaryosis to clonally related individuals or close kin (Cortesi and Milgroom 1988; Perkins and Turner 1988; Hoekstra 2001).

Several functional explanations of the evolution of somatic incompatibility have been suggested, of which defense against infection by deleterious genes is the most important (Caten 1972; Hartl et al. 1975; Nauta and Hoekstra 1994). Experiments indicate that somatic incompatibility is 100% effective in preventing nuclear mixing, but it cannot completely prevent the introgression of cytoplasmic elements, in particular mitochondrial plasmids and double-stranded RNA (dsRNA) viruses (Debets et al. 1994). The latter observation may be very relevant for explaining the apparent spread and stable existence of senescence-associated plasmids, particularly in *Podospora anserina*, where the whole species seems infected.

Genomic Anti-transposon Processes

A number of molecular genomic processes have been described that seem to function as anti-transposon devices. In the fungus *Neurospora*, Eric Selker discovered repeat-induced point mutation (RIP). During the sexual phase of the *Neurospora* life cycle, in the stage between fertilization and nuclear fusion, RIP efficiently detects duplicated DNA sequences (both linked and unlinked) and then modifies them by making G:C to A:T mutations. Cytosines remaining in a region mutated by RIP are typically methylated. As a result, the "ripped" sequence is inactivated. (Selker 1990). Similar processes have been described in a few other fungi as well.

Recently, RNA interference (RNAi) has been discovered and interpreted as a mechanism that (among other functions) suppresses transposon activity, specifically in the germ line, in nematodes, *Drosophila*, fungi and plants, and possibly also in mammalian cells (Tabara et al. 1999 ; Ketting et al. 1999). RNAi is the process where the introduction of dsRNA into a cell inhibits gene expression of the sequence that gave rise to the specific dsRNA molecules. Active transposons are expected to produce dsRNA. Transposons often insert into genes and are transcribed into RNA. Most transposons have inverted-repeat sequences at their ends that, when transcribed into RNA, will hybridize to form dsRNA.

SPECIFIC COUNTERMEASURES AGAINST GENOMIC PARASITES

Only a few genomic parasite systems have been investigated in sufficient detail to allow some insight into the complexities of the genetic regulation of their activity, for example the meiotic drive systems SD in *Drosophila* (Wu and Hammer 1990) and *t*-haplotypes in mouse (Lyttle 1991). A general conclusion from these reviews is that both systems are characterized by a high degree of genetic complexity with many different genes involved. Moreover, both have a considerable evolutionary age, in the order of several million years. The picture emerging then is that of a long-term coevolution between the parasitic genes and the host genome, resulting in ever-increasing complexity. In terms of individual fitness losses, the burden seems bearable to the species, but of course there is a danger here of ascertainment bias: we observe only those species that are able to cope with the genomic parasitism.

CONCLUSIONS AND DISCUSSION

Who suppresses the genomic outlaw? Returning to the central question of this chapter, we may draw several conclusions:

1. Genomic parasitism is a widespread phenomenon, affecting almost all species in a significant way.

2. Once established, a genomic parasite is not easily removed from the genome, often remaining for a very long time, although not always in active form. For example, comparing our own genome with a human society, we could say that about 50% of the buildings are in use as prisons to contain outlaws (inactive TEs).
3. Genetic systems are characterized by a number of features that make it difficult for genomic parasites to invade or become very deleterious to their hosts.
4. Every specific parasite that manages to invade a genome enters a coevolutionary process in which suppressive host mutations and subsequent parasite modifications that escape suppression are selected.

It has been suggested that this coevolutionary game will end in a victory to the host because in terms of numbers of genes, the host genome (the "parliament of the genes," Leigh 1977, 1991) vastly outnumbers the parasite, and the host genes are expected to cooperate in fighting the parasitic genes. Clearly, the short-term success of a parasitic gene depends on the competitive conditions at the replication level where it originates, which is often a (small) subset of the host's genome. It is questionable if the other genes at this level might form a sufficiently powerful "parliament." Perhaps a more fundamental problem with the concept of the parliament of the genes is how they are selected to cooperate. Any mutation enhancing individual fitness will be selected, but will not necessarily cooperate with other such mutations, except in the very limited sense that their effects are in the same direction, namely increased fitness.

The parliament of genes might, however, actually work in a different way, namely by drawing the parasitic genes themselves into its coalition. Consider the phenomenon of compensatory evolution, which has been well established experimentally in bacterial and viral systems (Burch and Chao 1999; Levin et al. 2000), and recently also in the fungus *Aspergillus* (Schoustra et al., unpublished). Following the introduction of a major deleterious mutation, strains will generally recover to their original fitness level *not* by a specific reversion mutation but through selection of a series of mutations that appear to be beneficial (i.e., enhance fitness) *conditional on the presence of the major deleterious mutation.* An interesting consequence of such compensatory evolution is that actually the removal of the deleterious element is prevented or retarded because fitness is only compensated for in the presence of the deleterious mutation. This conditional aspect points to epistasis (or genetic cooperation): the selected mutations "cooperate" with the parasitic gene to enhance individual fitness. In this way, the originally parasitic gene has become an essential component of a genetically interacting set of genes with a beneficial effect: a "criminal" converted into a "decent citizen" among the genes. Related to the genomic parasites discussed herein, such a case may be represented by mitochondrial male sterility in the wild beet (*Beta vulgaris*), in which female advantage of the male sterility has not been observed, whereas a cost appears to be associated with restorer alleles

(Laporte et al. 2001). It remains to be seen how often compensatory evolution can be identified in connection to genomic parasites.

REFERENCES

Arkhipova, I., and M. Meselson. 2000. Transposable elements in sexual and ancient asexual taxa. *Proc. Natl. Acad. Sci. USA* **97**:14,473–14,477.

Bertrand, H. 2000. Role of mitochondrial DNA in the senescence and hypovirulence of fungi and potential for plant disease control. *Ann. Rev. Phytopathol.* **38**:397–422.

Birky, C.W., Jr. 1996. Uniparental inheritance of mitochondrial and chloroplast genes: Mechanisms and evolution. *Proc. Natl. Acad. Sci. USA* **92**:11,331–11,338.

Birky, C.W., Jr. 2001. The inheritance of genes in mitochondria and chloroplasts: Laws, mechanisms, and models. *Ann. Rev. Genet.* **35**:125–148.

Burch, C.L., and L. Chao. 1999. Evolution by small steps and rugged landscapes in the RNA virus phi6. *Genetics* **151**:921–927.

Caten, C.E. 1972. Vegetative incompatibility and cytoplasmic infection in fungi. *J. Gener. Microbiol.* **72**:221–229.

Cortesi, P., and M.G. Milgroom. 1988. Genetics of vegetative incompatibility in *Cryphonectria parasitica*. *Appl. Envir. Microbiol.* **64**:2988–2994.

Cosmides, L.M., and J. Tooby. 1981. Cytoplasmic inheritance and intragenomic conflict. *J. Theor. Biol.* **89**:83–129.

Debets, A.J.M., X. Yang, and A.J.F. Griffiths. 1994. Vegetative incompatibility in *Neurospora*: Its effect on horizontal transfer of mitochondrial plasmids and senescence in natural populations. *Curr. Genet.* **26**:113–119.

Frank, S.A. 1989. The evolutionary dynamics of cytoplasmic male sterility. *Am. Nat.* **133**:345–376.

The Genome International Sequencing Consortium. 2001. Initial sequencing and analysis of the human genome. *Nature* **409**:860–921.

Gouyon, P.H., and D. Couvet. 1987. A conflict between two sexes, females and hermaphrodites. In: The Evolution of Sex and Its Consequences, ed. S.C. Stearns, pp. 245–260. Basel: Birkhäuser.

Griffiths, A.J.F. 1992. Fungal senescence. *Ann. Rev. Genet.* **26**:351–372.

Griffiths, A.J.F. 1995. Natural plasmid of filamentous fungi. *Microbiol. Rev.* **59**:673–685.

Hammer, M.F., J. Schimenti, and L.M. Silver. 1989. Evolution of mouse chromosome 17 and the origin of inversions associated with *t*-haplotypes. *Proc. Natl. Acad. Sci. USA* **86**:3261–3265.

Hartl, D., E.R. Dempster, and S.W. Brown. 1975. Adaptive significance of vegetative incompatibility in *Neurospora crassa*. *Genetics* **81**:553–569.

Hiraizumi, Y., L. Sandler, and J.F. Crow. 1960. Meiotic drive in natural populations of *Drosophila melanogaster*. III. Population implications of the segregation-distorter locus. *Evolution* **24**:415–423.

Hoekstra, R.F. 1990. Evolution of uniparental inheritance of cytoplasmic DNA. In: Organisational Constraints on the Dynamics of Evolution, ed. J. Maynard Smith and G. Vida, pp. 269–278. Manchester: Manchester Univ. Press.

Hoekstra, R.F. 2001. Functional consequences and maintenacc of vegetative incompatibility in fungal populations. In: Biotic Interactions in Plant–Pathogen Associations, ed. M.J. Jeger and N.J. Spence, pp. 27–34. Wallingford: CAB Intl.

Hurst, L.D., A. Atlan, and B.O. Bengtsson. 1996. Genetic conflicts. *Qtly. Rev. Biol.* **71**:317–364.

Hurst, L.D., and W.D. Hamilton. 1992. Cytoplasmic fusion and the nature of sexes. *Proc. Roy. Soc. Lond. B* **247**:189–194.

Ketting, R.F., T.H.A. Haverkamp, H.G.A.M. van Luenen, and R.H.A. Plasterk. 1999. mut-7 of *C. elegans*, required for transposon silencing and RNA interference, is a homolog of Werner Syndrome helicase and RnaseD. *Cell* **99**:133–141.

Laporte, V., F. Viard, M. Béna et al. 2001. The spatial structure of sexual and cytonuclear polymorphism in the gynodioecious *Beta vulgaris* ssp *maritima*. *Genetics* **157**: 1699–1710.

Laser, K.D., and N.R. Lersten. 1972. Anatomy and cytology of microsporogenesis in cytoplasmic male sterile angiosperms. *Bot. Rev.* **38**:425–454.

Law, R., and V. Hutson. 1992. Intracellular symbionts and the evolution of uniparental inheritance. *Proc. Roy. Soc. Lond. B* **248**:69–77.

Leigh, E.G.J. 1977. How does selection reconcile individual advantage with the good of the group? *Proc. Natl. Acad. Sci. USA* **74**:4542–4546.

Leigh, E.G.J. 1991. Genes, bees and ecosystems: The evolution of common interest among individuals. *Trends Ecol. Evol.* **6**:257–262.

Levin, B.R., V. Perrot, and N. Walker. 2000. Compensatory mutations, antibiotic resistance, and the population genetics of adaptive evolution in bacteria. *Genetics* **154**:985–997.

Lyttle, T.W. 1991. Segregation distorters. *Ann. Rev. Genet.* **25**:511–557.

Nauta, M.J., and R.F. Hoekstra. 1994. Evolution of vegetative incompatibility in filamentous ascomycetes. I. Deterministic models. *Evolution* **48**:979–995.

Perkins, D.D. 1992. *Neurospora*: The organism behind the molecular revolution. *Genetics* **130**:687–701.

Perkins, D.D., and B.C. Turner. 1988. Neurospora from natural populations: Toward the population biology of a haploid eukaryote. *Exp. Mycology* **12**:91–131.

Schnable, P.S., and R.P. Wise. 1998. The molecular basis of cytoplasmic male sterility and fertility restoration. *Trends Plant Sci.* **3**:175–180.

Selker, E.U. 1990. Premeiotic instability of repeated sequences in *Neurospora crassa. Ann. Rev. Genet.* **24**:579–613.

Tabara, H., M. Sarkissian, W.G. Kelly et al. 1999. The rde-1 gene, RNA interference, and transposon silencing in *C. elegans*. *Cell* **99**:123–132.

Van Damme, J.M.M. 1983. Gynodioecy in *Plantago lanceolata*. II. Inheritance of three male sterility types. *Heredity* **50**:253–273.

Van der Gaag, M., A.J.M. Debets, J. Oosterhof et al. 2000. Spore-killing meiotic drive factors in a natural population of the fungus *Podospora anserina*. *Genetics* **156**:593–605.

Werren, J.H., U. Nur, and C.-I. Wu. 1988. Selfish genetic elements. *Trends Ecol. Evol.* **3**:297–302.

Whitehouse, H.L.K. 1973. Towards an Understanding of the Mechanism of Heredity. 3d ed. London: Edward Arnold.

Wu, C.-I., and M.F. Hammer. 1990. Molecular evolution of ultraselfish genes of meiotic drive systems. In: Evolution at the Molecular Level, ed. R.K. Selander, T. Whittam, and A. Clark, pp. 177–203. Sunderland, MA: Sinauer.

15

The Transition from Single Cells to Multicellularity

Eörs Szathmáry[1] and Lewis Wolpert[2]

[1]Department of Plant Taxonomy and Ecology, Eötvös University, and Collegium Budapest (Institute for Advanced Study), 1014 Budapest, Hungary

[2]Department of Anatomy and Developmental Biology, University College, London WC1E 6BT, U.K.

ABSTRACT

Multicellularity requires overall coordination of at least some metabolic or informational processes among the cells of a colony. Developmentally differentiated cell types are a sufficient condition. Little new had to be invented by evolution in the transition from single cells to multicellular organisms. The basic processes required for development were already present in the eukaryotic cell. These included a program of gene activity as in the cell cycle, signal transduction, and cell motility. There are also processes similar to cell differentiation. Possible advantages of multicellularity included: more efficient predation (cooperative feeding), resistance to predation, division of labor, cannibalism and nutrient storage, and more efficient dispersal during starvation. The evolution of an egg was necessary to reduce the chance of selfish cells taking over, and it could have increased evolvability by rendering development coherent. The biggest hurdle could have been to down-regulate cell division at the appropriate time and place, which problem increases with organism size and complexity.

INTRODUCTION

What is multicellularity? We agree with Kaiser's (2001) view, that an overall co-ordination of function is a necessary and sufficient condition for a colony of cells to qualify as multicellular. According to this, most bacterial colonies are not multicellular, since they lack overall coordination, despite occasional self-organized patterned growth. We add to this that the occurrence of developmentally differentiated cell types in the colony is a sufficient condition of multicellularity. The interesting question is then this: Is it possible to have a multicellular organism with one cell type only? The two basic aspects of any living being are metabolism and informational operations (e.g., Maynard Smith

1986). We can thus say that if at least some parts of the metabolism or the information processing of the cells (confined to a single cell in unicellular organisms) are shared in a coordinated manner by all cells of the colony, we are dealing with a multicellular organism. Sharing must have an evolved genetic basis not found in unicellular organisms.

What then new had to be "invented" for the transition from single-cell organisms to multicellularity? What properties did these so-called simple organisms have, which were and are still fundamental for the development of multicellular organisms? How did embryonic development from an egg evolve? The key features required for embryonic development are a program of gene activity, cell differentiation, signal transduction so cells can communicate with each other, and cell motility to generate change in form. Single-cell organisms possessed all of these features in a primitive form and thus very little needed to be invented, although much had to evolve and be fine-tuned.

The basic organization and functions shared by all eukaryotic cells, but not prokaryotes, must have been present at least 2 billion years ago, before single-celled eukaryotes diverged. This conservation would include their large size — 1,000× the volume of the prokaryotic cell — their dynamic membranes capable of endocytosis and exocytosis, their membrane-bounded organelles (like the nucleus, mitosis, and meiosis), sexual reproduction by cell fusion, a cdk/cyclin-based cell cycle, actin- and tubulin-based dynamic cytoskeletons, cilia and flagella, and histone/DNA chromatin complexes. These ancient processes, which evolved in the single-celled prokaryotes and early eukaryotes long before Metazoa, constitute the core biochemical, genetic, and cell biological processes of Metazoa.

Eukaryotes presumably originated from bacteria (prokaryotes). It is uncertain when this actually happened. Molecular phylogeny tells us that Archaebacteria and eukaryotes are evolutionary sisters, but a molecular clock to pinpoint the event of origin simply does not exist. Some early fossil finds have been claimed as eukaryotic on the basis of having a large size, yet this is an error, since we know that huge bacterial cells occasionally arise. Thus, the older view holds that eukaryotes arose some 2 billion years ago and that animals, fungi, and plants diverged some 1.2 billion years ago. Cavalier-Smith (2002) has recently challenged this dating and suggests that archaebacteria and eukaryotes diverged from bacteria as late as 850 million years ago. Clearly, this issue has not been resolved. In any case, the spectacular radiation of the Metazoa happened at the beginning of the Cambrian, for which there is good fossil evidence.

What is puzzling is why the Cambrian explosion took place when it did. Two possible answers seem likely. One is that before complex multicellular organisms could evolve, some crucial invention or inventions in cell physiology or gene regulation had to be made; once made, there was rapid radiation into an ecologically empty world. The apparently monophyletic origin of the Metazoa, deduced from molecular data, is consistent with this view. In this chapter, we discuss what such inventions might be. An alternative is that it was not until some 600 million years ago that the physical conditions on the Earth made the Metazoan way of life practicable (Maynard Smith and Szathmáry 1995).

We first discuss some key features of multicellular organisms. Thereafter we ask how these features could have evolved and what could have been the selective advantages driving the transitions to multicellularity.

SOME KEY FEATURES OF MULTICELLULARITY

Cell Differentiation and Epigenetic Inheritance

Development requires controlled sequences of events. States must follow one another in progression: A–B–C–D–Š–Z. Intermediates in this sequence cannot be skipped; if they are, a teratological malformation is the result. This implies that there must be some means to monitor the completion of one step before the system moves to the next. The cell cycle is a nice illustration of how this problem can be solved. The system has two basic components: a "cell-cycle engine," and a set of feedback controls. The engine is a biochemical oscillator that can tick autonomously if left alone (although, as we shall see, usually it is not). The states of this biochemical clock trigger morphological events (Maynard Smith and Szathmáry 1999).

Cell differentiation depends on different genes being active in different cells. An understanding of how this can happen is shown by how the bacterium *Escherichia coli* can acquire the ability to use the sugar lactose. The mechanism is based on one gene producing a protein, which recognizes and binds to a specific DNA sequence at the start of a second gene (or sometimes of several linked genes), and so is able to regulate the activity of the second gene. In the particular case of the *lac* operon, regulation is negative; the regulatory protein switches off the second gene, unless it is rendered ineffective by binding to the "inducer," lactose. In other cases, however, regulation is positive; the regulated gene is inactive unless it is switched on by the regulator gene.

It turns out that such regulation is a universal feature of living cells. One feature worth emphasizing is that the "inducer" lactose binds to the regulator protein at a different site from that at which the regulator binds to the gene it is regulating. A consequence of this is that, in principle, any chemical substance can switch on any gene. That is, the "meaning" of an inducing signal is arbitrary, as the meanings of words are arbitrary. All complex communication depends on such arbitrary signals.

The activity of a particular gene, in a particular cell, can be under both positive and negative control from different sources and can depend on the stage of development and of the cell cycle, on the cell's tissue type, on its immediate neighbors, and so on. Gene regulation is complex and hierarchical. Yet the basic mechanism already exists in prokaryotes.

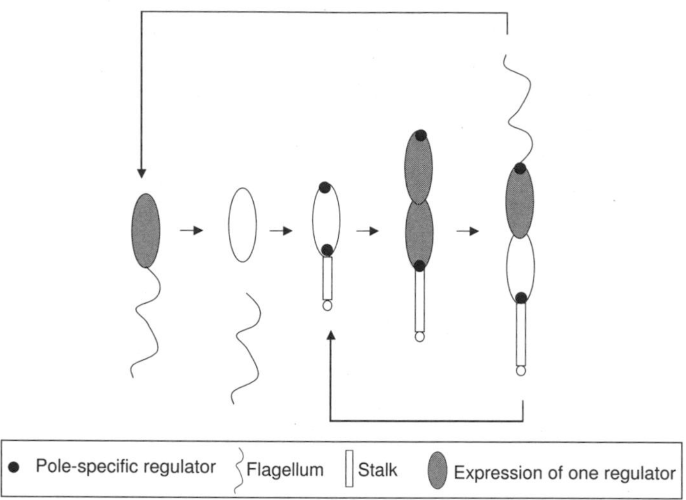

Figure 15.1 Schematic drawing of the *Caulobacter* life cycle. The development of the differentiated state follows a stem-cell strategy (after Martin and Brun 2000).

The development of the bacterium *Caulobacter* (Figure 15.1) sheds light on the origin of differentiation (Martin and Brun 2000; England and Gober 2001). The swarmer cell is flagellated before it becomes sessile; then it loses its flagellum and develops a stalk as it progresses through the cell cycle. Opposite the stalk, a new swarmer is produced; the stalked cell acts as a stem cell and divides again and again. Here the stalk is metabolically active and is thought to enhance nutrition while the cell is anchored to the nutritive substrate. A long stalk may also lift the cell above the level of most intense competition on the substrate surface. The swarmer is free to navigate into new habitats rich in nutrients.

There are two remarkable aspects of this development: (a) the stem-cell strategy works in a prokaryote and (b) development is strictly coupled to the cell cycle. Certain events in the cell cycle trigger flagellar development in the presumptive swarmer, whereas completion of this development gives the license for completion of the cell cycle. There is mutual feedback between these processes: checkpoints abound. It is crucial that the necessary asymmetry is generated by pole-specific localization of cell cycle regulators.

A good model is the budding yeast *Saccharomyces* (van Houten 1994), its mating system, and the behavior of its cytoskeleton. The yeast chemosensory system enables haploid cells of complementary mating type, known as **a** and α, to recognize each other. This then leads to changes in molecular activity and morphology, which eventually lead to cell fusion to produce a diploid cell. In both cell types, the mating pheromones bind to receptors on the surface that then leads to cell cycle arrest and changes in gene transcription — there is a typical signal transduction cascade. The ligand-activated receptor activates a trimeric G protein in a manner similar to that which occurs when muscarinic or adrenergic receptors are activated in Metazoa. Downstream is a kinase cascade that arrests the cells in G1 of the cell cycle and induces gene expression. This cascade has considerable similarities to that found in vertebrates.

Haploid-specific genes must be turned off in diploids, and meiosis must not be allowed to occur in haploids. Further, different mating type genes are expressed, depending on whether MAP or MATα is active. It is remarkable that both **a** and α alleles contain a region similar to the homeodomain. It seems that the homeodomain played a role in controlling the activities of other genes before the origin of multicellularity. It should be noted, however, that yeast may well be a secondarily simplified organism and could thus bear vestiges from its multicellular past.

The differentiated state arises in ontogeny, and in most cases it is passed on to offspring cells during cell division (a liver cell gives rise to liver cells). The DNA in the liver cell and the white blood cell is practically the same; it is the regulated (on-off) states of cells that is different and that is passed on. Therefore, multicellular organisms rest on a dual inheritance system (Maynard Smith 1990): genetic and epigenetic. There are three known types of epigenetic inheritance systems (Jablonka and Lamb 1995): (a) structural inheritance (such as in the cortical inheritance of ciliates), (b) steady-state systems (active gene regulatory networks), and (c) chromatin marking systems. For the long-term storage of epigenetic information, chromatin marking seems to be the most efficient. Of course, these systems may work together in the same organism. In *Drosophila* development, the activity of a hierarchical gene regulatory network sets the state of the chromatin marks that ensure long-term maintenance of the differentiated state. One candidate mechanism for chromatin marking is DNA methylation found in many, but not all, animals. Research over the past few years has focused on the interplay between DNA methylation, histone methylation, and acetylation (Rice and Allis 2001).

An emerging view is that histone methylation and acetylation can both serve to activate or silence genes, depending on the particular lysine residue in the N-terminal histone tail and the nature of the transcriptional regulator protein (Figure 15.2). These processes are widespread; they occur in *Drosophila* as well as yeast. Additionally, certain proteins target the core complex of nucleosomes, which affects the availability of DNA for transciption. Chromatin remodeling is known to be crucial for differentiation in animals (Müller and Leutz 2001) as well as in plants (Verbsky and Richards 2001). For example, chromatin remodeling in hematopoiesis is well documented, and deregulation of remodeling occurs in leukemia.

Signaling

Cell-cell signaling pervades all aspects of development, not just in vertebrates but in all animals (Metazoa) (Gerhart 1999). It is a typifying characteristic of the

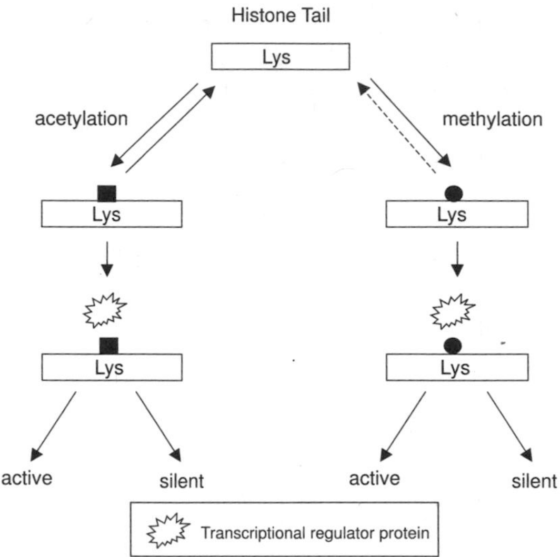

Figure 15.2 Epigenetic regulation by histone modifications. Depending on the transcriptional regulator, the effect of a particular lysine (Lys) modification can act as a positive or negative regulator. Methylation seems to be a more persistent epigenetic mark (after Rice and Allis 2001).

major multicellular life forms, animals, plants, and fungi, which diverged about 1.2 billion years ago from a common ancestor descended from a lineage of unicellular life forms. In Metazoa, at least 17 kinds of signal transduction pathways operate, each distinguished by its transduction intermediates. The pathways must have evolved and become conserved in pre-Cambrian times, before the divergence of basal members of most of the modern phyla. In Metazoan development and physiology, the responses of cells to intercellular signals include cell proliferation, secretion, motility, and transcription. These responses tend to be conserved among the Metazoa and shared with unicellular eukaryotes, and in some cases, even with unicellular prokaryotes. Protein components of the responses date back 2 billion years to ancestral eukaryotes or 3 billion to ancestral prokaryotes. Examples include the reversible phosphorylation of serine, threonine, and tyrosine residues by the interplay of protein kinases and phosphatases (Bakal and Davies 2000). Transmembrane signal transduction is a feature common to all eukaryotic and prokaryotic cells. The histidine-aspartate phosphorelays, signaling systems of eubacterial origin, are known to be widespread in eukaryotes outside the animal kingdom (Thomason and Kay 2000).

Lower eukaryotes, such as flagellates, slime molds, ciliates, and yeast cells, have many control mechanisms known from Metazoa (Christensen et al. 1998). This is true for regulatory systems, which have to do with fundamental features like cell survival, proliferation, differentiation, chemosensory behavior, and programmed cell death. Signaling in unicellular eukaryotes was believed to be confined to mating factors in, for example, ciliates and yeast cells. It is now evident that unicellular eukaryotes depend on extensive signaling systems for their mating, feeding, and limited differentiation.

There are also many similarities of the intracellular transduction systems in uni- and multicellular organisms. These similarities include G proteins, components of the inositol phospholipid pathway, phospholipase C and D, protein kinases, adenylate cyclase/cAMP systems, and guanylate cyclase/cGMP systems. cGMP is involved in cellular activities in unicellular organisms. It takes part in slug formation, cell cycle events, metabolism, and ciliary beating in ciliates. Enzyme activities are often controlled by protein phosphorylations in eukaryotes. Several systems are implicated in signal transduction pathways in mammalian cells in response to external stimuli, such as survival and growth factors. They include protein kinases which phosphorylate substrate enzymes at tyrosine and/or serine/threonine residues and similar systems seem to be at work in *Tetrahymena* with respect to survival and proliferation. The receptor tyrosine kinase pathway is found in all Metazoa, even sponges, but not outside Metazoa.

Ciliates respond to chemical cues in their environment; both mating and food searching make use of this mechanism. *Euplotes* mating involves small peptides, around forty amino acids, and there are about twelve possible alleles in the locus that codes for them. These pheromones act in an autocrine fashion and as a signal to a different mating type. In one species there are ten mating types. Folic acid and cAMP are used by *Paramecium* as signals for food. The ligands hyperpolarize the cell. Glutamate can cause a threefold increase of cAMP.

Embryos use a small number of signaling families of protein molecules that include TGFbeta, Hedgehog, Wnts, FGF. What is their origin? In general it is not known but one interesting case relates to the cystine knot three-dimensional structure, found in many extracellular molecules and conserved among divergent species (Vitt et al. 2001). In addition to the well-known members of the cystine knot superfamily, novel subfamilies of proteins (mucins, von Willebrand factor, bone morphogenetic protein antagonists, and slit-like proteins) were identified as putative cystine knot-containing proteins. Phylogenetic analysis revealed the ancient evolution of these proteins and the relationship between hormones (e.g., transforming growth factor-β, TGFβ) and extracellular matrix proteins (e.g., mucins). They are absent in the unicellular yeast genome but present in nematode, fly, and higher species, indicating that the cystine knot structure evolved in extracellular signaling molecules of multicellular organisms. The largest 10-membered cystine knot subfamily is the TGFb family that consists of transforming growth factors, bone morphogenetic proteins (BMP), growth differentiation factors, inhibins, and Müllerian inhibiting substance.

Two candidate proteins were found in yeast: the putative maltose permease (a transmembrane protein) and the metallothionein-like protein CRS5. Because their cystine knot signatures were not conserved in paralogs of the same species, they were excluded as cystine knot proteins. As no cystine knot structure can be identified in yeast, it is unlikely that this structure could have evolved in unicellular organisms. However, the use of four cysteines to form a cystine ring is widely practiced and has been described in protozoan as well as Metazoan species. The slime mold revealed three potential sequences, two prestalk proteins, and one predicted open reading frame, all with a signal peptide and without transmembrane helices.

These findings underscore the importance of the cystine knot structure for ligand-receptor interaction and cell-cell communication. This is corroborated by the absence of cystine knot structures in unicellular yeast and the presence of multiple subfamily members in the nematode, indicating that this ancient structure evolved parallel with the development of multicellular life. Note that a redox-sensitive device like this (with its disulfide bonds) is likely ancient, whereas the particular proteins we consider here may be more recently derived.

The cellular slime molds, which branched from the Metazoan line shortly before plants and fungi, have cell-cell signaling involving several components shared by Metazoa, such as cAMP, G-protein linked receptors, a variety of protein kinases, and JAK/STAT transcriptional control. From their unicellular past, early Metazoa had a lot to draw upon in the evolution of intercellular signaling.

The foregoing summary clearly shows the preadaptations for signaling in unicellular organisms. It also holds, however, that concomitant with the origin of multicellularity, these signaling systems have undergone radical evolution toward increased complexity. Comparison of the fully sequenced genomes of two species of cyanobacteria — the unicellular *Synechocystis* and the multicellular *Nostoc*— is a good case in point (Kaiser 2001). They have about 3200 and 7400 expressed proteins, respectively. Some of the extra genes code for a novel cell type in *Nostoc* (see below), but it is also clear that the number of signaling proteins has also dramatically increased; for example, there are 20 and 146 histidine kinase proteins in *Synechocystis* and *Nostoc*, respectively.

Motility

Single-cell organisms have molecular motors and these could provide the forces for morphogenesis. Chemotaxis in the slime mold *Dictyostelium* provides an important model for cytoskeletal organization and signal transduction, and chemotaxis is important in its own right. The chemotactic cell is polarized and polarity is fundamental to many developmental processes. Ligand binding (Chung et al. 2001) leads to rearrangement of the cytoskeleton; actin polymerization at the anterior end results in filopodial extension whereas myosin at the rear contracts to bring it forward. cAMP mediates chemotaxis through G protein-coupled/serpentine cAMP receptors, which are coupled to a G protein and guanylcyclase activation. This leads to activation, probably, of a PI3K kinase which could generate phospholipids leading to cytoskeletal regulation. Rac and Cdc 42 play a key role.

EVOLUTION OF MULTICELLULARITY

Multiple Origins of Multicellularity: Bacteria, Protists, and Algae

Attempts at various forms of multicellularity have independently been made by evolution at least twenty times (Bonner 2000; Kaiser 2001). Plants, animals, and fungi are just the most spectacular and complex achievements of this kind. There are essentially two ways to make a simple multicellular entity out of single cells: either the single cells divide and the offspring stick together, or a number of solitary cells aggregate to form the colony. With one exception, division-and-adhesion is characteristic of multicellular forms of aquatic origin, whereas aggregation prevails in colonies of terrestrial origin.

Cyanobacteria are prokaryotes that may stick together, forming a filament characteristic of the particular species. In the *Hormogonales* genus, division of labor is apparent: some cells (called heterocysts), spaced at regular intervals, are colorless and perform nitrogen fixation. One crucial condition for the division of labor is that optimal performance of different tasks should not disturb each other. This is satisfied here because oxygen is poisonous for the nitrogen-fixing enzymatic apparatus. (Unicellular cyanobacteria divide the labor *in time* rather than in space; photosynthesis is shut off while the cells fix nitrogen.) The positioning of the heterocysts is under strict developmental control, requiring cell–cell interaction (e.g., Kaiser 2001). RNA hybridization data suggest that vegetative cells and heterocysts differ in about 1000 expressed proteins.

Myxobacteria and slime molds provide one of the most spectacular cases of evolutionary convergence by their aggregative development (Bonner 2000; Crespi 2001). Initially, solitary cells come together when starved and the multicellular colony ultimately forms a fruiting body (Figure 15.3). The number of cells in the fruiting body ranges between 10^4 and 10^6, for all known wild types. The aggregation process produces striking transient dynamical patterns (such as ripping and swirling) which fade away upon fruiting body formation. Only spores formed in the fruiting body go to the next generation, whereas cells in the stalk die. Reproductive division of labor is apparent, with the formation of soma and germ. (Interestingly, in the case of myxobacteria, many nonreproductive cells lyse and their material is apparently eaten up by the spore-forming cells — a case of regulated cannibalism in a multicellular colony during starvation — an observation to which we return later.)

In the case of myxobacteria, two types of signal (called A and C) act at different stages of the life cycle, and many genes are partially or completely activated by these signals (see Kaiser 2001). It is thus perhaps not surprising that the 9.5 megabase pair genome of myxobacteria belongs to the largest known prokaryotic genomes.

What are the selective advantages of these lovely aggregative organisms? Aggregation by itself has its initial advantage in collective foraging: it is like collective hunting in lions. Fruiting body formation has an advantage in efficient

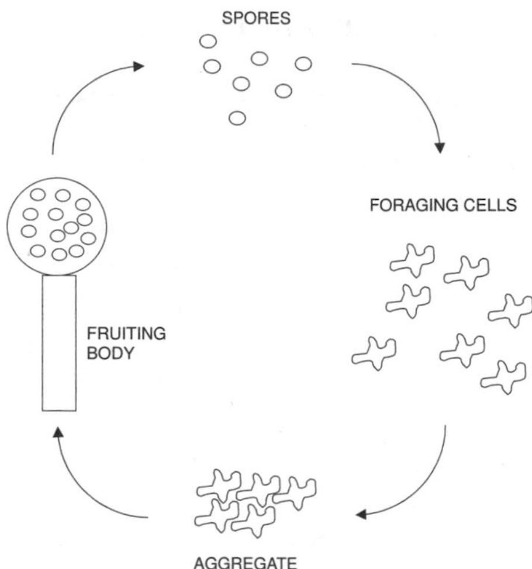

Figure 15.3 Schematic view of the slime mold life cycle. Multicellularity arises from aggregation on the soil. The fruiting body helps the dispersal of spores.

dispersal: the spores are normally lifted above the substrate and thus may disperse to better, and various, environments.

The volvocine algae provide an excellent test case for the evolution of multicellularity by division and adhesion (Kirk 1999). They belong to the chlorophyte (green) algae; their closest unicellular relative is *Chlamydomonas.* It is remarkable that reproductive division of labor (germ–soma differentiation) has occurred at least twice in this lineage. Here, sticking together is mediated by a common extracellular material (mostly glycoproteins and sulphated polysaccharides). Preadaptation to the multicellular life form is apparent in *Chlamydomonas*: they undergo "multiple fission." After growing to 2^n-fold in volume, they rapidly divide n times within the mother-cell wall, to produce 2^n offspring cells (n is a species-characteristic number). In the volvocaceans, cytokinesis is incomplete until the common envelope made of extracellular material is formed. This is a clear case of heterochrony that has led to a breakthrough in organization.

Another preadaptation to multicellularity in this lineage is the regulation of the flagellated state. *Chlamydomonas* and *Eudorina* are first motile, biflagellated ("somatic") cells that later lose their flagella, round up and form the fast dividing reproductive ("gonidia-like") cells. Thus in the common ancestor of multicellular volvocaceans, motility and reproduction are mutually exclusive. If an organism in that lineage wants to do both simultaneously, it must consist of at least two cells: one motile and one reproductive. The main advantage of multicellularity is presumably threefold: being able to divide while being motile, protection from predation by increased size (note that the developing small gonidia are in the protective interior of the organism), and enhanced capacity to store nutrients (such as polyphosphate in the extracellular matrix; Kaiser 2001).

In *Volvox*, what makes certain cells become somatic, others reproductive? Apparently, this needs asymmetric division, and the smaller cell invariably becomes somatic, by the action of a gene that shuts down the genes for reproductive development in it.

The Origin of Metazoa

One can imagine two models of the early Metazoa (Figure 15.4). The first depends on differences between cells that arise at the moment of cell division. The propagule divides by asymmetric cell division. The two cells remain attached to one another. One retains the capacity to divide, whereas the other is specifically adapted to feeding, at the expense of losing the ability to divide. When the "stem cell" divides again, one of the daughter cells remains attached to the feeding cell; the other must divide again to produce a new feeding cell. This illustrates a primitive differentiation between germ line and soma, with an associated division of labor. For the process to work, a stem cell must "know" whether it is attached to a feeding cell or not. This does not require cell-to-cell communication; it is sufficient that there be two kinds of cell division, one producing an attached feeding cell and the other an unattached stem cell.

An alternative process relies on cell-to-cell interactions. An imaginary two-celled organism reproduces by binary fission (vegetative reproduction), requiring two cell divisions. After the second division, the two central cells become decoupled, which triggers regulative differentiation of a new feeding cell in the offspring lacking one. Both cell polarity and cell-to-cell communication are needed: cells must be informed about the state of neighboring cells. Such organisms are expected to show regeneration upon injury. Although the model is

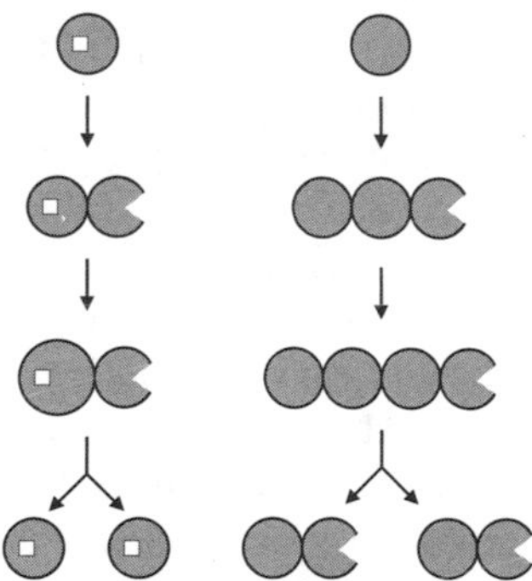

Figure 15.4 Two simple models of early cell differentiation (Wolpert 1990). There are only two cell types in this imaginary organism: one of them is specialized for feeding. One option is to have a stem cell (left) marked by a determinant (white block). The other option adopts cell polarity *as well as* cell-to-cell signaling.

imaginary, it is interesting that the very simple placozoan, *Trichoplax,* has only four cell types and reproduces by fission.

Another model for the origin of the Metazoa involves cooperation and cell death (Kerszberg and Wolpert 1998). A mutation in a protozoan resulted in the failure of the cells to separate following cell division. In addition, a cytoplasmic bridge may have persisted. A colony could develop by the repeated binary division of the constituent cells. Such a mutation could thus lead to the formation of colonies which were loose aggregates. These colonies could fragment when large and so provide a means of reproduction. But what was the selective advantage?

It could have been increase in size, which could have provided protection against predatory cells, but much more likely is what may have happened when conditions became unfavorable. When food was in short supply there would have been insufficient resources for the individual cells to grow and multiply, and death was imminent. Now the virtues of multicellularity become evident. Some cells gave up their lives for others. That is they were "eaten" by their neighbors. One possibility is that because of the cytoplasmic bridges between the cells, metabolites could move from one cell to its neighbor. Another possibility is that as some cells died their remains were taken up, phagocytosed, by adjacent cells. This last scenario has the advantage that it requires nothing new in terms of cell physiology.

Whereas a colony of independently reproducing cells could have been successful, mutations in all the individual lineages would have occurred and accumulated. This would have had two severe disadvantages. The first would have been at the level of how cells interacted in the colony (Michod, this volume). The cells would acquire different genetic constitutions and this would have led to competition rather than cooperation between the lineages. Second, it would have been difficult for the colonies to lose deleterious mutations or mutations in general, including those reverting to a unicellular state.

The solution to these problems lay in the evolution of the egg; if the various colonies arose from a few germ-like cells with low mutation rate, then the competition and the mutation problems would both disappear. It is not too difficult to imagine a series of mutations which would have given the inner cells an advantage with respect to eating their neighbors so that in hard times the outer cells died. Indeed, a redox gradient could have been utilized for this goal (Blackstone 2000). Small clusters of cells do form such gradients. There also seems to be an ancient link between programmed cell death and reactive oxygen species, which are often intermediates in redox signaling (Blackstone and Kirkwood, this volume). Note also that a morphogenic specification of the inner and outer layers of the fruiting body has evolved in myxobacteria (Kaiser 2001), where utilization of the dead cells by the presumptive spores is also known to happen.

There are a number of cases that provide strong support for cell death providing nutrition for adjacent cells. For example, in sponges, oocyte growth involves incorporation of nutrients from other cells, and the nutrient cells are phagocytosed by the growing oocyte. In starved hydra, epithelial cells are produced in excess such that about 20% die each day, and phagocytosis of the dead cells probably provides a survival mechanism. A similar situation seems to pertain in planaria.

Selective Advantages of Multicellularity and Levels of Selection

Division of labor could provide advantage once organisms were multicellular. This has been widely claimed to be a crucial step in the evolution of multicellular organisms, though the benefit has never been established. Bell and Koufopanou (1991), for example, suggest that the unexpectedly high rates of increase shown by colonial algae are made possible by the division of labor between somatic and germ cells. If the somatic cells are a source and the germ cells a sink, then there is the possibility that end product inhibition, which may act as a negative feedback mechanism for resources, could be reduced.

Michod and Roze (2000) argue that there was a trade-off between cell division and motility, such that dividing cells were less likely to be motile, and motile cells were less likely to divide. This condition cannot be general, but *Volvox,* discussed above, is a case in point. They assume that the transition to multicellularity was fueled by the benefits of cooperation and the advantages of large size but again the advantages are not universal. A positive association of the occurrence of a germ line with organism size is also predicted by an alternative hypothesis based on division of labor (Bell and Koufopanou 1991). This hypothesis states that the organism has a greater fitness when some cells specialize in reproduction (the germ cells) and other specialize in other functions (the somatic cells), because each task can be performed more efficiently. Furthermore, division of labor is assumed to be more fruitful in a big group than in a small one.

Their theory assumes that the march toward multicellularity is fueled by the advantages of cooperation and large size. Cooperation increases the fitness of the new higher-level unit, and, in this way, cooperation may create new levels of selection. The evolution of cooperation sets the stage for conflict, represented here by the increase of deleterious mutants within the emerging organism that tilt the balance of selection in favor of the lower level, cells in our case. The evolution of modifiers restricting the opportunity for selection among cells is the first higher-level function at the organism level.

There is an obvious question to be asked by an evolutionary biologist: How is it possible that evolution of the cell level (cheaters) does not destroy the higher-level unit (Maynard Smith and Szathmáry 1995) in aggregative organisms? This is all the more justified because during aggregation, unrelated cells may come together, annihilating the force of kin selection so effective when the colony is formed strictly from the offspring of a single cell, a zygote, for example. We know that colonies formed by unrelated cells tend to develop abnormally and that cheater genotypes can be found in the wild (reviewed by Pál and Papp 2000; Crespi 2001). The cheater phenotype may be induced by a single mutation. Cheaters refuse to go to the stalk (the soma); rather they prefer to become spores (germ). Moreover, they may induce wild-type cells to go preferentially to the stalk. All in all, a common outcome is that the stalk will be the shorter the more cheaters there are in the colony. There is an advantage to cheaters within the colony; this is presumably counterbalanced by selection against them at the colony level, as a result of inefficient dispersal (although there is evidence of reduced fertility at the colony level, also when the proportion of cheaters becomes excessive; we do not understand why this is so). Inefficient dispersal, due to shortening of the stalk, has two consequences: (a) the cheaters remain in the bad (exhausted) local environment; (b) they are less likely to infect distant, newly forming colonies.

Lenski and Velicer (2000) analyzed by simple game theory the fate of cheaters documented in myxobacteria (Velicer et al. 2000). The cheaters were obtained either through cultivation in nutrient-rich media or isolated as mutants defective in molecular signaling pathways. Cheaters invariably could invade a wild-type population and preferentially formed spores. The selective forces can be summarized by the following payoff matrix:

	P	C
P	1	0.5
C	10	0.1

P indicates the developmentally proficient genotype, whereas C denotes the cheater. This matrix is unusual in that it shows the fitness of the invader on the left in a population of residents on the top. It is shown that both types can mutually invade each other, resulting in a protected polymorphism. This case is formally similar to a model developed for the coexistence of standard and defective interfering (DI) viruses at high multiplicity of infection. In the latter model, while DI particles gain an advantage in each cell that is co-infected by standard viruses, they are unable to grow at all on their own (Szathmáry 1992).

There is a further twist, however. It seems that myxobacterial spores are quite sticky, which increases the chance of cheaters getting into the next colony; if colonies were founded by single spores, competition would happen mostly between, rather than within, colonies. Why are spores sticky? It seems the answer is that they germinate more successfully and they are much more efficient at collective predation if they start as a small herd from the beginning. This shows how conflicts between levels of selection are balanced and how they are linked to the evolution of development.

Evolution of Indirect Development

Could maximal indirect development and set-aside cells have evolved in some context other than planktotrophy (Blackstone and Ellison 2000)? As with all multicellular organisms, development of Metazoans must provide mechanisms to hold in check defecting cells, which selfishly favor their own replication rate over that of the multicellular group. The germ line is prominent among these mechanisms for conflict mediation. According to Davidson et al. (1995), a key evolutionary innovation was a group of undifferentiated "set-aside cells" on which novel patterns of gene expression could subsequently evolve to act. Blackstone claims the most important evolutionary novelty to have been the developmental use of yet undifferentiated set-aside cells, which retain indefinite division potential, as a substrate for the morphogenesis of large structures. However, Wolpert (1999) has argued that set-aside cells could not have evolved as suggested by Davidson, as metamorphosis is always intercalated into direct development.

Metamorphosis refers to a life cycle in which there is a free larva that is motile and may or may not be feeding, which undergoes a dramatic change in form to reach the adult state. The larval stage is for dispersal/feeding. In many cases, cells of the larva are replaced and die. A general principle of evolution of form is that there must be continuity and no hopeful monsters. That is, changes must be relatively small and each must be adaptive. The essence of metamorphosis is that there is no morphological continuity between the larva and the adult. For example, in the amphibian, one cannot go to the frog from the larva by small adaptive steps to form limbs. Similarly, with the sea urchin, the rudiment of the larva gives rise to key features of the adult and many larval cells die. Consider the frog: At some point in evolution the stage following somite formation became motile prematurely and this enabled the embryo to disperse a little — this was an advantage. It was from this that the larva, capable of feeding, eventually evolved. The trick was to get back to the normal developmental program and this change has adopted hormones as the trigger. For these reasons, the evolution of larva and metamorphosis must be due to intercalation of the larval stage into a directly developing animal. A stage of embryonic development or early growth becomes modified to form the larva, and metamorphosis is essentially a return to the original direct developmental program. It is not possible to imagine a scenario in which set-aside cells in a larval-like form could evolve to yield an adult by metamorphosis whose form is different to that of the larva. It is important to realize that imaginal disks of insects did not and do not arise from set-aside cells: in *Drosophila* they develop in the cellular blastoderm stage (3 hours after fertilization) to form small sheets of epithelial cells (Truman and Riddiford 1999). It is only by modification of the direct development that metamorphosis becomes possible.

Evolution of the Egg

As has been stated, reduction of the propagule to minimal size (i.e., one cell) is advantageous because it reduces conflicts due to increased kinship (Maynard Smith and Szathmáry 1995). Roze and Michod (2001) provide an insightful analysis of this issue, by considering how mutations affect the fate of the organism with varying propagule size. It turns out that within-organism selection is effective against uniformly deleterious mutations, i.e., those that have a negative effect at the cell as well as the organism level. This by itself would select for an *increase* in propagule size. Mutants that affect the organism but benefit the cell (such as those leading to cancer), called selfish mutants, cannot be effectively selected against in large propagules. If such mutants occur at a certain frequency, selection for a single-celled propagule follows.

We hypothesize that there may be another factor at play. Development from a single cell, an egg, is fundamental and essential for the evolution of complex multicellular organisms (Wolpert and Szathmáry 2002). A feature of evolution of multicellular organisms is the issue of evolvability, i.e., what cellular properties are necessary for organisms to evolve, as they have, an enormous variety of different forms. Could an asexual form of reproduction, involving budding (somatic embryogenesis) as in hydra, evolve to give complex new forms? Asked in this way, this is not a question of kin selection or competition between cells in the multicellular organism, but an issue that relates to developmental processes that generate the form of the organism. The main process for generating form is pattern formation, which specifies cell states in a group of cells so that they are different and thus can differentiate along different pathways. One mechanism for pattern formation is based on positional information in which cells acquire a positional identity with respect to boundaries and then interpret this positional value by a variety of cellular behaviors, such as differentiating into specific cell types or undergoing a change in shape and so exerting the forces required for morphogenesis (see Wolpert et al. 2002). Such a mechanism, and many other patterning processes, require cell signaling, all of which lead via signal transduction pathways to gene activation or inactivation. Such a process can only lead to reliable patterns of cell activities if all the cells have the same set of genes (they "speak the same language"). Essentially, they must obey the same set of rules. During evolution it is the change in genes that leads to new patterns forming during development. Once the pattern has been set up as in hydra, it is no longer possible to evolve significant changes for two reasons: (a) it is not possible to go through a developmental sequence and (b) mutations in individual cells mean that they all no longer have the same rules for behavior. It is only via a developmental program that organisms can evolve complex patterns, and this requires an egg. All larvae are intercalations in the developmental program (see above). There are multicellular organisms like the cellular slime molds which do not develop from an egg but by aggregation; however, their patterning for that reason has remained very simple for hundreds of millions of years—they could not evolve complex patterns of cell behavior.

There are a few obvious objections that can be raised against this reasoning. One is that we are confusing the disadvantage of asexuality as such, as dealt with by population geneticists, with those of having a large propagule. In fact, we are not doing this. Try to imagine a sexual life cycle combined with a large propagule size. This would mean that each of the cells in the propagule would have to be fertilized by a separate sperm cell, otherwise the propagule would still be one developing from a single fertilized egg. This would immediately generate a large amount of conflict as the result of the extra within-organism variation delivered through multiple fertilizations (Grosberg and Strathmann 1998).

Thus, the real question is whether one could evolve complex development with somatic embryogenesis that happens to be asexual. Sexual reproduction as such may have long-term advantages (e.g., Maynard Smith 1978), but we are discounting those from our budget and focusing instead on evolvability (another long-term effect) associated with a single-celled propagule as such.

Another objection is the example of insect colonies, often with multiple queens. To be sure, these are always evolutionary descendants of colonies with single single-mated queens, but then in the former worker, reproduction is completely suppressed. Moreover, unrelated queens, having founded a colony together, start to fight after maturation of the first workers until only one of them survives (Bernasconi and Strassmann 1999).

There are also examples of coadaptations between different species, such as in evolved mutualisms (e.g., the fig/fig-wasp system). Note that although the development of both partners may have undergone coordinated evolution in such cases, the majority and the integrity of the developmental processes still rest with organisms developing from an egg. We think it is *practically* impossible to have several-to-many asexual, partly differentiated, cell lineages mutating in all sorts of directions in genetic space and yet keep up the ability to evolve into viable novel forms. This may not be completely impossible; however, organisms developing from an egg would displace those without it in long-term evolution. This idea is open to modeling.

The Biggest Hurdles for the Origin of Multicellularity

Is then the evolutionary transition to multicellularity a difficult one or not? The blunt answer is, not at all, since multicellularity has arisen more than twenty times in evolution (see Bonner 2000). However, there are only three lineages that produced complex organisms: plants, animals, and fungi. Three hits in 3.5 billion years are not that many. So one is left with the feeling of some extrinsic or intrinsic difficulties. The most popular candidate for an extrinsic difficulty is the lack of sufficient oxygen in the atmosphere (discussed in Maynard Smith and Szathmáry 1995). This may in fact be part of the explanation, but there may have been intrinsic difficulties.

One possible intrinsic difficulty (maybe the biggest hurdle?) is *the appropriate down-regulation of cell division at the appropriate time and space in the organism.* This difficulty increases with the number of cell types and cells. As Szent-Györgyi once remarked, the fact that cancer cells divide like hell is not a miracle; the fact that most cells of the organism do not is amazing, however.

Another hurdle may be the complexity of development as such. This sounds tautological but we think it is deeper than that. What is required for a complex organism? It is complex development, which in turn requires complex regulation in the network of genes. Indeed, it is the connectivity of regularity gene interactions that seems to correlate well with intuitive feelings about organismic complexity (Szathmáry et al. 2001) or the number of cell types (Bonner 2000), rather than the mere number of coding genes, as previously thought (Maynard Smith and Szathmáry 1995). It is remarkable that Weismann favored a process of differentiation by gene elimination (to use our terms), rather than by gene silencing, because he could not imagine how the adequate signals could be generated and how the differentiated state could be maintained (see Maynard Smith 1986). It is likely that complex gene regulation requires *spatial separation of transcription from translation,* which would explain why all complex multicellular forms are eukaryotes.

ACKNOWLEDGMENT

We thank John Tyson, Neil Blackstone, Rick Michod, and Michael Lachmann for helpful comments on a draft version.

REFERENCES

Bakal, C.J., and J.E. Davies. 2000. No longer an exclusive club: Eukaryotic signalling domains in bacteria. *Trends Cell Biol.* **10**:32–38.

Bell, G., and V. Koufopanou. 1991. The architecture of the life cycle in small organisms. *Phil. Trans. Roy. Soc. Lond. B* **322**:81–89.

Bernasconi, G., and J.E. Strassmann. 1999. Cooperation among unrelated individuals: The ant foundress case. *Trends Ecol. Evol.* **14**:477–482.

Blackstone, N.W. 2000. Redox control and the evolution of multicellularity. *BioEssays* **22**:947–953.

Blackstone, N.W., and A.M. Ellison. 2000. Maximal indirect development, set-aside cells, and levels of selection. *J. Exp. Zool. (Mol. Dev. Evol.)* **104**:288–299.

Bonner, J.T. 2000. First Signals: The Evolution of Multicellular Development. Princeton, NJ: Princeton Univ. Press.

Cavalier-Smith, T. 2002. The neomuran origin of archaebacteria, the negibacterial root of the universal tree and bacterial megaclassification. *Intl. J. Syst. Evol. Microbiol.* **52**:7–76.

Christensen, S.T., V. Leick, L. Rasmussen, and D.N. Wheatley. 1998. Signalling in unicellular eukaryotes. *Intl. Rev. Cytol.* **177**:181–203.

Chung, C.Y., S. Funamoto, and R.A. Firtel. 2001. Signaling pathways controlling cell polarity and chemotaxis. *Trends Biochem. Sci.* **26**:557–566.

Crespi, B.J. 2001. The evolution of social behavior in microorganisms. *Trends Ecol. Evol.* **16**:178–183.

Davidson, E.H., K.J. Peterson, and R.A. Cameron. 1995. Origins of the bilateral body plans: Evolution of developmental regulatory mechanisms. *Science* **270**:1319–1325.

England, J.C., and J.W. Gober. 2001. Cell cycle control of cell morphogenesis in *Caulobacter. Curr. Op. Microbiol.* **4**:674–680.

Gerhart, J. 1999. Signalling pathways in development. *Teratology* **60**:226–239.

Grosberg, R.K., and R.R. Strathmann. 1998. One cell, two cell, red cell, blue cell: The persistence of a unicellular stage in multicellular life histories. *Trends Ecol. Evol.* **12**:112–116.

Jablonka, E., and M. Lamb. 1995. Epigenetic Inheritance. Oxford: Oxford Univ. Press.

Kaiser, D. 2001. Building a multicellular organism. *Ann. Rev. Genet.* **35**:103–123.

Kerszberg, M., and L. Wolpert. 1998. The origin of metazoa and the egg: A role for cell death. *J. Theor. Biol.* **193**:535–537.

Kirk, D.L. 1999. Evolution of multicellularity in the volvocine algae. *Curr. Op. Plant Biol.* **2**:496–501.

Lenski, R.E., and G.J. Velicer. 2000. Games microbes play. *Selection* **1**:89–95.

Martin, M.E., and Y.V. Brun. 2000. Coordinating development with the cell cycle in *Caulobacter. Curr. Op. Microbiol.* **3**:589–595.

Maynard Smith, J. 1978. The Evolution of Sex. Cambridge: Cambridge Univ. Press.

Maynard Smith, J. 1986. The Problems of Biology. Oxford: Oxford Univ. Press.

Maynard Smith, J. 1990. Models of a dual inheritance system. *J. Theor. Biol.* **143**:41–53.

Maynard Smith, J., and E. Szathmáry. 1995. The Major Transitions in Evolution. Oxford: W.H. Freeman.

Maynard Smith, J., and E. Szathmáry. 1999. The Origins of Life. Oxford: Oxford Univ. Press.

Michod, R.E., and D. Roze. 2001. Cooperation and conflict in the evolution of multicellularity. *Heredity* **86**:1–7.

Müller, C., and A. Leutz. 2001. Chromatin remodelling in development and differentiation. *Curr. Op. Genet. Dev.* **11**:167–174.

Pál, C., and B. Papp. 2000. Selfish cells threaten multicellular life. *Trends Ecol. Evol.* **15**:351–352.

Rice, J.C., and C.D. Allis. 2001. Histone methylation versus histone acetylation: New insights into epigenetic regulation. *Curr. Op. Cell Biol.* **13**:263–273.

Roze, D., and R. Michod. 2001. Mutation, multilevel selection, and the evolution of popagule size during the origin of multicellularity. *Am. Nat.* **158**:638–654.

Szathmáry, E. 1992. Natural selection and dynamical coexistence of defective and interfering virus segments. *J. Theor. Biol.* **157**:383–406.

Szathmáry, E., F. Jordán, and C. Pál. 2001. Molecular biology and evolution: Can genes explain biological complexity? *Science* **292**:1315–1316.

Thomason, P., and R. Kay. 2000. Eukaryotic signal transduction via histidine-aspartate phosphorelay. *J. Cell Sci.* **113**:3141–3150.

Truman, J.W., and L.M. Riddiford. 1999. The origins of insect metamorphosis. *Nature* **401**:447–452.

Van Houten, J. 1994. Chemosensory transduction in eukaryotic microorganisms: Trends for neuroscience? *Trends Neurosci.* **17**:62–71.

Velicer, G.J., L. Kroos, and R.E. Lenski. 2000. Developmental cheating in the social bacterium *Myxococcus xanthus. Nature* **404**:598–601.

Verbsky, M.L., and E.J. Richards. 2001. Chromatin remodelling in plants. *Curr. Op. Plant Biol.* **4**:494–500.

Vitt, U.A., S.Y. Hsu, and A.J.W. Hsueh. 2001. Evolution and classification of cystine knot-containing hormones and related extracellular signalling molecules. *Molec. Endocrinol.* **15**:681–694.

Wolpert, L. 1990. The evolution of development. *Biol. J. Linn. Soc.* **39**:109–124.

Wolpert, L. 1999. From egg to adult to larva. *Evol. Dev.* **1**:3–4.

Wolpert, L., R. Beddington, T. Jessell et al. 2002. Principles of Development. Oxford: Oxford Univ. Press.

Wolpert, L., and E. Szathmáry. 2002. Evolution and the egg: Multicellularity. *Nature* **420**:745.

16

Cooperation and Conflict Mediation during the Origin of Multicellularity

Richard E. Michod

Department of Ecology and Evolutionary Biology, University of Arizona, Tucson, AZ 85721, U.S.A.

ABSTRACT

The basic problem in an evolutionary transition is to understand how a group of individuals becomes a new kind of individual, having heritable variation in fitness at the new level of organization. We see the formation of cooperative interactions among lower-level individuals as a necessary step in evolutionary transitions; only cooperation transfers fitness from lower levels (costs to group members) to higher levels (benefits to the group). As cooperation creates a new level of fitness, it creates the opportunity for conflict between the new level and the lower level. Fundamental to the emergence of a new higher-level individual is the mediation of conflict among lower-level individuals in favor of the higher-level unit. We define a conflict mediator as a feature of the cell group (the emerging multicellular organism) that restricts the opportunity for fitness variation at the lower level (cells) and/or enhances the variation in fitness at the higher level (the cell group). There is abundant evidence that organisms are endowed with just such traits and numerous examples are reviewed here from the point of view of a population genetic model of conflict mediation. Our model considers the evolution of genetic modifiers that mediate conflict between the cell and the cell group. These modifiers alter the parameters of development, or rules of formation, of cell groups. By sculpting the fitness variation and opportunity for selection at the two levels, conflict modifiers create new functions at the organism level. An organism is more than a group of cooperating cells related by common descent and requires adaptations that regulate conflict within itself. Otherwise, its individuality and continued evolvability is frustrated by the creation of within-organism variation and conflict between levels of selection. Conflict leads to greater individuality and harmony for the organism through the evolution of adaptations that reduce it.

INTRODUCTION

Evolutionary individuals are units of selection and must satisfy Darwin's conditions of heritability and variation in fitness. Darwin's principles apply to different levels in the hierarchy of life, including genes, chromosomes, cells, cells within cells (eukaryotic cell), multicellular organisms, and social groups of organisms (Lewontin 1970). Because of the hierarchical nature of selection I take a multilevel selection approach to the origin of multicellularity and to evolutionary transitions. The multilevel selection approach to evolutionary transitions seeks to understand how a group of preexisting individuals becomes a new evolutionary individual, possessing heritable fitness variation at the group level and protected from within-group change by conflict mediators.

The transition to a new higher-level individual is driven by cooperation among lower-level individuals. Only cooperation trades fitness from the lower level (its costs) to the higher level (the benefits of cooperation for the group) (Table 16.1). Because cooperation exports fitness from lower to higher levels, cooperation is central to the emergence of new evolutionary individuals and the evolution of increased complexity. I believe this to be the case, even if the groups initially form via antagonistic interactions, as may have been the case during the origin of the eukaryotic cell (e.g., Maynard Smith and Szathmáry 1995; Michod and Nedelcu 2003a).

The flip side of cooperation is defection and selfishness leading to conflict among lower-level individuals between their effects at the cell and cell group levels; such conflicts must be mediated for heritable variation in fitness to increase at the cell group level (Michod 1999). We define a conflict mediator as a feature of the higher level (the group) that restricts the opportunity for fitness variation at the lower level (cells) and/or enhances the variation in fitness at the higher level (the cell group or organism).

The way in which the conflicts are mediated can influence the potential for further evolution (i.e., *evolvability*) of the newly emerged individual. We expect greater individuality to generally enhance evolvability by increasing the

Table 16.1 Cooperation and conflict among cells within organisms. The effects on size assume growth is indeterminate and that the sizes of adults vary depending upon composition of cells. The notation +/– means positive or negative effects on fitness at the cell or organism level.

	Level of Selection	
Cell Behavior	Cell	Group (organism)
Defection	(+) replicate faster or survive better	(+) larger (–) less functional
Cooperation	(–) replicate slower or survive worse	(–) smaller (+) more functional

potential for cooperation and restricting within-group change. However, evolution can be short sighted, and in *Volvox* it appears that conflict mediation led to a nonreplicative soma that, in turn, restricted the potential for further evolution (Nedelcu and Michod 2003; Michod et al. 2002).

A MODEL OF THE ORIGIN OF MULTICELLULARITY

Model Life Cycle

The model life cycle we have used to study cooperation and conflict mediation in the evolution of multicellularity is represented in Figure 16.1. Development of the multicellular group starts from an offspring or propagule group of *N* cells. This propagule may be formed in several ways, as discussed in the next section.

In the basic model, adult size is not fixed but rather depends on rates of cell division and time available for development (however, for consideration of fixed size, see section below on Determinant Growth). The fitness of the cell group or organism is the expected number of propagules it produces; this depends both on the size of the organism and on the frequency of mutant cells in the adult.

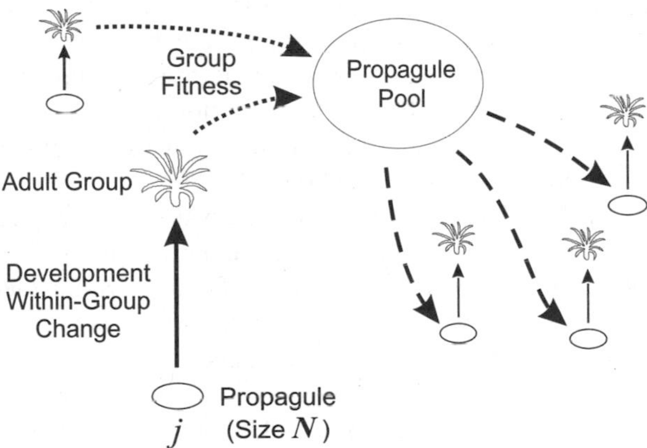

Figure 16.1 Model life cycle. The subscript j refers to a property of the propagule offspring group, typically its genotype or the number of mutant cells in the propagule; $j = 0, 1, 2, \ldots N$, where N is the total number of cells in the propagule, termed propagule size ($N = 1$ for single-cell reproduction). The fitness of group j is defined as the expected number of propagules produced by the group (dotted lines). Two components of group fitness are considered: the size of the adult group and its functionality. After production, the propagules form offspring groups of the next generation (dashed arrows). Sex may occur. The additional variables used in the model, but not specified here,include (a) the number and frequency of mutant and nonmutant cells in the adult cell group (that is after the propagule develops into an adult); (b) the change in mutant frequency within the cell group during development; and (c) fitness parameters at the cell and group level which stem from the interactions of the cell. For an application of the model to the case of the evolution of programmed cell death, see Table 16.3.

The complexity of the interactions among different cell types is represented by a single variable: cooperativity. We assume a single genetic locus controls the way in which cells interact. There are assumed to be two alleles, cooperate *C* and mutant-defecting cells *D*. Mutant-defecting cells (those carrying the *D* allele) no longer cooperate and this lowers the fitness of the cell group, as in Table 16.1. The fitness of the cell in terms of its replication and/or survival rates during development may be higher (selfish mutants) or lower (uniformly deleterious mutants) than nonmutant cells.

Mode of Propagule Formation

Concerning the formation of the propagule, we have considered three basic modes of reproduction: fragmentation, aggregation, and spore or zygote reproduction (with or without sex; Figure 16.2). In all three cases, the sequence of life cycle events involve the creation of a founding propagule or offspring group of N cells shown in Figure 16.1. This propagule could be a single cell if $N = 1$, as in the case of spore or zygote reproduction. Indeed, the case of spore reproduction can be seen as the limiting case of both fragmentation and aggregation modes (by setting $N = 1$). We have also considered the case of alternating fragmentation and spore reproduction every, say, v generations (Michod and Roze 1999). A fundamental difference between aggregation and the other reproductive modes is the opportunity for horizontal transfer of mutants to cell groups that contain no mutant cells. This is important because aggregation continually reestablishes mixed groups and concomitantly the opportunity for within-group selection and conflict between the two levels of selection.

Propagule size, N, influences fitness in several ways. First, propagule size affects the within- and between-group variance and opportunity for selection at the two levels, that is, it affects the opportunity for conflict. Smaller N may be seen as a conflict mediator, because smaller N increases the between-group variance and decreases the within-group variance. Second, propagule size has direct effects on fitness, because smaller N increases the number of possible fragments, but decreases adult size. As discussed below, we find that the direct effects of propagule size dominate the indirect effects in the evolution of reproductive mode, except when some mutations are selfish. When some mutations are selfish, the opportunity for selection at the two levels becomes the critical factor affecting the evolution of N.

Within-organism Change

As cells proliferate during the course of development, mutations occur leading to loss of cell function and cooperativity among cells. The mutants have a deleterious effect on the fitness of the group, while at the cell level, mutant cells may replicate slower (uniformly deleterious mutants) or faster (selfish cancer-like

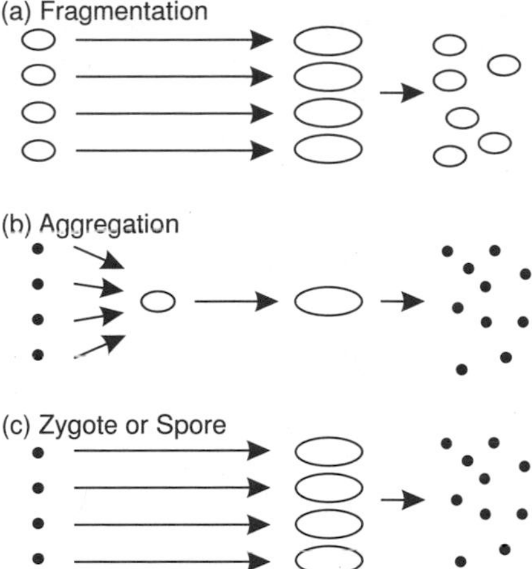

Figure 16.2 Modes of propagule formation (see also Figure 16.1). Small solid circles indicate single cells. Hollow ellipses indicate groups of cells. Small and large groups are shown. The small ellipses (of size N in the model) correspond to offspring propagule cell groups before cell division and development. The large ellipses correspond to adult groups. Under fragmentation (a), small offspring groups grow into larger adult cell groups, which produce offspring groups of the next generation. Under aggregation (b), single cells aggregate to form an offspring group which grows into an adult group which produce single cells of the next generation. Under zygote or spore reproduction (c), single cells divide and grow into adult cell groups which produce single cells of the next generation. If there is sex, fusion among single cells may occur in (c) prior to development into the multicellular form.

mutants) than nonmutant cells. A simple branching model of mutation from C to D and cellular selection has been considered, which extends previous work (Otto and Orive 1995) to include survival selection among cells in addition to differences in cell replication rate (Michod 1997). The mutation model allows the calculation of the expected number and frequency of mutant cells at the adult stage, and these variables are included in the recurrence equations for gene frequency change. Because of recurrent mutation from C to D, a mutation selection balance is achieved at the C/D locus. This balance takes into account selection at the cell and cell group levels. A mathematical description of the mutation selection model is given elsewhere (Michod 1997, 1999; Roze and Michod 2001).

Population Genetic Analysis

To study the evolution of conflict mediators, we employ a standard two-locus population genetic framework using genetic modifiers. As already discussed, the first locus controls cell behavior, that is, whether cells cooperate or not in their interactions with other cells. The recurrence equation for change in gene frequency at this first locus has been analyzed elsewhere in terms of the levels of cooperation and fitness variation and heritability maintained in the system (Michod 1997; Michod 1999). The level of cooperation among cells and fitness heritability at the cell group level depends on a variety of assumptions about development, mutation, and selection within the cell group. Thus, to study the consequences of development for the emergence of fitness heritability at the higher level, a second modifier locus is considered that changes these assumptions. For example, the modifier locus may create a germ line, allow for cell policing, change the propagule size, change the way in which cells are sampled to put in the propagule, or the modifier may limit the size of the group. Virtually any aspect of the development of the groups may be studied in this way to see if it serves to mediate conflict in favor of the multicellular group. The resulting two-locus population genetic model is analyzed using standard techniques.

The transition to multicellular individuals involves two general steps. First, cooperation must increase among cells in the group. Without cooperation the cells are independently evolving units. However, the increase of cooperation within the group is accompanied by an increase in the level of within-group change, and conflict as mutation and selection among cells leads to defection and a loss of cooperation. Organisms are more than cooperating groups of related cells. The second general step is the evolution of modifier genes that regulate this within-group conflict. Only after the evolution of modifiers of within-group conflict, do we refer to the group of cooperating cells as an "individual," because then the group possesses higher-level functions, conflict mediators, that protect its integrity.

One way of using the model to study the evolution of multicellular individuals is to investigate the model's equilibrium structure. The equilibria of the system with no linkage disequilibrium are described in Table 16.2. The evolution of cooperation in multicellular groups corresponds to the transition from equilibrium 1 to 3. The evolution of individuality supported by the spread of conflict mediators corresponds to a transition from equilibrium 3 to equilibrium 4. The question of the transition to individuality, then, boils down to the conditions for a transition from equilibrium 3 to equilibrium 4 in Table 16.2.

As discussed in more detail elsewhere (Michod and Roze 1999), conflict mediators increase by virtue of being associated with the more fit genotype and by increasing the heritability of fitness of that type. For example, at equilibrium 3 in Table 16.2, cooperating zygotes are more fit than defecting zygotes; the cooperating groups must be more fit, because for equilibrium 3 to be stable, the fitness of groups with cooperators must compensate for directional mutation toward defection (from C to D). The modifiers increase the heritability of fitness of the cooperating type and hitchhike along with these more fit chromosomes. They increase the heritability of fitness of the more fit type by decreasing the within-group change created by deleterious mutation.

The evolution of conflict mediators — functions that protect the integrity of the organism — are not possible, if there is no conflict among the cells in the first place. Conflict itself (the mutation selection balance at equilibrium 3 and the

Table 16.2 Equilibria for two-locus modifier model without linkage disequilibrium. The first locus controls cell behavior with two alleles cooperate, *C*, and defect, *D*. Recurrent mutation from *C* to *D* occurs during development. The second locus modifies aspects of development, group formation, or policing of mutant cells. The first stage in an evolutionary transition involves the increase of cooperation, the transition from Eq. 1 to Eq. 3. The second stage of an evolutionary transition involves the evolution of conflict mediation, the transition from Eq. 3 to Eq. 4. The effect of linkage disequilibrium and a mathematical description of the equilibria and eigenvalues are given elsewhere (Michod and Roze 1997; Michod 1999).

Eq.	Description of Loci	Interpretation of Equilibrium (Eq.)
1	No cooperation; no modifier	*Single cells, no organism*
2	No cooperation; modifier fixed	Not of biological interest, never stable
3	Polymorphic for cooperation and defection; no modifier	*Group of cooperating cells*: no higher-level functions
4	Polymorphic for cooperation and defection; modifier fixed	*Individual organism*: integrated group of cooperating cells with higher-level function mediating within-organism conflict

conflict between levels of selection) sets the stage for a transition between equilibrium 3 and 4 and the evolution of individuality.

CONFLICT MEDIATION

Kinds of Conflict Mediation

Let us consider briefly the kinds of conflict mediators studied to date using the models discussed in the last section. As already mentioned, we define a conflict mediator as a feature of the cell group that restricts the opportunity for fitness variation at the lower-level (cells) and/or enhances the variation in fitness at the higher level (the cell group or organism). Accordingly, one can think of two general classes of conflict mediators: those that restrict within-group change and those that increase the variation in fitness between groups, although both have the effect of increasing the heritability of fitness at the group level. It should be recognized that we focus on conflict mediation among cells and not on conflict at lower levels (e.g., among genes, chromosomes, and organelles). Conflict mediators that operate at these lower levels are also important to the origin of multicellularity and are discussed by Lachmann et al. (this volume).

Germ and Soma

By developing cell types specialized at vegetative and reproductive functions, the evolutionary opportunities of the majority of somatic cells are limited, because genes in somatic cells may spread in the population only if they cooperate with other genes in other cells, thereby doing something useful for the cell group or organism. There are four basic issues concerning the reproductively specialized germ cells: (a) how many cells are selected to form the propagule for the next generation, (b) the way in which these cells are sampled (two extremes would be cells selected randomly from all cells in the adult or selected from cells that are descendents of a single cell in the adult), (c) the time in development at which these cells are selected, and (d) the number of cell divisions between the propagules of two successive generations. Although these issues range on a continuum, the term "germ line" is often used for the special case in which a single cell (the spore or egg) is chosen from a distinct cell lineage set aside early in development. It is also often assumed there are fewer cell divisions in the germ line than in the soma.

When discussing the role of reproductive specialization as a conflict mediator, one must remember that other factors, such as division of labor, may have been important in the evolution of germ and soma. Nevertheless, specialization of cell types into reproductive and vegetative functions may still act to reduce conflict. For example, in the Volvocales, the soma likely evolved to lower the survival costs (due to compromised motility) of reproducing increasingly large groups (Koufopanou 1994; Michod and Nedelcu 2003b). Even in this case, the time of sequestration and number of cell divisions may be adjusted to reduce the opportunity for mutation (Michod et al. 2002).

Propagule Size

Multicellularity presumably evolved because of advantages for cells of group living (see Lachmann et al., this volume). However, most multicellular organisms begin their life cycle as a single cell. If group living is so advantageous, why return to a unicellular stage at the start of each generation?

A common hypothesis is that the unicellular bottleneck acts as a conflict mediator, by increasing kinship among cells in the organism, thereby aligning the interests of cells with the interest of the organism (Bell and Koufopanou 1991; Maynard Smith and Szathmáry 1995; Grosberg and Strathmann 1998). Smaller propagule size does increase between-group variation; however, propagule size has direct effects on the adult group size, in addition to its effects on conflict mediation. All things being equal, smaller propagules produce smaller adults. For this reason, we have studied the evolution of propagule size in simple cell colonies in the context of both selective factors: the direct effects on adult organism size and the more indirect effects on conflict mediation through the opportunity for selection on mutations at the cell and cell group levels (Michod and Roze 2000; Roze and Michod 2001). Our results show that evolution of propagule size is determined primarily by its direct effects on group size *except* when mutations are selfish. So long as some mutations are selfish, smaller propagule size may be selected, including single cell reproduction, even though smaller propagule size has a direct fitness cost by virtue of producing smaller organisms.

Time of Sequestration and Number of Cell Divisions in Germ Line

In our initial studies of the evolution of a germ line, we assumed for simplicity that the germ line was sequestered as a *single* cell set aside during the *first* cell division (Michod 1996; Michod and Roze 1997, 1999). However, most organisms depart from this ideal, and sequester cells later in development. For example, in the green alga, *V. carteri*, the precursors of the germ line are formed after five cell divisions, but the germ line is sequestered only after the ninth cell division. For this reason we have specifically modeled the selective forces acting on the time of sequestration, the number of cells sequestered, and the number of cell divisions in the germ line (Michod et al. 2002). Our results depend upon how the cost of germline sequestration is interpreted. We may interpret the cost of the germ line as stemming either from the *new* germ cells or the *missing* somatic-like cells (by that we mean, cells no longer available for somatic function). In the case where the germline cost is assumed to be proportional to the new germ cells, it is easier for a germ line to evolve ("easier" in the sense that the conditions on the parameters in the model are more relaxed) the earlier the germ line is sequestered, the lower the number of times it divides, and the fewer number of cells that are sampled. This is because there are only advantages to early segregation and low replication (in terms of a lower effective deleterious mutation rate resulting from the fewer number of cell divisions), and the cost of the germ line is smaller the fewer cells that are sampled. Organisms following this model should form a germ line by sequestering a single, nondividing cell during the first cell division.

What about when the cost of the germ line depends upon the number of missing cells unavailable for vegetative (somatic) function? The missing cells are those that would have been formed by the cells sequestered to form the germ cells. In this case, there is a cost to early sequestration of the germ line in terms of more missing somatic cells, and thus there is an intermediate optimum time for sequestration. Early sequestration is better in terms of coping with the threat of deleterious mutation; however, there is a greater penalty to pay in terms of missing cells unavailable for somatic function.

Mutation Rate

The vast majority of mutations are disadvantageous and therefore our models of germline sequestration considered mainly deleterious mutations, of either the uniformly deleterious or selfish varieties. Modifiers that lower the mutation rate are always selected for in our models because they reduce the opportunity for selfish mutations, which create conflict between the levels of selection.

Maynard Smith and Szathmáry (1995) suggest that germline cells may enjoy a lower mutation rate but do not offer a reason why. Bell (1985) interpreted the evolution of germ cells in the Volvacale as an outcome of specialization in metabolism and gamete production to maintain high intrinsic rates of increase while algae colonies got larger in size (see also Maynard Smith and Szathmáry 1995, pp. 211–213). I think there may be a connection between these two views.

As metabolic rates increase, so do levels of DNA damage. Metabolism produces oxidative products that damage DNA and lead to mutation. It is well known that the highly reactive oxidative by-products of metabolism (e.g., the superoxide radical O_2^-, and the hydroxyl radical •OH produced from hydrogen peroxide H_2O_2) damage DNA by chemically modifying the nucleotide bases or by inserting physical cross-links between the two strands of a double helix, or by breaking both strands of the DNA duplex altogether. Deleterious effects of DNA damage make it advantageous to protect a group of cells from the effects of metabolism, thereby lowering the mutation rate within the protected cell lineage.

This protected cell lineage — the germ line — may then specialize in passing on the organism's genes to the next generation in a relatively error-free state. Other features of life can be understood as adaptations to protect DNA from deleterious effects of metabolism and genetic error (Michod 1995): keeping DNA in the nucleus protects the DNA from energy-intensive interactions in the cytoplasm, nurse cells provision the egg so as to protect DNA in the egg, sex serves to repair genetic damage effectively while masking the deleterious effects of mutation. The germ line may serve a similar function of avoiding damage and mutation; by sequestering the next generation's genes in a specialized cell lineage, these genes are protected from the damaging effects of metabolism in the soma.

According to Bell (1985), the differentiation between the germ and the soma in the Volvocales results from increasing colony size, with true germ soma differentiation occurring only when colonies reach about 10^3 cells as in the *Volvox* section *Merillosphaera*. Assuming no cell death, this colony size would require a development time of approximately $t = 10$ in our model (in reality, because of cell death, larger t with more risks of within-colony variation would be needed to achieve the same colony size). Although Bell interpreted the dependence of the evolution of the germ line on colony size as an outcome of reproductive specialization driven by resource and energy considerations, this relation is also explained by the need for regulation of within-colony change (see panel F of Figure 6.1 in Michod 1999).

Determinant Growth

In our model, growth of the cell group was assumed to be indeterminate, and many factors influenced the number of cells in the adult organism. The main factors influencing the size of the adult were the replication and death rates of the cooperative and defecting cells, along with the time available for development.

Mutant-defecting cells are assumed to replicate faster and thus produce larger, though less functional, adults. Because organism fitness is assumed to depend upon the size of the adult, in addition to the level of cooperation, there is an advantage of defection at the organism level resulting from the organism's larger size, in addition to its advantage at the cell level (recall defecting cells replicate faster). One way of reducing the temptation of defection, that is conflict, is to control adult size, thereby removing the advantage of defection (cost of cooperation) at the organism level. Even if adult size is fixed, defecting cells still have a selection advantage within organisms; fixing adult size only removes the positive effect of defection at the organism level of selection.

Jie Li and I have considered an extension of the discrete generation model introduced above, in which a constant adult size is attained for all groups by assuming that the different kinds of zygotes develop for different periods of time (Li 1998, unpublished). For example, we allow *C* zygotes to develop for a longer period of time than *D* zygotes, so that both have the same number of cells in the adult stage (we maintain the assumption used here that *C* cells divide more slowly than *D* cells). We further assume there is a fixed time for reproduction, so that *D* zygotes reproduce for a longer period of time, since they reach adult size quicker. Because of the exponential nature of cell growth, only small differences in development time are needed to attain a fixed adult size. Consequently, there is little difference between organisms in the time available for reproduction. We have not yet considered a model with overlapping generations, although this is clearly in need of study.

Our results indicate that determinate growth acts as a conflict mediator. Constant adult size makes it much easier for cooperation to increase, and this effect is more pronounced for smaller mutation rates. In addition, much greater levels of harmony and cooperation are maintained within the organism if adult size is regulated. Cell death may have important effects with regard to organism size. Cell death increases the number of cell divisions required to reach a given adult size, and this has the additional consequence of increasing the opportunity for within-organism change and variation.

Policing

Another means of reducing conflict among cells is for the organism to actively police and regulate the benefits of defection (Boyd and Richerson 1992; Frank 1995). How might organisms police the selfish tendencies of cells? The immune system and programmed cell death are two examples. To model self-policing, we let the modifier allele affect the parameters describing within- and between-organism selection and the interaction among cells. Within-organism selection is still assumed to result from differences in replication rate, not cell survival. Cooperating cells in policing organisms spend time and energy monitoring cells and reducing the advantages of defection at a cost to the organism.

An explicit analysis is given elsewhere of immune system policing (Michod 1996, 1999; Michod and Nedelcu 2003b). In general, self-policing increases, if the cost of policing is not high.

As an explicit illustration of our study of conflict mediation and self-policing in the evolution of multicellularity, we consider the evolution of programmed cell death (PCD). PCD, sometimes termed apoptosis, is an evolutionarily conserved form of cell suicide that enables metazoans to regulate cell numbers and control the spread of cancerous cells that threaten the organism. It is best studied in *Caenorhabditis elegans* and mammals; however, similar traits have also been described in unicellular organisms. Presumably, in unicellular organisms, PCD is a form of kin-selected altruism, although there is little direct evidence of this.

We now illustrate how PCD may be viewed as a conflict mediator using our theory. The definitions of additional terms are given in Table 16.3. A PCD modifier lowers the rate of division (or survival) of the mutated cell (parameter *pcd*). We assume this occurs at some cost, δ, to the cell group, or organism. If there were no costs for the modifier, the modifier would always increase so long as it was introduced in a population in which cooperation was present (the role of cooperation is discussed further below).

In Figure 16.3 we report results for the evolution of PCD modifier alleles, assuming sexual reproduction ($N = 1$ with sex) and a single class of mutant cells D with fixed effect b (the replication rate of mutant cells without the PCD modifier allele; the replication rate of nonmutant cells is unity). Cells with the modifier allele express the PCD phenotype; mutant cells replicate at rate $pcd \times b$, instead of rate b in nonmodified cells. A perfect PCD phenotype would mean that no mutant cells replicate; in this case we would set $pcd = 0$. Of course, it is unlikely that the first PCD response was perfect, so we consider the entire range of possible values for the PCD phenotype ($0 \leq pcd < 1$). The cost of the PCD phenotype at the organism level is assumed to be δ—the benefit of cooperation is reduced in PCD cells to $\beta - \delta$, instead of β in non-PCD cells ($\beta = 3$ in Figure 16.3).

Table 16.3 Additional terms and variables for programmed cell death (PCD) modifier model.

b	Replication rate of mutant cells (relative to unity for nonmutant cells)
β	Effect of cooperation on group fitness
δ	Cost of PCD modifier to group fitness
pcd	Effect of PCD modifier on replication rate of mutant cells
r	Recombination rate between C/D locus and modifier locus
t	Development time
μ	Mutation rate per cell division
M, m	Alleles at modifier locus; M allele creates PCD phenotype

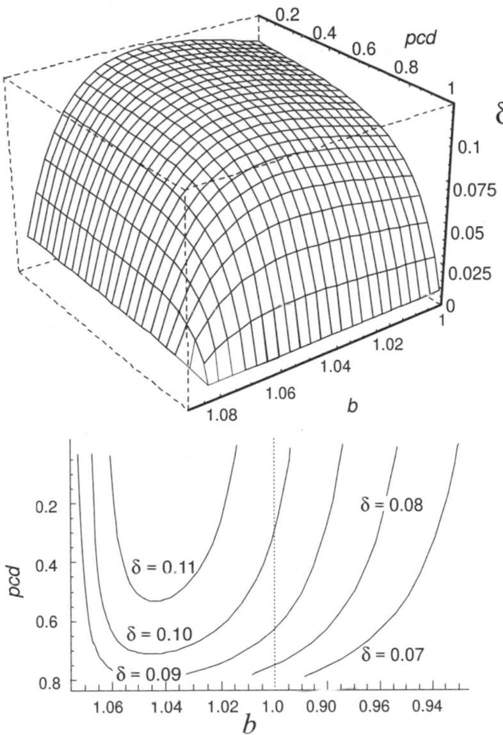

Figure 16.3 Evolution of apoptosis. Bottom panel gives 2D slices through the 3D surface in the top panel. PCD modifiers evolve for parameter values below the 3D surface and above the 2D curves. The parameter b is the replication rate of nonregulated mutants (relative to nonmutants) and pcd is the factor decrease in replication rate of PCD modified cells; modified cells replicate at rate $pcd \times b$. In the bottom panel five curves are plotted for different values of the cost to organisms of the PCD phenotype. Parameter values in the model (for specification of the model details, see Chapters 5 and 6 of Michod [1999]): offspring group size $N = 1$, time for development $t = 20$, benefit of cooperation $\beta = 3$, recombination rate between mutated locus and modifier locus $r = 0.2$, survival is incorporated in replication so $s_C = s_D = 1$, and mutation rate $\mu = 0.003$ (which is a value typical of the genome-wide rate in modern unicells [Drake 1974; Drake 1991]). In the bottom panel selfish mutations lie to the left of the vertical dotted line and uniformly deleterious mutations lie to the right.

An interesting feature of the results shown in Figure 16.3 is that uniformly deleterious mutations (ones that disrupt the functioning of the group and proliferate slower than normal cells, $b < 1$), may also select for PCD modifiers; however, to invade, the modifier requires lower costs of the PCD phenotype to the organisms than is the case for selfish mutations. It is common in the literature on PCD to assume that the risk of selfish mutations has lead to the evolution of the PCD phenotype. We see in Figure 16.3 that both uniformly deleterious and selfish mutations can select for PCD. We have also observed that both kinds of mutations select for the other kinds of modifiers that we have studied, such as germ line and self-policing modifiers.

Why do the curves in Figure 16.3 fall off rapidly as b increases up toward a value of approximately 1.07? As the proliferation advantage of mutants, b, increases, the equilibrium frequency of nonmutant cooperating cells decreases, eventually reaching zero at about 1.07 (when within-group change overpowers between-group selection for cooperation). Without variation at the cell interaction C/D locus, the PCD modifier, M, is disadvantageous, because when the modifier is introduced, the only genotypes are MD and mD (assuming haploidy for explanation purposes; where D is the mutant and M and m are the PCD and non-PCD alleles, respectively, at the modifier locus). Cell groups initiated by PCD cells (MD) end up being smaller than groups initiated by non-PCD cells (mD), because of the lower replication rate (or higher death rate) of PCD cells. However, when cooperating cells are maintained in the population before the PCD modifier is introduced, the significant competition is between groups initiated by CM and Cm cells. The cooperating groups carrying PCD modifiers (initiated by CM) end up being more functional and having fewer mutant cells in the adult stage; the associated fitness advantage can make up for the cost of PCD, δ (in the regions under the curves shown in the figure). The dependence of the evolution of PCD on the maintenance of cooperation reflects the need for a higher-level unit of selection (the cell group, or organism). The PCD modifier increases by virtue of tilting the balance in favor of the cell group, by enhancing its individuality and heritable fitness (Michod 1999).

SEX AND INDIVIDUALITY

Sex and individuality are in constant tension, because sex involves fusion and mixis of genetic elements and, so, naturally threatens the integrity of evolutionary units (see Lachmann et al., this volume). Yet, sex is fundamental to the continued well-being of evolutionary units too. Although sex seems to undermine individuality, sex has been rediscovered as each new level of individuality emerges in the evolutionary process. Sex holds the promise of a better future and a more whole and undamaged individual. Genetic redundancy and repair occur during the sexual cycle and are the key to greater wholeness and well-being for the individual (Michod 1995). Theories for the evolution of sex are discussed in three collections of papers (Stearns 1987; Michod and Levin 1988; Birky 1993).

Sex affects evolutionary transitions in our models in several ways. Sex affects the quantitative conditions for the evolution of conflict mediators: with recombination, it takes longer for the transition to occur (Michod 1999). The modifier increases by virtue of being more often associated with cooperating C alleles in gametes and recombination breaks apart this association. Although recombination can retard the transition between equilibrium 3 and 4 in Table 16.2, I do not see these quantitative differences as presenting any real barriers to the evolution of conflict modification and evolutionary transitions in sexual progenitors. More important, I think, is the way in which sex organizes variability and heritability of the traits and capacities that affect the fitness of the new emerging unit.

The effects of sex on fitness variation and heritability at the group level are studied in detail in the Appendix of Michod (1999), where it is shown that sex affects the level of conflict and variation within the emerging organism in profound ways (see also Michod 1997). Sex helps diploids maintain higher heritability of fitness under more challenging conditions especially when there is great opportunity for within-organism variation and selection. With sex, as the mutation rate increases and, concomitantly, the amount of within-organism change, more of the variance in fitness is heritable. Sex allows the integration of the genotypic covariances in a way not possible in asexual populations.

The increase in complexity during the evolution of multicellularity required new gene functions and an increasing genome size, which led to an increase in the deleterious mutation rate. It is often noticed that diploidy helps multicellular organisms tolerate this increase in mutation rate by masking recessive or nearly recessive deleterious mutations. However, once a diploid species reaches its own mutation selection balance equilibrium, the mutation load actually increases beyond what it was under haploidy (Haldane 1937; Hopf et al. 1988). There must be another factor that allows complex multicellular diploids to tolerate a high mutation rate and genetic error. This other factor may be sex.

Sex helps cope with genetic error in a variety of ways: by masking deleterious recessive mutations (Bernstein et al. 1985), by avoiding Muller's ratchet (Muller 1932), by removing deleterious mutations from the population (Kondrashov 1988), and through recombinational repair of DNA damage (Bernstein et al. 1985). To these we may add how sex maintains a higher heritability of fitness in the face of within-organism change resulting from somatic mutation.

As the mutation rate increases in sexual diploid organisms, the regression of fitness on zygote gene frequency actually increases (see Figure 9-2 of Michod 1999). In other words, as the mutation rate increases, and along with it the amount of within-organism change, more of the variance in fitness in sexual diploids is heritable than is explained by the alleles carried in the zygote.

How can this be? The greater mutation rate must result in greater levels of within-organism change. At equilibrium, this within-organism change must be balanced by a larger covariance of fitness with zygote frequency. This is what the Price equation states (see, e.g., Equation 5-2 of Michod 1999). In haploid and asexual diploid populations, this is accomplished by a greater variance in zygote gene frequency, whereas in sexual populations this can be accomplished by a greater regression of organism fitness on zygote frequency.

The fitness statistics we have studied (see Appendix of Michod 1999) apply before and after the transition. It is unclear whether these equilibrium statistics can be extended into the nonequilibrium realm of evolutionary transitions and if the results will hold up under more realistic genetic models. If so, the greater precision in the mapping of cooperative propensity onto group fitness should allow sexuals to make the transition from cells to multicellular organisms more easily under additionally challenging circumstances. This result is consistent with the view that the protist ancestor of multicellular life was likely sexual (Maynard Smith and Szathmáry 1995).

CONCLUSIONS

Multilevel selection theory predicts that for organisms to emerge from cooperating cell groups, they must acquire adaptations that reduce conflict so as to tilt the balance of selection away from the cell in favor of the multicellular group. There is abundant evidence that organisms are endowed with just such traits. Examples include a separate and sequestered germ line, passing the life cycle through a single cell stage, cell policing (including the immune system and programmed cell death), determinant growth, and a lowered mutation rate. In addition, sexual reproduction facilitates the maintenance of fitness heritability in the face of within-group change driven by high mutation rates.

ACKNOWLEDGMENTS

Many thanks to members of the "Cooperation and Conflict in the Evolution of Genomes, Cells, and Multicellular Organisms" Dahlem discussion group and to Denis Roze and Neil Blackstone for their comments on the manuscript.

REFERENCES

Bell, G. 1985. The origin and early evolution of germ cells as illustrated by the Volvocales. In: The Origin and Evolution of Sex, ed. H.O. Halvorson and A. Monroy, pp. 221–256. New York: A.R. Liss.

Bell, G., and V. Koufopanou. 1991. The architecture of the life cycle in small organisms. *Phil. Trans. Roy. Soc. Lond. B* **332**:81–89.

Bernstein, H., H.C. Byerly, F. Hopf, and R.E. Michod. 1985. DNA damage, mutation, and the evolution of sex. *Science* **229**:1277–1281.

Birky, W. 1993. American Genetics Society Symposium for the Evolution of Sex. *Journal of Heredity* **84**.

Boyd, R., and P.J. Richerson. 1992. Punishment allows the evolution of cooperation (or anything else) in sizable groups. *Ethol. Sociobiol.* **13**:171–195.

Drake, J.W. 1974. The role of mutation in bacterial evolution. *Symp. Soc. Gen. Microbiol.* **24**:41–58.

Drake, J.W. 1991. A constant rate of spontaneous mutation in DNA-based microbes. *Proc. Natl. Acad. Sci. USA* **88**:7160–7164.

Frank, S.A. 1995. Mutual policing and repression of competition in the evolution of cooperative groups. *Nature* **377**:520–522.

Grosberg, R.K., and R.R. Strathmann. 1998. One cell, two cell, red cell, blue cell, the persistence of a unicellular stage in multicellular life histories. *Trends Ecol. Evol.* **13**:112–116.

Haldane, J.B.S. 1937. The effect of variation on fitness. *Am. Nat.* **71**:337–349.

Hopf, F.A., R.E. Michod, and M.J. Sanderson. 1988. The effect of reproductive system on mutation load. *Theoret. Pop. Biol.* **33**:243–265.

Kondrashov, A.S. 1988. Deleterious mutations and the evolution of sexual reproduction. *Nature* **336**:435–440.

Koufopanou, V. 1994. The evolution of soma in the Volvocales. *Am. Nat.* **143**:907–931.

Lewontin, R.C. 1970. The units of selection. *Ann. Rev. Ecol. Syst.* **1**:1–18.

Maynard Smith, J., and E. Szathmáry. 1995. The Major Transitions in Evolution. San Francisco: W.H. Freeman.

Michod, R.E. 1995. Eros and Evolution: A Natural Philosophy of Sex. Reading, MA: Addison-Wesley.

Michod, R.E. 1996. Cooperation and conflict in the evolution of individuality. II. Conflict mediation. *Proc. Roy. Soc. Lond. B* **263**:813–822.

Michod, R.E. 1997. Cooperation and conflict in the evolution of individuality. I. Multilevel selection of the organism. *Am. Nat.* **149**:607–645.

Michod, R.E. 1999. Darwinian Dynamics: Evolutionary Transitions in Fitness and Individuality. Princeton, NJ: Princeton Univ. Press.

Michod, R.E., and B.R. Levin. 1988. Evolution of Sex: An Examination of Current Ideas. Sunderland, MA: Sinauer.

Michod, R.E., and A. Nedelcu. 2003a. Cooperation and conflict in the origins of multicellularity and the eukaryotic cell. In: Evolution: From Molecules to Ecosystems, ed. A. Moya and E. Font. Oxford: Oxford Univ. Press, in press.

Michod, R.E., and A. Nedelcu. 2003b. Individuality during evolutionary transitions. *Integ. Comp. Biol.,* in press.

Michod, R.E., A. Nedelcu, and D. Roze. 2002. Cooperation and conflict in the evolution of individuality. IV. Conflict mediation and evolvability in *Volvox carteri*. *BioSystems* **2190**:1–20.

Michod, R.E., and D. Roze. 1997. Transitions in individuality. *Proc. Roy. Soc. Lond. B* **264**:853–857.

Michod, R.E., and D. Roze. 1999. Cooperation and conflict in the evolution of individuality. III. Transitions in the unit of fitness. In: Mathematical and Computational Biology: Computational Morphogenesis, Hierarchical Complexity, and Digital Evolution, ed. C.L. Nehaniv, pp. 47–92. Providence, RI: American Mathematical Society.

Michod, R.E., and D. Roze. 2000. Some aspects of reproductive mode and the origin of multicellularity. *Selection* **1**:97–109.

Muller, H.J. 1932. Some genetic aspects of sex. *Am. Nat.* **66**:118–138.

Nedelcu, A., and R.E. Michod. 2003. Evolvability, modularity, and individuality during the transition to multicellularity in volvocalean green algae. In: Modularity in Development and Evolution, ed. G. Schlosser and G. Wagner. Chicago: Univ. of Chicago Press, in press.

Otto, S.P., and M.E. Orive. 1995. Evolutionary consequences of mutation and selection within an individual. *Genetics* **141**:1173–1187.

Roze, D., and R.E. Michod. 2001. Mutation load, multilevel selection and the evolution of propagule size during the origin of multicellularity. *Am. Nat.* **158**:638–654.

Stearns, S.C., ed. 1987. The Evolution of Sex and Its Consequences. Basel: Birkhauser.

17

Mitochondria and Programmed Cell Death

"Slave Revolt" or Community Homeostasis?

Neil W. Blackstone[1] and Thomas B. L. Kirkwood[2]

[1]Department of Biological Sciences, Northern Illinois University, DeKalb, IL 60115, U.S.A.

[2]Department of Gerontology, Institute for Ageing and Health, Wolfson Research Centre, University of Newcastle, Newcastle General Hospital, Newcastle upon Tyne NE4 6B, U.K.

ABSTRACT

Nucleotide sequence data and structural features suggest that eukaryotic mitochondria evolved from bacterial endosymbionts. Although mitochondria principally function in energy conversion, they also have a prominent role in programmed cell death, and this role may be a shared derived feature for eukaryotes. A key step in programmed cell death is the permeabilization of the outer mitochondrial membrane. Two models can explain this process: (a) Bax family proteins undergo conformational changes, oligomerize, and form large channels in the outer membrane, thus releasing the mitochondrial proteins that trigger cell death, or (b) in stressed mitochondria the permeability transition pore opens, solutes diffuse in, and the matrix swells, rupturing the outer membrane and releasing the mitochondrial proteins that trigger cell death. Inferences concerning the conflictual stages of the early mitochondrial symbiosis can be drawn from these models. If programmed cell death originated from host-parasite interactions, in which the mitochondrial symbionts killed their host prior to colonizing new ones, it would be expected that these cell death mechanisms would leave mitochondria healthy and intact. Nevertheless, both models suggest that considerable mitochondrial stress would have occurred during this process. It is perhaps more plausible that programmed cell death represents vestiges of conflictual stages that occurred after the symbiosis became obligate (e.g., efforts of a population of highly stressed symbionts not so much to kill the host, but to manipulate it). This latter view is reinforced by programmed cell death mechanisms that use reactive oxygen species (in plants, yeast, and animals) or their proxy, cytochrome *c* (in some animals). Reactive oxygen species can cause mutations that may have triggered host cell fusion and sexual recombination, ultimately restoring homeostasis to the mitochondrial community. Since programmed cell death functions principally to restrain the selfish replicatory potential of individual cells in multicellular groups, a tri-level (i.e., mitochondrial, cellular, and multicellular) view of this process emerges. Diverse features of extant multicellular organisms (e.g., cellular and mitochondrial damage, cell death and aging) can be illuminated by this tri-level view.

INTRODUCTION

Mitochondria, the powerhouses of eukaryotic cells, oxidize substrates (amino acids, carbohydrates, fatty acids) and reduce coenzymes NAD^+ (nicotinic adenine dinucleotide) and FAD (flavin adenine dinucleotide). Reoxidation of NADH and $FADH_2$ provides electrons to the electron transport chain. Electron flow between the major complexes of this chain drives the extrusion of protons, establishing a steep electrochemical gradient across the inner mitochondrial membrane (Figure 17.1). This gradient ultimately powers most cellular functions, particularly by allowing the formation of ATP (adenosine triphosphate) via ATP synthase (Scheffler 1999).

A number of putative homologies between mitochondria and bacteria stimulated the endosymbiont hypothesis (Margulis 1981). Subsequently, this hypothesis has been strongly supported by nucleotide sequence data (Gray et al. 1999). Mitochondria were likely derived from α-proteobacteria that formed associations with archaebacteria (or possibly primitive eukaryotes) as much as 2,000 million years ago. A wide variety of hypotheses concern the nature of this association and the initial capabilities of the symbiont and host cells (Martin et al. 2001; Michod and Nedelcu 2003). Nevertheless, it is generally accepted that the proto-mitochondria possessed a functional electron transport chain, whereas the host cells lacked this feature. Much of the early evolutionary dynamics between host and symbiont may have stemmed from this functional difference (Blackstone 1995).

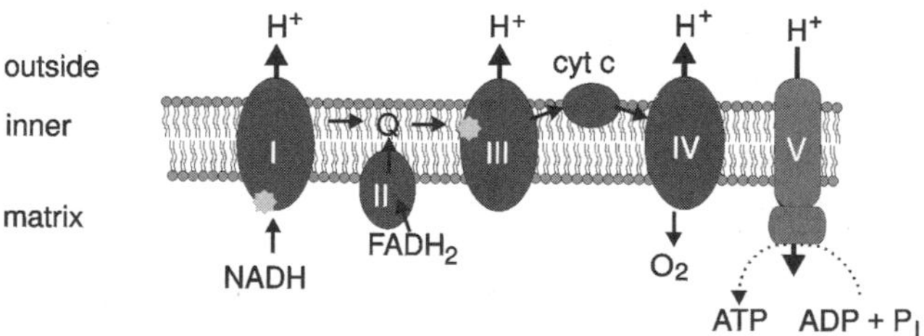

Figure 17.1 Schemata of the mitochondrial electron transport chain showing complexes I–V, coenzyme Q, and cytochrome *c*. Small arrows trace the flow of electrons from NADH and $FADH_2$ to oxygen. Large arrows show the extrusion of protons (H^+) by complexes I, III, and IV and the return of protons to the matrix via complex V, triggering the assembly of ATP (dashed arrow). Stars indicate the two major sites of reactive oxygen formation.

Recent studies have implicated mitochondria in a variety of eukaryotic signal transduction pathways, both normal and pathological (e.g., Bürkle 2000; Brownlee 2001). In this context, it is perhaps unsurprising that mitochondria have a prominent role in programmed cell death. In this process, termed "apoptosis" in metazoans, cells die as a result of an orderly, stereotyped, and tightly controlled cascade of events. Originally observed in mammalian systems (e.g., Kluck et al. 1997), a role for mitochondria in apoptosis has now been observed in all metazoans that have been studied. Indeed, programmed cell death in yeast and plants also involve mitochondria (Fröhlich and Madeo 2000; Lam et al. 2001). The relationship between eukaryotic and prokaryotic programmed cell death is not clear (Lewis 2000); plausibly, programmed cell death involving mitochondria is a shared derived feature of eukaryotes. Generally, programmed cell death provides a mechanism by which a higher-level evolutionary unit (e.g., multicellular organism, kin group of unicellular organisms, or population of unicellular parasites within a single host) can regulate the selfish replicatory tendencies of individual cells. Implication of mitochondria in this process provides a tantalizing suggestion that the evolutionary dynamics of the simple-to-complex cell transition may subsequently have influenced the dynamics of the unicellular-to-multicellular transition.

The role of mitochondria in programmed cell death inevitably suggests vestiges of early conflictual stages of the mitochondrial endosymbiosis. The obvious implications — symbionts kill their host — suggest that this process was derived from a parasitic or pathogenic stage in the symbiosis (Kroemer 1997; Frade and Michaelidis 1997; Mignotte and Vayssiere 1998). Accordingly, mitochondria participating in programmed cell death enact a vestigial "revolt" or "revenge" of an enslaved symbiont against the dominant host. Or do the data signify a more complex relationship? In this chapter we describe programmed cell death and consider the evolutionary implications of the role of the mitochondria. In the process, a "tri-level" (i.e., mitochondrial, cellular, and multicellular) perspective on programmed cell death and related features will be developed.

PROGRAMMED CELL DEATH

In metazoans, under a variety of circumstances, cells die as a result of an orderly, stereotyped cascade of cellular events (e.g., Zakeri et al. 2000). This cascade begins with extracellular signals, which initiate intracellular pathways and ultimately lead to the activation of caspases, a family of cysteine proteases. Caspases reside in the cytoplasm in an inactive form until mobilization, at which time they orchestrate the apoptotic phenotype. Because apoptosis likely has a crucial role in many human diseases, considerable efforts have focused on understanding the intracellular mechanisms that are necessary and sufficient for this process.

Although these mechanisms are complex and not completely characterized, it is nevertheless clear that critical parts of these pathways involve mitochondria. Members of the Bcl-2 protein family reside on the outer membrane of mitochondria and interact with each other to activate caspases. At least part of this activation process often involves the release of cytochrome *c* from mitochondria. Such release and the subsequent interactions of cytochrome *c* with Apaf-1 (apoptotic protease-activating factor 1) and caspase-9 in the presence of ATP seems to be a necessary step in this process in many cells. Mitochondria may also release other proteins which contribute to apoptosis (e.g., Smac/DIABLO, which is another caspase activator) and AIF (apoptosis-inducing factor, a nuclease activator). There is some suggestion that apoptotic genes may have once been mitochondrial (e.g., in the nucleus of the nematode *Caenorhabditis elegans* the *bcl-2* homologue is bicistronic with the cytochrome b_1 gene; Hengartner and Horvitz 1994).

A central, perhaps decisive, step in the process of release of mitochondrial proteins seems to be the permeabilization of the outer mitochondrial membrane. Two models of this process have been developed: the channel model and the permeability transition pore model (Figure 17.2; Martinou and Green 2001; Zamzami and Kroemer 2001). In the channel model, Bax and related proteins, which constitute a pro-apoptotic subfamily of Bcl-2 proteins, undergo posttranslational modifications leading to conformational changes, followed by their insertion into the outer mitochondrial membrane. Bax and Bcl-2 proteins, in general, show structural similarities to the pore-forming domains of diptheria toxin and bacterial colicins. Limited evidence suggests subsequent oligomerization, allowing these moderate-sized proteins to form large channels in the outer membrane, perhaps paralleling the large channels formed by pneumolysin, a toxin produced by *Streptococcus pneumoniae*. Ultimately, it is these channels that allow the release of the mitochondrial proteins that trigger cell death.

The second model focuses on the permeability transition pore, which is formed by a complex of the voltage-dependent anion channel (VDAC), the adenine nucleotide translocator (ANT), and several other proteins at the interface between the outer and inner mitochondrial membranes. When mitochondria are stressed, this pore opens and low-molecular weight solutes diffuse into the matrix. Osmotic swelling of the matrix ensues, rupturing the outer membrane, thus releasing the mitochondrial proteins that trigger cell death. Considerable debate concerns the role of Bcl-2 proteins in this process. There may be functional links between these proteins and VDAC or ANT, although the data do not yet provide a consensus in this regard.

These descriptions of programmed cell death focus largely on the well-studied metazoan systems. Nevertheless, at least some of these essential features are found in other organisms as well. Yeast lack *bcl-2* family genes, yet the expression of appropriate mammalian homologues in yeast cells results in the

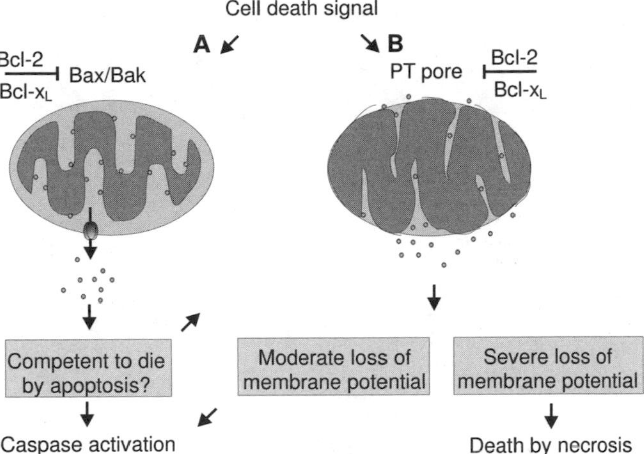

Figure 17.2 Models of the permeabilization of the outer mitochondrial membrane during apoptosis. In the channel model (A), Bax family proteins undergo conformational changes, oligomerize, and form large channels in the outer membrane, thus releasing the mitochondrial proteins, including cytochrome *c*, that trigger cell death. In the permeability transition (PT) pore model (B), these pores open, solutes diffuse in, and the matrix swells, rupturing the outer membrane and releasing the mitochondrial proteins that trigger cell death. The diagonal arrows in the lower figure indicate possible "crossover" pathways between these two models (modified from Martinou and Green 2001).

activation of a death program involving mitochondria (Fröhlich and Madeo 2000). In plants, the hypersensitive response usually functions to kill cells infected with parasites or pathogens; again, Bcl-2 family proteins and mitochondria seem to be involved in this process (Lam et al. 2001). Limited evidence thus suggests that some mechanisms of programmed cell death are conserved in eukaryotes, and it is assumed that these mechanisms in general (and the two models described above in particular) can provide insight into the evolution of the role of mitochondria in this process.

"SLAVE REVOLT"

The implication of mitochondria in programmed cell death has generally been interpreted as a vestige of an early host-parasite relationship (Kroemer 1997; Frade and Michaelidis 1997; Mignotte and Vayssiere 1998). By this view, free-living proto-mitochondria invaded and lived exploitatively inside the early host cell. Ultimately, these proto-mitochondria killed their host and resumed a free-living existence prior to invading other host cells. These conflictual stages of the endosymbiosis gave rise to modern mechanisms of programmed cell death, which is thus viewed as a vestigial "revolt" or "revenge" of the (eventually) enslaved symbiont against the (eventually) dominant host (Figure 17.3).

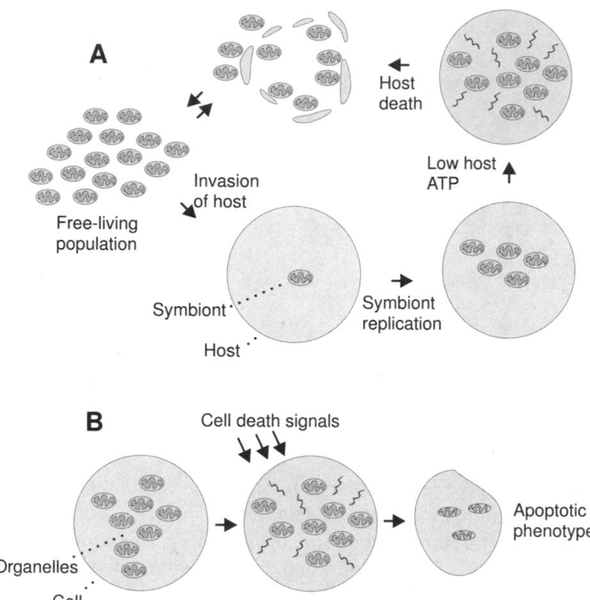

Figure 17.3 The evolution of programmed cell death from a host-parasite relationship. In the ancestral state (A), an individual from a free-living population of proto-mitochondria invades a host and replicates as a biotrophic pathogen. At some point the parasite load on the host becomes overwhelming and ATP levels in the cytoplasm drop. The parasites then permeabilize their outer membranes and trigger cell death by releasing or activating toxins in the cytoplasm. The host cell and possibly some of the parasites die; the remaining parasites feed on the remnants of the host as necrotrophic pathogens, perhaps in competition with free-living forms. Once higher evolutionary units were derived, programmed cell death (B) evolves from these mechanisms.

In the early conflictual stages, each parasitic invader is surrounded by a host vacuolar membrane. The parasite inserts porin-type channels into this membrane to monitor host ATP levels (high amounts of ATP in the fermentative host would indicate high levels of substrate for the parasite's electron transport chain). The parasite also releases caspase-like proteases into the host's cytoplasm. When levels of ATP drop, the porin channels open, triggering a decrease in the membrane potential of the host and a calcium flux that directly or indirectly activates the proteases. This cascade of events culminates with the death of the host, and its remnants are consumed by the proto-mitochondria. Subsequently, these proto-mitochondria continue a free-living existence or invade other host cells.

This view draws support from studies of *Neisseria gonorrhoeae* and related gram-negative bacteria (Rudel et al. 1996; Müller et al. 1999). Such bacteria have porins that are structurally and functionally similar to the mitochondrial VDAC. In gram-negative bacteria, porins typically form channels in the otherwise impermeable outer membrane to facilitate the diffusion of small solutes into the periplasmic space. However, when cells of *N. gonorrhoeae* interact with and enter human epithelial cells, their porins are capable of vectorially translocating into the host membranes. These porins apparently insert into both the plasma membrane and the vacuolar membrane surrounding the *N. gonorrhoeae* cell. Host ATP and other nucleoside triphosphates keep these porins closed. After prolonged infection, *N. gonorrhoeae* cells can trigger programmed cell death of the human host cell. Bacterial porins on the plasma membrane of the host cell participate in this process by allowing extracellular calcium to enter the cytoplasm. Generally, it is not unusual for parasites or pathogens to induce programmed cell death, for example, as in the hypersensitive response in plants (Lam et al. 2001).

In terms of the origins of the mitochondrial role in programmed cell death, the view that emerges is problematic in several respects. According to the permeability transition pore model, VDAC does indeed function in programmed cell death, and it does so in a manner quite reminiscent of porins: when the permeability transition pore opens, small solutes enter the mitochondrial matrix. However, none of the mitochondrial proteins that trigger cell death exit through this pore (they are too large). Rather, in this model, the matrix swells until the outer membrane ruptures, at which point the proteins are released. Is such swelling and rupture likely to be found in a successful intracellular parasite? Consider, for instance, that antimicrobial peptides of multicellular organisms likely target and destroy bacterial membranes (Zasloff 2002). Thus, osmotic stress and swelling, rupture of the external membrane, and loss of quantities of large proteins would likely constitute a significant fitness cost for the putative parasite. Even if it succeeded in killing the host prior to its own demise, subsequently the parasite would seem at a disadvantage relative to intact free-living forms.

Related hypotheses focus on the structural similarities between Bcl-2 family proteins and diptheria toxins and bacterial colicins (e.g., Muchmore et al. 1996). The diptheria toxin translocation domain is thought to dimerize and form a pH-dependent membrane pore. Colicins are used by various bacteria to weaken or kill competitors by forming channels in their membranes. Neither of these mechanisms seems to be a suitable precursor for the actions of Bcl-2 family proteins in programmed cell death, because during this process the channels form in what is functionally the outer membrane of mitochondria. These channels thus weaken the mitochondria, not the host or competitor cells.

Further difficulties with the parasitic hypothesis are apparent if the membrane system of proto-mitochondria is considered. Although the outer membrane of modern mitochondria is likely derived from the host, it cannot be primitive. Rather, proto-mitochondria likely exhibited the typical gram-negative condition: separate outer and cytoplasmic membranes with an intervening periplasmic space and cell wall. It is reasonable to expect that this membrane system had to be maintained as long as proto-mitochondria were capable of a free-living existence. Effecting mechanisms of programmed cell death would require significant disruption of the membrane system of such a symbiont.

Hypotheses deriving the role of mitochondria in programmed cell death from host-parasite interactions do not provide a good fit to the available data for either the channel model or the permeability transition pore model. Nor is it likely that other models could mitigate this discrepancy. Proteins released by mitochondria during cell death are relatively large (e.g., Smac/DIABLO is an end-to-end dimer that behaves as a 100 kDa protein), and their release requires major disruption of the outer mitochondrial membrane, whether this membrane is host-derived or that of the ancestral proto-mitochondria. Such disruptions, however accomplished, would seem to impose major fitness costs on the putative parasite (e.g., vis-à-vis healthy free-living competitors and predators). Nor would the symbiosis protect the parasite from these fitness costs, because the costs are incurred at the initiation of the free-living stage of the life cycle.

Other considerations also raise questions as to the benefits that accrue to intracellular parasites that actively kill their host, as opposed to parasites that merely passively allow their hosts to die. In the latter case, the liberated parasites certainly could still feed on the remnants of the host cell before colonizing new ones. The postulated trigger for the parasite action — a drop in ATP levels — would seem to presage the eventual demise of the host without any further action by the parasite. The rather evanescent benefits of active host killing coupled with the clear costs suggest additional difficulties for the host-parasite view of the evolution of cell death.

If the mitochondrial mechanisms of cell death are derived from a host-parasite relationship, it would seem necessary to invoke a population model, in which some subset of the population of proto-mitochondrial parasites in a cell would sacrifice themselves to kill the host for the benefit of surviving clone-mates. The selective dynamics of such "programmed proto-mitochondria death" would parallel those of programmed cell death; selection on the higher-level unit (the group of proto-mitochondria in the cell) may under some circumstances favor the altruistic sacrifice of some lower-level units. On the other hand, colonization of a host by multiple, unrelated invaders would likely not favor the evolution of such altruism. Even in a clonal population of invaders, selfish, loss-of-function variants could still arise. Ultimately, this hypothesis is not supported by the natural history of apoptosis; with the possible exception of nerve cells, the mitochondria in a cell behave uniformly during this process (Goldstein et al. 2000; D.R. Green, pers. comm.).

In summary, the essential features of the actions of mitochondria during cell death do not plausibly fit the host-parasite view — the costs to mitochondria seem too great and the benefits too small. Thus it seems unlikely that the current mechanisms of apoptosis evolved in the context of a parasite killing its host and resuming a free-living stage of the life cycle. Hypotheses that better fit the data can be more easily developed if programmed cell death is considered in the more general context of host-symbiont signaling, or in its modern form, nucleus-mitochondria "crosstalk."

SYMBIONT-HOST SIGNALING

Mechanisms of signaling are particularly likely to evolve in symbiotic relationships as the symbionts and host evolutionarily manipulate each other. Subsequently, as the symbiosis develops, these same mechanisms of signaling can be used to mitigate conflict in the relationship. Modern mitochondria certainly participate in a number of important signaling pathways (e.g., Scheffler 1999). Signaling frequently employs the electron transport chain, and reactive oxygen species (e.g., superoxide and hydrogen peroxide) are often intermediaries in mitochondrial signaling, either by themselves or in conjunction with nitric oxide. Generally, when metabolic demand is low and substrate is still available, mitochondria will enter the resting state ("state 4"). In this state, phosphorylation is minimal, electron carriers are highly reduced, and these carriers can act like a poorly insulated "wire," readily donating electrons to oxygen. On the other hand, when metabolic demand is high and sufficient substrate is available, mitochondria will enter "state 3." In this state, mitochondria are phosphorylating maximally, electron carriers are oxidized, and reactive oxygen formation is low. Finally, when there is metabolic demand but insufficient substrate, mitochondria will enter "state 2." In this state, phosphorylation is substrate limited, electron carriers are highly oxidized, and reactive oxygen formation is at a minimum. While these generalities may appear counter to the widely held notion that a high metabolic rate correlates with high levels of reactive oxygen formation, in fact, no such contradiction exists. Cells with a high metabolic rate develop many mitochondria and many electron carriers per mitochondrion; in such cells, the presence of many metal-containing macromolecules invariably leads to high levels of reactive oxygen formation. Nevertheless, the same cells will emit more reactive oxygen species when their mitochondria are in the resting state as compared to when their mitochondria are in states 2 or 3.

The role of mitochondrial signaling in human diabetes provides an instructive example (Brownlee 2001). All forms of diabetes are characterized by chronic hyperglycemia, leading to the development of microvascular pathogenesis. High levels of substrate cause mitochondria to enter the resting state 4, and the electron carriers become highly reduced. Increased production of reactive oxygen species ensues. Reactive oxygen, in turn, triggers several biochemical pathways (e.g., glucose-induced activation of protein kinase C, formation of advanced glycation end products, sorbitol accumulation, and NFκB activation). These biochemical pathways lead to the symptoms of the disease, yet diminishing mitochondrial reactive oxygen species blocks these pathways and alleviates the pathological effects.

Alternatively, mitochondria can function as a more subtle part of a complex signal transduction pathway. For instance, the damage-induced enzyme, poly(ADP-ribose) polymerase-1 (PARP-1), is activated within minutes following a genotoxic stress (Bürkle 2000). NAD^+ is the substrate of PARP-1, and the latter competes with mitochondria for this coenzyme. There is good evidence for a dual role of PARP-1 activation, i.e., a cytoprotective or an apoptosis-inducing function, depending on the degree of consumption of NAD^+ (Bürkle 2000). Apoptosis is associated with extreme NAD^+ depletion. On the other hand, a milder depletion resulting from a more moderate response of PARP-1 to DNA damage may transiently limit the formation of NADH (from NAD^+ during glycolysis and Krebs cycle). This decreases the entry of reducing equivalents via NADH into mitochondria and, as a result, decreases electron flow through the electron transport chain, thus inducing "state 2." Although this slows ATP synthesis to some extent, it also diminishes reactive oxygen formation and should thus spare both nuclear and mitochondrial DNA from oxidative attack.

Modern biomedicine has been greatly advanced by elucidating mitochondrial signaling. Such signaling can also inform investigations into the early evolution of the mitochondrial symbiosis. Consider obligately symbiotic mitochondria: such mitochondria can only interact with the environment through their host cells. Manipulation of their hosts is thus crucial to the evolutionary success of such mitochondria (but such "manipulation" should not be taken to imply "choice" or "purpose"; rather, it is a consequence of selection acting on the lower-level unit). In particular, mitochondrial manipulation has a dual focus: to obtain sufficient substrate from their hosts and to trigger rapid host growth and replication, thus increasing their habitat. Mitochondrial states 2, 3, and 4 correspond to major differences with respect to these ends. State 3 would seem to be the optimal state — substrate is plentiful and host metabolic demand is high suggesting a rapid growth rate. In state 2, starvation is occurring, and reactive oxygen is minimal. Low levels of reactive oxygen may trigger dormancy in nematodes (e.g., Larsen and Clarke 2002), and perhaps similar dormancy was triggered in the first complex cells. In state 4, substrate is plentiful, but the host has a low metabolic demand suggesting a low growth rate and low fitness. State 4 mitochondria release high levels of reactive oxygen species. Reactive oxygen likely triggers high levels of stress and high mutation rates in both host and symbionts. Under these conditions, fusion with another host cell followed by sexual recombination can ensue. Such recombination can produce genetically novel host cells, potentially with higher fitnesses (Blackstone and Green 1999). The biophysics of the electron transport chain can thus mediate fairly complex host-symbiont interactions (Blackstone 1995).

COMMUNITY HOMEOSTASIS

If programmed cell death is considered in view of the generality of mitochondria signaling, particularly as expected in an obligate symbiont, a different view of this process emerges. From the initiation of the mitochondrial symbiosis, mitochondria used their electron transport chain to manipulate their hosts' evolution. Nevertheless, selection on the higher and lower evolutionary units was generally synergistic (e.g., state 3 provides optimal growth for both units, state 2 favors dormancy for both units). In state 4, the greatest potential for conflict occurs. The key to interpreting the host-symbiont interaction in state 4 is to recognize that substrate alone has no evolutionary value. Rather, the value of substrate is that it permits replication, which is usually associated with fitness. In state 4, there is little metabolic demand so there can be no replication. As a consequence of biophysics, highly reduced electron carriers produce superoxide at two sites in the electron transport chain (Figure 17.1). Superoxide forms hydrogen peroxide spontaneously or via the enzyme superoxide dismutase. Neither of these reactive oxygen species is particularly dangerous; however, in the presence of metal-containing molecules, they can interact to form hydroxyl radicals. The latter are among the most reactive and mutagenic substances known. High mutation rates can trigger programmed cell death, and there is likely an ancient association between programmed cell death, reactive oxygen, and mutation (e.g., Fröhlich and Madeo 2000). Whereas state 4 likely produced considerable stress in both host and symbiont, the aerobic symbiont was likely better protected by antioxidant enzymes than the (formerly) anaerobic host (Blackstone 1995). Host cell death, however, may not have been the usual outcome; fusion between host cells followed by sexual recombination may have instead occurred. In many ways, sexual recombination in unicellular organisms provides an analogous mechanism to programmed cell death in multicellular organisms: both processes mitigate damage. The resulting genetically novel hosts may have provided enhanced fitness for themselves as well for their symbionts. Symbionts may also have been transmitted horizontally by this process.

By this view, the mitochondrial mechanisms of programmed cell death had their origins not in a host-parasite relationship, but in homeostatic mechanisms of the eukaryotic mitochondrial community (Figure 17.4). Indeed, limited data suggest that aspects of apoptosis are consistent with such metabolic signaling (e.g., Vander Heiden et al. 2001). The initial outcome of these mechanisms may have been host cell fusion and sexual recombination. Plausibly, this pathway was co-opted into programmed cell death only after higher-level units (e.g., multicellular organisms, kin groups of unicellular organisms, or populations of unicellular parasites within a single host) became the targets of selection. Reactive oxygen species would have stressed both host and symbiont. In derived forms, cytochrome *c* release may have functioned to upregulate reactive oxygen by blocking the electron transport chain and increasing the cytoplasmic levels of metal-containing molecules (Blackstone and Green 1999; Fröhlich and Madeo 2000; Lam et al. 2001). Subsequently, to diminish the harmful effects of reactive oxygen, cytochrome *c* itself became part of the signaling pathway. Similarly, other proteins released by stressed mitochondria also came to be part of the mechanism. Such pathways were further refined by crucial innovations, for example, recruitment of caspases as the terminal effectors of the apoptotic phenotype. In such a fashion, a pathway that initially led to sexual recombination became a mechanism of programmed cell death.

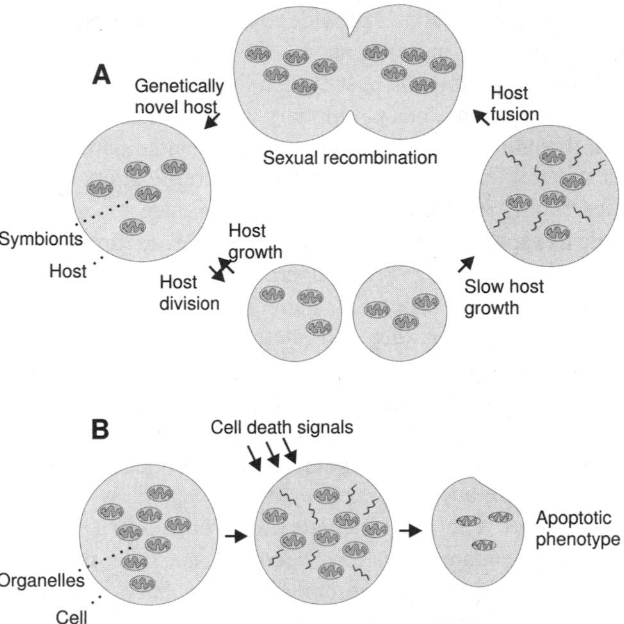

Figure 17.4 The evolution of programmed cell death from community homeostatic mechanisms. In the ancestral state (A), symbiotic proto-mitochondria multiply inside their host cells. Host cells grow and divide, maintaining the symbionts in "state 3." This cycle of growth and division can continue more-or-less indefinitely. However, endogenous or environmental factors may cause some host individuals to cease growing and dividing. This low metabolic demand shifts the symbionts into "state 4," and reactive oxygen forms, triggering cell fusion in the host and recombination in both the host and symbionts. Subsequently, reactive oxygen may have been replaced or supplemented by the release of proto-mitochondrial proteins. Once higher evolutionary units were derived, programmed cell death (B) evolves from these mechanisms.

CONCLUSIONS

The implication of mitochondria in mechanisms of programmed cell death provides a rare opportunity to develop insight into the conflictual stages of the early mitochondrial symbiosis. The common view of this interaction—as a vestige of a host-parasite relationship—seems compelling until the details of mitochondrial pathways in apoptosis are examined. A broader perspective, considering the widespread "crosstalk" between mitochondria and the nucleus, suggests that the signaling pathways in cell death perhaps arose in another context, that of metabolic signaling between symbiont and host. The electron transport chain and reactive oxygen species may have played a major part in this signaling. Subsequent evolution may have led to some subset of these pathways being co-opted into programmed cell death. This view still allows that mitochondria may have been parasitic at one time, but suggests that apoptosis did not arise from these parasitic interactions.

In addition to explaining the detailed mechanisms of cell death more effectively, this view may better correspond to general trends in host-parasite ecology. Modern endosymbiotic bacteria frequently manipulate their host's life cycle, including those aspects related to sexual reproduction (Hurst and Werren 2001). Many of these endosymbionts are α-proteobacteria (e.g., *Wolbachia*), the same group of bacteria that gave rise to modern mitochondria. Such biotrophic symbionts (i.e., those that need a living host to complete their life cycle) are perhaps more appropriate models for the evolution of mitochondria than the more necrotrophic pathogens.

Perhaps in departure from previous discussions of the evolution of complex cells, this view places emphasis less on host-symbiont interactions and more on interactions within the symbiont community. Properties of the eukaryotic cell (growth rate, dormancy, fusion, and sexual recombination) are viewed as emerging from the metabolic state of the mitochondrial community and the biophysics of the electron transport chain. Other features of eukaryotic cells can be interpreted in this context as well. For instance, the nucleus is seen not so much as a locus of control of the enslaved symbionts by the host, but rather as an innovation by a community of symbionts to alleviate conflict. Transfer of most of the mitochondrial genome to the nucleus greatly limited the potential for selfish variant mitochondria to arise. Such variants endanger not only the host but the community of lower-level units as well. Mitochondria-to-nucleus gene transfer was only feasible because the nuclear structure existed to begin with; no such structure, for instance, was available in the evolution of multicellularity. In this sense, the symbiosis and the host itself can be seen as a major innovation to diminish conflict within a population of symbionts (cf., Michod and Nedelcu 2003). In large part because of this innovation, levels-of-selection conflicts in eukaryotic cells are considerably diminished compared to, for instance, multicellular organisms. Multicellular organisms, in turn, exhibit more elaborate policing functions (see Michod, this volume; Lachmann et al., this volume).

Nevertheless, conflicts within the eukaryotic cell may still have an important role in the context of aging. It appears that the gradual loss of cellular function may be at least partly due to the build-up of defective mitochondria, raising further intriguing questions about why inter-mitochondrial selection within the host cell does not act to prevent such accumulation of defective mitochondria from occurring. Indeed it may be that the evolutionary transfer of mitochondrial genes to the nucleus has had, as a side effect, the consequence that the mitochondrial population is now vulnerable to clonal expansion of defective mitochondria, particularly if certain nuclear regulatory genes are damaged. Studies have shown that cells with abnormalities of the electron transport chain are apparently taken over by mitochondria of a single mutant mtDNA genotype (Brierley et al. 1998), suggesting that defective mitochondria somehow outcompete the wild type.

What might be the selective advantage for defective mitochondria? Several energy-dependent steps are needed for mitochondrial replication, and it is therefore hard to see how a defective mitochondrion could achieve an accelerated division rate. Although many mtDNA mutations involve deletion, there is little evidence that the smaller genome size itself confers any selection advantage. One suggestion is that mutant and wild-type mitochondria differ in their rates of degradation, intact mitochondria being turned over faster than defectives because of a greater rate of reactive oxygen-induced damage to the membranes of mitochondria with intact respiratory chains (de Grey 1997). If there is slower turnover of defective mitochondria but faster division of wild-type mitochondria, this might explain why an age-related accumulation of mutant mitochondria is seen in postmitotic tissues but less generally observed in dividing cell populations (Kowald and Kirkwood 2000). In proliferating cells, the wild-type mitochondrial population must double between successive cell divisions. Any cell falling short of its normal complement of wild-type mitochondria will be disadvantaged in its cell division rate, and therefore will be selected against. This replication advantage of intact mitochondria will keep mtDNA mutations at a low level within the cell population. However, in postmitotic or slowly dividing cells, the fact that intact mitochondria are turned over more rapidly than defective ones may cause the fraction of defective mitochondria within the population gradually to increase by a process that de Grey (1997) termed "survival of the slowest."

Given the crucial role of programmed cell death in regulating cell-level conflicts, implication of mitochondria in this process provides the tantalizing suggestion that the evolutionary dynamics of the simple-to-complex cell transition may subsequently have influenced the dynamics of the unicellular-to-multicellular transition. Likely, as groups of bacteria were evolving into complex eukaryotic cells, these emerging groups were also simultaneously evolving into multicellular communities and kin groups. The evolution of these highest-level communities may have been facilitated by the recruitment of lower-level (i.e., mitochondrial) pathways for the regulation of the intermediate level (i.e., the individual cell). In other words, mechanisms of conflict resolution within the eukaryotic cell may have been immediately co-opted into mechanisms of conflict resolution between eukaryotic cells within larger communities. Redox signaling mechanisms provided the first steps (Blackstone 2000), while subsequent innovations (e.g., caspases) led to more effective control of the cell's selfish replicatory potential via fully developed cell death pathways. Tri-level evolutionary dynamics may have enhanced the tendency of eukaryotes to form multicellular groups and ultimately contributed to the emergence of the crown groups: plants, fungi, and animals.

Vestiges of these tri-level evolutionary dynamics may still be apparent in extant multicellular organisms. An important feature of the relationship between mitochondria and programmed cell death in multicellular organisms involves their connection with damage. Mitochondria and their host cells have always been, and remain, vulnerable to random molecular damage, of which one of the most important categories is that inflicted by reactive oxygen. Damage from reactive oxygen affects all molecules and structures within the cell, but particularly significant is the damage to DNA. Oxidative damage to DNA increases the risk of neoplasia and contributes more generally to the impairment of cell function that underlies the aging process (von Zglinicki et al. 2001). Mitochondria are the primary source of reactive oxygen species within the cell. In view of the proximity of mtDNA to the site of reactive oxygen formation, mitochondria are also highly vulnerable as targets for reactive oxygen, as evidenced by tenfold higher mutation rate of mtDNA compared with nuclear DNA in some metabolically active animals. There is extensive evidence for the age-related accumulation of mtDNA mutations in a variety of postmitotic tissues (e.g., Brierley et al. 1998; Cottrell et al. 2000, 2001).

In adult tissues within multicellular organisms the primary function of programmed cell death appears to be the deletion of damaged cells, and damage to mitochondria plays an important role in triggering the cell death pathway. This response to damage is particularly evident in stem cells, such as those of intestinal epithelium which are highly sensitive to genotoxic stress and which readily undergo apoptosis following, for example, low dose (< 1Gy) γ-irradiation (Potten 1998). Thus, from the perspective of understanding the evolutionary forces that have acted to shape the role of mitochondria in triggering apoptosis, all three levels of selection need to be considered. The damage status of the mitochondrial population plays a key role in deleting host cells that otherwise might initiate the uncontrolled (selfish) proliferation of cancer cells which, in turn, might destroy the higher-level (multicellular) host.

ACKNOWLEDGMENTS

D. Green, S. Hill, A. Kowald, C. Proctor, P. Sozou, and D. Stenger contributed to discussion of these concepts and provided helpful comments. The National Science Foundation (IBN-00-90580) and the U.K. Biotechnology and Biological Sciences Research Council provided support.

REFERENCES

Blackstone, N.W. 1995. A units-of-evolution perspective on the endosymbiont theory of the origin of the mitochondrion. *Evolution* **49**:785–796.

Blackstone, N.W. 2000. Redox control and the evolution of multicellularity. *BioEssays* **22**:947–953.

Blackstone, N.W., and D.R. Green. 1999. The evolution of a mechanism of cell suicide. *BioEssays* **21**:84–88.

Brierley, E.J., M.A. Johnson, R.N. Lightowlers et al. 1998. Role of mitochondrial DNA mutations in human aging: Implications for the central nervous system and muscle. *Ann. Neurol.* **43**:217–223.

Brownlee, M. 2001. Biochemistry and molecular cell biology of diabetic complications. *Nature* **414**:813–820.

Bürkle, A. 2000. Poly(ADP-ribosyl)ation, genomic instability, and longevity. *Ann. NY Acad. Sci.* **908**:126–132.

Cottrell, D.A., E.L. Blakely, M.A. Johnson et al. 2001. Cytochrome c oxidase deficient cells accululate in the hippocampus and choroids plexus with age. *Neurobiol. Aging* **22**:265–272.

Cottrell, D.A., P.G. Ince, E.L. Blakely et al. 2000. Neuropathological and histochemical changes in a multiple mitochondrial DNA deletion disorder. *J. Neuropathol. Exp. Neurol.* **59**:621–627.

de Grey, A.D.N.J. 1997. A proposed refinement of the mitochondrial free radical theory of aging. *BioEssays* **19**:161–166.

Frade, J.M., and T.M. Michaelidis. 1997. Origin of eukaryotic programmed cell death: A consequence of aerobic metabolism. *BioEssays* **19**:827–832.

Fröhlich, K.-U., and F. Madeo. 2000. Apoptosis in yeast: A monocellular organism exhibits altruistic behaviour. *FEBS Lett.* **473**:6–9.

Goldstein, J.C., N.J. Waterhouse, P. Juin et al. 2000. The coordinate release of cytochrome *c* during apoptosis is rapid, complete, and kinetically invariant. *Nature Cell. Biol.* **2**:156–162.

Gray, M.W., G. Burger, and B.F. Lang. 1999. Mitochondrial evolution. *Science* **283**:1476–1481.

Hengartner, M.O., and H.R. Horvitz. 1994. *C. elegans* cell survival gene *ced-9* encodes a functional homolog of the mammalian proto-oncogene *bcl-2. Cell* **76**:665–676.

Hurst, G.D.D., and J.H. Werren. 2001. The role of selfish genetic elements in eukaryotic evolution. *Nature Rev. Gen.* **2**:597–606.

Kluck, R.M., E. Bossy-Wetzel, D.R. Green, and D.D. Newmeyer. 1997. The release of cytochrome *c* from mitochondria: A primary site for Bcl-2 regulation of apoptosis. *Science* **275**:1132–1136.

Kowald, A., and T.B.L. Kirkwood. 2000. Accumulation of defective mitochondria through delayed degradation of damaged organelles and its possible role in the ageing of post-mitotic and dividing cells. *J. Theor. Biol.* **202**:145–160.

Kroemer, G. 1997. Mitochondrial implication in apoptosis: Towards an endosymbiont hypothesis of apoptosis evolution. *Cell Death Differ.* **4**:443–456.

Lam, E., N. Kato, and M. Lawton. 2001. Programmed cell death, mitochondria, and the plant hypersensitive response. *Nature* **411**:848–853.

Larsen, P.L., and C.F. Clarke. 2002. Extension of life-span in *Caenorhabditis elegans* by a diet lacking in coenzyme Q. *Science* **295**:120–123.

Lewis, K. 2000. Programmed death in bacteria. *Microbiol. Mol. Biol. Rev.* **64**:503–514.

Margulis, L. 1981. Symbiosis in Cell Evolution. San Francisco: W.H. Freeman.

Martin, W., M. Hoffmeister, C. Rotte, and K. Henze. 2001. An overview of endosymbiotic models for the orgins of eukaryotes, their ATP-producing organelles (mitochondria and hydrogenosomes), and their heterotrophic lifestyle. *Biol. Chem.* **382**:1521–1539.

Martinou, J.-C., and D.R. Green. 2001. Breaking the mitochondrial barrier. *Nature Rev. Mol. Cell Biol*. **2**:63–67.

Michod, R.E., and A.M. Nedelcu. 2003. Cooperation and conflict during the unicellular-multicellular and prokaryotic-eukaryotic transitions. In: Evolution: From Molecules to Ecosystems, ed. A. Moya and E. Font. Oxford: Oxford Univ. Press, in press.

Mignotte, B., and J.-L.Vayssiere. 1998. Mitochondria and apoptosis. *Eur. J. Biochem.* **252**:1–15.

Muchmore, S.W., M. Sattler, H. Liang et al. 1996. X-ray and NMR structure of human Bcl-x, and inhibitor of programmed cell death. *Nature* **381**:335–341.

Müller, A., D. Günther, F. Düx et al. 1999. Neisserial porin (PorB) causes rapid calcium influx in target cells and induces apoptosis by the activation of cysteine proteases. *EMBO* **18**:339–352.

Potten, C.S. 1998. Stem cells in gastrointestinal epithelium: Numbers, characteristics and death. *Phil. Trans. Roy. Soc. Lond. B* **353**:821–830.

Rudel, T., A. Schmid, R. Benz et al. 1996. Modulation of *Neisseria* Porin (PorB) by cytosolic ATP/GTP of target cells: Parallels between pathogen accommodation and mitochondrial endosymbiosis. *Cell* **85**:391–402.

Scheffler, I.E. 1999. Mitochondria. New York: Wiley.

Vander Heiden, M.G., D.R. Plas, J.C. Rathmell et al. 2001. Growth factors can influence cell growth and survival through the effects of glucose metabolism. *Mol. Cell. Biol.* **21**:5899–5912.

von Zglinicki, T., A. Bürkle, and T.B.L. Kirkwood. 2001. Stress, DNA damage and ageing: An integrative approach. *Exp. Gerontol.* **36**:1049–1062.

Zakeri, Z., R.A. Lockshin, and C. Martinez-A., eds. 2000. Mechanisms of Cell Death II: The Third Annual Conference of the International Cell Death Society. *Ann. NY Acad. Sci.* **926**:1–238.

Zamzami, N., and G. Kroemer. 2001. The mitochondrion in apoptosis: How Pandora's box opens. *Nature Rev. Mol. Cell Biol.* **2**:67–71.

Zasloff, M. 2002. Antimicrobial peptides of multicellular organisms. *Nature* **415**: 389–395.

Standing, left to right: Neil Blackstone, Jack Werren, Lewis Wolpert, Rick Michod, Eörs Szathmáry, Axel Kowald, and Matthias Nöllenburg
Seated, left to right: Michael Lachmann, David Haig

18

Group Report: Cooperation and Conflict in the Evolution of Genomes, Cells, and Multicellular Organisms

Michael Lachmann, Rapporteur

Neil W. Blackstone, David Haig, Axel Kowald, Richard E. Michod, Eörs Szathmáry, John H. Werren, and Lewis Wolpert

INTRODUCTION

Most of us think of ourselves as individuals. The biological world around us contains a multitude of individuals, each of which is composed of many subunits that were separate in the evolutionary past. Some joined to form higher-level units, and others are still separately replicating but joined for life. This marriage harbored great benefits from mutualism and cooperation but also brought with it problems that arise from conflict of interest between the partners.

Conflicts between units arise when the selection pressures on some of the units favor one outcome, whereas those on other units favor another. The most basic conflict is between two units of the same species, when selection pressure on one of the units favors the survival of its own lineage over survival of the lineage of the other unit.

Conflict can exist when two units have influence over a common feature. The nucleus and mitochondrion can have an influence on the sex ratio, and thus a conflict over sex ratio can exist. A unit of a species might have another unit, but as long as it has no influence on that sex ratio, conflict will not arise. When evolution joins units into an association, then the tighter the association is, the more areas of common influence exist and the more potential for conflict exists. If the potential for conflict is lessened through the evolution of conflict mediation such as shared fate, selection pressures on the participants of the association will tend to favor similar outcomes, which will strengthen the association between the participants. Thus an association allows conflict, and lessening the conflict allows for stronger association. To understand this process, it is instructive to classify possible kinds of cooperation, source and types of conflicts that might arise in them, and the ways in which some of these conflicts are mediated (cf. Partridge and Hurst [1998] for further review and classification of such conflicts). In our discussions we tried to understand conflict in general, without separating conflict in interaction between units of the same species from conflict between interactions of units of different species. This was partly to understand the general features of conflict and conflict mediation, but primarily because the scenarios in which conflict arises usually involve both conflict between units of the same species and conflict between units of different species. Thus in the association between mitochondria and the cell nucleus, we have conflict between the mitochondria themselves and between the mitochondria and the nucleus.

Since organisms are composed of layers upon layers of cooperation, the formation of a new association could give rise to a conflict at one of the lower layers. Thus in the cooperative association between a multicellular parent and its internally carried offspring, conflicts at some of the lower levels that make up the multicellular partners (parent and offspring) can arise: a conflict between alleles at a locus, between organelles, or between cells.

Why have we not stated matters in terms of cooperation and conflict between genes? We could, for example, have talked about a conflict between a nuclear gene and a mitochondrial gene, but instead we talk about a conflict between the nucleus and the mitochondrion. As different levels of associations form, units have a shared fate, or a shared fate at a certain level of association. In those cases, the units will also have shared interests, and it is not necessary to separate them out into separate genes. For example, in a multicellular organism, a cell that replicates faster within the organism will have an advantage over other cells and will spread, often at a fitness cost to the whole organism — it will become a cancer. In this case, all the genes in the cell, including mitochondrial genes, autosomal genes, and sex-linked genes will usually have a shared interest in faster replication of the cell. In such a case, it is convenient to talk about the conflict of interest between cells in the multicellular organism.

We note that sexual reproduction plays an important role in cooperation and conflict. Many of the conflicts described would not exist at the population level without sexual reproduction. For example, conflict mediation seems a good explanation for why mitochondria are transmitted uniparentally, but not for why organisms are not asexual. We have tried, however, to not delve too deeply into questions that involve the reasons for sexual reproduction, since those are mostly unknown, or at least not agreed upon.

Classification of Cooperation

We classify cases of cooperation according to these types: (1) interchangeable vs. non-interchangeable units, (2) level of partner association, (3) asymmetry in transmission, (4) differences in replication rate, (5) mutational space or available strategies, and (6) type of benefit function. Using this classification, it is then possible to point out the areas of conflict that arise within associations of units. The factors that are known to be important for conflict, kinship, horizontal vs. vertical transmission, and shared fate are included in classification type (2) level of partner association.

Interchangeable vs. Non-interchangeable Units

Interchangeable units come from the same gene pool, and thus are in direct competition. Non-interchangeable units come from different gene pools and are not in direct competition. Here, partners are in conflict only as far as the selective forces that act on them are.

The symbiotic interaction between species is an example of interaction between non-interchangeable units, whereas cooperation between individuals within a species is an example of an interaction between interchangeable units.

When one type fixes in a population of interchangeable units, it also displaces all types available to the other partners in the association. On the other hand, when a type fixes in a population in which there is interaction between non-interchangeable units, then in the population of the other partners there are still different types.

Different alleles at one locus provide another example for a group of interchangeable units, whereas alleles at different loci are non-interchangeable: When a meiotic drive allele at locus **A** invades a population, it does not outcompete alleles at other loci, as it is not in direct competition with them. It may, however, have a conflict of interest with alleles at other loci: the selection process will cause a mutant to invade even if it lowers the total fitness of the organism, as long as its own drive at locus **A** increases, whereas the selection pressure on alleles at other loci favors those alleles that raise the total fitness of the organism. The selection pressure on alleles at unlinked loci is neutral with respect to meiotic drive at locus **A**. This is a conflict, since once such an allele, which lowers the total fitness of the organism and outcompetes other alleles at locus **A**, invades the population, mutants at other loci can invade if they reverse this effect, thereby raising the total fitness of the organism.

A single interaction can involve both interchangeable units and non-interchangeable units. For example, in the symbiosis between mitochondria and nucleus, we have both interchangeable units and non-interchangeable units: cooperation between the different mitochondria in a cell is a cooperation of interchangeable units, whereas cooperation between the mitochondria and the cell nucleus involves noninterchangeable units.

Even though the definition of interchangeable vs. non-interchangeable is clear cut in many cases, these are extremes taken from a continuum. Individuals in different species are non-interchangeable units and individuals from the same species are interchangeable units. It is obvious, however, that during speciation, there is a point at which individuals from the *same* species are non-interchangeable — for example, if they cannot drive the other to extinction, maybe because they already occupy different ecological niches. In the genome, genes at different loci are non-interchangeable, and genes at one locus are interchangeable. On the other hand, transposable elements can be seen as interchangeable units even when in different loci.

Partner Association and Kinship

One of the most important factors in the evolution of cooperation concerns the time span for which partners stay together. A well-known phenomenon in simple game theoretic examples, such as the Prisoners' Dilemma (PD) game vs. the repeated PD game, is that the length of partner association can play a role in the level of cooperation (see Axelrod 1984). Here we focus mainly on the length of association over evolutionary time.

Length of association — partner permanence vs. partner change (see Figure 18.1): In some cases partners are permanently joined and can never switch to other partners in the current population. Strict asexual reproduction provides an example of this: genes are in permanent association, and thus there is perfect alignment of transmission.

At the selective level that includes both partners, there is no long-term conflict of interest; the reproductive success of one partner is identical to that of the other. However, within the organism there could still be a short-term conflict of interest. For example, a transposable element in an asexual species has no conflict of interest with any other genes in the long term, but in the short term (within the lifetime of the lineage it is in) a transposable element might be selected for a high replication rate, even if it reduces the fitness of the organism.

In many cases, cooperation partners can be changed between generations. For example, genes at different loci in the genome are not in permanent association; recombination can change genetic partners. Partner permanence is the extreme case of a slow change of partners. In general, horizontal transfer causes partner change. When partner change occurs, one can talk about the level and fidelity of the association. This level is defined as the probability that partners in

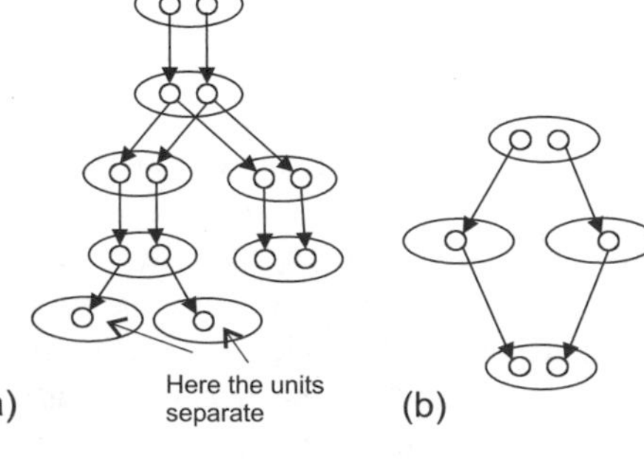

Figure 18.1 The length of association: (a) How long are partners expected to stay together? (b) How likely are they to meet again (compare also with kinship)?

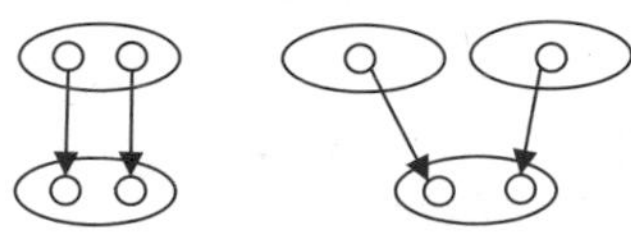

Figure 18.2 Aggregation vs. clonality: Do all units in the current association come from the same parent or from different parents?

the same cooperative group of one generation will be in the same group a certain number of generations in the future. Thus, genes on different chromosomes usually have a probability of 1/2 to stay together in the next generation after meiosis. In inbreeding populations, partners will have a high likelihood of meeting each other again, which will increase the expected length of association.

Type of formation — horizontal vs. vertical transfer (see Figure 18.2): When partners are changed, we can ask how the association was formed: through cloning or through aggregation. Aggregation is defined as the case in which partners that form a new association came from different associations in the past. Cloning is the case in which partners came from the same association. This is usually called horizontal vs. vertical transfer. When an association is formed by aggregation, horizontal transfer takes place. For example, the joining of genes from two mates during fertilization of the egg by the sperm is a case of aggregation, since the genes in the new cell come from two different cells. On the other hand, since the mitochondria in this fertilization come only from the egg, the association between the mitochondria is formed by cloning. When partner change is rare, the following is possible: an association can be formed by cloning and yet have nonpermanent partners. For example, plants can reproduce asexually by cloning, and yet the two alleles on the diploid chromosome in each cell are not in permanent association if sexual reproduction does occur from time to time.

Kinship (see Figure 18.3): In the case of cooperation between interchangeable units, we can ask not only if units that are in the same cooperative association have descended from units that were already in the same cooperative association, but also whether they have actually descended from one and the same unit, i.e., are identical by descent. If a cooperative association was formed through aggregation, the level of kinship between the units will have a strong influence on the level of conflict (Hamilton 1964).

Partner choice: Sometimes a unit can choose which other units to associate with, or can choose to leave an existing partner and find another. This choice does not have to be an active choice made by an individual. It can occur over evolutionary time. Partner choice creates the possibility of markets (see Bergstrom et al., this volume; Hammerstein, Chapter 5, this volume).

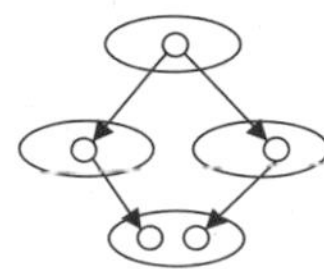

Figure 18.3 Kinship: How likely are two interchangeable units in an association formed by aggregation to be identical by descent?

Asymmetry in Transmission

Not all units that are currently in association will have the same future. It might be the case that two or more types of offspring are produced: a common example is the case of male and female offspring or offspring that are produced by unequal cell division. In the extreme case, only a subset of the units currently in association will be transmitted to the next generation — the units go through a bottleneck. The transmission of one of the two alleles at a locus to the egg nucleus and the relegation of the other allele to the polar bodies is a simple example of such a bottleneck. Asymmetric transmission can create a conflict between interchangeable units over who goes into which offspring. It can also create a conflict between non-interchangeable units as the fitness of the partners is dependent on the survival of different entities. (See Figure 18.4.)

Difference in Replication Rate

Whether partners are permanently associated or more loosely associated, non-interchangeable units can have a difference in replication rate. For example, different genes on the chromosome usually have the same replication rate. Transposable elements are one exception: the element itself replicates faster within the genome than other genes do. Such a difference in replication rate can create a conflict of interest between the units, since it allows for selection pressure that favors faster replicating units.

Mutational Space or Available Strategies

When a conflict occurs, the mutational space and strategies available to the units will strongly affect the outcome or resolution of the conflict. In a non-interchangeable association, units might have different available mutational spaces, different evolutionary rates, and different levels of phenotypic plasticity. For example, the mutational space of mitochondria and the range of influence they have on the organism is smaller than the space available to the nucleus and its influence because of the difference in the number of genes coded and because these genes effect only a limited part of cell function. On the other hand,

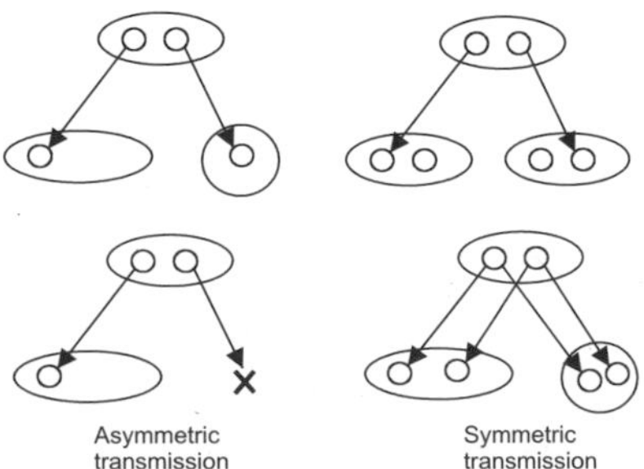

Figure 18.4 Transmission of partners in an association is called asymmetric if the offspring into which the different partners are destined can be distinguished.

mutation rates in mitochondria are sometimes higher than those in the nucleus but are, in general, highly variable (Wolfe et al. 1987; Pesole 1999). Both mutational space and mutational speed determine how fast a relevant mutation will arise when a conflict exists between the selection pressures acting on the mitochondria and on the nucleus. Mutational space is related to the concept of power, as known in economics (see Bowles and Hammerstein, this volume).

Type of Benefit Function

Fitness differences that result from the association and strategies available to the partners will affect the evolutionary outcome of the association. For example, in an interaction between interchangeable units, the benefit might be sublinear, linear (additive), or more than linear in the number of units that cooperate in an evolutionary interaction. In other cases, there might be very strong nonlinearities. Maynard Smith and Szathmáry (1995) give a good example of this effect: Imagine a group of people rowing a boat. If each person rows using two paddles, the increase in benefit, in terms of how fast the boat will get to its destination, is gradual in the number of rowers. On the other hand, if each person paddles on one side only, then removing one of the paddlers can have a catastrophic effect on this speed, since the boat will only go in circles. From this example we can see that the structure of the benefit or interaction function affects the conflict in the system, but that it can also be used as a conflict mediator: In one-sided paddling no single defector can invade since she will have a catastrophic effect on the fitness of the group and herself.

Conflict Mediation

The above classification will enable us to point to cases in which the type of cooperation is more susceptible to conflict. It also points to features that reduce conflict. Thus partner change increases the possibility for conflict, whereas partner permanence, or shared fate, reduces the possibility for conflict.

If units that have potential conflict have a large mutational space available to them, conflict is more likely; when this space is reduced, conflict is less likely. Thus, increased recombination rate reduces the mutational space available to conflict between alleles at the same locus because it reduces the total length of tightly linked loci.

In some cases specialized mechanisms for policing seem to have evolved. The immune system is such a policing agent, detecting cases of cancer in a multicellular organism. Another example is the detection and partial destruction of DNA sequences that appear twice in the genome of several fungi (*Neurospora crassa)* via mechanisms that induce hypermutation rates in repeated genes (RIP) or hypermethylation of such genes (MIP) (Selker 1999; see also Hurst and Werren 2001.) There are, however, many types of conflicts and conflict mediation, and we will expand on these throughout the rest of the chapter. For further discussion of conflict mediation, see the chapter by Michod in this volume.

ANALYSIS OF COOPERATION AND CONFLICT FOR GENES IN DIPLOID SEXUAL CELLS

Genes in the cell can be regarded as a cooperative association. In this association, many opportunities for conflict exist. We will first classify this cooperation based on the scheme described in the introduction and then discuss some possible areas of conflict and mechanisms for conflict mediation.

It should be noted that many or all of the cellular mechanisms are highly derived — thus genomes existed long before meiosis, and meiosis evolved under the background of genetic conflict. It is therefore hard to separate conflict from conflict mediation and its breakdown.

Interchangeable vs. non-interchangeable units: Within a genome, alleles at the same loci are interchangeable; alleles at different loci are non-interchangeable.

Partner association: During meiosis, partner association has a continuum from almost 1 for closely linked genes to 1/2 for genes far apart on the same chromosome or for genes on different chromosomes. The association level between alleles at the same locus between generations is close to 0. These two alleles will stay together only until the next meiosis.

Type of formation: Formation of the association is an aggregation between the genes in the sperm and the genes in the egg in the case of sexual reproduction and through cloning for asexual reproduction.

Asymmetry in transmission: Meiosis is usually a symmetric process; thus no asymmetry exists, though it is created in some cases. Sex-determining chromosomes have a different transmission pattern than the autosomal chromosomes. For example, the Y chromosome in humans is transmitted only through males.

Difference in replication rate: In cells in which all genes undergo coordinated replication, differences in replication rate do not usually exist. Such differences do exist for self-replicating units within the genome, such as transposable elements and microsatellites.

Mutational space or available strategies: When a conflict between alleles at a locus arises, the mutational space available to an allele that reduces the organism's fitness while increasing its own is limited only to the allele and to alleles tightly linked to it, whereas mutations in the whole rest of the genome could invade if they reduce this conflict and increase the organism's fitness. This idea has been termed the "parliament of genes": in cases in which there is a selection pressure on many genes to counter an effect caused by a few genes, the majority will win (Leigh 1977). One has to remember that for each particular case, the range of effects of the linked genes vs. the effect of the rest of the genome has to be considered. The mutation rate across genes is usually identical, though differences in mutation rate do exist. (One might predict that if there are mechanisms that enable local hot spots for mutations in certain genes, then "selfish genes" with a higher mutation rate would be more successful in countering the "parliament of genes.")

Benefit function: In the association between genes in the cell, there is a complex network of interactions. These include, in many cases, apparent high redundancy but also highly essential genes. In general, there could be cases in which turning off one gene would have huge consequences, and cases in which the consequences are very small.

Conflicts

From the above classification we can see that direct competition — where we would expect the highest level of conflict — exists between alleles at the same locus. Since the association between the alleles at meiosis is zero — they will end up in different organisms — an allele would be favored if it increased its own fitness even at the cost of reducing the fitness of its sister cell, while reducing the total fitness of the organism. This can happen only when meiosis is asymmetric. Although meiosis is generally fair, in that it results in equal transmission of both homologous chromosomes to the gametes, some genes have evolved into segregation distorters. These are overrepresented among the gametes in heterozygous individuals. Several such distorters are currently known (see Hurst and Werren 2001). Examples include the *t*-haplotype in mice, segregation distorters and sex-ratio distorting chromosomes in fruit flies and mosquitoes, as well as supernumerary chromosomes in a wide range of plants and animals. Current studies suggest that segregation distorters are likely to be more commonly found as we continue to investigate the genetics of organisms (Jaenike 1996).

When segregation distortion does occur, only genes that are tightly linked to the distorter have a shared interest in killing the sister cell. Unlinked genes will suffer reduced fitness as a result and thus are in conflict with the distorting genes. Recombination is a force that reduces the size of the linked loci, and thus the size of the group of genes that have shared interests. When a segregation distorter does evolve, mutational space of the genes which disfavor the drive is bigger and this is thought to cause the drive eventually to cancel. The concept of the "parliament of genes" then claims that in this conflict, the majority present in the unlinked genes eventually gains the upper hand and restores fairness to meiosis.

Because of asymmetry between the transmission patterns of sex chromosomes and autosomal chromosomes, conflicts exist between these non-interchangeable partners. For example the Y chromosome, which is transmitted only through males in an XY sex determination system, gets its fitness only through the male offspring of the male it is in, whereas the autosomal chromosomes get their fitness from both male and female offspring. Hamilton (1967) pointed out that segregation distortion, if it occurred on the Y chromosome, could drive a population to extinction, since eventually it would fix to have only males. Cosmides and Tooby (1981) extensively discuss these conflicts. Most known examples of segregation distortion are sex-ratio distorters.

A similar effect to segregation distortion is caused by converting elements: these alleles increase their representation among offspring, not by distorting segregation but by converting the sister allele. Homing endonucleases encode an endonuclease that introduces a double-stranded break at 15–20 bp recognition motifs. The break is not repaired by direct re-ligation, but by using the sequence that contains the homing endonuclease gene as a template. The end result is a conversion of the target sequence to one that contains the converting element. Repair also splits the recognition motif, thus preventing future self-cleavage. Thus, the homing endonuclease sequence is overrepresented among the gametes of heterozygous individuals and will increase in frequency, often to fixation (Gimble and Thorner [1992], taken from Hurst and Werren [2001]). In this case linked genes do not benefit from the overrepresentation, and thus the mutational space of the unit is limited to the sequence of the homing endonuclease itself. On the other hand, conflict with other genes is lower than in the case of segregation distortion: direct competition is only with the target sequence, not with any linked sites. Whereas segregation distorters usually destroy half the gametes, converters do not cause such a big direct fitness effect on the organism and thus are in less conflict with other genes in the genome.

Conflicts can arise between genes that have different replication rates in the genome. Gene replication is usually coordinated with each other and with cell division, so that the relative number of copies of the genes stays constant. When a certain gene overcomes this restriction, it can spread through the genome of the cell. A conflict with other genes will then arise, insofar as it reduces the total fitness of the cell.

When the organism is asexual and there is no horizontal transmission of genes, then, since all genes have a shared fate, there is no conflict of interest at the higher level between the copies of the replicating elements, and no conflict between the replicating elements and other genes in the cell. In this case, in the long-term, lineages in which the replicating elements are very harmful will be weeded out. At steady state the population will reach a mutation-selection balance between replication of the element within the cell and the disappearance of lineages in which it has a high copy number.

When the organism is sexual, the association between the replicated elements can be low and, since cells form by aggregation, the replicating elements can spread through the population. As a result, there is a lower selective pressure at the level of cell lineages for transposable elements to cooperate among themselves, and with other genes in the organism, to increase the fitness of their current lineage. In such a case, conflict between the transposable elements and the rest of the genome exists because of this difference in replication rate. Again, this conflict exists only insofar as the transposable elements reduce the fitness of their lineage. It is interesting to note that once a transposable element inserts itself into a position in the genome, at that position it will spread through the population of cells faster if it does not reduce the fitness of the cells it is in, even at the cost of losing its replicating ability. From the point of view of the population of transposable elements, such a mutant could be considered a "selfish element," even though from the point of view of the cell it is beneficial.

Examples of autonomous replicating elements are transposons and homing endonucleases. Sequences derived from transposons and other mobile elements make up over 45% of the human and 50% of the maize genome and are found in virtually all prokaryotes and eukaryotes. They are characterized by the ability to replicate and make additional copies of themselves so that they can accumulate within the genome.

Considerable evidence indicates that eukaryotic genomes are selected to repress autonomous-replicating elements or accommodate their presence. An example mentioned earlier is the detection and partial destruction of DNA sequences that appear twice in the genome of several fungi (*N. crassa)* via mechanisms that induce hypermutation rates in repeated genes (RIP) or hypermethylation of such genes (MIP; Selker 1999; cf. Hurst and Werren 2001).

ANALYSIS OF COOPERATION AND CONFLICT IN THE ASSOCIATION BETWEEN MITOCHONDRIA AND EUKARYOTIC CELL NUCLEUS

It is hypothesized that at some point in the evolution of the eukaryotic cell, a parasitic aerobic proteobacterium became an endosymbiont of an anaerobic host (Sagan 1967; Margulis 1981; Whatley et al. 1979; Cavalier-Smith 1981). Current knowledge suggests that such transitions, in which an endosymbiotic bacterium becomes an organelle, occurred only a handful of times. When such a tight association between the cell nucleus and endosymbiont occurs, many potential conflicts arise. A major force that seems to have reduced the number of potential conflicts is the transfer of many genes of the endosymbiont to the nucleus. In the hydrogenosome, which seems to have originated from a mitochondrion, all genes have been lost, which possibly removes all potential conflicts.

Many conflicts that occur between the nucleus and cytoplasmic elements have been discussed extensively by Cosmides and Tooby (1981), especially with respect to conflicts that arise in the production and fertilization of gametes. Here, we concentrate only on mitochondria as an example. (Rand [2001] also studies the various levels of conflicts between mitochondria in a population.) We will classify this cooperative system and then examine potential conflicts. As noted in the analysis of conflict between genes in diploid cells, it should be remembered that many of the features discussed here are highly derived.

Interchangeable units: In a single cell there are many mitochondria. These mitochondria within a single cell are interchangeable units in their association. The association between the mitochondria and the nucleus is an association between non-interchangeable units.

Partner association: During cell division the association between the different mitochondria in the cell is approximately 1/2, since that is the chance for two mitochondria to end up in the same daughter cell. The association between mitochondria and chromosomal genes is 1 during asexual cell division or mitosis and 1/2 during meiosis. Kinship depends on the mode of transmission. When mitochondria are uniparentally transmitted, kinship will depend on the relative probability that mitochondria have to advance to the next generation. If uniform, then all mitochondria will come from a common ancestor on average $1/2n$ generations ago, where n is the bottleneck size for mitochondria. If mitochondria are biparentally inherited, then the chance that two mitochondria descended from one mitochondrion in the same fashion (i.e., descended through the same individuals) is $a/(n+a)$, where n is again the bottleneck size and a is a constant that depends on the variance in replication rate of the mitochondria; a is 1 if mitochondria replicate randomly through a Poisson process and is 1/2 if each replicates exactly once per cell division.

Formation of association: In asexual reproduction and in sexual reproduction with uniparental inheritance of mitochondria, formation is through clonality. In sexual reproduction with biparental inheritance of mitochondria, during fusion of the gametes, the formation of the association of mitochondria is through aggregation.

Asymmetry of transmission: In uniparentally transmitted mitochondria, the mitochondria that end up in the fertilized egg of the sex that does not transmit the mitochondria will not continue to the next generation. In multicellular organisms, transmission bottlenecks of mitochondria do exist: not all mitochondria present in the fertilized egg will enter the germ line for several reasons: (a) programmed cell death of the oocytes, (b) nonreplication of mitochondria during division of the oocytes, and (c) high variance among mitochondria in different cells of the organism (Krakauer and Mira 1999).

An obvious asymmetry of transmission exists when mitochondria are transmitted uniparentally, between mitochondria that are in an offspring of the sex that transmits mitochondria to the next generation and those that do not.

Difference in replication rate: In single-celled organisms, mitochondria in general will replicate on average once per cell division (otherwise their number per cell would explode or dwindle), though the replication of these mitochondria is not coordinated with the replication of the nucleus, or with the replication of other mitochondria in the cell.

Mutational space: Mitochondria lost most of their original genes. For example, mammalian mitochondria retained only 13 of the protein-coding genes (Scheffler 1999) and thus have a limited range of mutations and strategies available to them. Post-transcriptional modification (e.g., RNA editing; Scheffler 1999) may further limit mitochondrial mutational space. The number of mitochondrial genes presumably has changed over evolutionary time, and thus the mutational spaces available to the mitochondria have also changed. In contrast, the mutational space available to the nucleus is large. Mutation rate in the mitochondria is sometimes much higher than that of the nuclear genome, although the ratio of mutation rate of mitochondria to those of the genome is variable (see Wolfe et al. 1987; Pesole 1999).

Type of benefit function: Fitness benefit with an increased number of mitochondria is gradual within the cell. Loss of a single mitochondrion is not catastrophic: if a single mitochondrion suffers a mutation that renders it nonfunctional, the adenosine triphosphate (ATP) supplied by the rest of the mitochondria will keep the cell as a whole mostly functional. This is further enhanced by the fact that many of the functions of the mitochondria are encoded by the nucleus, and thus the expression of those genes is not affected by the loss of a single mitochondrion in the cell.

Conflicts

As can be seen from this classification, the main conflict in this case is among mitochondria in the cell, since they are in direct competition, and between the mitochondria and nucleus when they are transmitted differently. There is a difference between the type of conflicts that arise in biparental transmission of mitochondria and in uniparental transmission. Therefore, we discuss these two cases separately.

Biparental Transmission

Since mitochondria are transmitted through both parents, several things occur: mitochondria in a fertilized egg form through aggregation, and thus there is horizontal transmission of mitochondria. The relatedness between mitochondria in that cell will also vary, being zero for mitochondria that came from different unrelated parents and higher between mitochondria from the same parent: between 1/2 and 1 on average, depending on factors such as variance in replication between mitochondria and inbreeding in the host population. This horizontal transmission allows for conflict between the mitochondria, even at the cost of lowering the fitness of the organism or the nuclear genes. Thus a conflict with the other genetic elements in the cell ensues.

Possible differences in replication rate between mitochondria within a cell provide one mechanism for such a conflict to take shape. A mitochondrion that replicates faster within a cell will have a higher chance to transmit to all offspring.

Energy allocation in the eukaryotic cell depends on the adenine nucleotide translocator (ANT), a common protein on the mitochondrial inner membrane which exchanges adenosine diphosphate (ADP) from the cytosol with ATP from the matrix. If ANT genes were mitochondrial, loss-of-function mutations would produce variants of mitochondria that could allocate all their ATP into their own replication. Consequently, it is not surprising that ANT genes are always found in the nucleus.

Mitochondria that destroy other mitochondria which do not carry a certain marker could also invade the population. To our knowledge, this has not been observed in mitochondria but has been in chloroplasts (Chiang 1976; Sears et. al. 1977).

Uniparental Transmission

Uniparental inheritance with developmental bottlenecks reduces heteroplasmy, so that the mitochondria within a cell are often clonally related and differ only by recent mutations. Therefore within a lineage, the above-mentioned conflicts will be restricted.

Since uniparental transmission creates an asymmetry in transmission between the eggs (ovules) and sperm (pollen), mitochondria that enhance their own transmission via eggs can invade a population, even if this transmission advantage is achieved at the cost of an even greater reduction of the fitness of offspring produced with the sperm. Cytoplasmic male sterility (CMS) caused by the mitochondria has evolved many times in flowering plants and provides the paradigmatic example of conflict between mitochondria and the nucleus. Mitochondria gain benefits through female fitness by causing the failure of pollen development. The effects of CMS genes in mitochondria are often countered by those of "restorer" genes in the nucleus. (See also below the discussion on parent–offspring interaction.) Thus, mitochondria can increase their own fitness in the female line even while decreasing the fitness of autosomal genes in the male and female line. Such mitochondria can increase in frequency or fix in the female population. Mitochondria could increase their fitness in female lineages by killing male offspring, by feminizing males, and by biasing the sex ratio toward females. These scenarios have been extensively analyzed by Cosmides and Tooby (1981). Mitochondria that increase their own replication rate at the cost of the female lineage they are in will be selected against at the population level. Such mitochondria should then be present in the population at a mutation-selection balance and should usually be the result of recent mutations.

During regular mitosis, no bottlenecks exist for mitochondria, since the cell division is symmetric. In the developmental process of multicellular organisms, however, such bottlenecks and asymmetries might exist. These would then select for selfish mitochondria within the soma of the organism, which again would be selected against at the population level. (See also below the discussion of conflicts between mitochondria in multicellular orginisms.)

Mitochondrial lineages in the mother go through a bottleneck before reaching the egg. These bottlenecks have been hypothesized to reduce the effect of Müller's ratchet, which can present a problem in mitochondria when they have limited amounts of recombination (cf. Kawano et al. 1995) and especially outcrossing in species with uniparental transmission (Bergstrom and Pritchard

1998; Krakauer and Mira 1999). When such bottlenecks exist, they cause mitochondria within an egg to compete to be the ones that survive the bottleneck.

In all the above-mentioned cases of conflict between mitochondria, a conflict between nuclear genes and the mitochondria would also arise, since in these cases the fitness of the organism that the mitochondria are in is lowered. Thus, mutations that could reduce the possibility of the conflicts occurring or reduce their effect would invade the chromosomal genes. Because of the asymmetry in the mutational space between mitochondria and the nucleus, the "parliament of genes" concept claims that such conflicts will usually be won by the nucleus.

In summary, both in biparental and uniparental inheritance, conflicts exist that are associated with the mitochondria. In biparental inheritance, strong conflict between mitochondria in the cell can exist; in uniparental inheritance, mitochondria are in conflict with the nucleus because of the different transmission patterns of nucleus and mitochondria. Mitochondria can outcompete other mitochondria in the population by increasing their own replication rate, by killing unmarked mitochondria or mitochondria that came from the different sex, and by increasing their own chance to go through bottlenecks. In biparental inheritance, a mitochondrion can lower the total fitness of the organism it is in but increase its own chance vs. the chance of other mitochondria to propagate to the offspring, thus increasing its own total fitness in the population. In uniparental inheritance, if a conflict between the mitochondria in the cell reduces the organism's fitness, they can be weeded out at the population level. However, this will not eliminate all effects of conflict between mitochondria in the cell. Imagine that the mitochondria in a population of organisms replicate at a slightly faster rate than would be optimal for these organisms. Now a mutation appears in one of the mitochondria that reduces its reproduction rate, so that the total fitness of the organism is higher. That mitochondrion will be outcompeted by other mitochondria in the cell, and thus there will be a low chance that the offspring of the organism will also carry a mitochondrion with the beneficial mutation. Of course, if this unlikely event happened and an organism appeared that has only mitochondria with the beneficial low rate of replication, then that organism will have a higher fitness and will most likely fix in the population. We see that the internal selection mechanism in the mitochondria will thus create a biased transmission profile: mutations that decrease the fitness of the mitochondrion they are in have a lower chance to be inherited to the next generation than ones that increase the fitness of the mitochondrion within the cell.

Conflict Mediation

Mechanisms that reduce conflict can minimize the causes or the means of conflict. Uniparental transmission of mitochondria reduces (but does not eliminate) conflict between mitochondria in the cell. Bottlenecks and segregation will do this as well. As we have seen above, these mechanisms also create opportunities for new conflict. Uniparental inheritance increases the asymmetry of transmission between nucleus and mitochondria. Transmission creates an asymmetry of transmission between the mitochondria themselves. Transfer of genes from the mitochondria to the nucleus can be a major contributor to the reduction of conflict. Out of the hundreds to thousands of protein-coding genes that existed in the original proteobacterium, only a very small number are present in the mitochondria of metazoans and only 13 are present in the mitochondria of mammals (Scheffer 1999). Many of the original genes have transferred to the nucleus; others might have simply been lost because they were no longer needed by the mitochondria inside a host. The fact that such a small number of genes are present in mitochondria probably reduces the frequency with which such conflicts arise, and this increases the ease with which the "parliament of genes" overcomes them. This does not necessarily mean that the genes have been transferred to the nucleus for that reason; mechanisms in which a direct selective pressure for such a transfer based on a reduction of conflict, are hard to envision. The transfer of the genes could have been beneficial in itself.

As mentioned above, the transition to becoming an organelle seems to have occurred only a handful of times. To explain why acquiring an organelle is so hard, Cavalier-Smith (2000) proposed that some membranes need pre-existing machinery in a membrane in order to target proteins into it. Therefore some of the membranes of cells can only be formed by splitting pre-existing membranes and are therefore called "genetic membranes." A "naked" membrane, one without the proteins necessary to incorporate proteins specific to a certain membrane type, can never become a membrane of this type. The cell membranes of bacteria belong to the category of genetic membranes, including the thylakoid membranes of cyanobacteria. In eukaryotes the endoplasmic reticulum-nuclear membrane complex belongs to this category, along with the double membranes of plastids and mitochondria. Nuclear genes code for most proteins of these organelles today, and the respective proteins are synthesized in the cytoplasm of the eukaryotic cell. For genes to transfer from the endosymbiont to the nucleus, a special mechanism for the gene products needs to evolve to target the gene products from the nucleus to the endosymbiont membrane. No nuclear-targeting gene targets that membrane, and no endosymbiont gene targets the outside of its membrane. Cavalier-Smith argues that because this targeting is hard to evolve, the evolutionary transition from endosymbiont to organelle is rare (see also Szathmáry 2000).

Why have not all mitochondrial genes been transferred to the nucleus? Changes in external electron sources and sinks (e.g., the food supply) perturb the redox state of electron carriers; if this perturbation can be transduced into gene activity, an adaptive response can ensue. Allen (1993) suggests that for an efficient functioning of this mechanism, the involved genes must reside spatially close to their gene products inside the mitochondrion (for a review, see Race et al. 1999). Others suggest that a generalized retargeting difficulty, because of size or hydrophobicity, is the cause (von Heijne 1986; Cavalier-Smith 2000). Notice that a eukaryotic organelle — the hydrogenosome — seems to have originated from mitochondria and subsequently lost all its genes (see Palmer 1997).

CLASSIFICATION OF MULTICELLULARITY

We define multicellularity as the spatial association between cells that occurs under genetic control. Unicellular organisms often appear in aggregations. We distinguish multicellularity as those cases in which the aggregation is under evolved genetic control of the individual cells. This type of control evolved many times over life's history (see Bonner 2000; Szathmáry and Wolpert, this volume). Recent discovery of a previously unknown multicellular fruiting body in such a well-studied organism, *Bacillus subtilis*, suggests that multicellular stages in the life cycle of bacteria may be more common than previously suspected (Branda et al. 2001; see also Table 18.1).

Since these associations occur at different levels of cooperation, we start by listing possible features of multicellular organisms that are relevant for cooperation, conflict, and conflict mediation within the multicellular organism.

1. Different cell types within the organism: Specialization and differentiation
   a. Reproductive division of labor: Do all cells in the organism reproduce, or do only some of the cells in the organism produce a new generation?
   b. Spatial differentiation

Table 18.1 Classification of multicellularity.

	Blue-green algae	Cellular slime molds	Plants	*Gonium*	*Porifera*
Different cell types and specialization					
• Reproductive division of labor	Yes	Yes	Yes	No	Yes
• Spatial differentiation	Yes	Yes	Yes	No	No
• Fate commitment	Yes	Yes	No	No	No
Aggregation vs. clonality	clonal, through splitting	aggregation	sexual/ asexual clonal	sexual/ asexual clonal	sexual/ asexual clonal
Size of propagules (How many cells from the original multicellular entity disperse together in the spore/ seed/embryo?)	splitting, i.e., 1/2 the organism	1 per spore	more than 1 cell	1 cell	1 egg for sexual reproduction; for asexual reproduction by fragmentation or gemmules.

c. Fate commitment: Do cells commit to their destiny so that their differentiated state cannot be reversed anymore?

2. Aggregation vs. clonality: Do the cells that form a new individual come from different parents, or from the same parent?
3. Size of propagules: How many cells from the multicellular parent(s) have offspring cells in the multicellular offspring?

Benefits and Detriments of Multicellularity

1. Reduced effective population size. For the same level of nutrients, a population of multicellular organisms will have a reduced population size, since multiple cells comprise a single organism and only the germ cells contribute to the effective population size.
2. Increase in generation time and reduction of mutation rate. Since multicellulars have many cells per organism, more cell divisions are required between generations. This results in an increase in generation time as defined by dispersal events. On the other hand, by controlling the rate of cell division of somatic vs. germ cells, multicellular organisms can control the number of cell divisions between generations, and thus reduce the effective mutation rate (relative to the total number of cell divisions in the organism).
3. Multicellular organisms are often larger than unicellular ones, or at least larger than the single cells that comprise them. This has several effects:
   - Dispersal: Larger size enables better dispersal in some multicellular organisms, e.g., through the creation of fruiting bodies, as occurs in myxobacteria.
   - Reduced ratio of surface area to volume: This has some drawbacks in transport of nutrients and disposal of waste, since there is less surface for exchange with the environment. The reduced ratio can also be advantageous when a slower exchange with the environment is desirable, e.g., for protection from heat loss or maintaining a high osmotic pressure. Furthermore, some cells may be internal and lose their contact with the external environment. This means that they have to rely on transport by other cells for nutrients, but it also means that they reside in a more protected environment.
   - Predation: Larger size provides an advantage in protection from predation as well as the ability to be a better predator, especially for engulfing larger prey.
   - Evading constraints of Reynolds number: Larger organisms can have a less random motion in watery solutions.
   - Survival advantages: Each multicellular organism can have a higher survival rate than a unicellular organism, since only the germ cells need to survive to produce progeny. The organism can increase the survival of the germ cells while reducing the survival chance for other

cells. Thus, initially the continuity of the lineage might have been assured through a strategy in which some cells "eat" other cells in time of nutrient deprivation (see Szathmáry and Wolpert, this volume).

4. Possibility of enclosing spaces within the multicellular organism: Three-dimensional topology provides an easier way of engulfing intercellular spaces for multicellulars. Thus, *Volvox* has a large space enclosed within the ball of cells in which it can store nutrients (Kirk 1998). An enclosing space is conducive to homeostasis through the regulation of the milieu interieur sensu Claude Bernard.
5. Division of labor: Specialization enables the different cells in the multicellular organism to invest only in the production of certain resources and cellular structures. It also reduces the potential for interference from the simultaneous execution of several tasks. For example, division of labor is thought to provide movement during cell division in *Volvox* (Kirk 1998).
6. Information sharing: Information about the environment gathered by the cells of the multicellular organism can be shared among them for zero or very low cost to enable better response to the environment (Zahavi 1971; Lachmann et al. 2000).

ANALYSIS OF COOPERATION IN A MULTICELLULAR ORGANISM

Interchangeable vs. non-interchangeable units: Cells are interchangeable units.

Partner association: We distinguish four cases.

1. The adult originates from a single cell, and this cell's components come from only one parent. *Volvox*, during their asexual life cycle, provide one example. Here, cells within an organism are permanently associated for the organism's lifetime, but this association is not transmitted to the next generation. Thus there is no partner change. Since all cells come from a common ancestor cell, and this comes from only one parent, kinship is high.
2. The adult originates from a single cell, and this cell's components come from multiple parents. This is the case, for example, in animal sexual reproduction. Here, the association between cells in the multicellular stage is permanent within the organism's lifetime, but there is a potential change of partners between generations. Kinship between cells depends on the number of parents. There is a potential for partner choice.
3. The adult originates from multiple cells, and these cells come from one parent. This is the case in asexual budding in plants. Here, the cells are in association for several generations if other types of reproduction (e.g., sexual reproduction) occur occasionally or in permanent association, if this is the only mode of reproduction. In this case there is the potential possibility that

the last common ancestor of the cells in the organism occurred many generations ago, since multiple parallel lineages of cells could exist within a single lineage of multicellulars. Thus kinship can be quite low.

4. The adult originates from multiple cells, and these come from multiple parents. This is the case in slime molds, since the organism is formed by aggregation of cells that potentially come from different parents. Here, cells are associated for the lifetime of the organism and change partners between generations. Kinship between cells in the organism depends on the kinship between the parents and the number of parents from which the cells originate. There is a potential for partner choice.

Asymmetry of transmission: Asymmetry of transmission occurs in several cases. First, an obvious asymmetry of transmission occurs in organisms with a germline–soma distinction. More generally, asymmetry will occur if some tissues in the multicellular organism have a higher chance to produce the next generation than other tissues. Second, in some organisms, different types of offspring can be produced, e.g., sperm and eggs, or flowers that are produced by different parts of the organism, or seeds with different dispersion strategies. All these will also produce an asymmetry.

Difference in replication rate: If replication of cells is not coordinated, then some cells could potentially reproduce faster than others.

Mutational space: In general, the mutational space available to all cells is identical. Some tissues might have an elevated mutation or epi-mutation rate.

Type of benefit function: Many different types exist.

Conflicts of Multicellularity

Cells in a multicellular organism are interchangeable units, i.e., in direct competition. This competition can arise within a single such organism, within the lineage of multicellular organisms, or at the population level. Below we will explain each of these levels further. The association of cells in a multicellular organism can also cause conflict between the interchangeable and non-interchangeable units that make up each of the cells.

Population Level

Conflicts at the population level can arise when there is vertical transmission between lineages. This happens when the organism is formed by aggregation or the cells that form the organism are formed by aggregation, e.g., in sexual reproduction. In this case, selfish elements that reduce the fitness of the organism they reside in, but increase their own fitness by increasing vertical transmission, can spread through the population. For a conflict to exist between cells, there needs to be a genetic variance between cells in the organism. Two examples follow:

Cellular slime molds form by aggregation. Usually, kinship between the cells is high because of the dispersal patterns. A mutant cell that decreases its own ability to become stalk (soma) and increases its own probability to become spore forming (germ) would decrease the total fitness of the organism (since the stalk is somewhat smaller the more stalk cells are in an organism), but would increase its own spore production. Some of these spores would then spread to aggregations of other genotypes, and the mutation could spread through the population (see Strassmann et al. 2000).

In organisms that originate from a single cell formed by fertilization (i.e., aggregation) and in which the mitochondria are inherited biparentally, the mitochondria in each cell in the organism are an aggregation of the mitochondria in the parents. Since the replication of mitochondria is not coordinated so that exactly one copy of each mitochondrion enters each of the daughter cells, there exist genetic differences between the mitochondria in different cells in the organism. A mitochondrion could invade that reduces the total fitness of the organism, but increases the chance of the cells that it resides in to become germ line. If one parent of the organism has such mitochondria, these will be overrepresented among the organism's offspring and thus increase their frequency in the population. In this case the conflict between mitochondria in different cells also creates a conflict between mitochondria and the nucleus, or other genetic elements. If the nucleus is genetically identical between cells and is highly related to the nuclei in other cells, there exists a selection pressure to negate the effects of the mutated mitochondria, increasing the organism's total fitness. Such conflicts can be lessened by increasing the relatedness of cells within the organism or forming new organisms by cloning instead of aggregation.

Lineage Level

Here a conflict can occur when there is genetic variance between cells in the organism, and this variance can be inherited between generations. In this case, cells that increase the representation of their offspring among the offspring of the organism will increase in frequency within the lineage, even if this comes at a cost of reduced fitness of the lineage as a whole. This lineage will then be selected against at the population level. The main difference between this type of conflict and the previous type is that when a conflict is limited to a conflict within a lineage, then population-level selection between lineages will select against lineages with selfish cell types. In the case of vertical transmission, population-level selection favors selfish cell types, and only higher-level structure selects against them. Conflicts within a lineage can arise through differences in replication rate or asymmetry of transmission. Since a lineage creates multiple sublineages, the population-level selection will also affect the frequency of the selfish individuals within a lineage. Below, three examples are given for these types of conflicts:

- In a multicellular organism that replicates by splitting, but in which the replication of cells and their distribution among the organism's offspring is not coordinated, a cell type that increases its own replication rate would increase in frequency within the lineage.
- In a multicellular organism that reproduces sexually and develops from a single egg and in which mitochondria are transmitted uniparentally, a mutant mitochondrion that increases the chance of the cell it is in to become the germ cell will increase its frequency within a lineage.
- In a multicellular organism that replicates by budding, a mutant cell type that increases its frequency in the buds would increase in frequency within the lineage. Thus in plants that undergo asexual reproduction through budding, a mutant cell type that reproduces faster or has a higher chance of producing new buds will spread through the lineage.

Organism Level

Conflict within an organism occurs when genetic variation within the organism is not transmitted between generations; the competition between cells is restricted to the organism's lifetime. At the population level, individuals with selfish cell types will be selected against. Since competition occurs within the individual, no conflict between cells to "take over" the germ cells will take place. On the contrary, a mutation in cells that invests less in producing germ and more in reproducing within the individual, will have a benefit within the organism for that cell type.

Conflict Mediation

A major hurdle in the evolution of multicellularity is the appropriate down-regulation of cell division at the right time and place. Multicellular organisms are made of cells. A proper functioning of the multicellular organism usually entails that not every cell that has enough resources to replicate will do so. Many cells in the organism have to give up their reproductive capability in the organism.

Linked with the three types of conflicts outlined above, we delineate three main areas of conflict mediation: (a) reduction of horizontal transfer and increase of kinship between cells in the multicellular organism, (b) reduction of the number of cell lineages within the organism that will produce offspring, (c) reduction of replication potential and detection of aberrant cells within the organism. Not all of these conflict-reducing mechanisms necessarily were selected for; in some cases, the life history of the organism results in fewer conflicts between the cells that compose the organism.

Germ Line

The first area of conflict mediation is the evolution of a germ line, where we distinguish three stages:

1. Propagule size: A smaller number of cells within the propagule increases kinship within the organism and thus reduces conflict. A larger number of cells in the propagule decreases kinship, and thus creates a conflict—a selection process that selects for cells that have a better ability to enter the propagule at a fitness cost to the organism. Such mutations can spread through the population if the organism is produced by aggregation but not if it is produced by clonality. Nevertheless, in clonality they reduce the overall fitness of the multicellular organism in which they occur.
2. Reproductive division of labor: Who proceeds to the next generation? A multicellular organism can evolve a reproductive division of labor, in which only some of the cells will produce the next generation of the multicellular. In a multicellular organism with a single-celled propagule, there is no selective process that selects for somatic cells that invade the propagule at a cost to the organism. A mutation like that could occur and could reduce the fitness of the organism it occurs in, but there would be no selective advantage to the mutation in the population.
3. Early sequestration: When is the germ line sequestered? Early sequestered germ line can reduce the number of cell divisions in a generation, and thus the mutational load that the organism experiences.

Soma

The second mechanism to reduce the conflict is the evolution of a soma. When some of the cells in the organism are forced to give up their ability to replicate, or replicate indefinitely within the organism, the potential for conflict between cells in the organism is reduced. Note that this is not the same as reproductive division of labor, in which some cells give up their ability to produce the next generation of the organism.

Other than reduction of conflict, there are several other benefits to the evolution of soma and germ. As mentioned above, an early sequestered germ can reduce the number of cell divisions between generations and thus the effective mutation rate. A disposable soma can have lower maintenance cost. It provides an efficient division of labor, since some cells can put all their resources into reproduction, whereas others do not need to maintain any reproductive ability.

It is important to note that soma and a full, early sequestered germ line are not present in all multicellular organisms. Even in metazoans it seems to be a relatively late evolutionary event. Plants have no real germ line. In sexual reproduction the propagule size of the fertilized seed can be one or more cells, and in vegetative growth, the propagule size is more than one cell. A terminally differentiated cell type does occur early in evolution.

Programmed Cell Death

A third mechanism of conflict mediation is programmed cell death (PCD). This is a slightly stronger mechanism of control of cell growth than simply preventing cell division. PCD is triggered by extracellular signals, following an orderly, stereotyped cascade of events regulated by an intrinsic pathway. There are two major types of such pathways: (a) extrinsic Fas pathways and (b) pathways in which mitochondria are central. Interestingly, there does not seem to be an equivalent pathway in which another organelle, the chloroplast, is in control. PCD has a couple of further functions. It is used in development for the formation of structures and for neuronal selection. It is used in infection control to eradicate infected cells and in oocyte selection. Finally, it is used in preventing uninhibited growth in the control of cancer.

PCD does occur in unicellular organisms, though we would expect such cases to occur only in kin groups or under similar selective scenarios.

BREAKDOWN OF CONFLICT MEDIATORS IN MULTICELLULARITY

When a conflict mediator exists, we can predict a pathological condition under which it will break down. In the absence of a mediator, such conditions will be even more common, as we see from the following examples.

In metazoans some somatic cells are unable to replicate indefinitely, which is one of the mechanisms to prevent conflict between cells within the organism. When this mechanism breaks down, we expect that a cell will start to divide indefinitely; PCD can then prevent further growth. When the mechanism of PCD breaks down, we expect cancer to occur. Cancer is somewhat more likely in cell types that have not lost the ability to replicate indefinitely.

The conflict between mitochondria in the same cell, as described earlier for diploid cells, is still present in the multicellular organism. This conflict is mediated in part by the small size of the mitochondrial genome and the reduced number of functions that it still controls. Although these mechanisms reduce the chance of conflict to occur, they do not eliminate it. A mutant mitochondrion could invade, resulting in cells that have a large number of mitochondria, each of which is not very functional for the cell. Only few cells like this should be observed, since there is no benefit for the cell. This phenomenon would mainly be expected in cells in which there is high turnover of mitochondria. Conflict between mitochondria in different cells is analogous to conflict between cells since there is no horizontal transfer of mitochondria inside the multicellular organism.

ANALYSIS OF COOPERATION AND CONFLICT IN THE ASSOCIATION BETWEEN PARENT AND FETUS

In some multicellular organisms a tight association between a parent (often, but not necessarily, the mother) and sexually produced offspring exists. This association provides increased survival chance for the offspring. Because of the highly asymmetric association between interchangeable units (see below), this association awakens many conflicts from lower levels of association in the organisms.

Interchangeable vs. non-interchangeable units: Parents and offspring are taken from the same gene pool and thus are, by our definition, interchangeable units. If a type invades the offspring population and takes over, it will also have wiped out all the genetic variation in the parent population. Because of the highly asymmetric nature of the association, this classification seems very nonintuitive. We can think of the association as a cooperation between pairs of individuals from the population, as might occur at early stages of the evolution of multicellularity: one cell divides to produce two daughter cells, which stay attached and cooperate for some time.

If we break down the association into the lower-level elements that make up each of the units, we encounter both interchangeable and non-interchangeable units. Thus the association between genes at different loci in these organisms are non-interchangeable units.

Partner association: Parent and offspring remain in association only for part of one generation, and thus the association between units in the parent vs. those in the offspring is 0.

Type of formation: Offspring are produced by an aggregation of genes from the parents, since they are produced sexually. Other elements in offspring are passed only through one parent. Kinship between different fetuses within the mother is usually 1/2, but can be lower for cases in which they arc the product of multiple matings.

Asymmetry in transmission: Parent and offspring have different functions and futures in this association. The expected fitness of mother and offspring can differ, which might cause a preference for elements to stay with the mother or continue to the offspring. If more than one offspring is carried, then those have usually a similar projccted future.

Difference in replication rate: Since neither parent nor offspring reproduce during the association, this difference does not exist.

Mutational space or available strategies: Mutational space is identical, although cells in the fetus undergo more replications. Differencc in available strategies can be created, e.g., if the genes of the fetus are not expressed up to a certain age, or if a tight control is kept over which gene products can be transferred from mother to fetus.

Type of benefit function: The parent provides all the benefit to thc offspring. The loss of fitness to the parent from terminating the association is usually smaller than the loss to the offspring. In this case many conflicts at lower levels of association are reawakened, each of which arc discussed separately below.

Conflicts between Cells

Mother and fetus are both multicellular. In a multicellular organism, the fact that the offspring is produced by only one cell prevents a conflict between cells in which a selection pressure exists favoring cells that become the germ cells. When the fetus is associated with the mother for a longer time, this restriction might be overcome by a cell in the mother that would also enter the fetus, so that more than one cell from the mother produces cells in the fetus.

A mutation in cells that would cause them to be transferred from parent to offspring, or from offspring to offspring could invade a lineage. If only one parent carries offspring, then such a mutant could not spread through the population (since, e.g., it is transmitted only by daughters), and would be selected out at the population level. Thus such mutants would be kept at a mutation-selection balance within lineages. This conflict is very similar to the conflict between mitochondria that can arise at the evolution of multicellular organisms; here, the fact that only one sex is carrying the fetus is equivalent to uniparental inheritance. In hermaphrodites such a mutation could spread through the population. One therefore expects fewer cases in which hermaphrodites carry fetuses internally, and in those cases this type of "parasitic cancer" could be present.

Conflicts between Autosomal Genes

Since the genomes of sexually produced offspring are produced by aggregation, there is a basis for conflict between alleles at the population level. Of the two alleles present in a diploid parent, only one is transmitted to each offspring. The randomizing process of meiosis provides protection against this conflict because the two alleles present in the parent cannot easily detect which one of them was transmitted to the offspring. In the absence of such information, the best they can do is to maximize the total number of surviving offspring and take their chances with the flip of the meiotic coin. This mechanism breaks down in some cases. Postsegregation distorters act by reducing the frequency of noncarrier individuals after fertilization. The distorter benefits if this increases the fitness of carriers (e.g., by reduced competition). Examples include spore killers in fungi and the Medea locus in flour beetles.

Similarly, alleles in the offspring are uninformed about whether they came from the mother or father. However, when such information is available, as is the case in genes that are imprinted (Haig 2000), then a conflict can result because alleles that come from the father could demand more investment from the mother, who does not carry those alleles. Alleles that come from the mother, on the other hand, have a higher interest in ensuring that the mother will have further offspring. If genes carry epigenetic marking that differs according to their sex of origin, then they can evolve to have a different behavior when marked or unmarked, and thus when transmitted through mother or father (Haig 2000).

Conflicts between Genes on Sex Chromosomes

The chromosome that is unique to the heterogametic sex is never present in the homogametic sex. Thus there will always be a conflict on parental investments between offspring of different sexes. Thus a gene on the Y chromosome that decreases parental investment in females but increases survival of male offspring will invade the population.

For parent-offspring conflict we should differentiate whether the heterogametic or homogametic sex carries the pregnancy. Many other modes of sex determination can exist and, in each, one could carry out the analysis of conflict as follows:

Pregnancy carried by homogametic sex (XX females): Alleles on the Y chromosome in a male fetus "know" that they did not originate from the parent carrying the pregnancy. Thus these alleles would increase investment in themselves even at a high cost to the mother (see Hurst 1994).

Pregnancy carried by heterogametic sex (ZW females): Alleles on the mother's W chromosome "know" that they are not present in any homogametic offspring, and thus W-linked alleles would invade the population if they increase the fitness of heterogametic offspring even at a high cost to the homogametic offspring. Thus pregnancy in a ZW sex determination system has more potential for conflict. A potential mechanism for conflict mediation would be a reduction of the size of the W chromosome and a mirroring of all genetic expression of the W in female fetuses by the Z and autosomal chromosomes in male fetuses.

All are similar to the conflict produced by segregation distorters, where genes at the other chromosomes would be in conflict with the genes causing the distorted investment, since they, on average, have a 1/2 chance of having an identical allele in the parent/offspring.

Conflicts between Cytoplasmic Elements

In uniparental inheritance, conflicts between parent and offspring created by elements of the cytoplasm are similar to those created by sex chromosomes. Obvious examples are for those elements that are transmitted in the cytoplasm of eggs but not via sperm (such as mitochondria and chloroplasts). There are many cases where such elements distort the sex ratio toward females by various mechanisms, including male killing, feminization of genetic males, and induction of asexuality (dispensing of males).

FURTHER REMARKS ON MULTICELLULAR ORGANISMS

Many further subjects arose in our discussions which did not fit into this chapter. We include a few of the interesting questions that came up in these discussions.

A major hurdle in the evolution of multicellularity is the appropriate down-regulation of cell division at the right time and place. This hurdle was overcome to different degrees by various separate evolutionary events in the evolution of multicellularity. Complex multicellularity, on the other hand, seems to have evolved only a very few times, which hints that there might be other "hurdles." One might be the evolvability of new cell types in the organism and linking them to the developmental plan of the organism. To address this question it would be instructive to expand our knowledge of the evolution of cell types. Can cell types be well defined? How often do new cell types evolve? How many cell types are there in multicellular organisms? In regard to these questions, a greater synthesis between evolutionary theory and molecular cell biology will further illuminate both fields.

Why are all complex multicellular organisms primitively sexual and eukaryotic? This question is linked to the question of why complex multicellularity evolved so late. What is the relationship between propagule size and evolvability of differentiation, under various assumptions about the type of control of differentiation that exists (i.e., differentiation through environmental signals or signals from other cells)?

Mitochondria and chloroplasts of most flowering plants are uniparentally inherited via ovules. Therefore, chloroplast genes could presumably benefit in the same manner as mitochondrial genes by causing male sterility. All known systems of CMS, however, involve mitochondrial genes; none involve the chloroplast. Why is this so? Do systems of chloroplast CMS exist but simply have not been recognized? Or does the chloroplast, unlike the mitochondrion, have little power to assert its interest during pollen formation?

Are there measurable costs to policing, e.g., for mechanisms that delete every gene duplication and prevent transposable elements? What are the measurable fitness costs of intragenomic conflict in real organisms? What is the balance sheet for meiosis in terms of increasing or reducing cooperation? How is fair meiosis maintained? Which features of eukaryotes evolved before the mitochondria became an endosymbiont of the eukaryotic cell?

What are the preadaptations of multicellularity? (See Szathmáry and Wolpert, this volume.) Under the right definitions, it should be far more widespread than currently thought. To understand the general principles of the evolution of multicellularity, we should seek and study those cases. It should be more widespread in prokaryotes—what are its distinctive features and advantages?

From our discussion on mitochondria, deleterious mutations, bottlenecks, and conflicts, we predict that germ cells should be subject to selection pressure that selects for least-loaded mitochondria in terms of mutational load that affects mitochondria performance.

SUMMARY

Every entity that we call "individual" in biology is made of many separate units. These units, if identical or different, will not strive toward (i.e., be selected for) the same goals. When we examine a certain biological system, we can classify the type of cooperation that exists according to the classification scheme we constructed. This will then point to the possible conflicts in the system. These conflicts should exist, or there should be mechanisms of conflict mediation that reduce the conflict. We could also make predictions of what is expected to happen when one of these mechanisms of conflict mediation breaks down.

It should not be expected that conflicts in the organisms will disappear. In some cases, mechanisms of conflict mediation reduce the conflicts; however, even if the conflict causes a fitness cost to the organism, and mechanisms for conflict mediation are directly selected for, conflict would still exist at some kind of mutation-selection balance (and in some cases a biased mutation-selection balance, as we pointed out in the discussion on conflicts between the mitochondria and the nucleus). In other cases a mechanism that reduces some of the conflicts will create others. Thus uniparental inheritance of mitochondria reduces the conflict between the mitochondria in the cell but increases the conflict between the mitochondria and the nucleus, since it causes the mitochondria to be transmitted on different lineages than the autosomal genes. Recombination reduces the mutational space available to the meiotic drive gene but makes it possible for transposons to spread through the population.

ACKNOWLEDGMENTS

We thank those conference participants who contributed to our interesting discussions, in particular Matthias Nöllenburg. We wish to thank Steve Frank, Franjo Weissing, and John Tooby, who contribued to this chapter, S. Ptak for reading and helping in its rewriting, and E. Jablonka for comments on the manuscript.

REFERENCES

Allen, J.F. 1993. Control of gene expression by redox potential and the requirement for chloroplast and mitochondrial genomes. *J. Theor. Biol*. **165**:609–631.

Axelrod, R. 1984. The Evolution of Cooperation. New York: Basic.

Bergstrom, C.T., and J. Pritchard. 1998. Germline bottlenecks and the evolutionary maintenance of mitochondrial genomes. *Genetics* **149**:2135–2146.

Bonner, J.T. 2000. First Signals: The Evolution of Multicellular Development. Princeton, NJ: Princeton Univ. Press.

Branda, S.S., J.E. González-Pastor, S. Ben-Yehuda et al. 2001. Fruiting body formation by *Bacillus subtilis*. *Proc. Natl. Acad. Sci. USA* **98**:11,621–11,626.

Cavalier-Smith, T. 1981. The origin and early evolution of the eukaryotic cell. *Symp. Soc. Gen. Microbiol*. **32**:33–84.

Cavalier-Smith, T. 2000. Membrane heredity and early chloroplast evolution. *Trends Plant Sci*. **5**:174–182.

Chiang, K.S. 1976. On the search for a molecular mechanism of cytoplasmic inheritance: Past controversy, present progress and future outlook. In: Genetics and Biogenesis of Chloroplasts and Mitochondria, ed. T. Bucher, W. Neupert, W. Sebald, and S. Werner, pp. 305–312. Amsterdam: North Holland.

Cosmides, L.M., and J. Tooby. 1981. Cytoplasmic inheritance and intragenomic conflict. *J. Theor. Biol*. **89**:83–129.

Gimble, F.S., and J. Thorner. 1992. Homing of a DNA endonuclease gene by meiotic conversion in *Saccharomyces cerevisiae*. *Nature* **357**:301–305.

Haig, D. 2000. The kinship theory of genomic imprinting. *Ann. Rev. Ecol. Syst.* **31**:9–32.

Hamilton, W.D. 1964. The genetical evolution of social behavior. *J. Theor. Biol.* **7**:1–52.

Hurst, G.D.D., and J.H. Werren. 2001. The role of selfish genetic elements in eukaryotic evolution. *Nature Rev. Gen.* **2**:597–606.

Hurst, L.D. 1994. Embryonic growth and the evolution of the mammalian Y chromosome. I. The Y as an attractor for selfish growth factors. *Heredity* **73**:223–232.

Jaenike, J. 1996. Sex-ratio meiotic drive in the *Drosophila quinaria* group. *Am. Nat.* **148**:237–254.

Kawano, S., H. Takano, and T. Kuroiwa. 1995. Sexuality in mitochondria: Fusion recombination, and plasmids. *Intl. Rev. Cytol.* **161**:48–110.

Kirk, D.L. 1998. Volvox: Molecular-Genetic Origins of Multicellularity and Cellular Differentiation. Cambridge: Cambridge Univ Press.

Krakauer, D., and A. Mira. 1999. Mitochondria and germ-cell death. *Nature* **400**:125–126.

Lachmann, M., G. Sella, and E. Jablonka. 2000. On the advantages of information sharing. *Proc. Roy. Soc. Lond. B* **267**:1287–1293.

Leigh, E.G. 1977. How does selection reconcile individual advantage with the good of the group? *Proc. Natl. Acad. Sci. USA* **74**:4542–4546.

Margulis, L. 1981. Symbiosis in Cell Evolution. New York: W.H. Freeman.

Maynard Smith, J., and E. Szathmáry. 1995. The Major Transitions in Evolution. Oxford: W.H. Freeman.

Palmer, J.D. 1997. Organelle genomes: Going, going, gone! *Science* **275**:790–791.

Partridge, L., and L.D. Hurst. 1998. Sex and conflict. *Science* **281**:2003–2008.

Pesole, G., C. Gissi, A. De Chirico, and C. Saccone. 1999. Nucleotide substitution rate of mammalian mitochondrial genomes. *J. Mol. Evol.* **48**:427–434.

Race, H.L., R.G. Herrmann, and W. Martin. 1999. Why have organelles retained genomes? *Trends Gen.* **15**:364–370.

Rand, D. 2001. Units of selection on mitochondrial DNA. *Ann. R. Ecol. Syst.* **32**: 415–448.

Sagan, L. 1967. On the origin of mitosing cells. *J. Theor. Biol.* **14**:225–276.

Scheffler, I.E. 1999. Mitochondria. New York: Wiley.

Sears, B., J.E. Boynton, and N.W. Gillham. 1977. Effect of increasing zygote age on transmission of chloroplast genes in green-alga *Chlamydomonas-reinhardtii*. *Genetics* **86**:S56–S57.

Selker, E.U. 1999. Gene silencing: Repeats that count. *Cell* **97**:157–160.

Strassmann, J.E., Z. Young, and D.C. Queller. 2000. Altruism and social cheating in the social amoeba *Dictyostelium discoideum*. *Nature* **408**:965–967.

Szathmáry, E. 2000. Evolution of replicators. *Phil. Trans. Roy. Soc. Lond. B* **355**:1669–1676.

von Heijne, G. 1986. Why mitochondria need a genome. *FEBS Lett.* **198**:1–4.

Whatley, J.M., P. John, and F.R. Whatley. 1979. From extracellular to intracellular: Establishment of mitochondria and chloroplasts. *Proc. Roy. Soc. Lond. B* **204**: 165–187.

Wolfe, K.H., W.-H. Li, and P.M. Sharp. 1987. Rates of nucleotide substitution vary greatly among plant mitochondrial, chloroplast, and nuclear DNAs. *Proc. Natl. Acad. Sci. USA* **84**:9054–9058.

Zahavi, A. 1971. The function of pre-roost gatherings and communal roosts. *Ibis* **113**: 106–109.

19

Cultural Evolution of Human Cooperation

Peter J. Richerson[1], Robert T. Boyd[2], and Joseph Henrich[3]

[1]Department of Environmental Science and Policy, University of California, Davis, CA 95616, U.S.A.

[2]Department of Anthropology, University of California, Los Angeles, CA 90095, U.S.A.

[3]Department of Anthropology, Emory University, Atlanta, GA 30322, U.S.A.

ABSTRACT

We review the evolutionary theory relevant to the question of human cooperation and compare the results to other theoretical perspectives. Then, we summarize some of our work distilling a compound explanation that we believe gives a plausible account of human cooperation and selfishness. This account leans heavily on group selection on *cultural* variation but also includes lower-level forces driven by both microscale cooperation and purely selfish motives. We propose that innate aspects of human social psychology coevolved with group-selected cultural institutions to produce just the kinds of social and moral faculties originally proposed by Darwin. We call this the "tribal social instincts" hypothesis. The account is systemic in the sense that human social systems are functionally differentiated, conflicted, and diverse. A successful explanation of human cooperation has to account for these complexities. For example, a tribal-scale cultural group selection process alone cannot account for human patterns of cooperation because, on one hand, much conflict exists within tribes and, on the other, people have proven able to organize cooperation on a much larger scale than tribes. We include multilevel selection and gene–culture coevolution effects to account for some of these complexities and discuss empirical tests of the resulting hypotheses. In particular, we argue that strong support for the tribal social instincts hypothesis comes from the structure of modern social institutions. These institutions have conspicuous "work-arounds" that shed light on the underlying instincts.

INTRODUCTION

Cooperation[1] is a problem that has long interested evolutionists. In both the *Origin* and *Descent of Man,* Darwin worried about how his theory might handle cases such as the social insects in which individuals sacrificed their chances to reproduce by aiding others. Darwin could see that such sacrifices would not ordinarily be favored by natural selection. He argued that honeybees and humans were similar. Among honeybees, a sterile worker who sacrificed her own reproduction for the good of the hive would enjoy a vicarious reproductive success through her siblings. Humans, Darwin (1874, pp. 178–179) thought, competed tribe against tribe as well as individually, and that the "social and moral faculties" evolved under the influence of group competition:

> It must not be forgotten that although a high standard of morality gives but slight or no advantage to each individual man and his children over other men of the tribe, yet that an increase in the number of well-endowed men and an advancement in the standard of morality will certainly give an immense advantage to one tribe over another. A tribe including many members who, from possessing in a high degree the spirit of patriotism, fidelity, obedience, courage, and sympathy, were always ready to aid one another, and to sacrifice themselves for the common good, would be victorious over most other tribes; and this would be natural selection.

More than a century has passed since Darwin wrote, but the debate among evolutionary social scientists and biologists is still framed in similar terms — the conflict between individual and prosocial behavior guided by selection on individuals versus selection on groups. In the meantime social scientists have developed various theories of human social behavior and cooperation — rational choice theory takes an individualistic approach while functionalism analyzes the group-advantageous aspects of institutions and behavior. However, unlike more traditional approaches in the social sciences, evolutionary theories seek to explain both contemporary behavioral patterns and the origins of the impulses, institutions, and preferences that drive behavior.

In this chapter we refer to "culture" as the information stored in individual brains (or in books and analogous media) that was acquired by imitation of, or teaching by, others. Because culture can be transmitted forward through time from one person to another and because individuals vary in what they learn from others, culture has many of the same properties as the genetic system of inheritance, but also of course many differences. The formal import of the analogies and disanalogies has been worked out in some analytical detail (e.g., Cavalli-Sforza and Feldman 1981; Boyd and Richerson 1985). We also subscribe to Price's approach to the concept of group selection. Heritable variation between entities can appear at any level of organization and any level above the individual merits the term group selection (Henrich 2003; Hamilton 1975; Price

[1] "Cooperation" has a broad and a narrow definition. The broad definition includes all forms of mutually beneficial joint action by two or more individuals. The narrow definition is restricted to situations in which joint action poses a dilemma for at least one individual such that, at least in the short run, that individual would be better off not cooperating. We employ the narrow definition in this chapter. The "cooperate" vs. "defect" strategies in the Prisoner's Dilemma and Commons games anchor our concept of cooperation, making it more or less equivalent to the term "altruism" in evolutionary biology. Thus, we distinguish "coordination" (joint interactions that are "self-policing" because payoffs are highest if everyone does the same thing) and division of labor (joint action in which payoffs are highest if individuals do different things) from cooperation.

1972; Sober and Wilson 1998). Here we focus on the more conventional notion that selection on variation between fairly large social units counts as group selection. In fact we have in mind, like Darwin and Hamilton, selection among tribes of at least a few hundred people, so we are referring to the *cultural analog* of what is sometimes called inter-demic group selection.

THEORIES OF COOPERATION

We draw evidence about cooperation from many sources. Ethnographic and historical sources include diverse religious doctrines, norms and customs, as well as folk psychology. Anthropologists and historians document an immense diversity of human social organizations, and most of these are accompanied by moral justifications, if often contested ones. Johnson and Earle (2000) provide a good introduction to the vast body of data collected by sociocultural anthropologists. Some important empirical topics are the focus of sophisticated work. For example, the cross-cultural study of commons management is already a well-advanced field (Baland and Platteau 1996), drawing upon the disciplines of anthropology, political science, and economics.

Human Cooperation Is Extensive and Diverse

Human patterns of cooperation are characterized by a number of features:

- *Humans are prone to cooperate, even with strangers.* Many people cooperate in anonymous one-shot Prisoner's Dilemma games (Marwell and Ames 1981) and often vote altruistically (Sears and Funk 1990). People begin contributing substantially to public goods sectors in economic experiments (Ostrom 1998; Falk et al. 2002). Experimental results accord with common experience. Most of us have traveled in foreign cities, even poor foreign cities filled with strange people for whom our possessions and spending money are worth a small fortune, and found risk of robbery and commercial chicanery to be small. These observations apply across a wide spectrum of societies, from small-scale foragers to modern cities in nation states (Henrich 2003).
- *Cooperation is contingent on many things.* Not everyone cooperates. Aid to distressed victims increases substantially if a potential altruist's empathy is engaged (Batson 1991). Being able to discuss a game beforehand and to make promises to cooperate affects success (Dawes et al. 1990). The size of the resource, technology for exclusion and exploitation of the resource, and similar gritty details affect whether cooperation in commons management arises (Ostrom 1990, pp. 202–204). Scientific findings correspond well to personal experience. Sometimes people cooperate enthusiastically, sometimes reluctantly, and sometimes not at all. People vary considerably in their willingness to cooperate even under the same environmental conditions.
- *Institutions matter.* People from different societies behave differently because their beliefs, skills, mental models, values, preferences, and habits have been

inculcated by long participation in societies with different institutions. In repeated play common property experiments, initial defections induce further defections until the contribution to the public good sector approaches zero. However, if players are allowed to exercise strategies they might use in the real world (e.g., to punish those who defect), participation in the commons stabilizes a substantial degree of cooperation (Fehr and Gächter 2002), even in one-shot (nonrepeated) contexts. Strategies for successfully managing commons are generally institutionalized in sets of rules that have legitimacy in the eyes of the participants (Ostrom 1990, Chapter 2). Families, local communities, employers, nations, and governments all tap our loyalties with rewards and punishments and greatly influence our behavior.

- *Institutions are the product of cultural evolution.*[2] Richard Nisbett's group has shown how people's affective and cognitive styles become intimately entwined with their social institutions (Cohen and Vandello 2001; Nisbett and Cohen 1996; Nisbett et al. 2001). Because such complex traditions are so deeply ingrained, they are slow both to emerge and decay. Many commons management institutions have considerable time depths (Ostrom 1990, Chapter 3). Throughout most of human history, institutional change was so slow as to be almost imperceptible by individuals. Today, change is rapid enough to be perceptible. The slow rate of change of institution means that different populations experiencing the same environment and using the same technology often have quite different institutions (Kelly 1985; Salamon 1992).
- *Variation in institutions is huge.* Already with its very short list of societies and games, the experimental ethnography approach has uncovered striking differences (Henrich et al. 2001; Nisbett et al. 2001). Plausibly, design complexity, coordination equilibria, and other phenomena generate multiple evolutionary equilibria and much historical contingency in the evolution of particular institutions (Boyd and Richerson 1992a); consider how different communities, universities, and countries solve the same problems differently.

Evolutionary Models Can Explain the Nature of Preferences and Institutions

These facts constrain the theories we can entertain regarding the causes of human cooperation. For example, high levels of cooperation are difficult to reconcile with the rational choice theorist's usual assumption of self-regarding preferences, and the diversity of institutional solutions to the same environmental problems challenges any theory in which institutions arise directly from universal human nature. The "second generation" bounded rational choice theory, championed by Ostrom (1998), has begun to addresses these challenges from within the rational choice framework. These approaches add a psychological

[2] We refer to cultural evolution as changes in the pool of cultural variants carried by a population of individuals as a function of time and the processes that cause the changes.

basis and institutional constraints to the standard rational choice theory. Experimental studies verify that people do indeed behave quite differently from rational selfish expectations (Fehr and Gächter 2002; Batson 1991). Although psychological and social structures are invoked to explain individual behavior and its variation, an explanation for the origins and variation in psychology and social structure is not part of the theory of bounded rationality.

Evolutionary theory permits us to address the origin of preferences. A number of economists have noted the neat fit between evolutionary theory and economic theory (Hirshleifer 1977; Becker 1976). Evolution explains what organisms want, and economics explains how they should go about getting what they want. Without evolution, preferences are exogenous, to be estimated empirically, but not explained. The trouble with orthodox evolutionary theory is that its predictions are similar to predictions from selfish rationality, as we will see below. At the same time, unvarnished evolutionary theory does do a good job of explaining most other examples of animal cooperation. To do a satisfactory job of explaining why *humans* have the unusual forms of social behavior depicted in our list of stylized facts, we need to appeal to the special properties of *cultural* evolution, and more broadly to theories of culture–gene coevolution (Henrich and Boyd 2001; Richerson and Boyd 1998, 1999; Henrich 2003).

Such evolutionary models have both intellectual and practical payoffs. The intellectual payoff is that evolutionary models link answers to contemporary puzzles to crucial long timescale processes. The most important economic phenomenon of the past 500 years is the rise of capitalist economies and their tremendous impact on every aspect of human life. Expanding the timescale a bit, the most important phenomena of the last 10 millennia are the evolution of evermore complex social systems and ever more sophisticated technology following the origins of agriculture (Richerson et al. 2001). A satisfactory explanation of both current behavior and its variation must be linked to such long-run processes, where the times to reach evolutionary equilibria are measured in millennia or even longer spans of time. More practically, dynamism of the contemporary world creates major stresses on institutions that manage cooperation. Evolutionary theory will often be useful because it will lead to an understanding of how to accelerate institutional evolution to better track rapid technological and economic change. Nesse and Williams (1995) provide an analogy in the context of medical practice.

Evolutionary Models Account for the Processes That Shape Heritable Genetic and Cultural Variation through Time

Evolutionary explanations are *recursive*. Individual behavior results from an interaction of inherited attributes and environmental contingencies. In most species, genes are the main inherited attributes; however, inherited cultural information is also important for humans. Individuals with different inherited attributes may develop different behaviors in the same environment. Every generation, evolutionary processes — natural selection is the prototype — impose environmental effects on individuals as they live their lives. Cumulated over the whole population, these effects change the pool of inherited information, so that the inherited attributes of individuals in the next generation differ, usually subtly, from the attributes in the previous generation. Over evolutionary time, a lineage cycles through the recursive pattern of causal processes once per generation, more or less gradually shaping the gene pool and thus the succession of individuals that draw samples of genes from it. Statistics that describe the pool of inherited attributes (e.g., gene frequencies) are basic state variables of evolutionary analysis. They are what change over time.

Note that in a recursive model, we explain individual behavior and population-level processes in the same model. Individual behavior depends, in any given generation, on the gene pool from which inherited attributes are sampled. The pool of inherited attributes depends in turn upon what happens to a population of individuals as they express those attributes. Evolutionary biologists have a long list of processes that change the gene frequencies, including natural selection, mutation, and genetic drift. However, no organism experiences natural selection. Organisms either live or die, reproduce or fail to reproduce, for concrete reasons particular to the local environment and the organism's own particular attributes. If, in a particular environment, some *types* of individuals do better than others, and if this variation has a heritable basis, then *we* label as "natural selection" the resulting changes in gene frequencies of populations. We use abstract categories like selection to describe such concrete events because we wish to build up some useful generalizations about evolutionary process. Few would argue that evolutionary biology is the poorer for investing effort in this generalizing project.

Although some of the processes that lead to cultural change are very different from those that lead to genetic change, the logic of the two evolutionary problems is very similar. For example, the cultural generation time is short in the case of ideas that spread rapidly, but modeling the evolution of such cultural phenomena (e.g., semiconductor technology) presents no special problems (Boyd and Richerson 1985, pp. 68–69). Similarly, human choices include ones which modify inherited attributes directly, rather indirectly, by natural selection. These "Lamarckian" effects are easily added to models and the models remain evolutionary so long as rationality remains bounded (Young 1998). Such models easily handle continuous (nondiscrete) traits, low-fidelity transmission, and any number of "inferential transformations" that might occur during transmission (Henrich and Boyd 2002; Cavalli-Sforza and Feldman 1981; Boyd and Richerson 1985). The degenerate case of omniscient rationality, of course, needs no recursion because everything happens in the first generation (instantly in a typical rational choice model). Viewed from the perspective of bounded rational choice, evolutionary models are a natural extension of the concept to study how the bounds genetically and culturally inherited elements impose on choice arise (Boyd and Richerson 1993).

Evolution Is Multilevel

Evolutionary theory is always *multilevel*; at a minimum, it keeps track of properties of individuals, like their genotypes, and of the population, such as the frequency of a particular gene. Other levels also may be important. Individual's phenotypes are derived from many genes interacting with each other and the environment. Populations may be structured (e.g., divided into social groups with limited exchanges of members). Thus, evolutionary theories are systemic, integrating every part of biology. In principle, everything that goes into causing change through time plays its proper part in the theory.

This in-principle completeness led Ernst Mayr (1982) to speak of "proximate" and "ultimate" causes in biology. Proximate causes are those that physiologists and biochemists generally treat by asking *how* an organism functions. These are the causes produced by individuals with attributes interacting with environments and producing effects upon them. Do humans use innate cooperative propensities to solve commons problems or do they have only self-interested innate motives? Or are the causes more complex than either proposal? Ultimate causes are evolutionary. The ultimate cause of an organism's behavior is the history of evolution that shaped the gene pool from which our samples of innate attributes are drawn. Evolutionary analyses answer *why* questions. Why do human communities typically solve at least some of the commons dilemmas and other cooperation problems on a scale unknown in other apes and monkeys? Human-reared chimpanzees are capable of many human behaviors, but they nevertheless retain many chimpanzee behaviors and cannot act as full members of a human community (Savage-Rumbaugh and Lewin 1994; Gardner et al. 1989). Thus we know that humans have different innate influences on their behavior than chimpanzees, and these must have arisen in the course of the two species' divergence from our common ancestor.

In Darwinian evolutionary theories, the ultimate sources of cooperative behavior are classically categorized into three evolutionary processes operating at different levels of organization (for a framework unifying these classical divisions, see Henrich 2003):

- *Individual-level selection*. Individuals and the variants they carry are obviously a locus of selection. Selection at this level favors selfish individuals who are evolved to maximize their own survival and reproductive success. Pairs of self-interested actors can cooperate when they interact repeatedly (Axelrod and Hamilton 1981; Trivers 1971). Alexander (1987) argued that such reciprocal cooperation can also explain complex human social systems, but most formal modeling studies make this proposal doubtful (Leimar and Hammerstein 2001; Boyd and Richerson 1989). Still, some version of Alexander's *indirect reciprocity* is perhaps the most plausible alternative to the cultural group selection hypothesis that we champion here. Most such proposals beg the question of how humans and not other animals can take massive advantage of indirect reciprocity (e.g., Nowak and Sigmund 1998). Smith (this volume) proposes to make language the key.[3]

- *Kin selection.* Hamilton's (1964) papers showing that kin should cooperate to the extent that they share genes identical by common descent are one of the theoretical foundations of sociobiology. Kin selection can lead to cooperative social systems of a remarkable scale, as illustrated by the colonies of termites, ants, and some bees and wasps. However, most animal societies are small because individuals have few close relatives. It is the fecundity of insects, and in one case rodents, that permits a single queen to produce huge numbers of sterile workers and hence large, complex societies composed of close relatives (Campbell 1983).
- *Group selection.* Selection can act on any pattern of heritable variation that exists (Price 1972). Darwin's model of the evolution of cooperation by intertribal competition is perfectly plausible, as far as it goes. The problem is that genetic variation between groups other than kin groups is hard to maintain unless the migration between groups is very small or unless some very powerful force generates between-group variation (e.g., Aoki 1982; Slatkin and Wade 1978; Wilson 1983). In the case of altruistic traits, selection will tend to favor selfish individuals in all groups, tending to aid migration in reducing variation between groups. Success of kin selection in accounting for the most conspicuous and highly organized animal societies (except humans) has convinced many, but not all, evolutionary biologists that group selection is of modest importance in nature (for a group selectionist's view of the controversy, see Sober and Wilson 1998). It is also important to note that the problem of maintenance of between-group variation applies *only* to altruistic/cooperative traits, not to social behavior in general. Nearly all evolutionary biologists would agree that group selection is likely to be important for any social interaction with multiple stable equilibria, such as those coordination situations mentioned by Smith (this volume).

We could make this picture much more complex by adding higher and lower levels cross-cutting forms of structure. Many examples from human societies will occur to the reader, such as gender. Indeed, Rice (1996) has elegantly demonstrated that selection on genes expressed in the different sexes sets up a profound conflict of interest between these genes. If female *Drosophila* are prevented from evolving defenses, male genes will evolve that seriously degrade female fitness. The genome is full of such conflicts, usually muted by the

[3] It is not obvious that language potentiates indirect reciprocity. Whereas superficially language may seem to promote the exchange of high-quality information required for indirect reciprocity to favor cooperation, this addition merely changes the question slightly to one of why individuals would cooperate in information sharing; language merely recreates the same public goods dilemma. Lies about hunting success, for example, are difficult to check, and often ambiguous. Among the Gunwinggu (Australian foragers), members of one band often lied to members of other bands about their success to avoid having to share meat (Altman and Peterson 1988).

fact that an individual's genes are forced by the evolved biology of complex organisms to all have an equal shot at being represented in one's offspring. Our own bodies are a group-selected community of genes organized by elaborate "institutions" to ensure fairness in genetic transmission, such as the lottery of meiosis that gives each chromosome of a pair a fair chance at entering the functional gamete (Maynard Smith and Szathmáry 1995; also Chapters 14–18, this volume).

Culture Evolves

In theorizing about human evolution, we must include processes affecting *culture* in our list of evolutionary processes along side those that affect genes. Culture is a system of inheritance. We acquire behavior by imitating other individuals much as we get our genes from our parents. A fancy capacity for high-fidelity imitation is one of the most important derived characters distinguishing us from our primate relatives (Tomasello 1999). We are also an unusually docile animal (Simon 1990) and unusually sensitive to expressions of approval and disapproval by parents and others (Baum 1994). Thus parents, teachers, and peers can rapidly, easily, and accurately shape our behavior compared to training other animals using more expensive material rewards and punishments. Finally, once children acquire language, parents and others can communicate new ideas quite economically. Our own contribution to the study of human behavior is a series of mathematical models of what we take to be the fundamental processes of cultural evolution (e.g., Boyd and Richerson 1985). Application of Darwinian methods to the study of cultural evolution was forcefully advocated by (Campbell 1965, 1975). Cavalli-Sforza and Feldman (1981) constructed the first mathematical models to analyze cultural recursions. The list of processes that shape cultural change includes:

- *Biases.* Humans do not passively imitate whatever they observe. Rather, cultural transmission is biased by decision rules that individuals apply to the variants they observe or try out. The rules behind such selective imitation may be innate or the result of earlier imitation or a mixture of both. Many types of rules might be used to bias imitation. Individuals may try out a behavior and let reinforcement guide acceptance or rejection, or they may use various rules of thumb to reduce the need for costly trials and punishing errors. Rules like "copy successful," "copy the prestigious" (Henrich and Gil-White 2001; Boyd and Richerson 1985) or "copy the majority" (Boyd and Richerson 1985; Henrich and Boyd 1998) allow individuals to acquire rapidly and efficiently adaptive behavior across a wide range of circumstances, and play an important role in our hypothesis about the origins of cooperative tendencies in human behavior (Henrich and Boyd 2001).
- *Nonrandom variation.* Genetic innovations (mutations, recombinations) are random with respect to what is adaptive. Human individual innovation is

guided by many of the same rules that are applied to biasing ready-made cultural alternatives. Bias and learning rules have the effect of increasing the rate of evolution relative to what can be accomplished by random mutation, recombination, and natural selection. We believe that culture originated in the human lineage as an adaptation to the Plio-Pleistocene ice-age climate deterioration which includes much rapid, high-amplitude variation of just the sort that would favor adaptation by nonrandom innovation and biased imitation (Richerson and Boyd 2000a, b).

- *Natural selection.* Since selection operates on any form of heritable variation and imitation and teaching are forms of inheritance, natural selection will influence cultural as well as genetic evolution. However, selection on culture is liable to favor different behaviors than selection on genes. Because we often imitate peers, culture is liable to selection at the sub-individual level, potentially favoring pathogenic cultural variants — selfish memes (Blackmore 1999). On the other hand, rules like conformist imitation have the opposite effect. By tending to suppress cultural variation within groups, such rules protect variation between them, potentially exposing our cultural variation to much stronger group selection effects than our genetic variation (Soltis et al. 1995; Henrich and Boyd 1998). Human patterns of cooperation may owe much to cultural group selection.

Evolutionary Models Are Consistent with a Wide Variety of Theories

Evolutionary theory prescribes a method, not an answer, and a wide range of particular hypotheses can be cast in an evolutionary framework. If population-level processes are important, we can set up a system for keeping track of heritable variation and the processes that change it through time. Darwinism as a method is not at all committed to any particular picture of how evolution works or what it produces. Any sentence that starts with "evolutionary theory predicts" should be regarded with caution.

Evolutionary social science is a diverse field (Borgerhoff Mulder et al. 1997; Laland and Brown 2002). Our own work, which emphasizes an ultimate role for culture and for group selection on cultural variation, is controversial. Many evolutionary social scientists assume that culture is a strictly proximate phenomenon, akin to individual learning (e.g., Alexander 1979), or is so strongly constrained by evolved psychology as to be virtually proximate (Wilson 1998). As Alexander (1979, p.80) puts it, "Cultural novelties do not replicate or spread themselves, even indirectly. They are replicated as a consequence of the behavior of vehicles of gene replication." We think both theory and evidence suggest that this perspective is dead wrong. Theoretical models show that the processes of cultural evolution can behave differently in critical respects from those only including genes, and much evidence is consistent with these models.

Most evolutionary biologists believe that individually costly group-beneficial behavior can only arise as a side effect of individual fitness maximization.

Above, we noted the problems with maintaining variation between groups in theory and the seeming success of alternative explanations. Many, but by no means all, students of evolution and human behavior have followed the argument against group selection forcefully articulated by Williams (1966).[4]

However, *cultural* variation is more plausibly susceptible to group selection than is genetic variation. For example, if people use a somewhat conformist bias in acquiring important social behaviors, variation between groups needed for group selection to operate is protected from the variance-reducing force of migration between groups (Boyd and Richerson 2002; Henrich and Boyd 2001; Boyd and Richerson 1985).

EVOLUTION OF COOPERATIVE INSTITUTIONS

Here we summarize our theory of institutional evolution, developed elsewhere in more detail (Richerson and Boyd 1998, 1999), which is rooted in a mathematical analysis of the processes of cultural evolution and is consistent with much empirical data. We make limited claims for this particular hypothesis, although we think that the thrust of the empirical data as summarized by the stylized facts above are much harder on current alternatives. We make a much stronger claim that a dual gene–culture theory of some kind will be necessary to account for the evolution of human cooperative institutions.

Understanding the evolution of contemporary human cooperation requires attention to two different timescales: First, a long period of evolution in the Pleistocene shaped the innate "social instincts" that underpin modern human behavior. During this period, much genetic change occurred as a result of humans living in groups with social institutions *heavily influenced by culture*, including cultural group selection (Richerson and Boyd 2001). On this timescale, genes and culture *coevolve*, and cultural evolution is plausibly a leading rather than lagging partner in this process. We sometimes refer to the process as "culture–gene coevolution." Then, only about 10,000 years ago, the origins of agricultural subsistence systems laid the economic basis for revolutionary changes in the scale of social systems. Evidence suggests that genetic changes in the

[4] Several prominent modern Darwinians, Hamilton (1975), Wilson (1975, pp. 561–562), Alexander (1987, p. 169), and Eibl-Eibesfeldt (1982), have given serious consideration to group selection as a force *in the special case* of human ultra-sociality. They are impressed, as we are, by the organization of human populations into units which engage in sustained, lethal combat with other groups, not to mention other forms of cooperation. The trouble with a straightforward group selection hypothesis is our mating system. We do not build up concentrations of intrademic relatedness like social insects, and few demic boundaries are without considerable intermarriage. Moreover, the details of human combat are more lethal to the hypothesis of genetic group selection than to the human participants. For some of the most violent groups among simple societies, wife capture is one of the main motives for raids on neighbors, a process that could hardly be better designed to erase genetic variation between groups, and stifle genetic group selection.

social instincts over the last 10,000 years are insignificant. Evolution of complex societies, however, has involved the relatively slow cultural accumulation of institutional "work-arounds" that take advantage of a psychology evolved to cooperate with distantly related and unrelated individuals belonging to the same symbolically marked "tribe" while coping more or less successfully with the fact that these social systems are larger, more anonymous, and more hierarchical than the tribal-scale systems of the late Pleistocene.[5]

Tribal Social Instincts Hypothesis

Our hypothesis is premised on the idea that selection between groups plays a much more important role in shaping culturally transmitted variation than it does in shaping genetic variation. As a result, humans have lived in social environments characterized by high levels of cooperation for as long as culture has played an important role in human development. To judge from the other living apes, our remote ancestors had only rudimentary culture (Tomasello 1999) and lacked cooperation on a scale larger than groups of close kin (Boehm 1999). The difficulty of constructing theoretical models of group selection on genes favoring cooperation matches neatly with the empirical evidence that cooperation in most social animals is limited to kin groups. In contrast, rapid cultural adaptation can lead to ample variation among groups whenever multiple stable social equilibria arise. At least two cultural processes can maintain multiple stable equilibria: (a) conformist social learning and (b) moralistic enforcement of norms. Such models of group selection are relatively powerful because they only require the social, not physical, extinction of groups. Formal theoretical models suggest that conformism is an adaptive heuristic for biasing imitation under a wide variety of conditions (Boyd and Richerson 1985, Chapter 7; Henrich and Boyd 1998; Simon 1990), and both field and laboratory work provide empirical support (Henrich 2001). Models of moralistic punishment (Boyd and Richerson 1992b; Boyd et al. 2003; Henrich and Boyd 2001) lead to multiple stable social equilibria and to reductions in noncooperative strategies if punishment is prosocial. As a consequence, we believe, a growing reliance on cultural evolution led to larger, more cooperative societies among humans over the last 250,000 years or so.

Ethnographic evidence suggests that small-scale human societies are subject to group selection of the sort needed to favor cooperation at a tribal scale. Soltis et al. (1995) analyzed ethnographic data on the results of violent conflicts among Highland New Guinea clans. These conflicts fairly frequently resulted in the social extinction of clans. Many of the details of this process are consistent with cultural group selection. For example, social extinction does not mean

[5] We are aware that much controversy surrounds the use of microevolutionary models to understand macroevolutionary questions. Our thoughts on the issues are summarized in Boyd and Richerson (1992a).

physical elimination of the entire group. Quite the contrary, most people survive defeat but flee as refugees to other groups, into which they are incorporated. This sort of extinction cannot support genetic group selection because so many of the defeated survive and because they would tend to carry their unsuccessful genes into successful groups, rapidly running down variation between groups. However, the effects of conformist cultural transmission combined with moralistic punishment makes between-group cultural variation much less subject to erosion by migration and within-group success of uncooperative strategies than is true in the case of acultural organisms.

The New Guinea cases had little information regarding the cultural variants that might have been favored by cultural group selection. Other examples are more informative in this regard. Kelly (1985) has worked out in detail the way bridewealth customs in the Nuer and Dinka, cattle-keeping people of the Southern Sudan, led to the Nuer maintaining larger tribal systems. These larger tribes, in turn, allowed the Nuer to field larger forces than Dinka in disputes between the two groups. As a result, the Nuer expanded rapidly at the expense of the Dinka in the 19th and early 20th centuries. Here, as in New Guinea, many Dinka lineages survived these fights and were often assimilated into Nuer tribes, a process, again, highly hostile to group selection on genes. The larger ethnographic corpus suggests that the sort of intergroup conflict described by Soltis and Kelly is very common, if not ubiquitous (Keeley 1996; Otterbein 1970). Darwin's picture of a group selection process operating at the level of competing symbolically marked tribal units with the outcome determined by differences in "patriotism, fidelity, obedience, courage, sympathy" and the like can work, but only upon cultural — not genetic — variation for such traits.

Consistent with this argument, evidence suggests that people in late Pleistocene human societies cooperated on a tribal scale (Bettinger 1991, pp. 203–205; Richerson and Boyd 1998). "Tribe" is sometimes used in a technical sense to include only societies with fairly elaborate institutions for organizing cooperation among distantly related and unrelated people. We apply the term to any institution that organizes interfamilial cooperation, even if it is rather simple and the amount of cooperation organized modest. Definitional issues aside, our claim is controversial because the archaeological record permits only weak inferences about social organization and because the spectrum of social organization in ethnographically known hunter-gatherers is very broad (Kelly 1995). At the simple end of the spectrum are "family-level" societies (Johnson and Earle 2000; Steward 1955), such as the Shoshone of the Great Basin and !Kung of the Kalahari. Because these two groups are so simply organized, some scholars used them as an archetypal model for Paleolithic societies (Kelly 1995, p. 2). However, such groups are likely poor examples of the "average" Paleolithic society because they inhabit and have adapted to marginal environments using subsistence strategies quite different from any known from the Paleolithic (R. Bettinger, pers. comm.). Also, we believe that the ethnographic societies used to exemplify the family level of organization actually have tribal institutions of some sophistication.

Much evidence suggests that typical Paleolithic societies were more complex than the Shoshone or the !Kung. Many late Pleistocene societies emphasized big game hunting, often in resource-rich environments, rather than the plant foods emphasized in the marginal environments inhabited by Kalahari foragers and the Shoshone. For example, the Kalahari foragers (along with the Aranda in the Australian Desert) anchor the low end of the distribution with respect to plant biomass found in regions of 23 ethnographically known nomadic foraging groups (Kelly 1995, p. 122). As Steward (1955) reports, big game hunting in ethnographic cases typically involves cooperation on a larger scale than plant collecting and small game hunting; thus we should expect societies in the late Pleistocene to be more, not less, socially complex than the !Kung and Shoshone. In any case we think it an error to try to identify an archetypal Pleistocene society; most likely last glacial societies spanned as large or larger a spectrum of social organization as ethnographically known cases. Art and settlement size (several hundred people) at upper Paleolithic sites in France and Spain suggest that these societies were toward the complex end of the foraging spectrum (Price and Brown 1985). In Czechoslovakia, the palisades and large housing structures look much more like the Northwest Coast Indians or Big-Men social forms of New Guinea than the !Kung or Shoshone (Johnson and Earle 2000).

Moreover, despite the marginality of their environment, the archetypal family-level societies do have tribal-scale institutions for dealing with environmental uncertainty (Wiessner 1984). For example, the Shoshonean peoples of the North American Great Basin foraged for most of the year in nuclear family units. Resources in the Basin were not only sparse but widely scattered, militating against aggregation into larger units during much of the year. Although such bands were generally politically autonomous, they were at least tenuously linked into larger units. In regard to the Shoshoneans, Steward (1955, p. 109) remarks the "... nuclear families have always co-operated with other families in various ways. Since this is so, the Shoshoneans, like other fragmented family groups, represent the family level of sociocultural integration only in a relative sense." Winter encampments of 20 or 30 families were the largest aggregations among Shoshoneans; however, these were not formal organizations but rather aggregations of convenience. Aside from visiting, some cooperative ventures, such as dances (fandangos), rabbit drives, and occasional antelope drives, were organized during winter encampments. The number of families that a given family might camp with over a period of years was also not fixed, although people preferred to camp with people speaking the same dialect (R. Bettinger, pers. comm.). Steward's picture of the simplicity of Shoshone has been challenged. Thomas (1986, p. 278) observes that, at best, Steward's characterization applied only to limiting cases, as, indeed, his frank use of them to imperfectly exemplify an ideal type suggests. Murphy and Murphy (1986), citing the case of the Northern Shoshone and Bannock, argue that the unstructured fluidity of Shoshonean society conceals a sophisticated adaptation to the sparse and uncertain resources of the Great Basin. The Shoshoneans maintained peace among themselves over a very large region, enabling families and small groups of families to move over vast distances in response to local feast and famine. When local resources permitted and necessity required, they were able to assemble considerable numbers of people for collective purposes. Murphy and Murphy cite the formation of war parties numbering in the hundreds to contest bison hunting areas with the Blackfeet. Indeed, the Shoshone and their relatives were relatively recent immigrants to the Great Basin who pushed out societies that were probably socially more complex but less well adapted to the sparse Great Basin environment (Bettinger and Baumhoff 1982). Murphy and Murphy summarize by saying "the Shoshone are a 'people' in the truest sense of the word." Compared to our great ape relatives, and presumably our remoter ancestors, Shoshonean families maintained generally friendly relations with a rather large group of other families, could readily strike up cooperative relations with strangers of their ethnic group, and organized cooperative activities on a considerable scale.

We believe that the human capacity to live in larger-scale forms of tribal social organization evolved through a coevolutionary ratchet generated by the interaction of genes and culture. Rudimentary cooperative institutions favored genotypes that were better able to live in more cooperative groups. Those individuals best able to avoid punishment and acquire the locally relevant norms were more likely to survive. At first, such populations would have been only slightly more cooperative than typical nonhuman primates. However, genetic changes, leading to moral emotions like shame and a capacity to learn and internalize local practices, would allow the cultural evolution of more sophisticated institutions that in turn enlarged the scale of cooperation. These successive rounds of coevolutionary change continued until eventually people were equipped with capacities for cooperation with distantly related people, emotional attachments to symbolically marked groups, and a willingness to punish others for transgression of group rules. Mechanisms by which cultural institutions might exert forces tugging in this direction are not far to seek. People are likely to discriminate against genotypes that are incapable of conforming to cultural norms (Richerson and Boyd 1989; Laland et al. 1995). People who cannot control their self-serving aggression ended up exiled or executed in small-scale societies and imprisoned in contemporary ones. People whose social skills embarrass their families will have a hard time attracting mates. Of course, selfish and nepotistic impulses were never entirely suppressed; our genetically transmitted evolved psychology shapes human cultures, and as a result cultural adaptations often still serve the ancient imperatives of inclusive genetic fitness. However, cultural evolution also creates new selective environments that *build cultural imperatives into our genes.*

Paleoanthropologists believe that human cultures were essentially modern by the Upper Paleolithic, 50,000 years ago (Klein 1999, Chapter 7) if not much earlier (McBrearty and Brooks 2000). Thus, even if the cultural group selection process began as late as the Upper Paleolithic, such social selection could easily have had extensive effects on the evolution of human genes through this process. More likely, Upper Paleolithic societies were the culmination of a long period of coevolutionary increases in a tendency toward tribal social life.[6]

We suppose that the resulting "tribal instincts" are something like principles in the Chomskian linguists' "principles and parameters" view of language (Pinker 1994). Innate principles furnish people with basic predispositions, emotional capacities, and social dispositions that are implemented in practice through highly variable cultural institutions, the parameters. People are innately prepared to act as members of tribes, but culture tells us how to recognize who belongs to our tribes, what schedules of aid, praise, and punishment are due to tribal fellows, and how the tribe is to deal with other tribes: allies, enemies, and clients. The division of labor between innate and culturally acquired elements is poorly understood, and theory gives little guidance about the nature of the synergies and tradeoffs that must regulate the evolution of our psychology (Richerson and Boyd 2000a). The fact that human-reared apes cannot be socialized to behave like humans guarantees that some elements are innate. Contrariwise, the diversity and sometimes-rapid change of social institutions guarantees that much of our social life is governed by culturally transmitted rules, skills, and even emotions. We beg the reader's indulgence for the necessarily brief and assertive nature of our argument here. The rationale and ethnographic support for the tribal instincts hypothesis are laid out in more detail in Richerson and Boyd (1998, 1999); for a review of the broad spectrum of empirical evidence supporting the hypothesis, see Richerson and Boyd (2001).

Work-around Hypothesis

Contemporary human societies differ drastically from the societies in which our social instincts evolved. Pleistocene hunter-gatherer societies were comparatively small, egalitarian, and lacking in powerful institutionalized leadership. By contrast, modern societies are large, inegalitarian, and have coercive leadership institutions (Boehm 1993). If the social instincts hypothesis is correct, our innate social psychology furnishes the building blocks for the evolution of complex social systems, while simultaneously constraining the shape of these systems (Salter 1995). To evolve large-scale, complex social systems, cultural evolutionary processes, driven by cultural group selection, take advantage of

[6] It would be a mistake to assume that complex technology is a prerequisite for tribal-level forms of social organization. At the time of European discovery, the Tasmanians had a technology substantially simpler than many upper Paleolithic peoples: they lacked bone tools, composite spears, bows, arrows, spear throwers, and fish hooks, etc. Yet, they lived in multiband groups, which controlled territories. Intertribal trade, warfare, and raiding were all commonplace (Jones 1995). The last 4,000 years of the Tasmania archaeological record do not look much different from many middle Paleolithic sites.

whatever support these instincts offer. For example, families willingly take on the essential roles of biological reproduction and primary socialization, reflecting the ancient and still powerful effects of selection at the individual and kin level. At the same time, cultural evolution must cope with a psychology evolved for life in quite different sorts of societies. Appropriate larger-scale institutions must regulate the constant pressure from smaller groups (coalitions, cabals, cliques) to subvert rules favoring large groups. To do this cultural evolution often makes use of "work-arounds." It mobilizes the tribal instincts for new purposes. For example, large national and international (e.g., great religions) institutions develop ideologies of symbolically marked inclusion that often fairly successfully engage the tribal instincts on a much larger scale. Military and religious organizations (e.g., Catholic Church), for example, dress recruits in identical clothing (and haircuts) loaded with symbolic markings, and then subdivide them into small groups with whom they eat and engage in long-term repeated interaction. Such work-arounds are often awkward compromises, as is illustrated by the existence of contemporary societies handicapped by narrow, destructive loyalties to small tribes (West 1941) and even to families (Banfield 1958). In military and religious organizations excessive within-group loyalty often subverts higher-level goals. If this picture of the innate constraints on current institutional evolution is correct, it is evidence for the existence of tribal social instincts that buttress the uncertain inferences from ethnography and archaeology about late Pleistocene societies. Complex societies are, in effect, grand natural social-psychological experiments stringently test the limits of our innate dispositions to cooperate. We expect the social institutions of complex societies to simulate life in tribal-scale societies in order to generate cooperative "lift." We also expect that complex institutions will accept design compromises to achieve such "lift," which would be unnecessary if innate constraints of a specifically tribal structure were absent.

Coercive Dominance

The cynics' favorite mechanism for creating complex societies is command backed up by force. The conflict model of state formation has this character (Carneiro 1970), as does Hardin's (1968) recipe for commons management.

Elements of coercive dominance are no doubt necessary to make complex societies work. Tribally legitimated self-help violence is a limited and expensive means of altruistic coercion. Complex human societies have to supplement the moralistic solidarity of tribal societies with formal police institutions. Otherwise, the large-scale benefits of cooperation, coordination, and division of labor would cease to exist in the face of selfish temptations to expropriate them by individuals, nepotists, cabals of reciprocators, organized predatory bands, greedy capitalists, and classes or castes with special access to means of coercion. At the same time, the need for organized coercion as an ultimate sanction creates roles, classes, and subcultures with the power to turn coercion to narrow advantage.

Social institutions of some sort must police the police so that they will act in the larger interest to a measurable degree. Indeed, Boehm (1993) notes that the egalitarian social structure of simple societies is itself an institutional achievement by which the tendency of some to try to dominate others on the typical primate pattern is frustrated by the ability of the individuals who would be dominated to collaborate to enforce rules against dominant behavior. Such policing is never perfect and, in the worst cases, can be very poor. The fact that leadership in complex systems always leads to at least some economic inequality suggests that narrow interests, rooted in individual selfishness, kinship, and, often, the tribal solidarity of the elite, always exert an influence. The use of coercion in complex societies offers excellent examples of the imperfections in social arrangements traceable to the ultimately irresolvable tension of more narrowly selfish and more inclusively altruistic instincts.

While coercive, exploitative elites are common enough, we suspect that no complex society can be based purely on coercion for two reasons: (a) coercion of any great mass of subordinates requires that the elite class or caste be itself a complex, cooperative venture; (b) defeated and exploited peoples seldom accept subjugation as a permanent state of affairs without costly protest. Deep feelings of injustice generated by manifestly inequitable social arrangements move people to desperate acts, driving the cost of dominance to levels that cripple societies in the short run and often cannot be sustained in the long run (Insko et al. 1983; Kennedy 1987). Durable conquests, such as those leading to the modern European national states, Han China, or the Roman Empire, leaven raw coercion with other institutions. The Confucian system in China and the Roman legal system in the West were far more sophisticated institutions than the highly coercive systems sometimes set up by predatory conquerors and even domestic elites.

Segmentary Hierarchy

Late Pleistocene societies were undoubtedly segmentary in the sense that supra-band ethnolinguistic units served social functions. The segmentary principle can serve the need for more command and control by hardening up lines of authority without disrupting the face-to-face nature of proximal leadership present in egalitarian societies. The Polynesian ranked lineage system illustrates how making political offices formally hereditary according to a kinship formula can help deepen and strengthen a command and control hierarchy (Kirch 1984). A common method of deepening and strengthening the hierarchy of command and control in complex societies is to construct a nested hierarchy of offices, using various mixtures of ascription and achievement principles to staff the offices. Each level of the hierarchy replicates the structure of a hunting and gathering band. A leader at any level interacts mainly with a few near-equals at the next level down in the system. New leaders are usually recruited from the ranks of subleaders, often tapping informal leaders at that level. As Eibl-Eibesfeldt (1989) remarks, even high-ranking leaders in modern hierarchies adopt much of the humble headman's deferential approach to leadership. Henrich and Gil-White's (2001) work on prestige provides a coevolutionary explanation for this phenomenon.

The hierarchical nesting of social units in complex societies gives rise to appreciable inefficiencies (Miller 1992). In practice, brutal sheriffs, incompetent lords, venal priests, and their ilk degrade the effectiveness of social organizations in complex societies. Squires (1986) dissects the problems and potentials of modern hierarchical bureaucracies to perform consistently with leaders' intentions. Leaders in complex societies must convey orders downward, not just seek consensus among their comrades. Devolving substantial leadership responsibility to subleaders far down the chain of command is necessary to create small-scale leaders with face-to-face legitimacy. However, it potentially generates great friction if lower-level leaders either come to have different objectives than the upper leadership or are seen by followers as equally helpless pawns of remote leaders. Stratification often creates rigid boundaries so that natural leaders are denied promotion above a certain level, resulting in inefficient use of human resources and a fertile source of resentment to fuel social discontent.

On the other hand, failure to articulate properly tribal-scale units with more inclusive institutions is often highly pathological. Tribal societies often must live with chronic insecurity due to intertribal conflicts. One of us once attended the *Palio,* a horse race in Siena in which each ward, or *contrada,* in this small Tuscan city sponsors a horse. Voluntary contributions necessary to pay the rider, finance the necessary bribes, and host the victory party amount to a half a million dollars. The *contrada* clearly evoke the tribal social instincts: they each have a totem — the dragon, the giraffe, etc., special colors, rituals, and so on. The race excites a tremendous, passionate rivalry. One can easily imagine medieval Siena in which swords clanged and wardmen died, just as they do or did in warfare between New Guinea tribes (Rumsey 1999), Greek city-states (Runciman 1998), inner city street gangs (Jankowski 1991), and ethnic militias.

Exploitation of Symbolic Systems

The high population density, division of labor, and improved communication made possible by the innovations of complex societies increased the scope for elaborating symbolic systems. The development of monumental architecture to serve mass ritual performances is one of the oldest archaeological markers of emerging complexity. Usually an established church or less formal ideological umbrella supports a complex society's institutions. At the same time, complex societies exploit the symbolic ingroup instinct to delimit a quite diverse array of culturally defined subgroups, within which a good deal of cooperation is routinely achieved. Ethnic group-like sentiments in military organizations are often most strongly reinforced at the level of 1,000–10,000 or so men (British and German regiments, U.S. divisions; Kellett 1982). Typical civilian symbolically marked units include nations, regions (e.g., Swiss cantons), organized tribal elements (Garthwaite 1993), ethnic diasporas (Curtin 1984), castes (Srinivas 1962; Gadgil and Guha 1992), large economic enterprises (Fukuyama 1995), and civic organizations (Putnam et al. 1993).

How units as large as modern nations tap into the tribal social instincts is an interesting issue. Anderson (1991) argues that literate communities, and the social organizations revolving around them (e.g., Latin literates and the Catholic Church), create "imagined communities," which in turn elicit significant commitment from members of the community. Since tribal societies were often large enough that some members were not known personally to any given person, common membership would sometimes have to be established by the mutual discovery of shared cultural understandings, as simple as the discovery of a shared language in the case of the Shoshone. The advent of mass literacy and print media — Anderson stresses newspapers — made it possible for all speakers of a given vernacular to have confidence that all readers of the same or related newspapers share many cultural understandings, especially when organizational structures such as colonial government or business activities really did give speakers some institutions in common. Nationalist ideologists quickly discovered the utility of newspapers for building of imagined communities, typically several contending variants of the community, making nations the dominant quasi-tribal institution in most of the modern world.

Many problems and conflicts revolve around symbolically marked groups in complex societies. Official dogmas often stultify desirable innovations and lead to bitter conflicts with heretics. Marked subgroups often have enough tribal cohesion to organize at the expense of the larger social system. The frequent seizure of power by the military in states with weak institutions of civil governance is probably a by-product of the fact that military training and segmentation, often based on some form of patriotic ideology, are conducive to the formation of *relatively* effective large-scale institutions. Wherever groups of people interact routinely, they are liable to develop a tribal ethos. In stratified societies, powerful groups readily evolve self-justifying ideologies that buttress treatment of subordinate groups, ranging from neglectful to atrocious. American White Southerners had elaborate theories to justify slavery, and pioneers everywhere found the brutal suppression of Indian societies legitimate and necessary. The parties and interest groups that vie to sway public policy in democracies have well-developed rationalizations for their selfish behavior. A major difficulty with loyalties induced by appeals to shared symbolic culture is the very language-like productivity possible with this system. Dialect markers of social subgroups emerge rapidly along social fault-lines (Labov 2001). Charismatic innovators regularly launch new belief and prestige systems, which sometimes make radical claims on the allegiance of new members, sometimes make large claims at the expense of existing institutions, and sometimes grow explosively. Contrariwise, larger loyalties can arise, as in the case of modern nationalisms overriding smaller-scale loyalties; sometimes for the better, sometimes for the worse. The ongoing evolution of social systems can develop in unpredictable, maladaptive directions by such processes (Putnam 2000). The worldwide growth of fundamentalist sects that challenge the institutions of modern states is a contemporary example (Marty and Appleby 1991). If T. Wolfe (1965) is right, mass media can be the basis of a rich diversity of imagined sub-communities using such vehicles as specialized magazines, newsletters, and web sites. The potential of deviant subgroups, such as sectarian terrorist organizations, to use modern media to create small but highly motivated imagined communities is an interesting variant on Anderson's theory. Ongoing cultural evolution is impossible to control wholly in the larger interest, at least impossible to control completely, and forbidding free evolution tends to deprive societies of the "civic culture" that spontaneously produces so many collective benefits.

Legitimate Institutions

In small-scale egalitarian societies, individuals have substantial autonomy, considerable voice in community affairs, and can enforce fair, responsive — even self-effacing — behavior and treatment from leaders (Boehm 1999). At their most functional, symbolic institutions, a regime of tolerably fair laws and customs, effective leadership, and smooth articulation of social segments can roughly simulate these conditions in complex societies. Rationally administered bureaucracies, lively markets, the protection of socially beneficial property rights, widespread participation in public affairs, and the like provide public and private goods efficiently, along with a considerable amount of individual autonomy. Many individuals in modern societies feel themselves part of culturally labeled tribal-scale groups, such as local political party organizations, that have influence on the remotest leaders. In older complex societies, village councils, local notables, tribal chieftains, or religious leaders often hold courts open to humble petitioners. These local leaders, in turn, represent their communities to higher authorities. To obtain low-cost compliance with management decisions, ruling elites have to convince citizens that these decisions are in the interest of the larger community. As long as most individuals trust that existing institutions are reasonably legitimate and that any felt needs for reform are achievable by means of ordinary political activities, there is considerable scope for large-scale collective social action.

Legitimate institutions, however, and trust of them, are the result of an evolutionary history and are neither easy to manage nor engineer. Social distance between different classes, castes, occupational groups, and regions is objectively great. Narrowly interested tribal-scale institutions abound in such societies. Some of these groups have access to sources of power that they are tempted to use for parochial ends. Such groups include, but are not restricted to, elites. The police may abuse their power. Petty administrators may victimize ordinary citizens and cheat their bosses. Ethnic political machines may evict historic elites from office but use chicanery to avoid enlarging their coalition.

Without trust in institutions, conflict replaces cooperation along fault lines where trust breaks down. Empirically, the limits of the trusting community define the universe of easy cooperation (Fukuyama 1995). At worst, trust does not extend outside family (Banfield 1958), and potential for cooperation on a larger scale is almost entirely foregone. Such communities are unhappy as well as poor. Trust varies considerably in complex societies, and variation in trust seems to be the main cause of differences in happiness across societies (Inglehart and Rabier 1986). Even the most efficient legitimate institutions are prey to manipulation by small-scale organizations and cabals, the so-called special interests of modern democracies. Putnam et al.'s (1993) contrast between civic institutions in Northern and Southern Italy illustrates the difference that a tradition of functional institutions can make. The democratic form of the state, pioneered by Western Europeans in the last couple of centuries, is a powerful means of creating generally legitimate institutions. Success attracts imitation all around the world. The halting growth of the democratic state in countries ranging from Germany to sub-Saharan Africa is testimony that legitimate institutions cannot be drummed up out of the ground just by adopting a constitution. Where democracy has struck root outside of the European cultural orbit, it is distinctively fitted to the new cultural milieu, as in India and Japan.

CONCLUSIONS

The processes of cultural evolution quite plausibly led to group selection being a more powerful force on cultural rather than genetic variation. The cultural system of inheritance probably arose in the human lineage as an adaptation to the increasingly variable environments of the recent past (Richerson and Boyd 2000a, b). Theoretical models show that the specific structural features of cultural systems, such as conformist transmission, have ordinary adaptive advantages. We imagine that these adaptive advantages favored the capacity for a system that could respond rapidly and flexibly to environmental variation in an ancestral creature that was not particularly cooperative. As a by-product, cultural evolution happened to favor large-scale cooperation. Over a long period of coevolution, cultural pressures reshaped "human nature," giving rise to innate adaptations to living in tribal-scale social systems. Humans became prepared to use systems of legitimate punishment to lower the fitness of deviants, for example. We believe that the cultural explanation for human cooperation is in accord with much evidence, as summarized by stylized facts about human cooperation with which we introduced our remarks. More detailed surveys of the concordance of our conjectures with various bodies of data may be found in Richerson and Boyd (1999, 2001) and Richerson et al. (2002).

Regardless of the fate of any particular proposals, we think that explanations of human cooperation have to thread some rather tight constraints. They have to somehow finesse the awkward fact that humans, at least partly because of our ability to cooperate with distantly related people in large groups, are a huge success yet quite unique in our style of social life. If a mechanism like indirect reciprocity works, why have not many social species used it to extend their range of cooperation? If finding self-reinforcing solutions to coordination games is mostly what human societies are about, why do not other animals have massive coordination-based social systems? If reputations for pairwise cooperation are easy to observe or signal (but unexploitable by deceptive defectors), why have we found no other complex animal societies based on this principle? By contrast, we do find plenty of complex animal societies built on the principle of inclusive fitness.

The unique pattern of cooperation of our species suggests that human cooperation is likely to derive from some other unique feature or features of human life. Advanced capacities for social learning are also unique to humans; thus culture is, *prima facie*, a plausible key element in the evolution of human cooperation. Our argument depends upon the existence of culture and group selection on cultural variation. Since sophisticated culture is unique to humans, we do not expect this mechanism to operate in other species. Ours is not the only hypothesis that passes this basic test. For example, E. Smith's (this volume) signaling hypothesis depends upon language, another unique feature of the human species. E. Hagen made a similar proposal in his comment on our background paper. He argued that the inventiveness of humans combined with language as a cheap communication device adapts us to solve problems of cooperation. We think that hypotheses in this vein, like Alexander's proposed indirect bias mechanism, cannot be decisively rejected, but they are far from completely specified. What is it that biases invention and cheap talk in favor of cooperative rather than selfish ends? The intuition that cheap talk, symbolic rewards, and clever institutions are in themselves sufficient to explain human cooperation probably comes from the common experience that people do find it rather easy to use such devices to cooperate (e.g., Ostrom et al. 1994). The difficult question is whether these are backed up by unselfish motives on the part of at least some people. A literal interpretation of experiments such as those of Fehr and Gächter (2002) and Batson (1991) suggests that unselfish motives play important roles. However, unselfish motives may be a proximal evolutionary result of an ultimate indirect reciprocity sort of evolutionary process rather than the result of a group selection mechanism. Those who attempt deception in a world of clever cooperators may simply expose their lack of cleverness, so that the best strategy is an unfeigned willingness to cooperate. The data that cultural group selection is an appreciable process (Soltis et al. 1995) is also not definitive, since it could be weak relative to some competing process of the indirect reciprocity sort.

Another complication is that hypotheses leaning on language, technology, and intelligence are appealing to phenomena with considerable cultural content. The evolution of technology and the diffusion of innovations are cultural processes that depend upon institutions and a sophisticated social psychology (Henrich 2001). Both the cultural and genetic evolution of our cognitive capacities (some of which gave rise to language) likely emerged from a culture–gene coevolutionary process (Henrich and McElreath 2002; Tomasello 1999). Thus, these hypotheses are not, we submit, clean alternatives to the cultural group selection hypothesis, absent further specification. In the future, we expect that competing hypotheses will be developed in sufficient detail that more precise comparative empirical tests will be possible.

For example, even if innatist linguists are correct that much of what we need to know to speak is innate, we wonder why more is not innate? Why is it that mutually unintelligible languages arise so rapidly? Would not we be better off if everyone spoke the some common entirely innate language? Not necessarily. Very often people from distant places are likely to have evolved different ways of doing things that are adaptive at home but not abroad. Similarly, avoiding listening to people is a wise idea if they are proposing a behavior deviant from locally prevailing coordination equilibria. Cultural evolution can run up adaptive *barriers* to communication quite readily if listening to foreigners makes you liable to acquire erroneous ideas (McElreath et al. 2003). Dialect evolution seems to be a highly nuanced system for regulating communication within languages as well as between them, although the adaptive significance of dialect is hardly well worked out (Labov 2001). Interestingly, in McElreath et al.'s model, using a symbolic signal to express a willingness to cooperate cannot support the evolution of a symbolic marker of group membership because defectors as well as potential cooperators will be attracted by the signal. A symbolic system can be used to communicate intention to cooperate only if potential cooperative partners can exchange trustworthy signals. Once symbolic markers became sufficiently complex as to be unfakable by defectors *and* a sufficiently large pool of relatively anonymous but trustworthy signalers exist, *then* cheap signals will be useful. Dialect is difficult to fake although cheap to use, and once some level of cooperation on a proto-tribal scale was possible, proto-languages might have come under selection to create unfakable signals of group membership that imply an intention to cooperate. We suspect that language could only have evolved in concert with a measure of trust of other speakers rather than being an unaided generator of trust. To the extent that cooperation is the game, one has no interest in listening to speakers whose messages are self-serving. Think of how annoying we find telemarketer's speech acts. Sociolinguists make much of the concept that speech is a cooperative system and argue that the empirical structure of conversation is consistent with this assumption (Wardhaugh 1992). Language seems to presuppose cooperation as much as it in turn facilitates cooperation.

That technology, like language, is one of the major components of the human adaptation is undeniable. It opens up opportunities to gain advantage to cooperation in hunting and defense, and to exploit the possibilities of the division of labor. What is less well understood is the extent to which technology is likely a product of large-scale social systems. Henrich (submitted) has analyzed models of the "Tasmanian Effect." At the time of European contact, the Tasmanians had the simplest toolkit ever recorded in an extant human society; it was, for example, substantially simpler than the toolkits of ethnographically known foragers in the Kalahari and Tierra del Fuego, as well as those associated with human groups from the Upper Paleolithic. Archaeological evidence indicates that Tasmanian simplicity resulted from both the gradual loss of items from their own pre-Holocene toolkit and the failure to develop many of the technologies that subsequently arose only 150 km to the north in Australia. The loss likely began after the Bass Strait was flooded by rising post-glacial sea levels (Jones 1995). Henrich's analysis indicates that imperfect inference during social learning, rather than stochastic loss due to drift-like effects, is the most likely reason for this loss. This suggests that to maintain an equilibrium toolkit as complex as those of late Pleistocene hunter-gatherers likely required a rather large population of people who interacted fairly freely so that rare highly skilled performances, spread by selective imitation, could compensate for the routine loss of skills due to imperfect inference. Neanderthals and perhaps other archaic human populations had large brains but simple toolkits. The Tasmanian Effect may explain why. Archaeology suggests that Neanderthal population densities were lower than the modern humans that replaced them in Europe and that they had less routine contact with their neighbors, as evidenced by shorter distance movement of high-quality raw materials from their sources compared to modern humans (Klein 1999).

The proposal that human intelligence is at the root of human cooperation is difficult to evaluate because of the ambiguity in what we might mean by intelligence in a comparative context (Hinde 1970, pp. 659–663). As the Tasmanian Effect illustrates, *individual* human intelligence is only a part of, and perhaps only a small part, being able to create complex adaptive behaviors. In fact, we think "intelligence" plays little role in the emergence of many of human complex adaptations. Instead, humans seem to depend upon socially learned strategies to finesse the shortcomings of their cognitive capabilities (Nisbett and Ross 1980). The details of human cognitive abilities apparently vary substantially across cultures because culturally transmitted cognitive styles differ (Nisbett et al. 2001). Although we share the common intuition that humans are individually more intelligent than even our very clever fellow apes, we are not aware of any experiments that sufficiently control for our cultural repertoires to be sure that it is correct. The concept of "intelligence" in individual humans perhaps makes little sense apart from their cultural repertoires: humans are smart in part because they can bring a variety of "cultural tools" (e.g., numbers, symbols, maps, various kinematic models) to bear on problems. A hunter-gatherer would seem an incredibly stupid college professor, but college professors would seem equally dense if forced to try to survive as hunter-gatherers (a few knowledgeable anthropologists aside). Even abilities as seemingly basic as those related directly to visual perception vary across cultures (Segall et al. 1966). Second, *intelligence* implies a means to an end, not an end in itself. Individual intelligence ought to serve the ends of both cooperation and defection. We suspect that actually defection, requiring trickery and deception, is better served by intelligence than cooperation. Game theorists assuming perfect, but selfish, rationality predict that humans should defect in the one-shot anonymous Prisoner's Dilemma, just as evolutionary biologists predict that dumb beasts using evolved predispositions will. Whiten and Bryne (1997) characterized our social intelligence as "Machiavellian," implying that it does indeed serve deception equally with honesty. However, just as humans punish altruistically, they seem also to exert their political intelligence altruistically (e.g., Sears and Funk 1990), biasing the evolution of institutions accordingly. On the basis of our brain size compared to other apes Dunbar (1992) predicts that human groups ought to number around 50. Hunter-gatherer co-residential bands do number around 50, but culturally transmitted *institutions* web together bands to create tribes typically numbering a few hundred to a few thousand people, as we have seen. Human political systems do seem to exceed in scale anything predicted on the basis of enhanced Machiavellian talents (supposing that such talents can on average increase social scale at all). The institutional basis of these systems is not far to seek. For example, Wiessner (1984) describes how institutions of ceremonial exchange of gifts knit the famous !Kung San bands into a much larger risk pooling cooperative. Australian aboriginal groups show similar functional patterns, which are built out of quite different and substantially more elaborate sets of cultural practices (Peterson 1979). Underpinning such individual-to-individual bond making is likely the kind of generalized trust that co-ethnics have for one another. If Murphy and Murphy (1986) are correct about the Northern Shoshone, a society of thousands constituted a functional "people" engaging in mutual aid in a hostile and uncertain environment on the basis of little more than a common language. In his classic ethnography of the Nuer, Evans-Pritchard (1940) describes how simple tribal institutions can knit herding people into tribes numbering tens of thousands, much larger than was possible among hunter-gatherers. The size of hunter-gatherer societies was evidently limited by low population density, not by their relatively unsophisticated institutions. Third, Henrich and Gil-White (2001) propose that human prestige systems are an adaptation to facilitate cultural transmission. Social learning means that the returns to effort in individual learning potentially result in gains for many subsequent social learners who do not have to "reinvent the wheel." If extra individual effort in acquiring better ideas pays off in prestige and if prestige leads to fitness advantages, then the social returns to effortful individual learning will in part be reflected in private returns to individual learners. Group selection on prestige systems may further enlarge the returns to investment individual learning and bring returns up to a level that reflects the group optimum amount of effort in individual learning. If this mechanism operates, human intelligence may have been enhanced by social selection emanating from institutions of prestige.[7]

[7] Similarly, as Smith (this volume) notes, Hawkes hypothesizes that men contribute to hunting success to "show off" and that showing off earns men reproductive success in terms of sexual favors from women. Contrary to what Hawkes supposes, this system is a possible focus of cultural group selection. In many hunter-gatherer groups, meat is

We propose that group selection on cultural variation is at the heart of human cooperation, but we certainly recognize that our sociality is a complex system that includes many linked components. Surely, without punishment, language, technology, individual intelligence and inventiveness, ready establishment of reciprocal arrangements, prestige systems, and solutions to games of coordination, our societies would take on a distinctly different cast, to say the least. Human sociality no doubt has a number of components that were necessary to its evolution and are necessary to its current functions. If such is the case, prime mover explanations giving pride of place to a single mechanism are vain to seek. Thus, a major constraint on explanations of human sociality is its systemic structure. Explanations have to have a plausible historical sequence tracing how the currently interrelated parts evolved, perhaps piecemeal. And explanations have to account for the current functional and dysfunctional properties of human social systems. We are far from having completed this task.

REFERENCES

Alexander, R.D. 1979. Darwinism and Human Affairs. The Jessie and John Danz Lectures. Seattle: Univ. of Washington Press.

Alexander, R.D. 1987. The Biology of Moral Systems: Foundations of Human Behavior. Hawthorne, NY: Aldine de Gruyter.

Altman, J., and N. Peterson. 1988. Rights to game and rights to cash among contempory Australian hunter-gatherers. In: Hunters and Gatherers, vol. 2: Property, Power, and Ideology, ed. T.R. Ingold, D. Riches, and J. Woodburn, pp. 75–94. Oxford: Berg.

Anderson, B.R.O'G. 1991. Imagined Communities: Reflections on the Origin and Spread of Nationalism. Rev. and extended ed. London: Verso.

Aoki, K. 1982. A condition for group selection to prevail over counteracting individual selection. *Evolution* **36**:832–842.

Axelrod, R., and W.D. Hamilton. 1981. The evolution of cooperation. *Science* **211**:1390–1396.

Baland, J.-M., and J.P. Platteau. 1996. Halting Degradation of Natural Resources: Is There a Role for Rural Communities? Oxford: Oxford Univ. Press.

Banfield, E.C. 1958. The Moral Basis of a Backward Society. Glencoe, IL: Free Press.

Batson, C.D. 1991. The Altruism Question: Toward a Social Psychological Answer. Hillsdale, NJ: Erlbaum.

Baum, W.B. 1994. Understanding Behaviorism: Science, Behavior, and Culture. New York: Harper Collins.

Becker, G.S. 1976. Altruism, egoism, and genetic fitness: Economics and sociobiology. *J. Econ. Lit.* **14**:817–826.

---

very widely shared and hunters often do not control its distribution. Personal favors granted to a successful hunter as recompense for effort will benefit all who share his kills. Showing that individuals who contribute heavily to the common good are rewarded is not evidence that group-selected effects are absent. In the end, group selection can succeed only if altruistic individuals on average do better than selfish ones. The fact that hunters are not allowed to bargain with consumers of their kills and yet are rewarded by consumers anyway is at least as consistent with the operation of group selection as with a competing individualist explanation.

Bettinger, R.L. 1991. Hunter-gatherers: Archaeological and Evolutionary Theory. Interdisciplinary Contributions to Archaeology. New York: Plenum.

Bettinger, R.L., and M.A. Baumhoff. 1982. The numic spread: Great Basin cultures in competition. *Am. Antiq.* **47**:485–503.

Blackmore, S. 1999. The Meme Machine. Oxford: Oxford Univ. Press.

Boehm, C. 1993. Egalitarian behavior and reverse dominance hierarchy. *Curr. Anthro.* **34**:227–254.

Boehm, C. 1999. Hierarchy in the Forest: The Evolution of Egalitarian Behavior. Cambridge, MA: Harvard Univ. Press.

Borgerhoff Mulder, M., P.J. Richerson, N.W. Thornhill, and E. Voland. 1997. The place of behavioral ecological anthropology in evolutionary science. In: Human by Nature: Between Biology and the Social Sciences, ed. P. Weingart, S.D. Mitchell, P.J. Richerson, and S. Maasen, pp. 253–282. Mahwah, NJ: Erlbaum.

Boyd, R., H. Gintis, S. Bowles, and P.J. Richerson. The evolution of altruistic punishment. *Proc. Natl. Acad. Sci. USA*. **100**:3531–3535.

Boyd, R., and P.J. Richerson. 1985. Culture and the Evolutionary Process. Chicago: Univ. of Chicago Press.

Boyd, R., and P.J. Richerson. 1989. The evolution of indirect reciprocity. *Social Networks* **11**:213–236.

Boyd, R., and P.J. Richerson. 1992a. How microevolutionary processes give rise to history. In: History and Evolution, ed. M.H. Nitecki and D.V. Nitecki, pp. 178–209. Albany: SUNY Press.

Boyd, R., and P.J. Richerson. 1992b. Punishment allows the evolution of cooperation (or anything else) in sizable groups. *Ethol. Sociobiol.* **13**:171–195.

Boyd, R., and P.J. Richerson. 1993. Rationality, imitation, and tradition. In: Nonlinear Dynamics and Evolutionary Economics, ed. R.H. Day and P. Chen, pp. 131–149. New York: Oxford Univ. Press.

Boyd, R., and P.J. Richerson. 2002. Group beneficial norms spread rapidly in a structured population. *J. Theor. Biol.* **215**:287–296.

Campbell, D.T. 1965. Variation and selective retention in socio-cultural evolution. In: Social Change in Developing Areas: A Reinterpretation of Evolutionary Theory, ed. H.R. Barringer, G.I. Blanksten, and R.W. Mack, pp. 58–79. Cambridge, MA: Schenkman.

Campbell, D.T. 1975. On the conflicts between biological and social evolution and between psychology and moral tradition. *Am. Psychol.* **30**:1103–1126.

Campbell, D.T. 1983. The two distinct routes beyond kin selection to ultrasociality: Implications for the humanities and social sciences. In: Nature of Prosocial Development: Theories and Strategies, ed. D.L. Bridgeman, pp. 71–81. New York: Academic.

Carneiro, R.L. 1970. A theory for the origin of the state. *Science* **169**:733–738.

Cavalli-Sforza, L.L., and M.W. Feldman. 1981. Cultural Transmission and Evolution: A Quantitative Approach. Monographs in Population Biology 16. Princeton, NJ: Princeton Univ. Press.

Cohen, D., and J. Vandello. 2001. Honor and "faking" honorability. In: Evolution and the Capacity for Commitment, ed. R.M. Nesse, pp. 163–185. New York: Russell Sage.

Curtin, P.D. 1984. Cross-cultural Trade in World History: Studies in Comparative World History. Cambridge: Cambridge Univ. Press.

Darwin, C. 1874. The Descent of Man. 2d ed. New York: American Home Library.

Dawes, R.M., A.J.C. van de Kragt, and J.M. Orbell. 1990. Cooperation for the benefit of us — not me or my conscience. In: Beyond Self-interest, ed. J.J. Mansbridge, pp. 97–110. Chicago: Univ. of Chicago Press.

Dunbar, R.I.M. 1992. Neocortex size as a constraint on group size in primates. *J. Hum. Evol.* **22**:469–493.

Eibl-Eibesfeldt, I. 1982. Warfare, man's indoctrinability, and group selection. *Zeitschrift Tierpsychol.* **67**:177–198.

Eibl-Eibesfeldt, I. 1989. Human Ethology: Foundations of Human Behavior. New York: Aldine de Gruyter.

Evans-Pritchard, E.E. 1940. The Nuer: A Description of the Modes of Livelihood and Political Institutions of a Nilotic People. Oxford: Clarendon.

Falk, A., E. Fehr, and U. Fischbacher. 2002. Approaching the commons: A theoretical explanation. In: The Drama of the Commons, ed. E. Ostrom, T. Dietz, N. Dolsak et al., pp. 157–191. Washington, D.C.: National Academy Press.

Fehr, E., and S. Gächter. 2002. Altruistic punishment in humans. *Nature* **415**:137–140.

Fukuyama, F. 1995. Trust: Social Virtues and the Creation of Prosperity. New York: Free Press.

Gadgil, M., and R. Guha. 1992. This Fissured Land: An Ecological History of India. Delhi: Oxford Univ. Press.

Gardner, R.A., B.T. Gardner, and T.E. Van Cantfort. 1989. Teaching Sign Language to Chimpanzees. Albany: SUNY Press.

Garthwaite, G.R. 1993. Reimagined internal frontiers: Tribes and nationalism — Bakhtiyari and Kurds. In: Russia's Muslim Frontiers: New Directions in Cross-cultural Analysis, ed. D.F. Eickelman, pp. 130–148. Bloomington: Indiana Univ. Press.

Hamilton, W.D. 1975. Innate social aptitudes of man: An approach from evolutionary genetics. In: Biosocial Anthropology, ed. R. Fox, pp. 115–132. New York: Wiley.

Hamilton, W.D. 1964. Genetic evolution of social behavior. I. II. *J. Theor. Biol.* **7**:1–52.

Hardin, G. 1968. The tragedy of the commons. *Science* **162**:1243–1248.

Hawkes, K. 1991. Showing off: Tests of an hypothesis about men's foraging goals. *Ethol. Sociobiol.* **12**:29–54.

Henrich, J. 2001. Cultural transmission and the diffusion of innovations: Adoption dynamics indicate that biased cultural transmission is the predominate force in behavioral change. *Am. Anthro.* **103**:992–1013.

Henrich, J. 2003. Cultural group selection, coevolutionary processes and large-scale cooperation. *J. Econ. Behav. Org.*, in press.

Henrich, J., and R. Boyd. 1998. The evolution of conformist transmission and the emergence of between-group differences. *Evol. Hum. Behav.* **19**:215–241.

Henrich, J., and R. Boyd. 2001. Why people punish defectors: Weak conformist transmission can stabilize costly enforcement of norms in cooperative dilemmas. *J. Theor. Biol.* **208**:79–89.

Henrich, J., and R. Boyd. 2002. On modeling cognition and culture: Why replicators are not necessary for cultural evolution. *Cult. Cogn.* **2**:67–112.

Henrich, J., R. Boyd, S. Bowles et al. 2001. In search of *Homo economicus*: Behavioral experiments in 15 small-scale societies. *Am. Econ. Rev.* **91**:73–78.

Henrich, J., and F.J. Gil-White. 2001. The evolution of prestige: Freely conferred deference as a mechanism for enhancing the benefits of cultural transmission. *Evol. Hum. Behav.* **22**:165–196.

Henrich, J., and R. McElreath. 2002. Are peasants risk-averse decision makers? *Curr. Anthro.* **43**:172–181.

Hinde, R.A. 1970. Animal Behaviour: A Synthesis of Ethology and Comparative Psychology. 2d ed. New York: McGraw-Hill.

Hirshleifer, J. 1977. Economics from a biological viewpoint. *J. Law Econ.* **20**:1–52.

Inglehart, R., and J.-R. Rabier. 1986. Aspirations adapt to situations — but why are the Belgians so much happier than the French? A cross-cultural analysis of the subjective quality of life. In: Research on the Quality of Life, ed. F.M. Andrews, pp. 1–56. Ann Arbor: Institute for Social Research, Univ. of Michigan.

Insko, C.A., R. Gilmore, S. Drenan et al. 1983. Trade versus expropriation in open groups: A comparison of two types of social power. *J. Pers. Soc. Psych.* **44**:977–999.

Jankowski, M.S. 1991. Islands in the Street: Gangs and American Urban Society. Berkeley: Univ. of California Press.

Johnson, A.W., and T.K. Earle. 2000. The Evolution of Human Societies: From Foraging Group to Agrarian State. 2d ed. Stanford: Stanford Univ. Press.

Jones, R. 1995. Tasmanian archaeology: Establishing the sequences. *Ann. Rev. Anthro.* **24**:423–446.

Keeley, L.H. 1996. War before Civilization. New York: Oxford Univ. Press.

Kellett, A. 1982. Combat Motivation: The Behavior of Soldiers in Battle. Intl. Series in Management Science/Operations Research. Boston: Kluwer.

Kelly, R.C. 1985. The Nuer Conquest: The Structure and Development of an Expansionist System. Ann Arbor: Univ. of Michigan Press.

Kelly, R.L. 1995. The Foraging Spectrum: Diversity in Hunter-gatherer Lifeways. Washington: Smithsonian Institution Press.

Kennedy, P.M. 1987. The Rise and Fall of the Great Powers: Economic Change and Military Conflict from 1500 to 2000. New York: Random House.

Kirch, P.V. 1984. The Evolution of the Polynesian Chiefdoms: New Studies in Archaeology. Cambridge: Cambridge Univ. Press.

Klein, R.G. 1999. The Human Career: Human Biological and Cultural Origins. 2d ed. Chicago: Univ. of Chicago Press.

Labov, W. 2001. Principles of Linguistic Change: Social Factors, ed. P. Trudgill, vol. 29, Language in Society. Malden, MA: Blackwell.

Laland, K.N., and G.R. Brown. 2002. Sense and Nonsense: Evolutionary Perspectives on Human Behaviour. Oxford: Oxford Univ. Press.

Laland, K.N., J. Kumm, and M.W. Feldman. 1995. Gene–culture coevolutionary theory: A test case. *Curr. Anthro.* **36**:131–156.

Leimar, O., and P. Hammerstein. 2001. Evolution of cooperation through indirect reciprocity. *Proc. Roy. Soc. Lond. B* **268**:745–753.

Marty, M.E., and R.S. Appleby. 1991. Fundamentalisms Observed: The Fundamentalism Project, vol. 1. Chicago: Univ. of Chicago Press.

Marwell, G., and R.E. Ames. 1981. Economist free ride: Does anyone else? *J. Publ. Econ.* **15**:295–310.

Maynard Smith, J., and E. Szathmáry. 1995. The major evolutionary transitions. *Nature* **374**:227–232.

Mayr, E. 1982. The Growth of Biological Thought: Diversity, Evolution, and Inheritance. Cambridge MA: Harvard Univ. Press.

McBrearty, S., and A.S. Brooks. 2000. The revolution that wasn't: A new interpretation of the origin of modern human behavior. *J. Hum. Evol.* **39**:453–563.

McElreath, R., R. Boyd, and P.J. Richerson. 2003. Shared norms can lead to the evolution of ethnic markers. *Curr. Anthro.* **44**:122–129.

Miller, G.J. 1992. Managerial Dilemmas: The Political Economy of Hierarchy. Cambridge: Cambridge Univ. Press.

Murphy, R.F., and Y. Murphy. 1986. Northern Shoshone and Bannock. In: Handbook of North American Indians: Great Basin, ed. W.L. d'Azevedo, pp. 284–307. Washington, D.C: Smithsonian Institution Press.

Nesse, R.M., and G.C. Williams. 1995. Why We Get Sick: The New Science of Darwinian Medicine. New York: Times Books.

Nisbett, R.E., and D. Cohen. 1996. Culture of Honor: The Psychology of Violence in the South. New Directions in Social Psychology. Boulder, CO: Westview Press.

Nisbett, R.E., K.P. Peng, I. Choi, and A. Norenzayan. 2001. Culture and systems of thought: Holistic versus analytic cognition. *Psychol. Rev.* **108**:291–310.

Nisbett, R.E., and L. Ross. 1980. Human Inference: Strategies and Shortcomings of Social Judgment. Englewood Cliffs, NJ: Prentice Hall.

Nowak, M.A., and K. Sigmund. 1998. Evolution of indirect reciprocity by image scoring. *Nature* **393**:573–577.

Ostrom, E. 1990. Governing the Commons: The Evolution of Institutions for Collective Action. Cambridge: Cambridge Univ. Press.

Ostrom, E. 1998. A behavioral approach to the rational choice theory of collective action. *Am. Pol. Sci. Rev.* **92**:1–22.

Ostrom, E., R. Gardner, and J. Walker. 1994. Rules, Games, and Common-pool Resources. Ann Arbor: Univ. of Michigan Press.

Otterbein, K.F. 1970. The Evolution of War: A Cross-cultural Study. New Haven, CT: Human Relations Area Files Press.

Peterson, N. 1979. Territorial adaptations among desert hunter-gatherers: The !Kung and Austalians compared. In: Social and Ecological Systems, ed. P. Burnham and R. Ellen, pp. 111–129. New York: Academic.

Pinker, S. 1994. The Language Instinct. New York: Morrow.

Price, G. 1972. Extensions of covariance selection mathematics. *Ann. Human Gen.* **35**:485–490.

Price, T.D., and J.A. Brown. 1985. Prehistoric Hunter-gatherers: The emergence of Cultural Complexity. Studies in Archaeology. Orlando: Academic.

Putnam, R.D. 2000. Bowling Alone: The Collapse and Revival of American Community. New York: Simon and Schuster.

Putnam, R.D., R. Leonardi, and R. Nanetti. 1993. Making Democracy Work: Civic Traditions in Modern Italy. Princeton, NJ: Princeton Univ. Press.

Rice, W.R. 1996. Sexually antagonistic male adaptation triggered by experimental arrest of female evolution. *Nature* **381**:232–234.

Richerson, P.J., and R. Boyd. 1989. The role of evolved predispositions in cultural evolution: Or sociobiology meets Pascal's Wager. *Ethnol. Socio.* **10**:195–219.

Richerson, P.J., and R. Boyd. 1998. The evolution of human ultrasociality. In: Indoctrinability, Ideology, and Warfare: Evolutionary Perspectives, ed. I. Eibl-Eibesfeldt and F.K. Salter, pp. 71–95. New York: Berghahn.

Richerson, P.J., and R. Boyd. 1999. Complex societies: The evolutionary origins of a crude superorganism. *Hum. Nat.: Inter. Biosoc. Persp.* **10**:253–289.

Richerson, P.J., and R. Boyd. 2000a. Climate, culture, and the evolution of cognition. In: The Evolution of Cognition, ed. C. Heyes and L. Huber, pp. 329–346. Cambridge, MA: MIT Press.

Richerson, P.J., and R. Boyd. 2000b. The Pleistocene climate variation and the origin of human culture: Built for speed. In: Evolution, Culture, and Behavior, ed. F. Tonneau and N.S. Thompson, pp: 1–45. Perspectives in Ethology 13. New York: Kluwer Academic/Plenum.

Richerson, P.J., and R. Boyd. 2001. The evolution of subjective commitment to groups: A tribal instincts hypothesis. In: Evolution and the Capacity for Commitment, ed. R.M. Nesse, pp. 186–220. New York: Russell Sage.

Richerson, P.J., R. Boyd, and R.L. Bettinger. 2001. Was agriculture impossible during the Pleistocene but mandatory during the Holocene? A climate change hypothesis. *Am. Antiq.* **66**:387–411.

Richerson, P.J., R. Boyd, and B. Paciotti. 2002. An evolutionary theory of commons management. In: The Drama of the Commons, ed. E. Ostrom, T. Dietz, N. Dolsak et al., pp. 403–442. Washington, D.C.: National Academy Press.

Rumsey, A. 1999. Social segmentation, voting, and violence in Papua New Guinea. *Cont. Pacific* **11**:305–333.

Runciman, W.G. 1998. Greek hoplites, warrior culture, and indirect bias. *J. Roy. Anthro. Inst.* **4**:731–751.

Salamon, S. 1992. Prairie Patrimony: Family, Farming, and Community in the Midwest. Chapel Hill: Univ. of North Carolina Press.

Salter, F.K. 1995. Emotions in Command: A Naturalistic Study of Institutional Dominance. Oxford: Oxford Univ. Press.

Savage-Rumbaugh, E.S., and R. Lewin. 1994. Kanzi: The Ape at the Brink of the Human Mind. New York: Wiley.

Sears, D.O., and C.L. Funk. 1990. Self interest in Americans' political opinions. In: Beyond Self-interest, ed. J. Mansbridge, pp. 147–170. Chicago: Univ. of Chicago Press.

Segall, M., D. Campbell, and M.J. Herskovits. 1966. The Influence of Culture on Visual Perception. New York: Bobbs-Merrill.

Simon, H.A. 1990. A mechanism for social selection and successful altruism. *Science* **250**:1665–1668.

Slatkin, M., and M.J. Wade. 1978. Group selection on a quantitative character. *Proc. Natl. Acad. Sci. USA* **75**:3531–3534.

Sober, E., and D.S. Wilson. 1998. Unto Others: The Evolution and Psychology of Unselfish Behavior. Cambridge, MA: Harvard Univ. Press.

Soltis, J., R. Boyd, and P.J. Richerson. 1995. Can group-functional behaviors evolve by cultural group selection: An empirical test. *Curr. Anthro.* **36**:473–494.

Squires, A.M. 1986. The Tender Ship: Governmental Management of Technological Change. Boston: Birkhäuser.

Srinivas, M.N. 1962. Caste in Modern India and Other Essays. Bombay: Asia Publ. House.

Steward, J.H. 1955. Theory of Culture Change: The Methodology of Multilinear Evolution. Urbana: Univ. of Illinois Press.

Thomas, D.H., L.S.A. Pendleton, and S.C. Cappannari. 1986. Western Shoshone. In: Handbook of North American Indians: Great Basin, ed. W.L. d'Azevedo, pp. 262–283. Washington, D.C.: Smithsonian Institution Press.

Tomasello, M. 1999. The Cultural Origins of Human Cognition. Cambridge, MA: Harvard Univ. Press.

Trivers, R.L. 1971. The evolution of reciprocal altruism. *Qtly. Rev. Biol.* **46**:35–57.

Wardhaugh, R. 1992. An Introduction to Sociolinguistics. 2d ed. Oxford: Blackwell.

West, R. 1941. Black Lamb and Grey Falcon. New York: Penguin.

Whiten, A., and R.W. Byrne. 1988. Machiavellian Intelligence: Social Expertise and the Evolution of Intellect in Monkeys, Apes, and Humans. Oxford: Oxford Univ. Press.

Wiessner, P. 1984. Reconsidering the behavioral basis for style: A case study among the Kalahari San. *J. Anthro. Arch.* **3**:190–234.

Williams, G.C. 1966. Adaptation and Natural Selection: A Critique of Some Current Evolutionary Thought. Princeton, NJ: Princeton Univ. Press.

Wilson, D.S. 1983. The group selection controversy: History and current status. *Ann. Rev. Ecol. Syst.* **14**:159–188.

Wilson, E.O. 1975. Sociobiology: The New Synthesis. Cambridge, MA: Belknap.

Wilson, E.O. 1998. Consilience: The Unity of Knowledge. New York: Knopf.

Wolfe, T. 1965. The Kandy-kolored Tangerine-flake Streamline Baby. New York: Farrar Straus and Giroux.

Young, H.P. 1998. Individual Strategy and Social Structure: An Evolutionary Theory of Institutions. Princeton, NJ: Princeton Univ. Press.

20

The Power of Norms

H. Peyton Young

Department of Economics, Johns Hopkins University, Baltimore, MD 21218, U.S.A.

ABSTRACT

Norms are basic building blocks of social and economic organization. This chapter proposes a framework for studying the evolution of norms based on the cumulative effect of many decentralized interactions by individuals. Predictions of the theory are illustrated for contractual norms between landlords and tenants in contemporary United States agriculture.

DEFINITION AND ROLE OF NORMS

A *norm* is a rule of behavior that is self-reinforcing: everyone wants to play their part given the expectation that others will play theirs. This definition encompasses simple rules that solve coordination problems, such as driving on a given side of the road, as well as more complex rules that involve sanctioning those who deviate from a first-order rule. (I express outrage if someone cuts in front of someone else in line; I refuse to associate with people who fail to tip in restaurants.) Norms structure our relations with others so completely that we often fail to recognize just how pervasive they are. The clothes I wear, the food I eat, the manner and time of day at which I eat it, the ways I address people, the obligations that I feel toward members of my family, the duties that I perform at work, and the amount that I earn are all determined to a significant degree by prevailing norms of behavior in the society.

Although sociologists and anthropologists have long understood the central role played by norms, economists have traditionally viewed them as being peripheral to economic decisions (notable exceptions are Schelling [1960], Akerlof [1997], and Sugden [1986]). The fact is, however, that norms influence terms of employment, the amount that people save and consume, attitudes toward debt, decisions about when to retire, and a host of other economic variables. Even property rights are governed to a considerable extent by social expectations about who is entitled to what (Hume 1739; Sugden 1986).

The social function of norms is to resolve problems of collective action and coordination (Ullman-Margalit 1977). Indeed, norms can be viewed as equilibria of appropriately defined games. These games often have a multiplicity of equilibria, as evidenced by the fact that solutions to a given coordination problem often differ substantially from one society to another, and also within a given society over time. However, although social norms are related to the notion of equilibrium, they are not the same thing.

To illustrate the distinction, consider two individuals who can divide a dollar provided they agree on *how* to divide it. Each makes a demand, and if the demands sum to at most one dollar their demands are met; otherwise they get nothing. If one demands 43 cents and the other 57 cents, the demands are in equilibrium: no one can gain by unilaterally changing his demand. Yet while this is an equilibrium, it is not a norm. The equilibrium is idiosyncratic to these particular individuals. Fifty-fifty division, by contrast, is a norm because it is a usual and customary solution in games of this kind. *More generally, a norm is an equilibrium behavior in a game played repeatedly by many different individuals in society where the behavior is known to be customary.* Note the importance of knowledge: behavior must not only *be* customary, it must be *known to be* customary or else behaviors are not in fact self-enforcing (Lewis 1969). Note also that this definition encompasses behaviors that require no sanctions by third parties to constitute an equilibrium. The latter are sometimes referred to as *conventions* rather than norms, though I shall not dwell on this distinction here.

THE EVOLUTION OF NORMS

How then do norms become established and what causes them to change? In Young (1993, 1998a), I suggest a general framework for investigating this question. The fundamental idea is that norms coalesce from the decentralized, uncoordinated choices of many interacting individuals. Roughly speaking, individuals are the particles of the system, and norms are the organizational forms that bind them together. Unlike particles, however, individuals make intentional choices based on perceived constraints and opportunities. We therefore need to explain how such individual choices can lead to the emergence of society-wide norms that promote coordinated behavior.

We model this situation as follows. Consider a given type of social interaction that regularly confronts different members of a society. For simplicity we shall think of this interaction as a two-person coordination game having multiple equilibria. At any given time t various solutions will have been tried, and there will be a distribution of behaviors in society. Roughly speaking this distribution defines the state of the system at time t. In the next period, a given pair (or pairs) of individuals interact with some probability. The information they have about the previous behavior of others leads them to expect some behavior (or distribution of behaviors) among their present opponents. These expectations lead them to make certain choices in the current period, which then become precedents for individuals in later periods. These choices may be purely rational, but more likely they involve elements of conformity, experimentation, inertia, and other forms of nonrational behavior. These other elements can be incorporated into a stochastic choice model that involves mostly rational behavior with some idiosyncratic elements.

The combination of these assumptions leads to a stochastic dynamical system that is built around the following feedback loop:

It can be shown that, under fairly general conditions, such a system converges to a situation that is close to being a norm in the sense that almost everyone is playing their part in a particular equilibrium, though idiosyncratic variations will also typically be present. In addition, however, the system occasionally shifts abruptly from one norm to another. These shifts can be induced either by system-wide shocks (e.g., changes in the underlying payoff structure, perhaps due to technological change) or by expectational "drift" that arises from the accumulation of idiosyncratic choices by a few individuals. Either one of these factors can push the system to a critical tipping point, beyond which expectations change and society converges to a new norm.

A central prediction of the theory is that the drift component tends, over time, to favor some norms over others (Young 1993, 1998b). In other words some norms are more *durable* than others in the sense that, once established, they tend to stay in place for longer periods of time. Surprisingly, the most durable norms are not necessarily efficient (i.e., Pareto optimal); in some situations, risk dominance is a better criterion of durability (Young 1993; Kandori et al. 1993).

Even when the evolutionary process does operate in favor of efficient norms, not all such norms are equally durable. In fact, the distributional properties of a norm also affect its durability to a significant extent. Specifically, in a pure coordination game between two players, the most durable norms are those that maximize the position of the worst-off party relative to the best outcome they could get. (That is, the outcome x is such that min $(u_1(x)/u_1^+, u_2(x)/u_2^+)$ is maximized over all x, where u_1^+ and u_2^+ are the maximum utilities 1 and 2 could get under some choice of x.) Thus, even when individuals always make choices that maximize their own utility, with no regard for the utility of others and no preference for fairness, the net effect of the evolutionary process is to favor norms that look *as if* people did have such preferences (Young 1998a).

EMPIRICAL STUDIES OF NORM FORMATION: AGRICULTURAL CONTRACTS

In this chapter I discuss how the theory can be brought to bear on situations where norms, and norm shifts, can be verified empirically. The identification of such situations poses a number of challenges. One difficulty is that norms sometimes become codified into laws, so that even though a norm might originally have emerged from the bottom up, it later becomes enforced from the top down, so that compliance is no longer voluntary. An example is the emergence of local ordinances in the southern U.S. requiring blacks to give up their seats to whites on public transportation. (Interestingly, even though this norm was codified, it was eventually overturned by the spontaneous [illegal] actions of a few individuals, notably Rosa Parks. This illustrates the general point that the viability of a norm ultimately rests on the shared expectation that people will conform to it, not on its legal status.)

Another difficulty in identifying empirical cases is that what appears to be a norm might be some form of spurious correlation. For example, some groups in the population smoke more than others (e.g., a much smaller proportion of black men are smokers than white women). Is this the result of different social norms operating in different groups, or is it the result of subtle differences in bodily responses to nicotine?

A study of norms by Young and Burke (2001; see also Burke and Young 2003) avoids some of these difficulties. Specifically, we studied whether social norms shape the terms of contracts between tenants and landlords in contemporary U.S. agriculture. Although this may seem like a somewhat unusual area in which to study the operation of norms, it has a number of important attributes. First, there is extensive data on contracts between tenants and landlords that extends over many years, gathered by agricultural economists and research branches of the Department of Agriculture. Second, modern U.S. agriculture is a highly competitive and sophisticated business in which both tenants and landlords have a great deal of scope for making choices. (It is quite unlike the situation where southern sharecroppers or European peasants were trapped on the land through debt and lack of alternative employment.)

Third, there is little reason to think that norms are the spurious by-product of associational preferences: farmers do not move to an area because they like the other farmers there, but in most cases because they were born there. (This applies to landowners as well as tenants.) In other words, interactions are determined by geography and happenstance of birth; they are not contaminated by the endogenous sorting of people into like-minded groups — a difficulty that has bedevilled other norm studies (see Manski 1993).

Fourth, there is extensive data on the underlying quality of factor inputs, so that the standard refuge of the skeptic — the existence of common unobservables — can be eliminated by appropriate statistical tests. Finally, the choice of contract is a purely voluntary act between principal and agent that is not constrained by law. Hence, standard competitive market forces should, in theory, determine the outcome.

PREDICTIONS OF STANDARD THEORY

Let us quickly review what standard competitive theory would predict in this situation. Each year a given landowner (the principal) and a prospective tenant farmer (the agent) negotiate a contract for the coming year. The main factors governing the outcome of the bargain are: quality of the land, quality of the labor, the cost of monitoring the tenant's performance, the parties' attitude toward risk, and their opportunity costs, that is, the value of their next-best alternatives. (For the tenant this includes the option of alternative employment, say in a nearby factory.) This problem has been extensively studied from a theoretical point of view (Cheung 1969; Stiglitz 1974; Hayami and Otsuka 1993). Basically the monitoring cost and attitudes toward risk determine the form of the contract, that is, whether the parties prefer a share contract, a cash lease, a wage contract, or some hybrid form. In our study we restrict attention to the subset of farms that adopted the share format, thus effectively sidestepping the impact of monitoring costs and risk attitudes, which we cannot observe in any event. (Theory says that those agents who choose share contracts tend to be those who are more risk averse and for whom monitoring costs are low.)

Among those agents who opt for share contracts, the empirical question is what division of the crop they negotiate. In principle they could agree to different divisions of each crop (corn, wheat, and soybeans) as well as to different divisions of each input (e.g., fertilizer, seed, and equipment). In practice, however, almost all share contracts are expressed in terms of a *single* share for all the outputs and a *single* share for all inputs except equipment, which is the sole responsibility of the tenant.

Assume for the moment that all labor is similar in quality and that the reservation wage in a given area is fixed, say, by the going wage in factory employment. Then, in competitive equilibrium, laborers will earn the reservation wage and the residual surplus will go to the landowner as pure rent. In particular, soils of a given quality should earn the same rent across different farms, and higher-quality soils should command higher rents. This implies that the negotiated share to the tenant should be lower on higher-quality farms. Furthermore, the two parties know how to adjust the share appropriately because the quality of soils on a given farm is rated according to a scheme that gives expected productivity per acre of different crops, holding labor and other inputs constant, and both parties know the ratings, which are a matter of public record.

We turn now to the data to see if these predictions hold. These data come from a sample survey conducted by the Illinois Cooperative Agricultural Extension Service (1995). Of the 1704 responses in the 1995 survey, cropsharing contracts were the most frequent (55%) and land rent contracts the next most frequent (41%); all other contract forms (mostly livestock and pasture leases) constituted less than 4%. We restrict ourselves to an analysis of the 935 cropshare contracts.

Figure 20.1 shows the frequency distribution of shares of corn output, which is virtually the same as the frequency distribution of shares for soybeans and wheat. Note that 1/2–1/2 is by far the most common division, and virtually all contracts use either 1/2–1/2, 3/5–2/5, or 2/3–1/3. (Here and elsewhere we list the share to the tenant first.) These data are difficult to reconcile with the standard

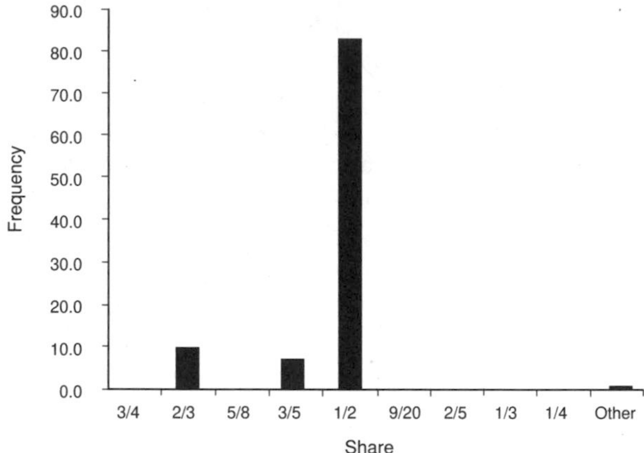

Figure 20.1 Crop share frequencies in Illinois: tenant's share of the corn crop. Illinois Cooperative Agricultural Extension Service (1995).

competitive account, because they show so little variation in contract terms. (It is also quite suspicious that the shares concentrate solely on fractions with small denominators.)

The situation is further illuminated by looking at the distribution of shares in different parts of the state. Illinois exhibits considerable variation in its soil characteristics. In the north, the land is mostly flat and the soils are on average highly productive, whereas in the south the land tends to be hillier, the topsoil is not as thick, and on average it is less productive. (This north-south division corresponds roughly to the southern boundary of the last major glaciation.) When we compare contract frequencies in the northern and southern parts of the state, substantial differences appear (see Figure 20.2). In the north, contract terms are almost exclusively 1/2–1/2, whereas in the south the predominant contracts are 2/3–1/3 and 3/5–2/5.

These differences make sense from an economic point of view. Because the land in the south is, on average, inherently less productive than the land in the north, the share for the tenant must be higher in the south if net returns to the tenants in the two regions are to be comparable. Viewed in this way, the data seem to vindicate standard competitive theory.

The data only vindicate standard theory in the crudest sense, however. In the first place, there should not be three shares but a spectrum of shares reflecting the soil qualities of the farms in question. Are we to believe, for example, that all of the farms in the north have the same soil quality, which justifies a share of 1/2, while all the farms in the south have one of two soil qualities, namely, those that justify 2/3 or 3/5 to the tenant?

This hypothesis seems absurd on the face of it. Moreover, it is completely contradicted by the actual distribution of soil qualities. In both the north and the south there is a wide range of qualities, and the highest produces over twice as much per acre as the lowest (holding labor and other inputs constant). In fact,

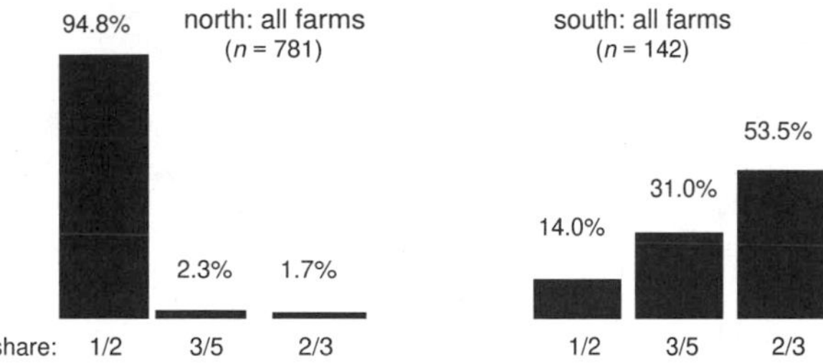

Figure 20.2 Distribution of share contracts by region. Illinois Cooperative Agricultural Extension Service (1995).

this is true within virtually each individual *county* in the north and the south. According to standard theory the negotiated shares should reflect these differences. Instead, the shares in a given county cluster around a small number of values independently of the soil quality. Figure 20.3 illustrates this effect for two representative counties, one in the north and one in the south.

PREDICTIONS OF A NORMS-BASED THEORY

To understand a model of norm formation that can account for these phenomena, three facts need to be explained. First, we have seen that there is a tendency for contract terms to cluster on a few simple fractions that have a priori focal power. This is the *quantum effect*. Second, there is much less heterogeneity locally than competitive theory would predict. This is the *local conformity effect*. Notice that the second does not follow from the first: even though the quantum effect limits the number of distinct contracts that might be observed, contract terms could still vary substantially from one farm to the next, resulting in a great deal of local diversity. This is not confirmed by the data even though local differences in fundamentals might call for it. The third fact to be explained is that the contractual norm differs between the two regions in a way that *is* broadly consistent with the competitive model. This is the *regional diversity effect* (Young 1998b).

Consider the following dynamic model. In each period, landlords propose contracts to tenants. In proposing a contract, the landlord takes into account both economic and psychological factors. First, he obviously cares about the expected returns from the contract. Second, he wants to conclude an agreement expeditiously and without a lot of haggling, which argues for keeping the terms simple. Third, he values his relationship with the tenant and his standing among his neighbors. Hence he wants to conclude an agreement that is perceived by the tenant to be fair and that adheres to general standards within the local community. In particular, if he were to offer the tenant less than the going share, he risks antagonizing someone he must work with and he may be seen as greedy or exploitative by his neighbors. If he offers the tenant more than the going share, the tenant may be happy but the neighboring landlords will not be; moreover, this goes against his interest in maximizing returns. The tradeoffs between these

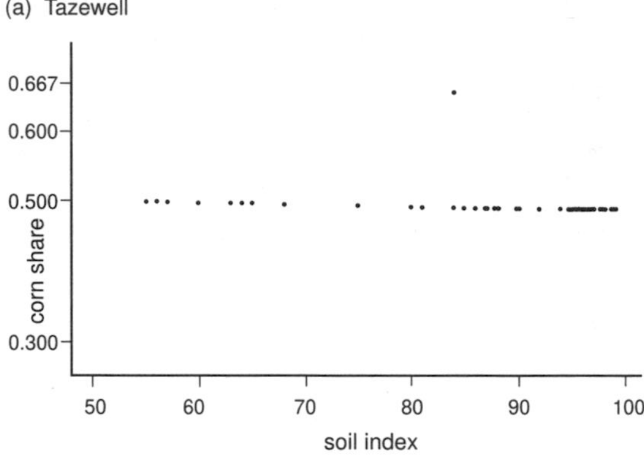

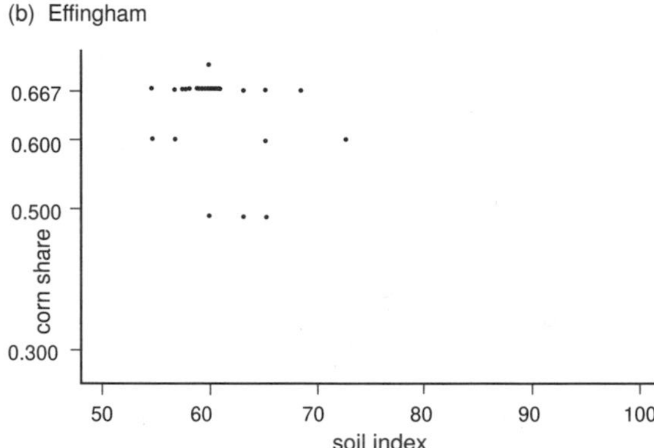

Figure 20.3 Distribution of shares in (a) Tazewell Co. (north) and (b) Effingham Co. (south).

considerations are difficult to estimate and will no doubt vary substantially among regions and among particular principal-agent pairs. Here we shall posit a model that could, in principle, be estimated empirically from event histories that specify the temporal sequence of contract adoptions by location. (The Illinois data do not allow us to make this estimation because they do not contain a sufficiently large panel of farms, and the locations are aggregated so as not to reveal the identity of particular respondents.)

Assume that each landlord chooses from a menu of simple contracts that apply a single, easy-to-calculate fraction to all inputs and outputs. The probability of choosing a given contract in the menu is given by a logit function in which the log probability of a contract increases linearly in its expected returns and the degree to which it conforms with contracts used on nearby farms. The relative weighting between these two factors is a parameter that can be estimated. This type of setup is standard in discrete choice analysis (McFadden 1974; Brock and Durlauf 2001) and is relatively easy to estimate from microlevel data (if we had such data).

The basic predictions of the model are as follows. Consider any distribution of soil qualities that is heterogeneous locally and exhibits substantial shifts in *average* quality between geographic regions. It can be shown that, for a wide range of parameter values, the model converges with high probability to a situation characterized by *regional customs*. That is, there is near-uniformity of the contractual custom within each region, but substantial differences between regions that reflect underlying differences in average quality. Boundaries between regional customs form endogenously and are quite sharp: just over the border, people do things differently. However, there may be nothing that marks the boundary line per se: it suffices that average land quality be somewhat different on opposite sides. This is just what we see in the Illinois data.

The existence of these regional norms has important implications for the economic returns realized by labor and land. To see this, suppose that the share to the tenant is uniform in a given region, irrespective of soil quality. Then, on soils of higher quality the tenant will earn higher net returns because he captures a fixed fraction of the increased yield per acre that the higher-quality soil produces. Thus, unless the landlord enforces a higher level of labor input per acre on the higher-quality land, labor will effectively capture a portion of the land rent because of the rigidity of the contract.

It can be shown, however, that labor input per acre does not increase on higher-quality land (Burke and Young 2001). (This exploits the detailed knowledge we have of the productivity of different soils when inputs remain fixed.) In other words, tenants succeed in capturing part of the rent that should accrue to land. Moreover, the amount of rent capture is quite sizable. Our estimate is that the tenant captures about *one-third* of the rent that ought to accrue to higher soil quality, and this is attributable mainly to the rigidities induced by contractual custom. On a farm of several hundred acres, the rent capture by the tenant may amount to several thousand dollars per year, which is a nonnegligible fraction of his income.

Of course, this argument would be undermined if we could identify hidden costs to the tenant or hidden benefits to the landlord that justify the higher payments to labor on higher-quality land. One possibility, for example, is that yields on higher-quality soils are more variable than on lower-quality soils. This would imply that risk-averse tenants must receive higher expected payments to be willing to farm high-quality soils. (As pointed out earlier, there is reason to believe that the tenants in our sample are risk averse because they opted for a share contract instead of a land rent contract.) In fact, however, the data show that variability in yield is not significantly related to soil quality.

A second possibility is that higher returns to the tenant result in lower turnover rates, thus reducing transaction costs for the landlord. Although the data do not allow us to test for this possibility directly, the rate of tenant turnover is so low that only a small portion of the labor premium could be explained by this consideration (if in fact it explains any of the premium).

A third possibility is that higher-quality tenants migrate to higher-quality land. In other words, equilibrium is achieved in the labor market through assortative matching rather than by adjusting contract terms. If this explanation is to have any force, however, then higher-quality tenants must generate an increase in yield that goes beyond the increase attributable to higher land quality alone, holding other inputs fixed. The evidence strongly suggests that this is not the case; indeed, if anything, there appears to be some slacking off in the quality or quantity of labor input as land quality rises.

Finally, we consider the possibility that the increase in tenant net income is the result of spurious correlation effects. It could be, for example, that reservation wages happen to be higher in regions with higher land quality. In this case, the return to tenant labor would rise with land quality, but the relationship would be purely spurious. We test for this and other local fixed effects and find that they do not account for the observed increase in tenant income.

The question remains as to how this kind of behavior can be sustained in a competitive market environment. Are norms really so powerful that they can distort economic returns to this extent? I conjecture that this phenomenon is quite general and that similar distortions would be uncovered in many other parts of the economy if people would only look for them. Among the prime candidates are decisions about when to retire, how much to save for retirement, how much it is prudent to borrow, how much to pay your lawyer to defend against a lawsuit, how much to pay senior faculty in comparison to junior faculty, how many stock options to award the CEO of a corporation, and so on. In all of these cases, my guess is that social norms substantially alter the decisions that would be predicted by the standard competitive model and that economists need to start taking these effects seriously.

REFERENCES

Akerlof, G. 1997. Social distance and social decisions. *Econometrica* **65**:1005–1027.

Brock, W., and S.N. Durlauf. 2001. Discrete choice with social interactions. *Rev. Econ. Stud*. **68**:235–260.

Burke, M.A., and H.P. Young. 2003. Contract uniformity and factor returns in agriculture. Baltimore: Dept. of Economics, Johns Hopkins Univ.

Cheung, S.N.S. 1969. The Theory of Share Tenancy. Chicago: Univ. of Chicago Press.

Hayami, Y., and K. Otsuka. 1993. The Economics of Contract Choice: An Agrarian Perspective. Oxford: Clarendon.

Hume, D. 1739. A Treatise of Human Nature. Oxford: Clarendon.

Illinois Cooperative Agricultural Extension Service. 1995. Cooperative Extension Service Farm Leasing Survey. Urbana-Champaign: Dept. of Agricultural and Consumer Economics, Univ. of Illinois.

Kandori, M., G. Mailath, and R. Rob. 1993. Learning, mutation, and long-run equilibria in games. *Econometrica* **61**:29–56.

Lewis, D. 1969. Convention: A Philosophical Study. Cambridge, MA: Harvard Univ. Press.

Manski, C. 1993. Identification of endogenous social effects: The reflection problem. *Rev. Econ. Stud.* **60**:531–542.

McFadden, D. 1974. Conditional logit analysis of qualitative choice behavior. In: Frontiers in Econometrics, ed. P. Zarembka, pp. 105–142. New York: Academic.

Schelling, T. 1960. The Strategy of Conflict. Cambridge, MA: Harvard Univ. Press.

Stiglitz, J.E. 1974. Incentives and risk-sharing in sharecropping. *Rev. Econ. Stud.* **41**: 219–255.

Sugden, R. 1986. The Economics of Rights, Cooperation and Welfare. Oxford: Blackwell.

Ullman-Margalit, E. 1977. The Emergence of Norms. New York: Oxford Univ. Press.

Young, H.P. 1993. The evolution of conventions. *Econometrica* **61**:57–84.

Young, H.P. 1998a. Conventional contracts. *Rev. Econ. Stud.* **65**:776–792.

Young, H.P. 1998b. Individual Strategy and Social Structure. Princeton, NJ: Princeton Univ. Press.

Young, H.P., and M.A. Burke. 2001. Competition and custom in economic contracts: A case study of Illinois agriculture. *Am. Econ. Rev.* **91**:559–573.

21

Human Cooperation

Perspectives from Behavioral Ecology

Eric A. Smith

Department of Anthropology, University of Washington,
Seattle, WA 98195–3100, U.S.A.

ABSTRACT

Humans, like all social species, face various collective action problems (difficulties achieving potential benefits from cooperating when coordination is required or individuals have incentives to defect). Humans solve these problems through various means: communication, monitoring, enforcement, and selective incentives. This chapter summarizes the theory and evidence on human cooperation found in the field of human behavioral ecology, categorized topically: resource sharing, cooperative production, aid-giving, and coalition-based conflict. A more speculative question is then addressed, "Why are humans so cooperative?" The suggested answers revolve around linguistic communication, technology, and coalitional behavior. In particular, language clearly increases the likelihood of solving coordination games and appears to lower the cost of monitoring and enforcement in other payoff environments. Language is also likely to enhance signaling and reputation effects. Technology and complex division of labor increase fitness interdependencies between individuals, and the potential payoffs to coalition members; these in turn provide new opportunities for development of norms and institutions to solve collective action problems. The chapter closes with some caveats about the limits to human cooperation.

INTRODUCTION

All social species face various collective action problems, i.e., various opportunities for cooperation that can yield benefits, but which can be thwarted by free riding and other forms of selfishness, as well as by coordination failures. In comparison to other vertebrates, humans appear to be remarkably good (though not perfect) at solving these collective action problems. They do so through a variety of means, including communicating about options and preferences, socially transmitting norms and other codified information, monitoring the behavior of others, imposing punishment for selfish behavior, and dispensing selective incentives for cooperative or prosocial behavior.

The means by which people manage to capture the benefits of cooperation, and the conditions under which such solutions are more or less likely to occur, are studied by analysts using several different theoretical approaches. Here I survey the approach known as human behavioral ecology, one that is complementary to but distinct from other prominent approaches to studying the evolution of human cooperation, such as evolutionary psychology and cultural inheritance theory (Smith 2000).

I begin with definition of some key terms and then outline the main features of the research strategy employed in human behavioral ecology. Next, I present summaries of behavioral ecology research in various domains of human cooperative behavior: resource sharing, cooperative production, aid-giving, and coalition-based conflict. The second major section offers rather speculative answers to the question, "Why are humans so cooperative?" The answers proffered (and interrogated) revolve around linguistic communication, technology, coalitional behavior, and kinship. I close with some caveats about the limits to human prosociality.

THE BEHAVIORAL ECOLOGY OF COLLECTIVE ACTION

Defining the Problem

In accord with the usual meaning in behavioral ecology as well as some areas of social science, I define *cooperation* as collective action for mutual benefit (Clements and Stephens 1995; Dugatkin 1997). By *collective action*, I mean whenever two or more individuals must interact or coordinate their actions to achieve some end. This end is generally to provide a *collective good*, meaning any material good or service that is then available (though not necessarily in equal amounts) for consumption by the members of some collective (e.g., a family, a village, an organization, a nation), whether or not consumption by some reduces the amount available to the remainder. Note, as defined here, cooperation does not necessarily entail (nor does it exclude) altruism, either temporary (Trivers's "reciprocal altruism") or in terms of expected lifetime fitness.

The simplest form of cooperation involves *coordination*; this applies when actors share preferences on the rank ordering of each strategy pair in the interaction and thus always mutually benefit from cooperation.[1] In behavioral ecology, coordination interactions are usually labeled "mutualism," and a distinction is often made between *by-product mutualism* (Brown 1983) where A benefits from B's action but B would perform the action and gain benefits regardless (e.g., evading predators via the "selfish herd" effect), and *synergistic mutualism*

[1] In some fields, coordination problems are considered to lie outside the domain of cooperation, i.e., cooperation must by definition solve collective action problems where interactors have conflicts of interest (see Bowles and Gintis, this volume).

(Maynard Smith and Szathmáry 1995), where coordination yields increased per capita benefits (e.g., coordinated efforts to deter predators). A *collective action problem* (CAP) arises (a) when coordination is difficult (e.g., due to imperfect information about the actions others will take) or (b) when cooperation is individually costly but collectively beneficial (as in games of Chicken, Prisoner's Dilemma, etc.). *Free riding* consists of benefiting from a collective good without paying the costs of providing that good. A *second-order collective action problem* arises whenever the means needed to solve one CAP (e.g., monitoring, teaching, enforcement) itself poses a CAP (e.g., because it provides a collective good on which some could free ride).

Human Behavioral Ecology: Research Strategy

The adaptationist program in contemporary evolutionary biology proposes that natural selection has designed organisms to respond to environmental conditions in fitness-enhancing ways. With this as a starting point, behavioral ecologists formulate and test formal models incorporating specific optimization goals, currencies, and constraints, and use these to study evolution and adaptive design of animal behavior in ecological context. Some researchers in anthropology and cognate disciplines have adapted this approach, in conjunction with theory and method from the home discipline, as tools to analyze human behavior. Human behavioral ecology emerged in the 1970s and grew rapidly in the 1990s (Winterhalder and Smith 2000). Because it incorporates material from the much older tradition of ecological anthropology and pays some attention to the roles of intentionality and cultural evolution, it is not quite as radical a departure from standard social science as it might first appear (Smith and Winterhalder 1992).

Focusing as they do on behavior, and particularly social behavior with a strong cultural component, human behavioral ecologists must analyze a very labile and causally complex set of phenomena. They generally attempt to explain such complex patterns of behavioral variation as forms of *phenotypic adaptation* to varying social and ecological conditions. The focus is on testing predictions about the match between environmental conditions or payoffs and behavioral variation, without worrying too much about developmental or learning mechanisms that create or maintain this match. The link between such phenotypic adaptation and genetic evolution is provided by positing that the former is guided by "decision rules." These decision rules are presumed panhuman cognitive adaptations that have evolved by natural selection (or recurrent cultural evolution) and guide behavioral variation in ways sensitive to environmental context. In the language of game theory, decision rules are usually conditional strategies that take the general form "In context X, adopt one behavioral tactic; in context Y, switch to an alternative tactic," and so on. Strategies can be conditional on the actor's phenotype (e.g., "I will signal only if I am high quality") or on aspects of the social and nonsocial environment (e.g., "Ally with Joan only if she reciprocates" or "Pursue a given prey type only if it raises my mean return rate"). Behavioral variation arises as individuals match their conditional strategies to their particular socioecological settings and endowments.

Forms of Cooperation

I briefly discuss various forms of cooperation that human behavioral ecologists have analyzed, summarizing key models and representative empirical studies.

Resource Sharing

People can, of course, share a large variety of resources: land, unharvested resources, dwellings and other durable goods, labor, and so on. Research in human behavioral ecology has dealt with many of these; however, the greatest amount of research has concentrated on sharing of food in subsistence economies. Unlike most other primates, humans often harvest resources of sufficient "package size" (e.g., large game) or in sufficient bulk (e.g., an agricultural crop) that some combination of transfer to those without the resource or storage for later use is likely. A variety of behavioral ecology models are employed to analyze these phenomena (Winterhalder 1996), each making somewhat different assumptions about the socioecological circumstances specified (e.g., group size, conditionality of transfer decisions, the nature of the resource) and the evolutionary mechanism invoked (e.g., individual, kin, sexual, or trait-group selection).

Possible benefits of food sharing include risk reduction (buffering variation in individual or household food income through pooling of asynchronous and unpredictable harvests), obtaining resources without working for them (a benefit to the recipient only!), gains to trade (I produce food X more efficiently and you produce food Y more efficiently, and we mutually benefit through exchanging some of our production), and advertising the producer's quality. These hypothesized benefits correspond to distinct explanatory models from behavioral ecology: risk-reduction reciprocity, scrounging (also known as "demand sharing" or "tolerated theft"), trade, and costly signaling. Possible costs of food sharing, corresponding with the same set of explanations, include nonreciprocation (defection in a delayed-reciprocity system), exploitation (by scroungers), transaction costs (in arranging and carrying out trade), and signal costs (e.g., food income foregone or choice of a production strategy with high display value but low production efficiency).

Risk-reduction reciprocity. The Aché Indians of Paraguay have perhaps the best-documented food-sharing behavior. When studied some fifteen years ago, Aché hunters shared game evenly and without kin bias with all members of the band, regardless of foraging success. Kaplan et al. (1990) calculated that on average Aché families produced less than 1,000 calories per member on 27% of the 412 days in their sample, but after sharing only 3% days resulted in food intake below this threshold. They estimate that without food sharing, an Aché family experiencing average foraging success and variance would fail to obtain at least 50% of its caloric needs for 3 weeks running about once per 17 years. Further calculations suggest that sharing of meat increases average family nutritional status most, honey an intermediate amount, and collected foods the least; these correspond to observed rankings of sharing frequency, with meat being shared evenly in the band while collected foods are shared to a moderate degree (honey again being intermediate). While various resource qualities are correlated with sharing frequency, package size predicts much more of the resource-by-resource variance in sharing (54%) than does standard deviation in harvest success across families (23%), suggesting that Aché use package size as a robust rule of thumb for sharing decisions and/or that declining marginal value for acquirers of retaining large packages is more important that the marginal value of shares for the recipients. The net result is that "there is no discernable relationship between the amount of calories produced and the amount eaten," and the best predictor of family food consumption is number of dependents (Kaplan et al. 1990, p. 128).

Although Aché food sharing reduces consumption risk, the system is clearly not based on dyadic reciprocity. As Hawkes (1993) points out, Aché hunters do not directly control the distribution of their catch, and sharing is unconditional on foraging effort. Higher producers (i.e., better hunters) might still obtain a net nutritional gain (risk reduction that outweighs lost food income, which in case of large kills is in excess of the producer's needs); alternatively, they might be rewarded with social benefits (see below). In either case, the system of food sharing does not conform to the pattern of dyadic Tit-for-Tat conditional reciprocity envisioned in standard reciprocity theory. The suggestion that better hunters are rewarded by other band members for their production efforts and sharing raises a second-order CAP, since these rewards (e.g., greater sexual access, deference in disputes, greater solicitude for their offspring by unrelated individuals) would seem to entail private costs to those who grant them, yet provide a public good — securing the continued production of better hunters who share unconditionally with all. It is also important to note that the Aché case appears to be very unusual cross-culturally, with many other well-documented cases (e.g., Hadza, Hiwi, !Kung, Meriam, Yanomamo) lacking this extreme degree of resource pooling. Indeed, such pooling is even absent in contemporary Aché settlements (see below).

Tolerated theft. When food is acquired unpredictably, asynchronously, and in relatively large packets, at any one point in time there are likely to be "haves" and "have-nots." Given that food is likely to be characterized by diminishing marginal value to any one possessor/consumer (Figure 21.1), transfer from haves to have-nots will increase the fitness of the latter far more than it will reduce the fitness of the former. Blurton Jones (1987) suggested that under these conditions we might expect "tolerated theft" to occur, since have-nots should be willing to pay greater costs than haves in contesting resource possession. Such interactions are known as "scrounging" in the social foraging literature, and

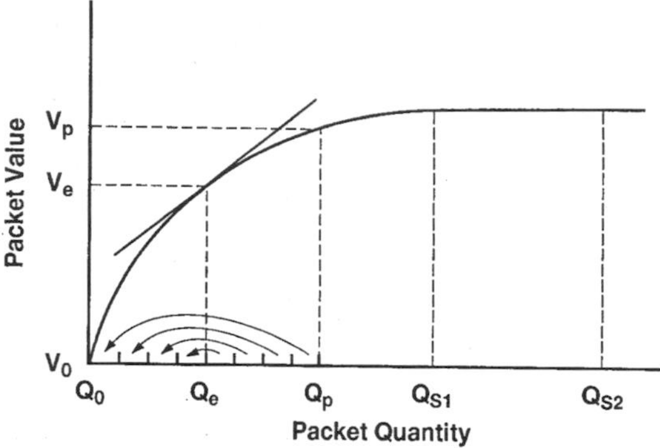

Figure 21.1 Demand sharing and declining marginal resource value in the producer-scrounger game. In the two-player case illustrated, a producer acquires a packet of size Q_p and value V_p, and relinquishes it in small portions to the scrounger (who initially has nothing, at Q_0/V_0). With equal competitive abilities (and in this simple graph, costless transactions), transfers will cease when both players attain equal marginal value and possess the equilibrium quantity Q_e. After Winterhalder (1996).

"demand sharing" in social anthropology. If contestants are of equal competitive ability, possess no other (or equal) food stores, and are characterized by equal marginal utility curves for food consumption, and if detection of harvests is immediate (or consumption sufficiently delayed), the equilibrium outcome is plausibly an equal division of the catch (Figure 21.1). If these assumptions are relaxed, the tolerated theft model will of course yield more complex predictions (Winterhalder 1996). Behavioral ecologists have modeled this process in some detail, using game theory (Giraldeau and Caraco 2000) as well as trait-group selection (Wilson 1998). Jones and others have noted that if the various relevant parameter values (e.g., competitive ability) are common knowledge, both parties may benefit by conventional solutions ("tolerating" transfers from haves to have-nots) rather than engaging in physical combat or the like — a form of mutualism nested within directly conflicting interests.

Fieldworkers disagree strenuously over the empirical relevance of tolerated theft. Whereas Hawkes (1993) suggests it is the main dynamic at work in food sharing among the Hadza (savannah hunter-gatherers of Tanzania), and Bliege Bird and Bird (1997) argue that it is better supported than alternative explanations (such as risk-reduction reciprocity) among Meriam turtle hunters of northern Australia, others find no evidence of it in the peoples they study. Thus, Kaplan et al. (1990) argue that Aché evidence contradicts tolerated theft hypotheses, in that hunters (a) actually consume less of their own production than do others, (b) return solitary kills to camp without first consuming any themselves, (c) and often call for aid upon encounter of game or honey (thus reducing their personal return rate, though often enhancing the group return rate).

Signaling. A very different explanation of sharing invokes costly signaling: by successfully harvesting and then distributing difficult-to-capture resources, individuals may reliably signal various socially important qualities, thereby benefiting themselves as well as potential allies, mates, or competitors who gain both food and useful information about the provider (Smith and Bliege Bird 2000; Gintis et al. 2001). The advantage of the costly signaling explanation is that it does not raise the collective action problems posed by reciprocity models (the threat of unilateral defection) or tolerated theft (free riding on the production efforts of others by scrounging resources). In a stable signaling system, observers will confer social benefits on signalers not as reciprocation, but because doing so is their best move: signaling indicates qualities that make it advantageous to preferentially mate with, ally with, or defer to the signaler. The weakness of signaling explanations for sharing is that the resource transfers themselves may be somewhat incidental to the signaling equilibrium; for this reason, while signaling may be a necessary component of the explanation, it is not sufficient (Gintis et al. 2001).

This apparent weakness, however, can be mitigated or eliminated under one of several conditions. First, sharing resources may serve to attract an audience, hence increasing the "broadcast efficiency" (observer per unit signal) and making the sharing a vehicle for signaling. Smith and Bliege Bird (2000) argue this is why Meriam hunters are willing to pay the entire cost of harvesting large marine turtles which they donate in toto to communal feasts hosted by unrelated clans.

Second, sharing may be somehow integral to the quality being signaled. This could happen in one of two ways: the quality being signaled might refer to the ability to generate a production surplus, or it might refer to ongoing commitment to the recipient group. Ability to generate a surplus (because of productive prowess, skilled management, or control over labor and/or resources) appears to be the key quality being signaled in many systems of communal feasting, potlatching, give-aways, and the like described in the ethnographic literature on myriad small-scale foraging, horticultural, and pastoral societies (Boone 1998). It may also play a role in the production of public goods in archaic state societies as well as contemporary electoral politics (for further discussion, see section on *Coalitions and Conflict* below). Signaling commitment to a social group by unconditionally sharing resources with its members (providing a public good) is a possibility suggested by Schelling's (1960; Nesse 2001) theory of strategic commitment, but has not yet been formally modeled or tested (Smith and Bliege Bird, submitted). The basic idea is simple enough: if I wish to convince you of my sincere ongoing commitment to a common project, I can honestly signal this commitment by contributing to the common good at levels that would not be beneficial to me were I planning on defecting over the next time period. Extended courtship (and the associated opportunity costs of time) and economic transfers such as bridewealth or dowries are straightforward examples of the phenomenon, but more subtle forms are possible (e.g., voluntarily yielding first authorship on a chapter to signal ongoing commitment to collaborative research).

Why do people share "windfall" resources more readily? A variety of lab experiments as well as anecdotes from naturalistic settings indicate that so-called "windfall resources" (those obtained by chance rather than as a result of concerted effort) are more readily shared than earned resources. For example, both Japanese and American subjects of both sexes answering hypothetical scenario questions were statistically more willing to share money (hypothetically) obtained by lottery than the same amount when it was (hypothetically) earned for participation in lab exercises (Kameda et al. 2002; see also Camerer and Thaler 1995). In effect, it appears that windfall resources are viewed as common property subject to communal sharing rules, whereas earned resources are viewed as private property that will be shared only under more stringent conditions set by the earner.

One possible explanation of this windfall-resource psychology is that it is a convention to minimize conflict costs (or more generally transaction costs) involved when resources are acquired in an unpredictable and asynchronous fashion. Kameda et al. (2003) have constructed a model for such a context, marrying the logic of Hawk–Dove games to tolerated theft. This model considers four strategies: Egoist (never share own harvest, demand a share of Other's harvest), Bourgeois (never share or demand), Communalist (always share, always demand), and Saint (always share, never demand). There is, of course, a resource of value V, contest costs of C, and group size is allowed to vary. Kameda et al. show that the Communalist strategy is evolutionarily stable under a wide range of parameter values. However, this result depends on certain assumptions, including pairwise contests followed by equal partitioning of the resource among all contestants (should the acquirer lose any contest), asymmetric conflict costs (winners pay none), and the elimination of both first- and second-order free riders via punishment.

As Kameda et al. suggest, the uncertainty involved in harvesting large game (as compared to small game or sessile resources such as plant foods) could give it the characteristics of a windfall resource, and they suggest this as the reason why a windfall psychology (or communal-sharing norm) might have evolved in *Homo*. In any case, experimental results suggest that any experiments where resources are provided by the experimenter to the subjects arbitrarily may invoke a greater propensity to share than would be the case for earned resources. This should be considered in interpreting the results of experiments utilizing the Ultimatum and Dictator games, since the stakes here are inherently unearned (windfall) resources. As Camerer and Thaler (1995, p. 216) state:

> Subjects are handed \$10 in manna from experimental heaven and asked whether they would like to share some of it with a stranger who is in the room. Many do. However, if the first player is made to feel as if he earned the right to the \$10, or the relationship with the other player is made less personal, then sharing shrinks.

On the other hand, windfall psychology cannot account for the much greater propensity for sharing found among hunter-gatherers than among social carnivores, nor the wide cross-cultural (and intra-cultural) variation in sharing rates for Ultimatum game players documented by Henrich et al. (2001).

How much inertia do sharing systems possess? Extant studies of food sharing in subsistence economies reveal some salient patterns: resources associated with higher production variance (e.g., big game) tend to be more widely shared, most food sharing is not structured as conditional reciprocity, demand sharing is common, and those who produce more and share more often have enhanced prestige. However, data do not unambiguously support any one of the explanations sketched above for perhaps the following reasons: (a) each food sharing system may be shaped by a different set of causal factors (risk reduction here, tolerated theft there, etc.); (b) none of the current models may be causally relevant; or (c) these systems may not be at a local optimum as a result of cultural inertia, bounded rationality, or stochastic factors. To address this last possibility, we will need detailed comparative studies, preferably diachronic ones. The best candidate to date again comes from the Aché study team.

Gurven et al. (2001) studied food sharing among Aché resident in a recently formed village of 117 people, which is comparable in size to many seminomadic or village-dwelling hunter-gatherers and horticulturalists but about twice as large as the median size of the nomadic pre-contact Aché residential groups and several times larger than the groups studied on forest treks in the 1980s.[2] Sharing patterns show a marked difference from those observed on treks: (a) any given household directs almost all its food sharing to just 2 or 3 other households (usually close relatives), and little or nothing is given to any of the other 22 households in the village; (b) there is a strong element of contingency (dyadic reciprocity) in sharing patterns — those to whom you give food are much more likely to give food to you; (c) despite the continuing egalitarian sociopolitical organization, there is no tendency for foraged foods to be shared preferentially with those who lack them. These same patterns have been documented for other settled forager-horticulturalists (Hiwi, Yanomamo, and Yora), yet they have developed very recently among the Aché, with their movement into permanent settlements. Gurven et al. argue that this dramatic shift can be attributed to a few key changes: larger group size, which increases the difficulty of detecting free riders; decreased risk (variance or unpredictability in daily food income), which reduces the payoff from pooling before consumption; and increased privacy due to home construction, which increases the ability to hide food from others and thus reduces the effectiveness of demand sharing. Although more detailed analyses are needed, this case does suggest that patterns of food sharing are very sensitive to socioecological context and can shift rapidly in response to changed conditions even in small, relatively isolated societies.

---

2 It is important to note that the Aché observed on treks in the 1980s were residing in settlements when not on trek, and group size and composition on these treks differed from that reconstructed before contact/settlement. Thus, trek data do not necessarily mimic aboriginal patterns any more closely than the data collected later in settlements.

Cooperative Production

Cooperative production, ranging from group hunting or fishing to construction of buildings or facilities (e.g., fish weirs), is a universal feature of human societies. It may offer several advantages: increased per capita resource harvest rate, reduced variation in harvest rates, reduced losses to competitors, and increased vigilance and predator detection. Cooperative production, however, can also increase resource depletion and competition; even where cooperation is beneficial, optimal group size itself may be unstable as a result of conflicts of interest between existing members and potential joiners. In any case, once groups form they provide the context for complex social dynamics, including economies of scale as well as competition and conflict over labor contributions and division of the product.

One form of cooperative production given great prominence in scenarios of hominid evolution is group hunting; behavioral ecologists have given this corresponding attention in ethnographic studies. The standard expectation has been that cooperative hunting occurs when there are economies of scale: per capita return rate R increases as a function of group size n, so that $R_1 > R_n$ for some range of n (Smith 1985). Suppose the per-capita return rate curve reaches a maximum at some intermediate group size n_{opt}, the optimal group size, and then declines gradually as n increases (Figure 21.2). Then members of a group of size n_{opt} have an interest in preventing additional individuals from joining the group, whereas potential joiners would increase their returns as long as $R_{n+1} > R_1$. This simple model thus predicts a conflict of interest between n members and a prospective joiner whenever $R_n > R_{n+1} > R_1$ (Smith 1985). The model, of course,

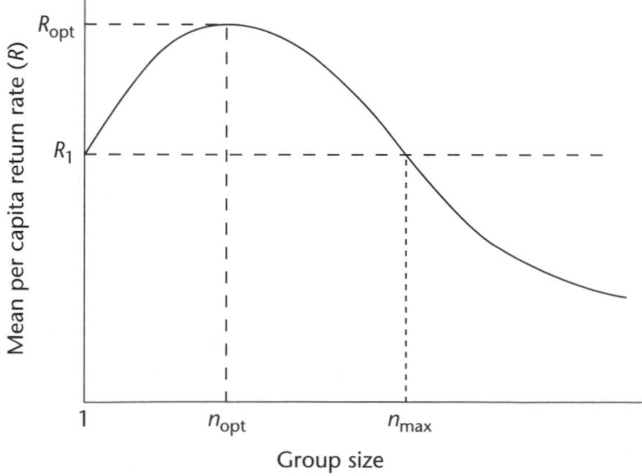

Figure 21.2 Optimal group size and member-joiner conflict. When per capita return R (a member's share of group production) reaches a maximum R_{opt} at $n_{opt} > 1$, members will benefit by restricting further entry, but potential joiners have an incentive to join as long as their share will exceed their return rate from solitary production R_1, up to the equilibrium group size of n_{max}. After Smith (1985).

Table 21.1 Group size of Inuit hunt types (data from Smith 1991, p. 316).

	Optimal group size = 1	Optimal group size > 1	Totals
Optimal group size = modal	6 hunt types	2 hunt types	8
Optimal group size < modal	2 hunt types	2 hunt types	4
Optimal group size > modal	0 hunt types	4 hunt types	4
Totals	8 hunt types	8 hunt types	16

says nothing about how such a conflict will be resolved. If members do not exclude joiners beyond n_{opt}, perhaps because exclusion presents a collective action problem, then presumably the group size will exceed the optimum (at the limit, equilibrating at n_{max}, when per capita returns are equal to R_1).

In my study of Inuit (Canadian Eskimo) hunters, I found that for hunt types where the highest per capita return rate (measured in calories per hunter hour) came from solitary hunting, the modal group size was indeed 1 in most cases (Table 21.1). For hunt types characterized by some payoff to cooperative hunting, results were quite mixed and seemed to reflect both coordination failure (where the modal observed group size was less than n_{opt}) and joiner "crowding" (where modal $n > n_{opt}$). An instructive case of the latter is beluga whale hunting, which usually occurs a day's travel from the settlement at an estuary. Hunters arrive at the hunting site in boats containing 2–3 hunters which have made the trip independently (or in coordination with perhaps one other boat) and stay one or more days. Groups smaller than 5 did not ever capture belugas, presumably because at least two boats are needed to coordinate pursuit; per capita return rate declined monotonically above $n = 6$, yet mean group size was 10.7 and groups were as large as 16. The likely explanation is twofold: lack of information makes it difficult to predict how many other hunters will be at the hunting site on any given day, and once having made the journey, hunters have no foraging options that will yield higher per capita returns even when the site is crowded (Smith 1991) — a combination of coordination failure and joiner crowding. (It is worth noting that behavioral ecologists studying cooperative hunting in nonhuman species have also had difficulty demonstrating anything that goes beyond by-product mutualism [review in Dugatkin 1997; cf. Boesch 2002].)

Alvard and Nolin (2002) report on an extensive study of cooperative hunting in the Indonesian community of Lamalera. They describe a quite complex system, involving corporate kin groups (subclans) that own traditional paddle-and-sail-powered vessels and field crews organized into specific roles (e.g., harpooner, helmsman, bailers). A set of rules specifies a very precise division of the catch, not only among the members of a successful boat crew, but also among other designated recipients (sail maker, boatwright, boat manager, and various individuals or groups with hereditary rights). Given this division of the catch (e.g., for sperm whales, 14 named shares assigned to at least 40 designated recipients) and observed hunting behavior, Alvard and Nolin calculate that whaling provides significantly greater mean returns to each crew member (ca. 0.55 kg/hr) than does the next best alternative of net fishing (0.34 kg/hr). They thus conclude that Lamalera whaling is a case of synergistic mutualism, with a payoff schedule matching the classic "stag hunt" coordination game.

Although some might view coordination games and mutualistic equilibria as not all that interesting or difficult to achieve, Alvard and Nolin (2002, p. 547) argue otherwise:

> Substantial coordination is required to subsist on cooperatively acquired resources. Behaviors must be synchronized, rules must be agreed to (even if tacitly), and assurance, trust and commitment must be generated among participants for the collective benefits of cooperative hunting to be realized.

Note, however, that in the Lamalera case the complex system of sharing rules, designated roles (e.g., boat manager), and division of labor are all necessary to ensure successful coordination of effort and hence a Pareto-efficient mutualistic payoff. These elements, in particular the sharing rules, go well beyond a simple coordination game, and in fact must have been produced by a long process of cultural evolution.

Sharing rules in hunter-gatherer societies are not often as precisely specified as in the Lamalera case. However, there are certainly norms that vary within societies (e.g., with respect to different resource types) as well as between them. The sharing rule implied in the joiner-member model outlined above is an equal division of the catch among members of the production group, with those excluded being on their own. At the opposite extreme would be a communal-sharing rule, where all producers in the band or village — whether their own production efforts were cooperative or solitary — pool and equally divide the product. The joiner-member model can be modified accordingly (Smith 1985). For simplicity, assume that the village contains N producers who can each decide whether to engage in cooperative production in a single group of size n, or to be one of m individuals engaged in solitary production ($n + m = N$). Given communal sharing, an individual share (regardless of production tactic) equals $(nR_n + mR_1)/N$. It follows that any individual will increase her production share by becoming the nth member of the cooperative-production group as long as

$$nR_n + (m-1)R_1 > (n-1)R_{n-1} + mR_1 ,$$

which simplifies to

$$nR_n - (n-1)R_{n-1} > \mathrm{R}_1.$$

This last inequality states that under communal sharing, the decision rule for production is to participate in cooperative production so long as the marginal gain in the group production rate (the left side of the inequality) exceeds the rate that can be obtained from solitary production (the right side). This stylized model illustrates how a change in sharing rules can significantly alter incentives and, in this case, dissolve the joiner-member conflict.

Ethnographic data indicate that the behavior of forest-dwelling Aché (see discussion above) closely approximates the communal-sharing rule. Consistent with the prediction just made, Aché engage in cooperative foraging (e.g., calling for aid in prey capture, pointing out resources for others to harvest, helping with prey tracking and capture) even when the act of doing so leaves the donor's return rate unchanged or lower, so long as it raises the band's overall return rate (Hill et al 1987; Hill 2002). This is particularly true for hunting and honey harvesting, but less so for gathering; again, it is well documented that Aché-gathered foods are shared much less communally than are meat and honey. Whereas Hill (2002) interprets the extensive cooperation while foraging as "altruistic" (at least in the short term), the arguments just given suggest it is simply mutualistic, *as long as the harvest is communally shared.* Of course, such communal sharing is itself a remarkable example of cooperation-in-need explanation.

While the studies summarized above provide many insights, they are couched primarily in terms of average individuals (sometimes differentiated by sex). They thus offer little insight into individual differences in constraints and opportunities that might affect decisions to participate in cooperative production. An interesting effort in this direction is a study by Sosis et al. (1998), which used bargaining theory to explain differences in individual participation in cooperative fishing production on Ifaluk atoll in Micronesia. They showed that fishing effort was lower among older men, those from higher-status clans, those with more education, and those with adult sons residing in their household, but correlated positively with need for food (as measured by household stores and numbers of dependents). These matched their predictions regarding the factors that will enhance bargaining power in interactions determining individual contribution to cooperative fishing efforts on Ifaluk. (Of course, a variety of other explanations could account for these observations.)

Aid-giving Behavior

Although aiding unrelated adult conspecifics who are seriously ill or incapacitated is reported for some dolphin species, such behavior is, in degree if not in kind, uniquely developed in humans. Darwin felt it so notable as to single it out to illustrate the unique moral evolution of our species, and paleoanthropological evidence suggests it first arose in Neanderthals and early *Homo*.

Sugiyama and Chacon (2000) studied the effect of illness and injury on foraging returns and aid-giving among two Amazonian village peoples: Yora (Peru) and Shiwiar (Ecuador). They estimate that injuries reduced foraging effort by at least 10.6% and that if a hunter of average skill is incapacitated, this reduces protein intake by 18% whereas if he were the best hunter in the group then protein intake would drop 32–37%. They note that reciprocal altruism fails to offer a convincing explanation for aid-giving (feeding of incapacitated individuals and their dependents), since the more ill one is (a) the lower the probability of survival, (b) the longer until recovery, (c) the lower the ability to punish defectors, and hence (d) the greater the temptation to defect (fail to aid the incapacitated). As Sugiyama and Chacon (2000, p. 384) put it, "in a world where only the logic of kin selection and reciprocal altruism operate, there comes a point at which abandonment of a sick or injured individual becomes the adaptive choice." Although ethnographic anecdotes indicate that such abandonment does sometimes occur, Sugiyama and Chacon suggest that the threshold for abandonment can be increased by strategies such as costly signaling of willingness to provide public goods (e.g., sustained hunting effort and widespread sharing of the catch) and social niche differentiation to position oneself as providing irreplaceable benefits (see also Tooby and Cosmides [1996] on the "banker's paradox"). Evidence that this occurs among Aché has been provided by Gurven et al. (2000). However, neither study provides any direct tests of hypotheses concerning mechanisms by which such a system could evolve or be stabilized. This is one area in which models incorporating partner choice (e.g., Cooper and Wallace 1998; Bshary and Noë, this volume) would seem to have much promise.

Coalitions and Conflict

Humans are arguably unique among vertebrates in the size, importance, and diversity of their coalitions. Although much coalitional behavior in small-scale societies is kin based, even this presents challenging problems for evolutionary analysis (see below). In any case, nonkin coalitions are important vehicles for within- and between-community competition in all human societies. Recent theory (Gil-White 2001; McElreath et al. 2003) and experimental data (Bornstein et al. 2002) suggest that culturally defined in-group identity, such as ethnicity, allows people to predict the presence of hard-to-observe norms and behavioral propensities, thus facilitating the solution of coordination problems. Yet what about more costly forms of cooperation? Perhaps the most striking acts of self-sacrificial cooperation in both humans and social insects occur in the context of coalition-based violent conflict, including warfare. In social insects, within-colony relatedness is usually very high, but this is not normally the case for raiding and warfare among humans.

Patton (2000) studied warfare ("intercoalitional violence"), male status, and reciprocity in an Indian community in the Ecuadorian Amazon. He found that male status is strongly correlated with warrior status (with status of both types scored by the independent rankings elicited through interviews). Unpublished data (Patton, pers. comm.) indicate that these status measures are positively correlated with reproductive success. This parallels evidence for the Yanomamo of Venezuela of a strong relationship between reproductive success and *unokai* status (marked by a public ceremony given to those who have killed an enemy on a raid): men who are *unokai* average over twice as many wives and over three times as many offspring as do other men (Chagnon 1988). Patton (2000, p. 420) argues that these social benefits of participating in coalitional violence are underpinned by a system of indirect reciprocity and reflect "an evolved strategy for the use of violence for status gain within an coalitional context." Given the doubts about how such a system of indirect reciprocity might work (Boyd and Richerson 1989; Leimar and Hammerstein 2001), we need to consider alternative hypotheses.

One such alternative is costly signaling. Proven ability in lethal fighting with enemies should be a reliable signal of the physical, emotional, and cognitive qualities that would make someone a formidable competitor. Such individuals might often be desirable allies, and competitors with lower competitive ability might find it wise to defer to them. Warriors might also be desirable mates, if their proclivity for violence is not too generalized and protection from other males has high adaptive value; in any case, they might have an easier time using alliances with, and intimidation of, other males to gain more mating opportunities. Of course, in systems (such as the Yanomamo) where some men gain wives through raiding and abduction, there can be a fairly direct link between coalitional violence and reproductive success (tempered of course by tradeoffs involving increased mortality risk).

It is important to distinguish social systems where participation in lethal conflict is voluntary and unpaid from those where military service in the rank-and-file is coerced (by conscription, threat of imprisonment, etc.) or is a source of income and upward mobility for relatively impoverished classes. The former includes the vast majority of small-scale societies, whereas all states and some chiefdoms fall in the latter category. Hierarchical societies may use various institutional means of encouraging morale and commitment (ideology, intimate face-to-face relations in modular combat units; see Richerson et al., this volume). However, given the direct incentives (threats and rewards) that motivate enlistment, the evolutionary explanation of lethal risk taking in combat among members of stratified societies is simpler (or rather, deflected to accounting for the social institutions that carry out third-party enforcement of military and political unity). In contrast, among small-scale societies, military conflict is primarily organized at the level of voluntary raiding parties led by charismatic leaders, and the adaptive payoffs include booty, captive females, and the local status enhancement noted above. (Exceptions occur in very densely populated but still small-scale societies, such as highland New Guinea.) In these systems, there is little evidence of self-sacrificial devotion to the military success of the entire society, and within-group factionalism (resulting quite often in homicide) is often common, though controlled to some extent by various institutions (e.g., adjudication by elders) as well as by threats of revenge.

Coalition-based conflict need not be violent to be important; much of political life in any society is dominated by more restrained forms of conflict. Models of political microdynamics usually assume that politicians gain power as part of a reciprocal exchange: a politician promises goods to his constituents in return for the favor of their support. Given the delayed return here ("I support you now, you return the favor by providing collective goods in the future"), defection is always a distinct possibility. Costly signaling might not eliminate the risk of defection, but it could help predict which individuals are less likely to do so: if a candidate can reliably signal a superior ability to obtain resources for redistribution, he should have a higher probability of actually doing so when elected. Here, costly signaling does not guarantee honesty of intent to deliver collective goods, but it may guarantee honest advertisement of ability to do so.

A variety of political systems, ranging from the semi-egalitarian "big man" systems of Melanesia to the stratified chiefdoms of the Northwest Coast Indians, appear to display various elements of this costly signaling dynamic of garnering political support through magnanimity (Boone 1998). In these cases, and arguably in many instances of electoral politics in modern industrialized democracies, political candidates use distributions of goods to signal honestly their ability to benefit supporters in the future. The big man, chief, or congressional candidate encourages others to donate wealth or labor in his support by displaying honest signals of his skill in accumulating resources for redistribution, thus ameliorating the most problematic aspect of delayed reciprocity, risk of default.

Whereas these arguments concern power plays within a political system, signaling may also play an important role in competition between systems. The archaeologist Fraser Neiman (1998) proposes that the florescence of monumental architecture (particularly flat-topped pyramids) among Classical Maya city-states was a form of costly signaling serving to advertise honestly the political and economic (and hence military) power of competing kingdoms. Whereas warfare was certainly common enough in these and other archaic states, Neiman argues that such provisioning of public goods served the interests of elites in competing polities by deflecting costly conflict in cases where the architectural signals indicated equally matched opponents, while simultaneously signaling the power of the elite class to the commoners within their polity.

WHY ARE HUMANS SO COOPERATIVE?

Having surveyed some relevant research in human behavioral ecology, I move now to the more speculative part of this chapter. A key question our discussion group at Dahlem sought to address is why cooperation in large groups with low relatedness is so common in humans. Various answers to this question have been proposed in the literature, including:

- a hominid population structure directly favoring the evolution of cooperation via genetic group selection (Alexander 1974; Hamilton 1975; Boehm 1997);
- genetic group selection among alternative (individually selected) equilibria (Boyd and Richerson 1990);
- cultural group selection facilitated by conformist transmission (Boyd and Richerson 1985).

As discussed by Richerson et al. (this volume), the case for the first alternative is weak. The latter two have a sounder theoretical basis (if yet untested empirically) and are ably discussed in other chapters in this volume (see Richerson et al., and Bowles and Gintis). Here I focus on alternative or complementary accounts giving central explanatory roles to language (symbolic communication) and technology.

Language and Collective Action

Symbolic communication using various linguistic media (spoken, written, signed) is a specialty of *Homo sapiens.* Most anthropologists consider it the most significant derived feature of our lineage, one that enables the transfer of large volumes of cultural information. Although communication is certainly possible without language, language allows people to communicate about events remote in time and space (including imagined futures), to express (however imperfectly) high-level cognitive abstractions and internal subjective states, and to create collectively cultural webs of meaning. Language also allows people to flatter, lie, dissemble, mislead, and obfuscate — all of which can be quite adaptive for those doing so (if not for those listening). These various aspects of linguistic communication have important implications for the forms and extent of cooperation.

Cheap Talk and Coordination

Coordination games are a relatively straightforward but underappreciated context for cooperation and the first place to look for effects of linguistic communication on cooperation. It seems almost certain that language greatly facilitates the several aspects of solving coordination problems: defining options, specifying players' preferences, and agreeing on the solution. This communication need not even be direct: I can tell Sam that Rob said he would meet him at the *Alexanderplatz* TV tower at noon, or a traffic sign can tell me which way to drive on a one-way street.

Ample evidence from lab experiments (e.g., Crawford 1998) and the real world indicates that pregame communication can significantly enhance the probability of attaining efficient solutions to coordination problems (as well as several other game forms). This suggests that "cheap talk" can be quite valuable in cases where agents share common interests:

> ... solutions that include pregame negotiations are often considered trivial by economists because all humans can easily communicate in this way. From a comparative evolutionary perspective such a solution is far from trivial. The adaptive value of being able to communicate honest cooperative intent with a statement such as "I will hunt whales tomorrow with you if you hunt whales tomorrow with me" is hard to overestimate. (Alvard and Nolin 2002, p. 549)

In addition, language could be critical for defining conventions to minimize transaction costs and stabilize Pareto-superior solutions to coordination games (Alvard and Nolin 2002). When coordination games are repeated over generations, it is plausible that the locally prevalent solutions will become codified as written or orally transmitted norms that come to seem "natural" and exogenous to participants.

Signaling and Reputation Effects

Many social interactions (and the games that model these) look very different once we consider the signaling value of alternative strategies. Thus, a one-shot public goods game with a single Nash equilibrium of defection (failure of any player to provide the good) can be transformed into a signaling game where the strategy pair "signal only if high quality, provide social benefits only to signalers" is an equilibrium with a large basin of attraction (Gintis et al. 2001). As discussed above, such an analysis may explain a range of public-goods provisioning, from big-game hunting with unconditional sharing to charity galas in capitalist societies.

The role of signaling in favoring cooperation is a relatively new topic that is as yet poorly studied but is likely to be of great importance, particularly in the human case. Because of this novelty, the role of signaling effects is often overlooked. Even the venerable game of Chicken derives its name from a form of human behavior that makes little sense unless one realizes that the situation referred to (whether or not to yield to an opponent in a ritualized public contest of nerves) is nested within a larger game involving reputation effects. Such contests are not limited to 1950s American teenagers but are culturally widespread, ranging from various forms of dueling to male initiation rites to military maneuvers of state societies. (Of course, Chicken games need not entail an underlying signaling context, e.g., this is usually absent in the Hawk–Dove version analyzed in behavioral ecology.)

When signalers can derive social benefits from a number of individuals (not just a single partner), the payoffs from signaling can be greatly enhanced by some means of efficient broadcasting. Linguistic communication provides a vehicle for very low-cost (hence efficient) signaling. Instead of having to direct signals physically to observers, signalers can rely on observers to spread the word to others (e.g., the fact that Toma killed a giraffe can become known to distant parties that never tasted a morsel of that giraffe).

Of course, there is the important issue of what (if anything) ensures honesty. Much current work is aimed at understanding how low-cost linguistic communication can be linked to costly signaling theory. One proposed answer turns on social enforcement: dishonest statements can be discovered and punished (Lachmann et al. 2001). While this argument is certainly correct, it would limit the signaling value of language to situations where receivers can use other means to verify signal honesty (as well as coordination contexts where there is no incentive to be dishonest). A second proposal turns on reputation effects: if I pass unreliable information too often, you will come to discount what I say, and then I will find it difficult to influence your behavior. This is also plausible, but it can be costly for receivers while they are building up information on others' honesty, and the payoffs and dynamics here scarcely differ from nonlinguistic signaling.

With language (unlike a peacock's tail or a sparrow's status badge), you can learn something about my track record for honesty from third parties. The means by which such third-party information is transmitted ranges from gossip to testimony at public hearings to media accounts. None of these forms of linguistic communication are necessarily honest themselves, but I doubt that any have zero reliability. As with other forms of information accrual (trial-and-error learning, observation of others' behavior, etc.), individuals face a problem of statistical evaluation that they may or may not be able to solve in any given case. I would expect individuals to give greater weight to first-hand accounts of direct experience with individual X (e.g., "Sally brought me food when I was sick," "John lied to me"), and to multiple independent first-hand accounts, than to vague or second-hand accounts (e.g., "I hear Sally is a nice person," "Jane told me that John can't be trusted"). By marrying models of many-sided cultural transmission to the problem of establishing reputations for cooperation and honesty, we ought to be able to put the ideas of third-party reputation and indirect reciprocity ("standing" or "image score") on more solid footing.

Monitoring, Assortment, and Enforcement

The arguments just given focus primarily on how linguistic communication can improve outcomes in dyadic interactions, but what about multiplayer interactions that involve trust, public goods, potential for defection, and the like? I see at least three ways in which language can enhance the possibility of cooperative outcomes. First, linguistic communication might significantly lower the cost of monitoring selfish behavior in a Prisoner's Dilemma or public goods payoff environment. It is widely recognized that as group size increases beyond a very small number, the difficulty of each agent observing the behavior of all other agents makes free riding and other forms of selfish behavior much more likely to proliferate (e.g., Boyd and Richerson 1988). Language, however, allows individuals to learn about defection from other group members without having to observe it themselves directly.

Second, if language can be used to communicate information about honesty and cooperative history, then it can facilitate positive assortment of groups of cooperators. It is well known that such positive assortment can be very effective in enhancing the evolution and stability of cooperation. The problem, of course, is how to ensure reliability of the information or markers used for assortment. Language alone cannot do this (it is too easy to pretend to be a cooperator, even to oneself), but other means do exist (including the costly signaling avenue sketched above). What language can do is make it much easier to find out (with admittedly imperfect but presumably nonzero accuracy) an individual's past track record of cooperative behavior. These reputations, amplified through linguistic communication, should significantly ease the task of forming groups composed of cooperators. Again, some explicit models of this process, incorporating both linguistic communication and assortment dynamics, are sorely needed to evaluate such plausibility arguments.

Third, and perhaps most speculatively, I propose that linguistic communication can help reduce the cost of punishing defectors. The lowest-cost form of punishment is simply the third-party communication about behavior and reputation just discussed. Again, I expect this information to be of intermediate reliability and thus better than no information. Many forms of human cooperation, particularly those involving larger or variable-membership groups, rely on rules and norms that define both the rules of cooperation and modes of enforcement (including punishment). Language plays an indispensable role in formulating and transmitting these rules and norms. At the higher end of punishment cost, when punishers must directly confront defectors and impose penalties upon them, linguistic communication can at least play a role in coordinating a cooperative form of punishment. Cases where members of a hunter-gatherer band secretly plotted the abandonment or even assassination of incorrigible offenders are described in the ethnographic literature; such coordinated actions greatly reduce the per-capita cost incurred by the punishers and would be essentially impossible without linguistic communication.

Commitment

Many forms of human cooperation rely on commitments, including both secured forms such as enforceable contracts and less secured forms such as public or private promises and codes of honor (Nesse 2001). Language certainly must greatly facilitate the making of commitments, in which individuals agree in advance to a prescribed course of action, operating perhaps under a Rawlsian veil of uncertainty about what the future outcome might be. Thus, Carl and I might agree to take turns buying a lottery ticket (or going hunting), with the explicit agreement that whoever happens to succeed will share the proceeds with the other. The facilitating role of language should be particularly important for commitments to involving multiparty collective action, where nonlinguistic communication about future contingencies would be difficult if not impossible.

In addition, it seems obvious that linguistic communication also greatly expands the possibilities for advertising (and monitoring) commitments, for the reasons described above with regard to monitoring, assortment, etc. Commitments that are advertised widely (through linguistic communication) may offer advantages to the one making the commitment, and they can then be monitored by a larger audience.

Technology and Collective Action

Language may also play a critical role in making both technology and complex division of labor possible (though certainly not inevitable). By technology, I mean more than just tools; I mean a combination of tools, culturally transmitted knowledge about tool manufacture, and the use of tools in various realms, particularly in economic production.

The issue here is how technology can increase the payoffs from cooperative production. Examples from small-scale societies, even ones with "stone-age" technology, are plentiful: nets and brush or stone surrounds for game drives, fish weirs, multiperson (or multiply-deployed) watercraft, etc. Higher payoffs from cooperative production mean a greater incentive to solve collective action problems, to ensure any needed coordination, and counter free riding. Once cooperative production and other forms of (nonkin *n*-person) fitness interdependence mediated by technology and language have a foothold, they generate incentives to develop supporting social institutions and norms (Kaplan et al. submitted).

A single ethnographic example can illustrate my argument. The horse-mounted nomadic bison-hunting Indians of the North American Great Plains region are the stereotypical Indian culture of cinema and popular writing. Prior to ca. 1700, however, no such culture existed. As horses became available (after the Pueblo Revolt drove the Spanish colonists temporarily out of New Mexico), various Indian peoples migrated out onto the plains and rapidly developed a new way of life: a coadapted economy, residence pattern, set of political and religious institutions, kinship system, and so on (Oliver 1962). Within less than a century, Indians from various regions (mostly outside the plains) and with no common language or shared set of social institutions had converged on a new and distinct way of life. This lifeway, recorded in great detail by travelers and ethnographers, was remarkably adapted to the exigencies of using horses and bows and arrows (later rifles) to hunt bison, an extremely abundant (ca. 60 million) but heretofore difficult to locate and harvest nomadic herd animal.

Of particular interest here is the collective action problem posed when a tribe of several thousand people aggregated together for the summer months. The Cheyenne case is representative:

> From the time of the performance of the great ceremonies [around summer solstice] to the splitting up of the tribe at the end of the summer, no man or private group may hunt alone. During the early summer months the bison are gathered in massive herds, but distances between herds may be great. A single hunter can stampede thousands of bison and spoil the hunt for the whole tribe. To prevent this, the rules are clear, activity is rigidly policed [by a formal warrior's association], and violations are summarily and vigorously punished. (Hoebel 1978, p. 58)

A payoff matrix could hardly be clearer. Hoebel goes on to provide several detailed accounts of cases in which the rule barring selfishly "jumping the gun" was violated, and the prescribed punishment meted out (including killing the violators' horses and smashing their weapons). Again, note that these rules and institutions, brought to bear to ensure that the potential gains from collective action not be eroded by selfish behavior, had come into existence in just the few decades that elapsed from the Cheyenne abandoning horticultural villages in

Minnesota and becoming nomadic equestrian bison hunters on the Plains. They were clearly a response to a new economic opportunity afforded by the technology of mounted bison hunting and could not have existed without a symbolic cultural system based on language.

What about Kinship?

In the heady early days of sociobiology, many thought that explanations based in kin selection would unlock the mysteries of human sociality. After all, kinship — real or metaphorical — is a key organizing principle in all societies, and a linchpin for collective action in many. Inclusive fitness, however, has not proved to be the universal acid that dissolves the problems of human cooperation (nor even insect sociality). One problem is that coefficients of relatedness drop off rapidly outside a narrow orbit of close kin, whereas much of the puzzle of human sociality concerns the high amount of cooperation between members of different families (though they may often belong to the same large corporate kin group, such as a clan). Another is that kinship is often defined culturally in ways that do not line up well with the calculus of inclusive fitness. Thus, in many societies we find that unilineal kin groups (e.g., clans or lineages defined either patrilineally or matrilineally) are important foci of cooperation and within-group factionalism. Such systems seem peculiar from the standard perspective of kin selection, as they arbitrarily define half of one's genetic kin as closer cooperators than the other half.

An alternative view is that kinship is simply one of many possible conventions people use for defining in-groups in order to compete with out-groups. Yet if all kinds of arbitrary distinctions can be stable in complex games, why do people settle on kinship as the convention so often? One possible answer: given that so much of the social system in small-scale societies is based on kinship, it is a very convenient preadaptation on which to hang your coalition structure. In any case, it is a fair generalization that unilineal kin groups occur only where there are economically defendable forms of property that cannot be effectively managed or inherited in family lines (e.g., cattle herds, complex agricultural holdings, salmon streams, positions on a council of chiefs) and where formal bureaucratic structures for solving conflicts over such property rights (i.e., state systems) do not exist. In effect, unilineal kin groups are a means of forming coalitions to compete with other coalitions. If coalitions were based solely on genetic relatedness, each Ego would have a different set of preferred coalition partners (except in the limiting case of full siblings), so group boundaries would be ambiguous at best; at worst, conflicts would erupt between kin along lines defined by Hamilton's rule, and it would be difficult or impossible to hold large coalitions together (van den Berghe 1979; Alvard 2003).

Defining coalition boundaries on the basis of unilineal descent (e.g., every Ego belongs from birth to the clan of his/her mother) may solve the ambiguity problem, but in itself this does not vanquish the problems posed by cross-cutting loyalties (based on true genealogical relatedness, or other shared interests) or free-rider problems. Thus, using kinship (or ethnicity, or a variety of other conventional markers) to define group boundaries might be relatively straightforward when solving coordination problems (McElreath et al. 2003), but what is to stop a defector from free riding on the collective goods provided by kinsmen? My (highly speculative) answer is that it might be possible to extend kin-based cooperation to contexts where individual and group interests conflict if group affiliation is sufficiently costly (e.g., you won't be recognized as a member of the Turtle clan unless you undergo ritual scarification with risk of infection or donate sufficient quantities of goods to clan feasts). Under these conditions, it might not pay to fake one's affiliation, the cost only being worth paying if one is committed for the long haul. Still, this proposal is vulnerable to the question of who will enforce the cost-paying rule, as well as what to do about collective goods that are nonexcludable.

Doug Jones (2000) has developed an interesting variant on the kinship-as-group-nepotism argument. Using a combination of explicit population genetics involving multilevel selection and *n*-player game theory, he derives results that amplify kin selection and extend it to groups of various size. These results depend, however, on an exogenous solution to large-group collective action problems; in effect, they explore the implications for kin selection of having solved *n*-player collective action problems by some other means.

Limits to Cooperation

Much of the recent literature on the evolution of human cooperation extols the ascendance of prosocial norms, pro-community institutions, and innate cooperative preferences. Even allowing for the fact that some evolutionary models of the evolution of such "prosociality" are based on chronic and lethal inter-group conflict (e.g., Richerson and Boyd 2001), I suggest this picture is rather simplistic. Cooperation in human groups is far from perfect. Many social institutions and practices are grossly unfair to segments of the society (e.g., women, the poor, subjugated castes, and ethnic groups). Free riding, socioeconomic exploitation, and other inequalities with major fitness consequences are well-known features of state societies. Ethnographic evidence indicates that at least some of these are also common in small-scale (nonstate) societies, though at arguably lower levels. For one thing, monitoring of and sanctions against antisocial behavior are universal, which in turn suggests selfish behavior is also ubiquitous. Within-group homicide rates can be very high in stateless societies (or in areas where the state is weak), and these often concern disputes over adultery, theft, "honor," or alleged witchcraft (rather than enforcement of prosocial norms).

According to some accounts, conformist cultural transmission and/or enforcement of prosocial norms act to reduce fitness differences drastically within groups, thus facilitating group selection (Boehm 1997; Wilson 1998; Bowles and Gintis, this volume). However, quantitative evidence from various societies with egalitarian or semi-egalitarian sociopolitical structure (Aché, Achuar, Hadza, Hiwi, !Kung, Meriam, Yanomamo) reveals substantial differences in at least male reproductive success (Smith et al. 2003). This suggests to me that resource sharing and other egalitarian elements in small-scale societies may have less impact on fitness differentials than some have proposed. Indeed, various explanations of resource sharing — risk reduction, costly signaling, tolerated scrounging, as well as bargaining dynamics in dominant-subordinate relations — indicate that giving away some portion of one's resources may offer higher marginal fitness returns than hoarding them. Interpreting resource (and power) sharing as prosocial "leveling mechanisms" (Bowles and Gintis, this volume) may mask the prime evolutionary forces that shape such behavior, as well as their fitness consequences.

In sum, conflict, exploitation, free riding, and reproductive skew appear to be much more pervasive in small-scale societies than is commonly realized and large-scale collective action much less common. (In state societies, exploitation and inequality is generally more institutionalized, but conflict management and large-scale cooperation are facilitated by segmentation into smaller groups where trust and enforcement is more likely, as well as by third-party enforcement with selective incentives for enforcers.) Humans may be much more cooperative than baboons or chimpanzees, but the evidence suggests to me that the gap is not so vast as portrayed in some accounts.

CONCLUSION

As Richerson and Boyd (2001, p. 212) note, the unique features of human sociality "cast into question explanations that should apply widely to many other species.... If a cheap, honest, cooperative signaling system evolves in a straightforward way, then we should expect many species to use it, and cooperation on the human pattern should be relatively common." The point is well taken, and the challenge is to provide evolutionary explanations for human cooperation that are powerful enough to explain the empirical evidence without being so broad as to predict identical outcomes in other species.

Currently there are several plausible accounts for the evolution of human cooperation. I have nominated symbolic communication and the fitness interdependencies arising from technologically mediated complex division of labor as species-specific elements that shift human behavioral ecology toward more intensive and larger-scale cooperation. These elements arose in the context of yet poorly understood evolutionary transitions creating our species (and its immediate predecessors), a transition in which natural selection favored the ability to produce surplus resources and expand the scale of social interaction, which in turn required solving collective action problems that other species have not managed to overcome. Several participants at this Dahlem Workshop propose a crucial role for group selection (cultural and/or genetic) in generating the intensified cooperation of our species; however, given the lack of development of alternatives, I would argue this remains an open question. Our theoretical possibilities are rich, but meaningful evaluation of these will require expanded model-building and empirical testing.

ACKNOWLEDGMENTS

I am very grateful to Michael Alvard, Carl Bergstrom, Hillard Kaplan, and Peter Richerson for helpful comments on an earlier draft.

REFERENCES

Alexander, R.D. 1974. The evolution of social behavior. *Ann. Rev. Ecol. Syst.* **5**:325–383.

Alvard, M.S. 2003. Kinship, lineage identity, and an evolutionary perspective on the structure of cooperative big game hunting groups in Indonesia. *Hum. Nat.*, in press.

Alvard, M.S., and D. Nolin. 2002. Rousseau's whale hunt? Coordination among big-game hunters. *Curr. Anthro.* **43**:533–559.

Bliege Bird, R., and D.W. Bird. 1977. Delayed reciprocity and tolerated theft: The behavioral ecology of food sharing strategies. *Curr. Anthro.* **38**:49–78.

Blurton Jones, N.G. 1987. Tolerated theft: Suggestions about the ecology and evolution of sharing, hoarding and scrounging. *Soc. Sci. Info.* **26**:31–54.

Boehm, C. 1997. Impact of the human egalitarian syndrome on Darwinian selection mechanisms. *Am. Nat.* **150(Suppl.)**:S100–S121.

Boesch, C. 2002. Cooperative tactics in hunting among Tai chimpanzees. *Hum. Nat.* **13**:27–46.

Boone, J.L. 1998. The evolution of magnanimity: When is it better to give than to receive? *Hum. Nat.* **9**:1–21.

Bornstein, G., U. Gneezy, and R. Nagel. 2002. The effect of intergroup competition on intragroup coordination: An experimental study. *Games Econ. Behav.* **41**:1–25.

Boyd, R., and P.J. Richerson. 1985. Culture and the Evolutionary Process. Chicago: Univ. of Chicago Press.

Boyd, R., and P.J. Richerson. 1988. The evolution of reciprocity in sizable groups. *J. Theor. Biol.* **132**:337–356.

Boyd, R., and P.J. Richerson. 1989. The evolution of indirect reciprocity. *Social Networks* **11**:213–236.

Boyd, R., and P.J. Richerson. 1990. Group selection among alternative evolutionarily stable strategies. *J. Theor. Biol.* **145**:331–342.

Brown, J.L. 1983. Cooperation: A biologist's dilemma. *Adv. Study Behav.* **13**:1–37.

Camerer, C., and R.H. Thaler. 1995. Anomalies: Ultimatums, dictators and manners. *J. Econ. Persp.* **9**:209–219.

Chagnon, N.A. 1988. Life histories, blood revenge, and warfare in a tribal population. *Science* **239**:985–992.

Clements, K.C., and D.W. Stephens. 1995. Testing models of non-kin cooperation: Mutualism and the prisoner's dilemma. *Anim. Behav.* **50**:527–535.

Cooper, B., and C. Wallace. 1998. Evolution, partnerships and cooperation. *J. Theor. Biol.* **195**:315–328.

Crawford, V. 1998. A survey of experiments on communication via cheap talk. *J. Econ. Theory* **78**:286–298.
Dugatkin, L.A. 1997. Cooperation among Animals: An Evolutionary Perspective. New York: Oxford Univ. Press.
Gil-White, F. 2001. Are ethnic groups biological "species" to the human brain? Essentialism in our cognition of some social categories. *Curr. Anthro.* **42**:515–554.
Gintis, H., E.A. Smith, and S.L. Bowles. 2001. Cooperation and costly signaling. *J. Theor. Biol.* **213**:103–119.
Giraldeau, L.-A., and T. Caraco. 2000. Social Foraging Theory. Monographs in Behavior and Ecology. Princeton, NJ: Princeton Univ. Press.
Gurven, M., W. Allen-Arave, K. Hill, and A.M. Hurtado. 2000. "It's a wonderful life": Signaling generosity among the Aché of Paraguay. *Evol. Hum. Behav.* **21**:263–282.
Gurven, M., W. Allen-Arave, K. Hill, and A.M. Hurtado. 2001. Reservation food sharing among the Aché of Paraguay. *Hum. Nat.* **12**:273–297.
Hamilton, W.D. 1975. Innate social aptitudes of man: An approach from evolutionary genetics. In: Biosocial Anthropology, ed. R. Fox, pp. 133–155. London: Malaby.
Hawkes, K. 1993. Why hunter-gatherers work. *Curr. Anthro.* **34**:341–362.
Henrich, J., R. Boyd, S. Bowles et al. 2001. Cooperation, reciprocity and punishment in fifteen small-scale societies. *Am. Econ. Rev.* **91**:73–78.
Hill, K. 2002. Altruistic cooperation during foraging by the Aché, and the evolved human predisposition to cooperate. *Hum. Nat.* **13**:105–128.
Hill, K., H. Kaplan, K. Hawkes, and A.M. Hurtado. 1987. Foraging decisions among Aché hunter-gatherers: New data and implications for optimal foraging models. *Ethol. Sociobiol.* **8**:1–36.
Hoebel, E.A. 1978. The Cheyennes: Indians of the Great Plains. 2d ed. New York: Holt, Rinehart and Winston.
Jones, D. 2000. Group nepotism and human kinship. *Curr. Anthro.* **41**:779–809.
Kameda, T., M. Takezawa, and R. Hastie. 2003. The logic of social sharing: An evolutionary game analysis of adaptive norm development. *Pers. Soc. Psych. Rev.*, in press.
Kameda, T., M. Takezawa, R.S. Tinsdale, and C.M. Smith. 2002. Social sharing and risk reduction: The psychology of windfall gains. *Evol. Hum. Behav.* **23**:11–33.
Kaplan, H., K. Hill, and M. Hurtado. 1990. Fitness, foraging and food sharing among the Aché. In: Risk and Uncertainty in Tribal and Peasant Economies, ed. E. Cashdan, pp. 107–144. Boulder, CO: Westview Press.
Lachmann, M., S. Szamado, and C.T. Bergstrom. 2001. Cost and conflict in animal signals and human language. *Proc. Natl. Acad. Sci. USA* **98**:13,189–13,194.
Leimar, O., and P. Hammerstein. 2001. Evolution of cooperation through indirect reciprocity. *Proc. Roy. Soc. Lond. B* **268**:745–753.
Maynard Smith, J., and E. Szathmáry. 1995. The Major Transitions in Evolution. San Francisco: W.H. Freeman.
McElreath, R., R. Boyd and P. J. Richerson. (2003). Shared norms and the evolution of ethnic markers. *Curr. Anthro.* **44(1):**122–129.
Neiman, F.D. 1998. Conspicuous consumption as wasteful advertising: A Darwinian perspective on spatial patterns in Classic Maya terminal monument dates. In: Rediscovering Darwin: Evolutionary Theory and Archeological Explanation, ed. C.M. Barton and G.A. Clark, pp. 267–290. Archeological Papers of the American Anthropological Association 7. Washington, D.C.: American Anthropological Assn.
Nesse, R.M., ed. 2001. Evolution and the Capacity for Commitment. New York: Russell Sage.
Oliver, S.C. 1962. Ecology and cultural continuity as contributing factors in the social organization of the Plains Indians. In: Univ. of California Publications in American Archaeology and Ethnology 48, No. 1, pp. 1–90. Berkeley: Univ. of California.

Patton, J.Q. 2000. Reciprocal altruism and warfare: a case from the Ecuadorian Amazon. In: Adaptation and Human Behavior: An Anthropological Perspective, ed. L. Cronk, N. Chagnon, and W. Irons, pp. 417–436. Hawthorne, NY: Aldine de Gruyter.
Richerson, P.J., and R. Boyd. 2001. The evolution of subjective commitment to groups: A tribal instincts hypothesis. In: Evolution and the Capacity for Commitment, ed. R.M. Nesse, pp. 186–220. New York: Russell Sage.
Schelling, T.C. 1960. The Strategy of Conflict. Cambridge, MA: Harvard Univ. Press.
Smith, E.A. 1985. Inuit foraging groups: Some simple models incorporating conflicts of interest, relatedness, and central-place sharing. *Ethol. Sociobiol.* **6**:27–47.
Smith, E.A. 1991. Inujjuamiut Foraging Strategies: Evolutionary Ecology of an Arctic Hunting Economy. Hawthorne, NY: Aldine de Gruyter.
Smith, E.A. 2000. Three styles in the evolutionary study of human behavior. In: Human Behavior and Adaptation: An Anthropological Perspective, ed. L. Cronk, N. Chagnon, and W. Irons, pp. 27–46. Hawthorne, NA: Aldine de Gruyter.
Smith, E.A., and R.L. Bliege Bird. 2000. Turtle hunting and tombstone opening: Public generosity as costly signaling. *Evol. Hum. Behav.* **21**:245–261.
Smith, E.A., R. Bliege Bird, and D.W. Bird. 2003. The benefits of costly signaling: Meriam turtle-hunters. *Behav. Ecol. Sociobiol*. **14**:116–126.
Smith, E.A., and B. Winterhalder, eds. 1992. Evolutionary Ecology and Human Behavior. Hawthorne, NY: Aldine de Gruyter.
Sosis, R., S. Feldstein, and K. Hill. 1998. Bargaining theory and cooperative fishing participation on Ifaluk atoll. *Hum. Nat.* **9**:163–204.
Sugiyama, L., and R. Chacon. 2000. Effects of illness and injury on foraging among the Yora and Shiwiar: Pathology risk as adaptive problem. In: Adaptation and Human Behavior: An Anthropological Perspective, ed. L. Cronk, N. Chagnon, and W. Irons, pp. 371–395. Hawthorne, NY: Aldine de Gruyter.
Tooby, J., and L. Cosmides. 1996. Friendship and the banker's paradox: Other pathways to the evolution of adaptations for altruism. *Proc. Brit. Acad*. **88**:119–143.
van den Berghe, P. 1979. Human Family Systems: An Evolutionary View. New York: Elsevier.
Wilson, D.S. 1998. Hunting, sharing, and multilevel selection: The tolerated theft model revisited. *Curr. Anthro.* **39**:73–97.
Winterhalder, B. 1996. Social foraging and the behavioral ecology of intragroup resource transfers. *Evol. Anthro.* **5**:46–57.
Winterhalder, B., and E.A. Smith. 2000. Analyzing adaptive strategies: Human behavioral ecology at twenty-five. *Evol. Anthro.* **9**:51–72.

22

Origins of Human Cooperation

Samuel Bowles and Herbert Gintis

Santa Fe Institute, Santa Fe, NM 87501, U.S.A.

ABSTRACT

Biological explanations of cooperation are based on kin altruism, reciprocal altruism, and mutualism, all of which apply to human and nonhuman species alike. Human cooperation, however, is based in part on capacities that are unique to, or at least much more highly developed in, *Homo sapiens*. In this chapter, an explanation of cooperation is sought that works for humans but does not work for other species, or works substantially less well. Central to this explanation will be human cognitive, linguistic, and physical capacities that allow the formulation of general norms of social conduct, the emergence of social institutions regulating this conduct, the psychological capacity to internalize norms, and the formation of groups based on such nonkin characteristics as ethnicity and linguistic behavior, which facilitates highly costly conflicts among groups. Agent-based modeling shows that these practices could have coevolved with other human traits in a plausible representation of the relevant environments. The forms of cooperation to be explained are confirmed by natural observation, historical accounts, and behavioral experiments and are based on a plausible evolutionary dynamic involving some combination of genetic and cultural elements, the consistency of which can be demonstrated through formal modeling. Moreover, the workings of the models developed account for human cooperation under parameter values consistent with what can be reasonably inferred about the environments in which humans have lived.

INTRODUCTION

> The Americans ... are fond of explaining almost all the actions of their lives by the principle of self interest rightly understood; they show with complacency how an enlightened regard for themselves constantly prompts them to assist one another and inclines them willingly to sacrifice a portion of their time and property to the welfare of the state. In this respect I think they frequently fail to do themselves justice; in the United States as well as elsewhere people are sometimes seen to give way to those disinterested and spontaneous impulses that are natural to man; but the Americans seldom admit that they yield to emotions of this kind.
>
> — Alexis de Tocqueville
>
> (*Democracy in America*, 1830, Book II, chapter VII)

Cooperation among humans is unique in nature, extending to a large number of unrelated individuals and taking a vast array of forms. By cooperation we mean an individual behavior that incurs personal costs to engage in a joint activity that confers benefits exceeding these costs to other members of one's group. This applies, for example, to contributing in a public goods game.[1] Although the absence of this unique type of cooperation in other species could be an evolutionary accident, a more plausible explanation is that human cooperation is the result of human capacities that are unique to our species.

Common explanations of cooperation in other species based on genetic relatedness (kin altruism) and repeated interactions (e.g., reciprocal altruism) certainly apply to cooperation in humans as well. However, the capacities underlying these mechanisms are not unique to humans: repeated interactions and interactions among kin are common in many species. We do not seek to diminish the importance of these familiar modes or to suggest that extensions of them to account for uniquely human aspects of cooperation are uninteresting. Rather we suggest that it would be fruitful to seek an explanation of cooperation that works for humans but, because it centrally involves attributes unique to humans, does not work for other species, or works substantially less well.

Central to our explanation will be human cognitive, linguistic, and physical capacities that allow the formulation of general norms of social conduct, the emergence of social institutions regulating this conduct, the psychological capacity to internalize norms, and the basing of group membership on such nonkin characteristics as ethnicity and linguistic behavior, which facilitates highly costly conflicts among groups. Of course, it will not do to posit these rules and institutions a priori. Rather, we must show that these could have coevolved with other human traits in a plausible representation of the relevant environments.

Our thinking, while necessarily speculative, has been disciplined in three ways. First, the forms of cooperation we seek to explain are confirmed by natural observation, historical accounts, and behavioral experiments. Second, we require that our account be based on a plausible evolutionary dynamic involving some combination of genetic and cultural elements, the consistency of which can be demonstrated through formal modeling. Third, the workings of the models we develop must account for human cooperation under parameter values consistent with what can be reasonably inferred about the environments in which humans have lived. When the models in question resist analytical solution (because they are complicated and highly nonlinear), this third requirement entails computer simulation under plausible parameter values.

The chapter is structured as follows:

- We support our assertion that explanations based on kin and reciprocal altruism are incomplete.
- We characterize key individual behavioral traits that we think account for much of human cooperation. We term this *strong reciprocity*.

[1] This definition of cooperation excludes mutually beneficial interactions (mutualisms), the evolutionary explanation of which is relatively simple; nonproductive forms of altruism (in which the benefit received does not exceed the cost to the altruist); and those lacking the common benefits of joint activity that are characteristic of the behaviors we wish to explain.

- We explain why multilevel selection among human groups operating on both cultural and genetic variability must play an important role in our explanation.
- We show that some common human institutions create the conditions under which multilevel selection is especially powerful. This provides a reason why group-level institutions, such as resource sharing as well as warfare, may have coevolved with the individual behaviors we call strong reciprocity.
- We explain why strong reciprocators may have been favored evolutionarily under conditions where their actions constituted a difficult-to-fake (costly) signal of their otherwise unobservable qualities as a mate, coalition partner, or opponent.

We argue that the maintenance of group boundaries through the parochial exclusion of "outsiders" may have contributed to the evolutionary success of cooperative behaviors. This, in turn, may provide part of the explanation of the salience of group membership as a determinant of the scope of cooperative relationships.

We develop the idea that human capacities to internalize norms and mobilize emotions in support of cooperative behavior have attenuated the conflict between individual-interest and group-benefit, and have thus supported cooperative interactions even under conditions when multilevel selection and the cooperation-inducing effects of costly signaling are weak.

In the spirit of this gathering, we concentrate on expressing a point of view, without giving the full attention to the more nuanced and formal arguments that a more extended presentation would allow. Nor do we take note of our immense debt to the work of other scholars, many of them joining us in this workshop, except to say that what follows is the result of a sustained collaboration in recent years with Ernst Fehr, Simon Gächter, Armin Falk, Urs Fischbacher, and their coauthors, as well as with Robert Boyd, Marcus Feldman, Joe Henrich, Peter Richerson, and Eric Alden Smith. Our own contributions to the ideas expressed here are summarized in our recent synthetic works (Gintis 2000a; Bowles 2003).

WHY EXPLANATIONS BASED ON KIN AND RECIPROCAL ALTRUISM ARE INCOMPLETE

We do not doubt that relatedness is an important part of the explanation of human cooperation, as it is among other animals, and that cooperation among kin may have been a template whose gradual extension contributed to cooperation among nonkin. However, to explain human cooperation among large numbers of unrelated individuals in this way is implausible.

Similarly, repeated interactions allowing retaliation against antisocial actions undoubtedly contribute to sustaining cooperation among humans and perhaps among some other animals. Some have suggested that the evolution of cooperation among entirely self-interested humans is explained in this manner.

This, however, is false. First, much of the experimental evidence about human behaviors contributing to cooperation comes from nonrepeated interactions, or from the final round of a repeated interaction. We do not think that subjects are unaware of the one-shot setting, or unable to leave their real-world experiences with repeated interactions at the laboratory door. Indeed, evidence is overwhelming that humans readily distinguish between repeated and nonrepeated interactions and adapt their behavior accordingly. Nonexperimental evidence is equally telling: common behaviors in warfare as in everyday life are not easily explained by the expectation of future reciprocation.

Second, conditions of early humans may have made the repetition–retaliation mechanism an ineffective support for cooperation. Members of mobile foraging bands could often escape retaliation by relocating to other groups. Moreover, in many situations critical to human evolution, repetition of an interaction was quite unlikely, as when groups faced dissolution as the result of group conflict or an adverse environment.

Third, the conditions under which repetition and retaliation can explain why self-regarding individuals would cooperate are not met in settings where large numbers interact. The celebrated "folk theorem," which is frequently invoked to show that repeated interactions among self-regarding individuals can support seemingly other-regarding behaviors, does not extend plausibly from two-person to n-person groups for large n. Critical differences between dyadic and n-person interactions in this respect are that (a) the number of accidental defections or perceived defections increases with n, and such "trembles" dramatically increase the cost of punishing defectors; (b) probability that a sufficiently large fraction of a large group of heterogeneous agents will be sufficiently forward-looking to make cooperation profitable decreases exponentially as n rises; and (c) coordination and incentive mechanisms required to ensure punishment of defectors by self-regarding group members become increasingly complex and unwieldy as n increases.[2] Although many important human interactions are dyadic (e.g., mutualistic exchange of goods), many important examples of cooperation (e.g., risk reduction through co-insurance, information sharing, maintenance of group-beneficial social norms, and group defense) are large group interactions. For these cases, the folk theorem provides no reason to expect cooperation to be common and durable rather than rare and ephemeral.

PSYCHOLOGICAL AND BEHAVIORAL ASPECTS OF ALTRUISM: PROSOCIAL EMOTIONS AND STRONG RECIPROCITY

Prosocial emotions are physiological and psychological reactions that induce agents to engage in cooperative behaviors as we have defined them above. Some

[2] A well-known theorem showing that repetition among a large number of agents can support efficient cooperative equilibria (Fudenberg and Maskin 1990) effectively requires group members to be infinitely lived and does not apply even approximately to human groups under the most optimistic assumptions concerning longevity and future orientation.

prosocial emotions, including shame, guilt, empathy, and sensitivity to social sanction, induce agents to undertake constructive social interactions; others, such as the desire to punish norm violators, reduce free riding when the prosocial emotions fail to induce sufficiently cooperative behavior in some fraction of members of the social group (Frank 1987; Hirshleifer 1987).

Without prosocial emotions, we would all be sociopaths, and human society would not exist, however strong the institutions of contract, governmental law enforcement, and reputation. Sociopaths have no mental deficit except that their capacity to experience shame, guilt, empathy, and remorse is severely attenuated or absent. They comprise 3–4% percent of the male population in the United States (Mealey 1995), but account for approximately 20% of the United States' prison population and between 33% and 80% of the population of chronic criminal offenders.

Prosocial emotions are responsible for the host of civil and caring acts that enrich our daily lives and render living, working, shopping, and traveling among strangers feasible and pleasant. Moreover, representative government, civil liberties, due process, women's rights, respect for minorities, to name a few of the key institutions without which human dignity would be impossible in the modern world, were brought about by people involved in collective action, pursuing not only their personal ends but also a vision for all of humanity. Our freedoms and comforts alike are based on the emotional dispositions of generations past.

Whereas we think evidence is strong that prosocial emotions account for important forms of human cooperation, there is no universally accepted model of how emotions combine with more cognitive processes to affect behaviors. Nor is there much agreement on how best to represent the prosocial emotions that support cooperative behaviors, although we (Bowles and Gintis 2002) have attempted one in this direction. It is uncontroversial, however, to assert that there are many civic-minded acts that cannot be explained by self-regarding preferences, including why people vote, why they give anonymously to charity, and why they sacrifice themselves in battle. In dealing with these areas of social life, a suggestive body of evidence points to a behavior that we call *strong reciprocity*. A strong reciprocator comes to a new social situation with a predisposition to cooperate, is predisposed to respond to cooperative behavior on the part of others by maintaining or increasing his level of cooperation, and responds to free-riding behavior on the part of others by retaliating against the offenders, even at a cost to himself, and even when he cannot reasonably expect future personal gains from such retaliation. The strong reciprocator is thus both a *conditionally altruistic cooperator* and a *conditionally altruistic punisher* whose actions benefit other group members at a personal cost. We call this reciprocity "strong" to distinguish it from such forms of "weak" reciprocity as reciprocal altruism, indirect reciprocity, and other such interactions that posit individually self-regarding behavior sustained by repeated interactions or positive assortation (see Fehr et al. 2002).

MULTILEVEL SELECTION

In populations composed of groups characterized by a markedly higher level of interaction among members than with outsiders, it has long been recognized that evolutionary processes may be decomposed into between-group and within-group selection effects (Price 1970). Where the rate of replication of a trait depends on the composition of the group, and where group differences in composition exist, group selection contributes to the pace and direction of evolutionary change. Until recently, however, most who modeled evolutionary processes under the joint influence of group and individual selection have concluded that the former cannot offset the latter, except where special circumstances (small group size, limited migration) heighten and sustain differences between groups relative to within-group differences.

Thus, group selection models are widely judged to have failed to explain evolutionary success of individually costly forms of group-beneficial sociality. But group selection operating on genetic and cultural variation may be of considerably greater importance among humans than other animals. Among the distinctive human characteristics that enhance the relevance of group selection is our capacity to suppress within-group phenotypic differences (e.g., via resource sharing, co-insurance, consensus decision making), conformist cultural transmission, ethnocentrism (which supports positive assortation within groups and helps maintain group boundaries), and the high frequency of intergroup conflict.

In Gintis (2000b) we develop an analytical model showing that under plausible conditions strong reciprocity can emerge from reciprocal altruism, through group selection. The paper models cooperation as a repeated n-person public goods game in which, under normal conditions, when agents are sufficiently attentive to future gains from group membership, cooperation is sustained by trigger strategies, as asserted in the folk theorem. However, when the group is threatened with extinction or dispersal, say through war, pestilence, or famine, cooperation is most needed for survival. Probability of one's contributions being repaid in the future, however, decreases sharply when the group is threatened, since the probability that the group will dissolve increases and hence the incentive to cooperate will dissolve. Thus, *precisely when a group is most in need of prosocial behavior, cooperation based on reciprocal altruism will collapse*. Such critical periods were common in the evolutionary history of our species. A small number of strong reciprocators, who punish defectors *without regard for the probability of future repayment*, can dramatically improve the survival chances of human groups. Moreover, humans are unique among species that live in groups and recognize individuals, in their capacity to inflict heavy punishment at low cost to the punisher, as a result of their superior tool-making and hunting ability. Indeed, and in sharp contrast to nonhuman primates, even the strongest man can be killed while sleeping by the weakest, at low cost to the punisher. A simple argument using Price's equation then shows that under these conditions, strong reciprocators can invade a population of self-regarding types and can persist in equilibrium.

Our joint work with Boyd and Richerson (Boyd et al. 2003) shows, through agent-based simulations, that for some cooperative behaviors — notably punishing those who violate cooperative norms — group selection on culturally transmitted traits can be decisive even for very large groups and for substantial rates of migration. The reason for this surprising result is that if most members of a group are adhering to the norm, the costs incurred by those predisposed to punish violators are very small for the simple reason that violations are infrequent. Thus while within-group selection against the cooperative behavior exists, it is very weak in the neighborhood of the cooperative equilibrium. This supports the persistence over long periods of substantial between-group differences in composition, some with virtually all cooperative agents predisposed to cooperate and punish those who do not, and other groups composed of virtually all self-regarding individuals Additional between-group variance is provided by intergroup conflicts following which winning groups absorb losers and then divide.

One particularly attractive property of these models is that they predict a heterogeneous equilibrium with a considerable fraction of both self-regarding and strong reciprocator types, as is often found in the experimental literature (Fehr and Gächter 2002).

COEVOLUTION OF INSTITUTIONS AND BEHAVIORS

If group selection is part of the explanation of the evolutionary success of cooperative individual behaviors, then it is likely that group-level characteristics (e.g., relatively small group size, limited migration, or frequent intergroup conflicts) that enhance group selection pressures coevolved with cooperative behaviors. Thus group-level characteristics and individual behaviors may have synergistic effects. This being the case, cooperation is based in part on the distinctive capacities of humans to construct institutional environments that limit within-group competition and reduce phenotypic variation within groups, thus heightening the relative importance of between-group competition and allowing individually costly but in-group-beneficial behaviors to coevolve with these supporting environments through a process of interdemic group selection.

The idea that the suppression of within-group competition may be a strong influence on evolutionary dynamics has been widely recognized in eusocial insects and other species. Alexander (1979), Boehm (1982), and Eibl-Eibesfeldt (1982) first applied this reasoning to human evolution, exploring the role of culturally transmitted practices that reduce phenotypic variation within groups. Examples of such practices are leveling institutions, such as resource sharing among nonkin, namely those which reduce within-group differences in reproductive fitness or material well-being. These practices are leveling to the extent that they result in less pronounced within-group differences in material well-being or fitness than would have obtained in their absence. Thus, the fact that good hunters who are generous toward other group members may experience higher fitness than other hunters and enjoy improved nutrition (as a result of consumption smoothing) does not indicate a lack of leveling unless these practices also result in lesser fitness and worse nutrition among less successful hunters (which seems highly unlikely).

By reducing within-group differences in individual success, such practices may have attenuated within-group genetic or cultural selection operating against individually costly but group-beneficial practices, thus giving the groups adopting them advantages in intergroup contests. Group-level institutions are thus constructed environments capable of imparting distinctive direction and pace to the process of biological evolution and cultural change. Hence, evolutionary success of social institutions that reduce phenotypic variation within groups may be explained by the fact that they retard selection pressures working against in-group-beneficial individual traits and the fact that high frequencies of bearers of these traits reduces the likelihood of group extinctions.

We have modeled an evolutionary dynamic along these lines with the novel features that genetically and culturally transmitted individual behaviors as well as culturally transmitted group-level institutional characteristics are subject to selection, with intergroup contests playing a decisive role in group-level selection (Bowles 2001; Bowles et al. 2003). We show that intergroup conflicts may explain the evolutionary success of both (a) altruistic forms of human sociality toward nonkin and (b) group-level institutional structures such as resource sharing that have emerged and diffused repeatedly in a wide variety of ecologies during the course of human history. In-group-beneficial behaviors may evolve if they inflict sufficient costs on outgroup individuals and group-level institutions limit the individual costs of these behaviors and thereby attenuate within-group selection against these behaviors.

Our simulations show that if group-level institutions implementing resource sharing or nonrandom pairing among group members are permitted to evolve, group-beneficial individual traits coevolve along with these institutions, even where the latter impose significant costs on the groups adopting them. These results hold for specifications in which cooperative individual behaviors and social institutions are initially absent in the population. In the absence of these group-level institutions, however, group-beneficial traits evolve only when intergroup conflicts are very frequent, groups are small, and migration rates are low. Thus the evolutionary success of cooperative behaviors in the relevant environments during the first 90,000 years of anatomically modern human existence may have been a consequence of distinctive human capacities in social institution building.

STRONG RECIPROCITY AS A SIGNAL OF QUALITY

Cooperative behaviors may be favored in evolution because they enhance the individual's opportunities for mating and coalition building. This would be the case, for example, if sharing valuable information or incurring dangers in defense of the group were taken by others as an honest signal of the individual's otherwise unobservable traits as a mate or political ally. Much of the literature on costly signaling and human evolution explains such behaviors as good hunters contributing their prey to others, but the same reasoning applies to cooperative behaviors. Cooperative behaviors would thus result in advantageous alliances for those signaling in this manner, and the resulting enhanced fitness or material success would then account for the proliferation of the cooperative behaviors constituting the signal. With Eric Alden Smith (Gintis et al. 2001), we have modeled this process as a multiplayer public goods game that involves no repeated or assortative interactions, so that noncooperation would be the dominant strategy if there were no signaling benefits. We show that honest signaling of underlying quality by providing a public good to the rest of the group can be evolutionarily stable and proliferate in a population in which it is initially rare, provided that certain plausible conditions hold. Behaviors conforming to what we call strong reciprocity could have thus evolved in this way.

Our signaling equilibrium alone, however, does not require that the signal confer benefits on other group members. Antisocial behaviors could perform the same function: beating up one's neighbor can demonstrate prowess just as much as behaving bravely in defense of the group. If signaling is to be an explanation of group-beneficial behavior, the logic of the model must be complemented by a demonstration that group-beneficial signaling is favored over antisocial signaling. We supply this by noting that the level of public benefit provided may be positively correlated with the individual benefit the signaler provides to those who respond to the signal. For instance, the signaler who defends the group is more likely to confer a benefit (say, protection) on his partner or allies than the signaler who beats up his neighbor. Group-beneficial signals may attract larger audiences than antisocial signals. Finally, group selection among competing groups would favor those at group-beneficial signaling equilibria over those either at nonsignaling equilibria or those at antisocial signaling equilibria.

As this last reason suggests, the effects of signaling and group selection on cooperation may be synergistic rather than simply additive. Group selection provides a reason why signaling may be prosocial, whereas signaling theory provides a reason why group-beneficial behaviors may be evolutionarily stable in a within-group dynamic, thus contributing to between-group variance in behavior and thereby enhancing the force of group selection.

PAROCHIALISM AND RECIPROCITY

The predisposition of individuals to behave cooperatively often depends on the identities of the individuals with whom they are interacting: "insiders" are favored over "outsiders." Insider-outsider distinctions play a critical role in the above models. In our group selection models, cooperative behaviors conferring benefits on fellow group members allowed highly cooperative groups to prevail in intergroup conflicts. Thus the very behaviors that are beneficial to one's own group are costly or even lethal to members of other groups. In addition, as we have seen, maintenance of group boundaries to limit the extent of migration and the frequency of intergroup conflict contribute substantially to the force of group selection in promoting cooperation within groups. Thus it seems likely that within-group cooperation and hostility toward "outsiders" coevolved.

In-group favoritism is often supported by the cultural salience of physical, linguistic, and other behavioral markers identifying insiders and outsiders in conjunction with exclusionary practices, which we call parochialism. We have modeled parochialism as a filter on the ascriptive traits of those with whom one might interact, a particular filter excluding those with "objectionable" traits (Bowles and Gintis 2000). Members of groups benefit in two ways from the adoption of more parochial filters: equilibrium group size and cultural heterogeneity of group members is thereby reduced, and this enhances the effectiveness of mutual monitoring and reputation-building in supporting high levels of within-group cooperation. More parochial groups forego the economies of scale, gains from exchange, and possible collective cognition benefits of larger and more diverse membership. The degree of parochialism observed in a population will depend on the balance of these benefits and costs of exclusionary practices. As these have evolved over time with the effects of changing environments and technologies, our analysis of "optimal parochialism" may provide a way of modeling the coevolution of cooperation and out-group hostility, though we have not attempted this ambitious project.

PROSOCIAL EMOTIONS: MODELS AND EXPERIMENTAL EVIDENCE

As we have argued above, adherence to social norms is underwritten not only by the cognitively mediated pursuit of self-interest but also by emotions. Shame, guilt, empathy, and other visceral reactions play a central role in sustaining cooperative relations. The puzzle is that prosocial emotions are at least *prima facie* altruistic, benefiting others at a cost to oneself. Thus, under any payoff-monotone dynamic in which the self-regarding trait tends to increase in frequency, prosociality should atrophy.

Pain is a presocial emotion. Shame is a social emotion: a distress that is experienced when one is devalued in eyes of one's consociates because of a value that one has violated or a behavioral norm that one has not lived up to.

Does shame serve a purpose similar to that of pain? If being socially devalued has fitness costs, and if the amount of shame is closely correlated with the level of these fitness costs, then the answer is affirmative. Shame, like pain, is an aversive stimulus that leads the agent experiencing it to repair the situation that led to the stimulus and to avoid such situations in the future. Shame, like pain, replaces an involved optimization process with a simple message: whatever you did, undo it if possible, and do not do it again. Of course, the individual can override the unpleasurable shame sensation if the benefits are sufficiently great, but the emotion nevertheless, on average, will reduce the frequency of shame-inducing social behaviors.

Since shame is evolutionarily selected and is costly to use, it must on the average confer a selective advantage on those who experience it. Two types of selective advantage are at work here. First, shame may raise the fitness of an agent who has incomplete information (e.g., as to how fitness-reducing a particular antisocial action is), limited or imperfect information-processing capacity, and/or a tendency to undervalue costs and benefit that accrue in the future. Probably all three conditions conspire to react suboptimally to social disapprobation in the absence of shame, and shame brings us closer to the optimum. The role of shame in alerting us to negative consequences in the future, of course, presupposes that society is organized to impose those costs on rule violators. Shame may have coevolved with the emotions motivating punishment of antisocial actions (the reciprocity motive in our model).

The second selective advantage to those experiencing shame arises through the effects of group competition. Where the emotion of shame is common, punishment of antisocial actions will be particularly effective and, as a result, seldom used. Thus groups in which shame is common can sustain high levels of group cooperation at limited cost and will be more likely to spread through interdemic group selection. Shame thus serves as a means of economizing on costly within-group punishment.

Shame can be investigated in the laboratory. In Bowles and Gintis (2002) we consider a public goods game where agents maximize a utility function that captures five distinct motives: personal material payoffs, one's valuation of the payoffs to others, which depend both on one's altruism and one's degree of reciprocity, and one's sense of guilt or shame when failing to contribute one's fair share to the collective effort of the group. Shame is evident if players who are punished by others respond by behaving more cooperatively than is optimal for a material payoff-maximizing agent. We present indirect empirical evidence suggesting that such emotions play a role in the public goods game. However, direct evidence on the role of emotions in experimental games remains scanty.

INTERNALIZATION OF NORMS

An *internal norm* is a pattern of behavior enforced in part by internal sanctions, including shame and guilt as outlined above. People follow internal norms when they value certain behaviors for their own sake in addition to, or despite, the effects these behaviors have on personal fitness and/or perceived well-being. The ability to internalize norms is nearly universal among humans. While widely studied in the sociology literature (socialization theory), it has been virtually ignored outside this field (but see Caporael et al. 1989 and Simon 1990).

Socialization models have been strongly criticized for suggesting that people adopt norms independent of their perceived payoffs. In fact, people do not always blindly follow the norms that have been inculcated in them; instead, at times, they treat compliance as a strategic choice (Gintis 1975). The "oversocialized" model of the individual presented in the sociology literature can be counteracted by adding a phenotypic copying process reflecting the fact that agents shift from lower to higher payoff strategies (Gintis 2003b).

All successful cultures foster internal norms that enhance personal fitness, such as future orientation, good personal hygiene, positive work habits, and control of emotions. Cultures also universally promote altruistic norms that subordinate the individual to group welfare, fostering such behaviors as bravery, honesty, fairness, willingness to cooperate, and empathy with distress of others.

Given that most cultures promote cooperative behaviors, and if we accept the sociological notion that individuals internalize norms that are passed to them by parents and other influential elders, it becomes easy to explain human cooperation. If even a fraction of society internalized the norms of cooperation and punished free riders and other norm violators, a high degree of cooperation could be maintained in the long run. Thus we are left with two puzzles: why do we internalize norms, and why do cultures promote cooperative behaviors?

We provide an evolutionary model in which the capacity to internalize norms develops because this capacity enhances individual fitness in a world in which social behavior has become too complex and multifaceted to be fruitfully evaluated piecemeal through individual rational assessment (Gintis 2003a). Internalization moves norms from constraints that one can treat instrumentally toward maximizing well-being to norms that are then valued as ends rather than means. It is not difficult to show that if an internal norm enhances fitness, then for plausible patterns of socialization, the allele for internalization of norms is evolutionarily stable.

We (Gintis 2003a) use this framework to model Herbert Simon's (1990) explanation of altruism. Simon suggested that altruistic norms could "hitchhike" on the general tendency of internal norms to be fitness enhancing. However, Simon provided no formal model of this process and his ideas have been widely ignored. This chapter shows that Simon's insight can be analytically modeled and is valid under plausible conditions. A straightforward gene–culture coevolution argument then explains why fitness-reducing internal norms are likely to be prosocial as opposed to socially harmful: groups with prosocial internal norms will outcompete groups with antisocial, or socially neutral, internal norms.

CONCLUSION

Two themes run through our account of the origin of cooperation among humans: (a) the importance of groups in human evolution and the power of multilevel selection; (b) the underlying dynamic of gene–culture coevolution. We close with comments on what we consider two mistaken approaches: the tautological extension of self-interest to the status of the fundamental law of evolution and the representation of culture as an epiphenomenal expression of the interaction of genes and environments.

Like de Tocqueville's "Americans," a distinguished tradition in biology and the social sciences has sought to explain cooperative behavior "by the principle of self-interest, rightly understood." From J.B.S. Haldane's quip that he would risk his life to save eight drowning cousins to the folk theorem of modern game theory, this tradition has clarified the ways that relatedness, repeated play, and other aspects of social interactions among members of a group might confer fitness advantages on those engaging in seemingly unselfish behaviors. The point is sometimes extended considerably by noting that if the differential replication of traits by selection operating on either culturally or genetically transmitted traits is monotonic in payoffs, only traits that on average have higher payoffs will be evolutionarily successful. If selfish behaviors are then *defined* as those that on average have higher payoffs, the principle of self-interest becomes the fundamental law of evolution.

Some prominent researchers in evolutionary biology have taken precisely this tack. Richard Dawkins (1989), for instance, states in the course of the first four pages of *The Selfish Gene* that "a predominant quality to be expected in a successful gene is ruthless selfishness. This gene selfishness will usually give rise to selfishness in individual behavior .... Let us try to *teach* generosity and altruism, because we are born selfish."[3] Similarly, drawing out the philosophical implications of the evolutionary analysis of human behavior, Richard Alexander (1987) says, "ethics, morality, human conduct, and the human psyche are to be understood only if societies are seen as collections of individuals seeking their own self-interest ... That people are in general following what they perceive to be their own interests is, I believe, the most general principle of human behavior." (pp. 3, 35).

Like de Tocqueville, we object to the tautological extension of the principle of self-interest. Our concern is not with the fitness-based or other payoff monotonic dynamic process assumed in this approach. It goes without saying that traits that experience lower fitness in a population will be handicapped in any plausible evolutionary dynamic: even cultural evolution may be strongly biased toward proliferation of behaviors leading to individual material success. Rather, our concern is with the distortion of the term "self-interest." Those who, in Darwin's words, were "ready to warn each other of danger, to aid and defend each other" would tautologically be deemed "selfish" if, as Darwin (1871/1973) suggested, tribes in which these behaviors were common would "spread and be victorious over other tribes." We have eschewed the terms "selfish" or "self-interested" to avoid confusions and have instead defined cooperative behaviors in terms of their costs to the individual and their beneficial consequences for group

[3] Note the tendency in this last sentence to identify self-interest with on-average higher payoffs, but then the use of the term in its everyday sense, which is completely unwarranted.

members. Our models and simulations show that these behaviors may proliferate under plausible conditions as the result of the group structure of human populations and success of groups in which cooperators are common.

Turning to our second point, we note that reduction of culture to an effect of the interaction of genes and natural environments is a common, if rarely explicit, aspect of accounts from such diverse authors as Karl Marx and some modern-day sociobiologists. Like the principle of self-interest, the hypothesis that the interaction of natural environments and genes affects the evolution of cultures has yielded numerous insights. It is also true, however, that culture affects the natural and social environments in which the relative fitness of genetically transmitted behavioral traits is determined. Cavalli-Sforza and Feldman (1981), Boyd and Richerson (1985), Durham (1991) and others have provided compelling examples of these cultural effects on genetic evolution. Our own models of the coevolution of genetically transmitted individual behaviors and culturally transmitted group-level institutions are but some of the many models of this process. In one of our models, for example, we have seen that the presence of a culturally transmitted convention (resource sharing) is essential to the evolution of a genetically transmitted altruistic trait governed by natural selection. It may be helpful to represent human cultures, especially the institutional structures they support, as a case of niche construction, i.e., the creation of a particular environment such that genetic evolution is affected (Laland et al. 2000; Bowles 2000).

The challenge of explaining the origins of human cooperation has led us to the study of the social and environmental conditions of life of mobile foraging bands and other stateless simple societies which arguably made up human society for most of the history of anatomically modern humans. The same quest has made noncooperative game theory (which assumes the absence of enforceable pre-play agreements) an essential tool. But as several authors have pointed out, most forms of contemporary cooperation are supported by incentives and sanctions based on a mixture of multilateral peer interactions and third party enforcement, often accomplished by the modern nation state. It would be modest and perhaps even wise to resist drawing strong conclusions about cooperation in the 21st century on the basis of our thinking about the origins of cooperation in the Late Pleistocene.

ACKNOWLEDGMENTS

We are grateful to Eric Alden Smith for helpful comments and to the Santa Fe Institute and the John D. and Catherine E. MacArthur Foundation for support of this research.

REFERENCES

Alexander, R.D. 1979. Biology and Human Affairs. Seattle: Univ. of Washington Press.

Alexander, R.D. 1987. The Biology of Moral Systems. Hawthorne, NY: de Gruyter.

Boehm, C. 1982. The evolutionary development of morality as an effect of dominance behavior and conflict interference. *J. Soc. Biol. Struct.* **5**:413–421.

Bowles, S. 2000. Economic institutions as ecological niches. *Behav. Brain Sci.* **23**.

Bowles, S. 2001. Individual interactions, group conflicts, and the evolution of preferences. In: Social Dynamics, ed. S.N. Durlauf and H.P. Young, pp. 155–190. Cambridge, MA: MIT Press.

Bowles, S. 2003. Microeconomics: Behavior, Institutions, and Evolution. Princeton, NJ: Princeton Univ. Press.

Bowles, S., and H. Gintis. 2000. Persistent parochialism: The dynamics of trust and exclusion in networks. Santa Fe Institute Working Paper 00-03-017. Santa Fe, NM: Santa Fe Institute.

Bowles, S., and H. Gintis. 2002. Prosocial emotions. Santa Fe Institute Working Paper 02–07–028. Santa Fe, NM: Santa Fe Institute.

Bowles, S., J.-K. Choi, and A. Hopfensitz. 2003. The coevolution of individual behaviors and group level institutions. *J. Theor. Biol.*, in press

Boyd, R., H. Gintis, S. Bowles, and P.J. Richerson. 2003. Evolution of altruistic punishment. *Proc. Natl. Acad. Sci. USA* **100**:3531–3535.

Boyd, R., and P.J. Richerson. 1985. Culture and the Evolutionary Process. Chicago: Univ. of Chicago Press.

Caporael, L., R. Dawes, J. Orbell, and J.C. van de Kragt. 1989. Selfishness examined: Cooperation in the absence of egoistic incentives. *Behav. Brain Sci.* **12**:683–738.

Cavalli-Sforza, L.L., and M.W. Feldman. 1981. Cultural Transmission and Evolution. Princeton, NJ: Princeton Univ. Press.

Darwin, C. 1871/1973. The Descent of Man. New York: Appleton Press.

Dawkins, R. 1989. The Selfish Gene. 2d ed. Oxford: Oxford Univ. Press.

Durham, W.H. 1991. Coevolution: Genes, Culture, and Human Diversity. Stanford: Stanford Univ. Press.

Eibl-Eibesfeldt, I. 1982. Warfare, man's indoctrinability and group selection. *J. Comp. Ethnol.* **60**:177–198.

Fehr, E., and S. Gächter. 2002. Altruistic punishment in humans. *Nature* **415**:137–140.

Fehr, E., U. Fischbacher, and S. Gächter. 2002. Strong reciprocity, human cooperation and the enforcement of social norms. *Nature* **13**:1–25.

Frank, R.H. 1987. If *Homo economicus* could choose his own utility function, would he want one with a conscience? *Am. Econ. Rev.* **77**:593–604.

Fudenberg, D., and E. Maskin. 1990. Evolution and cooperation in noisy repeated games. *Am. Econ. Rev.* **80**:275–279.

Gintis, H. 1975. Welfare economics and individual development: A reply to Talcott Parsons. *Qtly. J. Econ.* **89**:291–302.

Gintis, H. 2000a. Game Theory Evolving. Princeton, NJ: Princeton Univ. Press.

Gintis, H. 2000b. Strong reciprocity and human sociality. *J. Theor. Biol.* **206**:169–179.

Gintis, H. 2003a. The hitchhiker's guide to altruism: Gene–culture coevolution and the internalization of norms. *J. Theor. Biol.* **220**:407–418.

Gintis, H. 2003b. Solving the puzzle of human prosociality. *Ration. Soc.* **15**.

Gintis, H., E.A. Smith, and S. Bowles. 2001. Costly signaling and cooperation. *J. Theor. Biol.* **213**:103–119.

Hirshleifer, J. 1987. Economics from a biological viewpoint. In: Organizational Economics, ed. J.B. Barney and W.G. Ouchi, pp. 319–371. San Francisco: Jossey-Bass.

Laland, K., F.J. Olding-Smee, and M. Feldman. 2000. Group selection: A niche construction perspective. *J. Consc. St.* **7**:221–224.

Mealey, L. 1995. The sociobiology of sociopathy. *Behav. Brain Sci.* **18**:523–541.

Price, G.R. 1970. Selection and covariance. *Nature* **227**:520–521.

Simon, H. 1990. A mechanism for social selection and successful altruism. *Science* **250**:1665–1668.

Standing, left to right: Karl Sigmund, Franjo Weissing, John Tooby, Pete Richerson, Astrid Hopfensitz
Seated, left to right: Eric Smith, Sam Bowles, Peyton Young, Joe Henrich, Rob Boyd

23

Group Report: The Cultural and Genetic Evolution of Human Cooperation

Joseph Henrich, Rapporteur

Samuel Bowles, Robert T. Boyd, Astrid Hopfensitz, Peter J. Richerson, Karl Sigmund, Eric A. Smith, Franz J. Weissing, and H. Peyton Young

INTRODUCTION

Whereas many aspects of human sociality show the clear imprint of our primate phylogeny (e.g., mother-offspring kinship, dominance hierarchies), human cooperation is unique in a number of respects. At a macro level, human cooperation varies substantially from nonhuman primates in both its scale and the nature of its variability. Although the scale of cooperation in other primates rarely exceeds two or three individuals (e.g., in grooming and coalitions), humans in some societies cooperate on scales involving hundreds or even thousands of individuals.[1] Curiously, unlike all other primates, the scale of human cooperation varies across social groups, from societies that are economically independent at the family level — showing little cooperation outside the extended kin circle (e.g., Johnson and Earle 2000: Machiguenga, Shoshone) — to the vast scales found in chiefdoms and modern states. Further, substantial degrees of variation in the scale of cooperation can be observed among social groups inhabiting identical environments (e.g., Kelly 1985: the Nuer and Dinka; Atran et al. 1999: Itza, Ladinos, and Kekchi Maya). Finally, although primate species typically show little variation in behavioral domains of cooperative behavior, human social groups vary substantially in their domains of cooperation. Some groups

1 "Cooperation" is used in the game theoretic sense to include only situations in which a potential "free-rider problem" exists. We exclude a range of other situations encompassed in the broader definition of cooperation that includes games with multiple equilibria (e.g., coordination games).

cooperate in fishing but not house-building or warfare, whereas other groups cooperate in house-building and warfare but not fishing.

At the macro level, variation in cooperation across both social groups and behavioral domains is immense, whereas at a micro level, human behavior and psychology exhibit a number of panhuman patterns that must be addressed if evolutionary theory is to provide a more complete account of the emergence of human cooperation. We provide a general sketch of the relevant behavioral and psychological patterns gleaned from a combination of field and experimental data from across the social sciences, and suggest that the existing set of theoretical tools from evolutionary biology are insufficient to explain these patterns.[2]

First, all human societies have regularized patterns and expectations of conduct that reflect some more or less shared notions of what is appropriate behavior for individuals within that group; of course, these expectations may vary among social categories (e.g., married vs. unmarried females) or with individual status. Violations of these normative expectations will likely provoke affective responses in the first party (violators may experience shame or regret) and in the injured second parties (who may experience anger or moral outrage at the first party). Even third-party observers often experience anger and moral outrage toward the contextually defined "violator," and empathy toward the injured party, i.e., the normative rules determine with whom the third party empathizes. These affective responses may lead to punishments, which are individual actions taken by second or third parties against the first-party violator.[3] They may take a range of forms, from violence to gossip. Under some circumstances, despite such punishments, seemingly costless acts of contrition and apologies can atone for violations (Silk, this volume). Most importantly, while this general pattern is pervasive across human societies, the domains of behavior, expectations, and details of the locally relevant beliefs and behaviors to which this panhuman constellation of cognitive processes, emotions, and affective reactions is linked varies tremendously from society to society. Interestingly, this constellation applies to a wide range of behaviors (e.g., clothing choice) that are not merely cooperative circumstances. Building on these observations, we use the word "norm" to refer to regularized behavioral patterns that carry the kinds of expectations and affective responses described above.

Incest prohibitions provide an interesting example of the peculiarity of certain aspects of these normative phenomena from an evolutionary point of view. Many societies have strong prohibitions against sex between brothers and sisters. When this does occur, and the incestuous pair is discovered, they sometimes experience so much shame that they commit suicide or run away. Both second parties (e.g., the violators' parents, who suffer a fitness loss when their

[2] As with all simplifications, our goal is to capture some of the underlying patterns, not to account for all the contextually specific variations.

[3] The emotions of shame and regret may be construed as forms of self-punishment by the norm violator.

kids engage in incest) *and* unrelated third parties often show extreme levels of moralistic outrage that may result in assault, murder, or ostracism (putting aside the gossip, etc.). The puzzling feature here is the outrage of third parties since one might predict that fitness-maximizing strategists would, if they took any interest at all, actively encourage other individuals (nonrelatives) to engage in incest, thereby increasing their own relative fitness. Of course, human behavioral ecologists have only just begun to attack the challenges of explaining third-party punishments and norm-based behavior, so serious alternatives may yet answer the above challenges.

Other examples of cooperative behavioral patterns that seem to reflect the basic norm structure sketched above include food sharing, cooperative hunting and fishing, territorial defense, and cooperative labor (e.g., communal gardening and house-building). The Aché provide a well-studied example of highly egalitarian sharing of certain specified kinds of food (big game and honey). Successful Aché hunters are expected to deliver the meat to the group quietly (no boasting allowed), where it is divided into remarkably equal portions and distributed across all parties (both relatives and nonrelatives alike). Interestingly, the hunter is not allowed to eat the meat himself (Kaplan and Hill 1985). Also interesting is the fact that Aché sharing norms are highly context specific (see Smith, this volume); the above description applies in the context of small trekking parties, whereas when residing at the larger permanent settlement sharing (even of game) is much more restricted (Gurven 2001).

A wide range of experimental work confirms the patterns suggested by ethnography. In particular, such patterns indicate that individuals are willing to punish other individuals at a cost to themselves, even in one-shot encounters, but only if some culturally specific notion of fairness has been violated (Henrich 2000; Henrich et al. 2001; Henrich and Smith 2003). In two-person bargaining games, individuals in many societies are willing to pay a cost to inflict punishment on individuals who violate their expectations of fairness. However, people in societies without expectations of fairness (based on postgame interviews and ethnographic observations) vis-à-vis these experimental games show little or no willingness to punish (Henrich et al. 2001). Similarly, experiments designed to test the willingness of third-party observers to punish unfairness in first parties show that third parties are quite willing to inflict costly punishment on unfairness, even at a cost to themselves (Fehr and Henrich, this volume). N-person public goods games with and without punishment options show that individuals are willing to punish, and this punishment promotes cooperation; individuals also do not stick to a costly practice if others are violating it with impunity.

Punishment is not the only means of sustaining a norm, however. People can anticipate the negative emotions and loss of self-esteem that result from norm violations, which act as a built-in deterrent not requiring the intervention of others. Furthermore, both ethnographic and historical material suggest that people in all human societies worry, at least to some degree, about their reputations — about what others think about them. People want good reputations, but most of all they want to avoid bad reputations; self-esteem may provide a gauge of an individual's assessment of his/her own reputation. Information about reputation is likely transmitted principally through gossip, which can be extremely intense in many small-scale societies. Again, experimental evidence is consistent with the ethnography. Individuals increase their prosocial behavior when reputation-building becomes possible, and individuals use reputation information to make decisions about cooperative interactions (Fehr and Henrich, this volume).

Some researchers have proposed that genetic evolutionary processes operating through selection pressure have influenced the psychological processes that sustain cooperation (e.g., Simon 1990; Richerson and Boyd 1998). For example, by attaching emotions, affect, and moralistic sentiments to deviations from learned behavioral patterns, individuals' subjective assessments of the payoffs may change (punishing a norm violator feels good) such that "cooperating" and "punishing" no longer reduces an individual's subjective payoff. Second, natural selection may have altered human cognitive information processes related to, for example, attention, memory, and the calculation of future expectations such that individuals under particular stimuli systematically miscalculate the immediate (within-group) consequences of their decisions to cooperate or punish. Longer-run interactions or multilevel culture–gene coevolutionary processes may have favored psychological mechanisms in which individuals overestimate their chances of being "caught" or underestimate the cost of seeking revenge. Finally, humans are literally creatures of habit, which means certain learning or training processes (drill) may inculcate behavior programs that are merely executed without any consideration of the costs and benefits. Firefighters, for example, are trained in nonrisky situations into a rigid program of actions and contingency actions. This training allows them to charge into burning buildings in situations that cause the untrained to run the other way. Military training performs a similar function.

KEY ASPECTS OF THE HUMAN PHENOTYPES

Before embarking on our review and discussion of theoretical approaches, we should first note that other coevolving aspects of human societies and psychology substantially influence both the genetic and cultural evolutionary pathways to the observable patterns of human cooperation. Undoubtedly, from the psychological angle, our capacities of learning, language, analogical reasoning, and metaphorical extension have substantial and important impacts on the evolutionary processes that were likely to have been important in shaping our psychology. Below, we highlight the importance of specific forms of social learning in understanding the evolution of cooperation in one-shot n-person interactions.

Economic specialization, social segmentation, and communal decision making undoubtedly all played a role in the dramatic "scaling up" of cooperation and societal scale that has taken place during the last 10,000 years (with the rise of chiefdoms and nation-states), and may have had consequences for genetic evolution over considerably longer timescales. For example, Richerson and Boyd (1998, 2000) and Richerson et al. (this volume) argue that by subdividing larger groups (e.g., military organizations, lineage groups, and corporations), culturally evolved institutions create small-group environments that make effective use of our ancient social instincts by embedding them in a larger context. Boehm (1996) has argued that, because many small-scale human societies likely engaged in various forms of communal decision making in emergency situations, group members partake in a shared fate that levels the fitness variation within groups and enhances between-group selection processes, thereby generating a species-specific circumstance that favors genetic group selection. Because many aspects of cooperative social behavior are transmitted culturally from one generation to the next, we first consider the different mechanisms that may be responsible for this transmission.

Transmission Mechanisms

Learning mechanisms allow for the nongenetic transmission of behavioral repertoires (or information) from one generation to the next. These mechanisms can be partitioned into asocial learning (e.g., trial and error, Bayesian updating) and social learning (imitation, teaching). Available empirical data suggest that the following types of categories must be learned in some fashion:

- Social rules/norms: "All large game must be shared with the group"; "the ideal sharecropping contract is 50/50."
- Social roles: Elders, potential marriage partners, male-female responsibilities.
- Belief systems: "Eating snakes will cause you to vomit and die."
- Tastes: Do we like to eat insects, snakes, snails, bone marrow, and/or the brains of dead relatives?
- Techniques: What are the tasks for processing bitter (toxic) manioc?

Having set out these categories, we realize that one of the key areas for future research is to figure out what is actually transmitted between individuals and what is inferred from limited information. Young (1998), Boyd and Richerson (1985), and Cavalli-Sforza and Feldman (1981) have begun to develop theories for how social norms and rules are transmitted within and between social groups by learning and imitation. We suspect that combining these formal approaches with lines of research in cognitive anthropology and psychology (e.g., Fiske 1991) may produce further important advances; Atran et al. (2002) provides an interesting effort in that direction.

Another important aspect of learning is timescale. Short-term learning processes allow for frequent updating. Hunters may update their information on prey encounter rates as they hunt, and farmers may imitate the crop choices of their successful neighbors from year to year. In contrast, longer-term processes may occur only rarely or only during a specified learning window (e.g., dialect acquisition). Such longer-term processes likely influence the acquisition of moral rules and social categories. For example, in many societies males are forbidden from talking to (let alone having incestual sex with) their mother's brother's daughter (these girls are "classificatory sisters"), while they are strongly encouraged to marry their father's sister's daughter. Similarly, certain difficult skills, such as kayak manufacturing or knowing which cycad seeds are cyanide-free by inspection, may be sufficiently complex that substantial modifications are unlikely to be introduced by later experience.

In considering cooperative situations specifically, individuals must be able to acquire through some process of observation, sanctioning, inference, and teaching the details of when and how much to cooperate and punish (Is one defection enough? Does a person's intent matter when they defect?). Individuals must learn about the costs of being punished, including social disapproval (shame), material losses (fines), and physical violence. Individuals must also learn, in some sense, about the relationship between context and emotional experience. That is, should they feel embarrassed any time they are naked in public (as in Chicago), or only when they are not near some body of water (as in Berlin). Somehow individuals are able to learn about all this even when norm violations are rare; if most people obey local norms most of the time, then individuals will have little direct experience with seeing norm violators getting punished. Teaching may enhance this process in some circumstances: Inuit parents give young children opportunities to hoard food and then chastise them if they do so. In considering the importance of teaching, we should remember that available evidence suggests that direct teaching of the kind observed in Inuit and Western middle-class parents is not typically observed in most small-scale societies (Toren 1990; Lancy 1996; Fiske 1998). Moreover, people clearly show moral outrage over norm violations about which they were never explicitly taught.

Undoubtedly, much of human learning is facilitated by analogical reasoning and metaphorical extension. Individuals learn approaches, solutions, and "proper" behavior in one context, and they extend them in "similar circumstances." Unfortunately, little is known about how individuals do this in relation to figuring out norms of conduct, emotional attachments, etc., among similar situations. This is an important area of empirical research for anthropologists and social psychologists.

Social Learning Capacities

Understanding the details of human social learning may be critically important for studying the evolution of cooperation. Humans, unlike other animals, possess extremely refined capacities for imitation and other forms of social learning. In fact, humans are unique in their degree of reliance on social learning and the sophistication of their imitation-based inferential capacities. Many aspects of human social learning capacities can best be understood as cognitive adaptations for acquiring adaptive behaviors and strategies at low costs in information-poor environments (Bloom 2000). Both theory and evidence suggest that, instead of merely imitating a random individual from the population or learning from one's parents, individuals use both model-based cues and integrative algorithms to extract adaptive information from their social world. Model-based cues allow individuals to focus their attention (a scarce resource) on those individuals most likely to have information that is worth acquiring (in terms of individual fitness for the acquirer), e.g., novice hunters might imitate the practices of the most successful hunter in their group. Integrative algorithms allow information from multiple individuals to be combined in a way that increases the likelihood of adopting the most adaptive strategies, behaviors, etc. for the current environment; in many ambiguous situations, a rule like "copy the majority" gives individuals the best chance of acquiring an adaptive behavior.

By using model-based cues related to individuals' skills, success, achievement, and payoffs, and by combining these with cues about self-similarity, this cognitive adaptation allows individuals to focus their social learning efforts on those individuals who are mostly likely to possess adaptive behavior/information that is appropriate for the individual's particular circumstances. Similarity cues may be numerous, although sex and ethnicity are especially salient. Finally, because more experienced individuals will bestow material benefits (gifts) and displays (public praise) on particularly skilled or knowledgeable individuals, naive individuals can use the pattern of these deferential behaviors to assist them in figuring out who is likely to possess useful information (Henrich and Gil-White 2001).

A substantial amount of empirical work from psychology, economics, and anthropology confirms a variety of predictions derived from the above theory (summarized in Henrich and Gil-White 2001). For our purposes, however, these findings confirm that people preferentially imitate the ideas, opinions, beliefs, strategies, and behaviors of prestigious, skilled, and successful individuals across a wide range of domains (even in domains outside the expertise of the prestigious individual). For example, in a synthesis of the diffusion of innovations literature, Rogers (1995) argued that the spread rate of novel technologies and new economic practices into different social groups depends on how quickly prestigious local "opinion leaders" adopt these innovations. In the laboratory, experimental economists have found that MBA students tend to mimic the decisions of successful players in multi-round market games (Kroll and Levy 1992). In a different experiment, Offerman and Sonnemans (1998) showed that subjects making investment decisions tended to copy the beliefs of successful individuals (about the current environment), even when players clearly knew that these individuals had the same information about the current situation as they did (see also Pingle 1995; Pingle and Day 1996).

In contrast to model-based cues, an integrative learning rule like conformist transmission represents a set of mental mechanisms that allows individuals to extract adaptive information by integrating the behavioral observations of several individuals. In its simplest form, conformist transmission can be glossed as a rule to "copy the majority" or "ignore the outliers" preferentially. By biasing individuals in favor of copying common behaviors, preferences, or behavioral strategies, this transmission bias tends to homogenize social groups. There are both theoretical and empirical reasons to believe that humans possess cognitive capacities for conformist transmission. Theoretically, Henrich and Boyd (1998) have shown that genes favoring a heavy reliance on social learning and conformist transmission (copying the majority) can outcompete genes favoring individual learning in both spatially and temporally varying environments. This model predicts two important things: (a) that individuals should increase their reliance on social learning when individual (or environmental) information becomes less certain or as the difficulty of the problem increases and (b) that individuals should rely on copying the majority (conformist transmission) under a wide range of conditions (Boyd and Richerson 1985; see also Ellison and Fudenberg 1993).

Independent experimental work in psychology supports both of these predictions as well as a number of other predictions arising from this model. Psychologists studying conformity have shown that, as a task's difficulty and financial rewards rise, individuals *increase* their reliance on imitation (vs. individual analysis) even though others do not know how they behave (reducing any fear of social sanctions; Insko et al. 1985; Baron et al. 1996). Further, with real money on the line, other experiments show that individuals rely on copying the majority in *social dilemmas*, both when self-interest conflicts with group interest and when self-interested choices correspond to group-interested choices (Smith and Bell 1994; Wit 1999). Finally, Henrich (2001) shows that the slow take-offs and "critical mass tipping-point" observed in many empirical studies of the diffusion of innovations are quite consistent with effects of conformist transmission, although alternative explanations have been proposed (Valente 1995).

THEORETICAL DIRECTIONS

In this section, we will discuss four theoretical approaches to human cooperation: (a) indirect reciprocity and reputation, (b) signaling, (c) models of *n*-person cooperation, and (d) culture–gene coevolution. The first two are closely linked, as are the latter two, so these categories are delineated primarily for expositional purposes. In each case, we briefly describe the basic underlying idea and summarize the relevant conclusions vis-à-vis the empirical patterns that characterize human cooperation.

Indirect Reciprocity and Reputation

Theoretical work on indirect reciprocity has shown that reputational information about an individual's past behavior will allow dyadic cooperation to evolve and, in some cases, to remain stable. Work by Leimar and Hammerstein (2001), Nowak and Sigmund (1998), and Boyd and Richerson (1989) shows that strategies that use information about the reputation for past cooperation of the other individual in the interaction can be evolutionarily stable.[4]

This line of theory suggests that humans should both actively seek out and remember information about the behavior of other individuals in past interactions and information about the behavior of the other individuals with whom they have interacted. Such theoretical findings may help us understand why humans are so interested in gossip and rumors (e.g., N. Smith 2001). It also illuminates a range of experimental findings on the importance of reputation (Fehr and Henrich, this volume).

On the empirical side (McElreath et. al., this volume), the available evidence indicates that people remember some portion of interactions with friends and in other kinds of reciprocal relationships. The limited experimental evidence on these matters suggests, however, that people may forget or not even bother to store most of this information. For example, Milinski et al. (2001) performed an experiment designed to investigate the use of two different bookkeeping strategies in an iterated Prisoner's Dilemma (PD) setting. The first one, Pavlov (Nowak and Sigmund 1993), attends to both its own and its partner's previous round payoffs to determine how to behave in the present. The second strategy, Generous Tit-for-Tat (GTFT) (Nowak and Sigmund 1992), simply copies what its partner did in the last round but occasionally cooperates when its partner defected. Since Pavlov requires more memory than GTFT, Milinski and Wedekind introduced a memory constraint into the game by requiring subjects to play a memory game in which they had to match symbols cards. After each round of the game with the same partner, each subject was allowed to turn over two cards. If they did not match, the cards were turned back over. Subjects were paid the *product* of their scores in the iterated PD and the memory game (and they knew this), so that they would not ignore either game.The findings indicate that (a) under the memory constraint, people's behavior better fit the GTFT strategy, whereas (b) when their memory was unconstrained, their behavior better fit a Pavlovian strategy.

In a similar vein, Bendor et al. (1991) questioned whether Tit-for-Tat bookkeeping is a good strategy in all reciprocal interactions. Like Axelrod (1981, 1984), Bendor solicited strategies for a series of computer tournaments. Strategies were paired at random and played a repeated game. During each round of

---

4 Information about "standing" incorporates both what the current interactant did in their last interaction (help/cooperation vs. not-help/defect) and what the standing was of the individual they interacted with in their last interaction (Sugden 1986). We restrict our discussion of indirect reciprocity to "standing" approaches, leaving out Nowak and Sigmund's "image scoring." We do this based on Leimar and Hammerstein's (2001) comparison of the relative success of these two approaches in explaining indirect reciprocity.

the game, each player picked a number between zero and one. Larger numbers cost the player more and benefited its partner more. Individuals could not find out the other player's number exactly. Instead, they "observed" the other player's number with a normally distributed random error added. Strategies that maintained ongoing accounts, and attempted to return as much on average as they received, performed poorly. Tit-for-Tat also performed poorly. The best-performing strategies were the ones that played a number that was some modest percentage larger than the one that they "observed" their opponents using in the prior round. Bendor argues that account-keeping strategies did poorly because the errors in perception caused them to walk randomly through the space between zero and one. Such strategies overfit their noisy observations, and "overresponded" to deviations. In contrast, strategies that were a bit more generous than their partner tended to push up toward the maximum payoff without too much risk of exploitation, and they resist the temptation of overresponding to errors in their information. Of course, the success of a strategy depends on the mixture of other strategies in the population, so these findings need qualification. Nevertheless, they do indicate that we should be skeptical of the intuition that only very complete bookkeeping strategies can be successful and avoid exploitation.

In general, these results suggest that memory space really is a scarce resource and that under the right conditions, strategies that use simple bookkeeping can outperform strategies that maintain detailed books. This result suggests that it may be impractical for individuals to maintain detailed accounts of long-term relationships. Instead, people may only track recent interactions, or only interactions with substantial costs and benefits. Currently, we know of no experimental evidence that addresses these issues.

Although this work contributes to our understanding of human sociality, two puzzles present themselves: If these models are so robust, why don't we observe reputation-based cooperation throughout nature, as is the case with kin-based cooperation? Why are people cooperative even when they lack information about their interactants?

In addressing the first, let us set aside the possibility that such reputation-based cooperation is actually quite common in nature (but we just haven't noticed it yet) and consider other possibilities. One is that syntactical language allows for large amounts of reputational information to spread through populations rapidly (Smith, this volume). While language is clearly important to reputation effects in humans (as noted above in our discussion of gossip), why gossipers pass on true rather than false but self-serving information is less clear (cf. Lachmann et al. 2001).

Although reputation-based explanations do illuminate some puzzling phenomena, they suffer from a number of drawbacks that suggest that they are insufficient to explain large swaths of human prosocial behavior. First, these models have been restricted to cooperation in dyads, and work on n-person cooperation (warfare, food sharing, etc.) strongly suggests that the results from these dyadic models will not generalize to the n-person situation (Boyd and Richerson 1988; see below). Second, these cooperative strategies are quite susceptible to at least two kinds of errors: (a) if individuals do not know the "standing" of some fraction of the individuals they may interact with, the basin of attraction of cooperation shrinks dramatically — if the "fraction known" drops below 80%, cooperation is extremely unlikely to evolve or be maintained; (b) if the "standing ranking" is even moderately noisy, cooperation is unlikely to evolve or be maintained (Panchanathan and Boyd 2002; Boyd and Richerson 1989; Leimar and Hammerstein 2001). Given that humans cooperate in n-person dilemmas in societies ranging from small-scale foraging groups (Lee 1979; Kaplan and Hill 1985) to large-scale nation-states, information quality is likely to be poor in many situations. On the other hand, in many cases of large-scale cooperation, participants are subdivided into much smaller groups (e.g., kin groups, combat units) that are often embedded in informational hierarchies (Richerson and Boyd 2000), which may substantially reduce the problems noted above.

Signaling

Signaling provides another mechanism that facilitates cooperation. This suggests that individuals may be willing to provide collective goods in the absence of reciprocity or punishment when the provision of such goods can act as an honest signal to other individuals of the provisioner's high quality. As a consequence of receiving this advertisement, other individuals are more likely to provide fitness-enhancing benefits to the signaler in the form of coalitional alliances and mating opportunities. There are two key conditions required for evolutionary stability of such signaling. First, signals must convey information about underlying qualities of the signaler that would be advantageous for observers to know. Second, signals must impose a cost on the signaler that is linked to the quality being advertised, so as to validate the accuracy of the signal. This link can take one of two forms: either lower-quality signalers pay higher marginal costs for signaling, or they reap lower marginal benefits. These two conditions are related, since quality-dependent cost (the second condition) serves to ensure that the signal honestly advertises the relevant underlying qualities of the signaler (the first condition). Because a "good signal" has high broadcast value (lots of people receive the signal), the provision of collective goods provides one effective type of signal.

To explore an application of this theory to a specific case of human cooperation, Bliege Bird et al. (2001) studied turtle hunting on the island of Mer in the Torres Straits (northern Australia). In response to a request from feast organizers, Meriam hunters provide marine turtles (live weight ca. 100–150 kg) for consumption at a previously announced feast. A hunt leader and his crew travel by boat to distant reefs in hopes of locating turtles, even when turtles can be easily collected on beaches during the nesting season. Compared to collecting, hunting is more costly (in time, energy, and risk), provides meat less efficiently (due to higher travel, search, and pursuit costs), and is associated with much wider distributions of meat. Hunters keep little or no meat for themselves and take on a variety of costs for which they are not materially compensated (e.g., time and energy in hunting, money for fuel, time organizing and preparing the hunting team and its equipment prior to the hunt). The ability to bear such costs appears to be linked to hunter quality: because a hunt leader is an organizer and decision maker, his abilities peak as he gains skill and experience. Signals sent by hunting also are efficiently broadcast: hunts are associated with larger numbers of consumers and thus a broader audience (Smith and Bliege Bird 2000). Quantitative data indicate that hunters do not receive increased shares of collected turtle or other foods, as we might predict if risk-reduction reciprocity were structuring the payoffs for hunting (Bliege Bird et al. 2002).

The signaling explanation of collective goods provisioning as applied to the Meriam turtle-hunting case proposes that turtle hunters benefit from unconditional sharing because their harvesting success sends honest signals about their quality to the community in which they will play out their lives as mates, allies, and competitors. Paying attention to such signals can benefit observers because the costs and potential for complete failure inherent in the signal guarantees that it is an honest measure of the underlying qualities at issue: only those endowed with the skills necessary will succeed and hence be asked repeatedly to serve on crews or as hunt leaders. Benefits accruing to signalers (hunters) will depend upon the specific signal and audience; for hunt leaders, benefits might consist of being deferred to by elders or gaining the benefits of a hardworking wife's labor (Smith and Bliege Bird 2000); for jumpers this might include a means of establishing prestige among peers and hence preferential access to various social resources, including enhanced mating success. Whatever the pathways, hunters can be shown to have much higher age-specific (and lifetime) reproductive success than other Meriam men (Smith et al. 2003).

The game theoretic model of Gintis et al. (2001) provides one mechanism for the evolution of such a system. However, honest signaling of quality need not be beneficial to the signaler's group and, indeed, this model applies equally well to socially neutral or harmful forms of costly signaling. This raises the question of why costly signaling should ever take the form of providing collective goods. After all, in other species such signaling generally involves displays such as peacock's tails, roaring contests between red deer, or ritualized struggles between male elephant seals, which provide no overall group benefits. Furthermore, there appear to be numerous human examples of such socially wasteful displays as foot-binding, headhunting, various forms of conspicuous consumption, duels, violent brawling, and even the conspicuous flouting of social norms.

There are several possible approaches to explaining why signals might tend to be prosocial (Gintis et al. 2001). One explanation is *cultural group selection* among alternative evolutionarily stable equilibria, as described below, but now applied to various signaling equilibria (Boyd and Richerson 1990). A second involves the aforementioned "broadcast value," i.e., providing collective goods can attract a larger audience, thus enhancing broadcast efficiency. There may, however, be other potential signals that do not involve collective goods but have equal or better broadcast value. The cultural group selection explanation predicts that only humans will have stabilized signals at prosocial equilibria which vary among social groups, whereas the broadcast value explanation implies that nonhumans should also utilize group provisioning as a way amplifying the broadcast value of their signal At this point, further data will be required to determine if proposed cases of either human or nonhuman group provisioning can be definitively ascribed to costly signaling; proposed nonhuman instances include several bird species, particularly ravens (Heinrich and Marzluff 1991; Mesterton-Gibbons and Dugatkin 1999). Finally, a third explanation of the provision of public goods may apply if providing collective benefits serves as an honest signal that responders who partner with such a signaler will gain future private benefits (e.g., more resources).

N-person Cooperative Models

Indirect reciprocity, reputation, and signaling certainly contribute to explaining some aspects of cooperation in humans. However, much of human cooperation involves large-scale, simultaneous interaction among unrelated individuals; cooperation in warfare, food sharing and help during natural disasters provide classic examples. Interestingly, little theoretical work has been done on modeling this kind of n-person cooperation, and the limited work that has been done strongly suggests that lessons learned from two-person models of dyadic cooperation and indirect reciprocity cannot readily extend the n-person situation. Boyd and Richerson (1988) show that repeated interactions, even when some amount of genetic relatedness is added, are unlikely to explain cooperation in groups larger than about ten. The size of the basin of attraction for the "cooperation equilibrium" in this model goes down with the power $1/n$. Next, we examine another effort to explain n-person cooperation using the logic of repeated interaction. After this, we describe two alternative coevolutionary approaches that have been proposed for solving n-person cooperative dilemmas.

Single-punisher Repeated Game Equilibria

Boyd and Richerson (1992) studied models of the evolution of punishment. In the simplest model, there are three types: defectors who defect until punished and then cooperate, punishers who cooperate and punish defectors, and "contributors" who cooperate but do not punish. Groups of size n are sampled from a global population and interact for a lengthy period. The fitness of each type is the average over all groups. This model has two kinds of equilibria:

1. If punishment by a single individual can induce all others in the group to cooperate, and if the increase in that individual's long-run payoff due to

such cooperation is sufficient to compensate for costs of punishment, then there is a stable polymorphic equilibrium in which punishers and defectors coexist. At this equilibrium there is, on average, about one punisher per group. There is no second-order free-rider problem because punishment creates sufficient private benefits to compensate a punisher.

2. If the costs of being punished are greater than the costs of cooperating, mutant defectors cannot invade a population of all punishers. Such a population can be invaded by contributors, but the selection pressure in favor of such individuals is typically weak. Stability of punishment does not depend on there being any long-run benefit of cooperation. Defectors are eliminated because they are punished.

Other models in Boyd and Richerson (1992) study the effect of kinship, defectors who resist punishment for a number of time periods, and punishers who do not cooperate. Taken together this work suggests that in *small groups*, in which a single punisher can have a substantial effect, the evolution of cooperation induced by punishment is much like other forms of reciprocity, and it is enhanced by kinship and long periods of interaction. But for larger groups, kinship reduces the stability of cooperation and long-term benefits are irrelevant. In such groups, punishment is self-stabilizing and thus punishment can stabilize cooperation, but it can stabilize anything else as well — even maladaptive behaviors.

This model generates a number of predictions about the expected pattern of human cooperation and prosocial psychology. First, it predicts that cooperation should exist only among small groups. Second, it fails to explain the variation in the scale of cooperation across human societies and predicts that the private interests of a few punishers will determine domains of cooperation in local small social groups. Third, it predicts that cooperative norms will periodically collapse, as rare punishers eventually die and are not likely to be replaced immediately. Fourth, it predicts that most people are neither cooperators nor punishers and only cooperate through punishment and threat of punishment. This implies that most humans will lack both moral outrage and any willingness to punish third parties (even if it is in their long-term private interest to do so). In contrast, a small fraction of punishers will have a strategic taste for punishment but lack any shame or fear of being punished. All of these predictions seem wrong, given what we know about human social behavior from ethnography, economic experiments, and social psychology.

Before we discuss evidence for a role for cultural group selection in driving culture–gene coevolution, it is important to realize that many evolutionary social scientists think that large-scale cooperation arises in our species, as it seems to do in nonhuman species, by mechanisms other than group selection. The most plausible alternative to cultural group selection processes discussed below is, we believe, the hypothesis that (a) human social intelligence is sufficient to make reciprocity work in fairly large groups, perhaps larger than most current theoretical models suggest (Dunbar 1998), and (b) that people are also clever private-return punishers (Ruttan and Borgerhoff Mulder 1999). Some combination of reciprocity, reputation, and coercion by self-interested leaders could easily account for the ancestral band scale of human social organization. This would be especially true if the models of reciprocity so far conceived are rather too conservative, or if supplemented by some mechanism like using altruism as a costly signal. Two empirical challenges for this position are whether the scale of social organization in the late Pleistocene was restricted to the band scale, and how to explain the relative rapidity with which the scale of social organization increased in the Holocene. Some felt that the cultural group selection position was overrepresented in our discourse and that the alternatives described here should be given serious consideration.

Cultural Group Selection and Culture–Gene Coevolution

Humans rely heavily on various forms of learning, in particular social learning. These social learning mechanisms create a second system of inheritance that interacts with genetic inheritance. Interestingly, because the dynamics of cultural transmission are fundamentally different from genetic transmission, the usual impediments to multilevel selection processes in genetic systems do not have nearly the same impact on cultural evolution. Building on this idea, we describe a line of theoretical reasoning showing three things. First, if humans rely primarily on "payoff-biased updating" (copying the successful) but admit even a small amount of conformist transmission, the one-shot n-person cooperation–punishment dilemma is transformed into a problem with two stable equilibria: one at full cooperation, the other at the usual full defection. Second, the interaction between groups at the cooperative equilibrium with those at noncooperative equilibria can lead to the rapid spread of cooperation because individuals from cooperative groups (receiving a higher average payoff) will be selectively imitated by individuals from noncooperative groups (but not vice versa). Creating a second equilibrium fundamentally changes the problem, because selection among alternative equilibria does not suffer from the same problem as cooperative dilemmas in multilevel genetic transmission scenarios. This is because with multiple equilibria, the within-group selective processes favoring a local equilibrium resist the mixing processes produced by migration that usually erode the variation between groups (Boyd and Richerson 1990). Third, once cooperative groups have spread widely via cultural group selection, within-group selection will favor genes that support and reinforce cooperative equilibrium.[5]

Conformist Transmission Can Stabilize Cooperation by Stabilizing Punishment

Henrich and Boyd (2001) have shown that if human social learning psychology contains both a transmission bias to copy successful individuals ("payoff-

[5] Unlike models of genetic group selection, cultural group selection models have received little direct criticism; cf. Palmer et al. (1997).

biased" transmission) and a bias to copy high-frequency behaviors (conformist transmission), and there are an arbitrary number of "punishing levels," then highly cooperative equilibria can exist even if conformist transmission is only a weak component of human cultural transmission. Without any transmittable punishing strategies, conformist transmission can only stabilize costly cooperative strategies without punishment, but only if this conformist transmission is quite strong compared to payoff-biased transmission. All other things being equal, payoff-biased transmission causes higher payoff variants to increase in frequency. Thus cooperation is not evolutionarily stable under plausible conditions because not cooperating leads to higher individual relative payoffs (within groups) than cooperating. On its own, payoff-biased transmission suffers from the same problem as natural selection in genetic evolution. However, if our social learning psychology contains a combination of conformist and other social learning mechanisms, then, if cooperation becomes common, conformist transmission will oppose payoff-biased transmission and favor cooperation. When cooperation is not too costly, conformist transmission will maintain cooperative strategies in the population at high frequency. However, because both theory and evidence (Henrich 2001) suggest that conformist transmission is relatively weak compared to payoff-biased transmission (and the costs of cooperation are probably substantial), it seems unlikely that conformist transmission alone will be able to maintain cooperation.

A quite different logic applies to the maintenance of punishment. Suppose that culturally transmitted punishing and cooperating strategies are both common and that being punished is sufficiently costly such that cooperators have higher payoffs than defectors. Rare invading second-order free riders who cooperate but do not punish will achieve higher payoffs than punishers because they avoid the costs of punishing. Since defection does not pay, the only defections will be due to rare mistakes, and thus the *difference* between the payoffs of punishers and second-order free riders will be relatively small compared to that between first-order free riders and cooperators. Hence, conformist transmission is more likely to stabilize the punishment of noncooperators than of cooperation itself. As we ascend to higher-order punishing, the difference between the payoffs to punishing versus nonpunishing decreases geometrically toward zero because the occasions that require the administration of punishment become increasingly rare. Second-order punishing is required only if someone erroneously fails to cooperate, and then someone else erroneously fails to punish that mistake. For third-order punishment, yet another failure to punish must occur. As the number of punishing stages (i) increases, conformist transmission, no matter how weak, will at some stage overpower payoff-biased imitation and stabilize common i^{th}-order punishment. Once punishment is stable at the ith stage, payoffs will favor strategies that punish at the $i-1$ order, because common punishers at the ith order will punish nonpunishers at stage $i-1$. Stable punishment at stage $i-1$

means that payoffs at stage $i - 2$ will favor punishing strategies, and so on down the cascade of punishment. Eventually, common first-order punishers will stabilize cooperation/altruism at stage zero.[6]

Selection among Alternative Equilibria

Once stable culturally transmitted differences arise between groups, at least three different forms of cultural group selection may influence the evolution of practices, beliefs, ideas, and values: demographic swamping, intergroup competition, and prestige-biased group selection. (For the latter, see Henrich 2003.)

Demographic swamping produces changes in the frequency of cultural traits in an overall metapopulation because some social groups (perhaps just one) reproduce new individuals faster than other groups as a result of some set of culturally transmitted ideas or practices that are relatively stable in those groups; this is natural selection acting on between-group cultural variation. Demographic swamping may explain the spread of early agriculturalists into regions once dominated entirely by hunter-gatherers. Agriculturalists gradually replace foragers, increasingly compressing them into tracts of nonarable land. Such empirical cases of demographic swamping suggest that this is probably the slowest kind of cultural group selection, operating on timescales of millennia (Young and Bettinger 1992; Cavalli-Sforza et al. 1994; Diamond 1997).

In intergroup competition, different cultural groups may also compete directly for access to resources through warfare and raiding. Cultural practices and beliefs that provide a competitive edge to groups in warfare will proliferate at the expense of traits that make groups less effective in competition (and more likely to be defeated, absorbed, or dispersed). Such cultural traits might relate to beliefs about patrilocality, heroism, patriotism, economic cooperation (leading to surplus production), the villainy of foreigners, and the proper forms of social or political organization (or all of these). In exploring cultural group selection resulting from intergroup competition, Soltis et al. (1995) calculated evolutionary rates using a model based on group "extinctions" ("extinction" only implies that the group must be disbanded and its members scattered, not necessarily killed). Using empirical data from New Guinean horticultural groups, Soltis et al. estimated that a group-beneficial cultural trait could spread to fixation on timescales of 500 to 1000 years.

One of the best-documented cases of cultural evolution via intergroup competition occurred during the 18th century among the anthropologically famous ethnic groups of the Nuer and Dinka. Before 1820, the Nuer and Dinka occupied adjacent regions in the southern Sudan (Kelly 1985). Despite inhabiting similar

6 Boyd et al. (2003) have combined the fact that the payoff difference between punishing and not punishing is smaller than that between cooperating and not cooperating with group selection to show that the conditions for the evolution of cooperation and punishment are much more plausible for cultural evolution than genetic evolution.

environments and possessing identical technology, the two groups differed in significant ways. Economically, both raised cattle; the Dinka maintained smaller herds of approximately nine cows per bull, whereas the Nuer maintained larger herds with two cows per bull. The Nuer ate mostly milk, corn, and millet and rarely slaughtered cows, whereas the Dinka frequently ate beef. Politically, the Dinka lived in small groups, the largest of which corresponded to their wet season encampment. In contrast, the Nuer organized according to a patrilineal kin system that structured tribal membership across much larger geographic areas. Consequently, the size of a Dinka social group was limited by geography, whereas the Nuer system could organize much larger numbers of people over greater expanses of territory. Despite the similarity of their environments, these two groups showed substantial economic and political differences. Over about 100 years, starting in about 1820, the Nuer dramatically expanded their territory at the expense of the Dinka, who were driven off, killed, or captured and assimilated. As a result, Nuer beliefs and practices spread fairly rapidly across the landscape relative to Dinka beliefs and practices, despite the fact that the Nuer were soon living in the once "Dinka environment" and the fact that many Nuer were formerly Dinka who had adopted Nuer customs.

Genes Respond in the Wake of Cultural Group Selection

Genes may respond to the changed social environment created by cultural evolution. By systematically altering the selective environment faced by genes, cultural evolution via cultural group selection may lead to the subsequent spread of prosocial genes by purely within-group selection processes — genes that would not otherwise be favored without the action of cultural processes (Laland 1994; Henrich and Boyd 2001). This ongoing coevolutionary process between cultural group selection and within-group genetic selection processes could in theory favor genes that allow individuals to learn cooperative strategies rapidly and punishment, to avoid mistaken defections (and thereby avoid the risk of being punished) as well as to learn to punish (thereby avoiding the punishment of nonpunishers). More specifically, by driving down the probability that individuals commit mistaken defections, natural selection might act to expand the size of the group (the n-person group) at which the cooperative equilibrium is stable. Thus, the interaction of genes and culture might ratchet up the maximum size of stable cooperative groups.

These theoretical results account for several observations we made earlier about the nature of human cooperation. First, at a micro level, because these prosocial genes would have evolved in a world with substantial amounts of culturally evolved between-group variation, such genes would foster prosocial psychologies adapted to cue off local behavioral patterns. To avoid being punished for defection or to enhance the rate of learning from mistaken defections, this suggests that shame may have evolved. To avoid being punished for not punishing a defector, moral outrage and anger toward defectors (norm breakers) may have evolved. To enhance learning by observing, empathic abilities may have developed: "If I feel some of your shame for a norm violation, I may be able to avoid a similar situation." Most important, because different human groups may end up at different cooperative equilibria in different behavior domains (e.g., house-building, fishing) all these emotions and affective responses, as well as the learning of social norms/rules, will be designed to adapt themselves to the local equilibria, because they coevolved in a culturally structured world (contrast this with fear of snakes, the dark, etc., which do not vary cross culturally). At a macro level, this suggests that human groups will vary in both their domains and scale of cooperation.

More specific adaptations, such as an in-group psychology, may be favored when stable cooperation in specific domains is particularly favored in cultural group selection. For example, if cooperation in intergroup conflicts and territorial defense were a particularly favorable culturally transmitted equilibrium, then within-group selective processes could have favored in-group psychologies that would have allowed individuals to acquire rapidly cooperative traits that would allow them to attack neighboring groups and defend their group territory. By "favorable" we mean a cooperative equilibrium that would have consistently provided competitive advantage over groups at different stable equilibria across a wide range of ecologies.

Unlike purely genetic approaches to the evolution of human cooperation, this culture–gene coevolutionary approach is applicable only to highly cultural species that live in large groups — and as far as we know, that is a human-only club. Consequently, it provides one explanation of why humans stand alone among mammals in both the scale and nature of their cooperation. For further discussion on the question of why human-like cultural transmission capacities are so rare in nature, see Boyd and Richerson (1996).

Culturally Transmitted Conventions as a Form of Niche Construction

Where group selection processes are at work favoring the proliferation of a genetically transmitted altruistic trait, but within-group selection operates against the altruistic trait, the outcome depends on the relative strengths of within-group as opposed to between-group selection pressures. Humans are distinctive in the extent to which we engage in cultural practices, such as resource sharing or monogamy that retard the within-group pressures and thus favor the evolution of altruistic traits. These practices may be transmitted culturally in the sense that adherence to a particular practice (e.g., sharing some types of food) is in the interest of each member as long as most members do the same. A group's institutions thus constitute a niche, i.e., a modified environment capable of imparting distinctive direction and pace of evolutionary change (Laland et al. 2000; Bowles 2000). Accordingly, the evolutionary success of variance-reducing social institutions may be explained by the fact that they retard selection pressures working against in-group-beneficial individual traits coupled with the fact that high frequencies of bearers of these traits reduces the likelihood of group extinctions (or increases the likelihood of a group's expanding and propagating new groups).

Although the idea that suppression of within-group competition may have a strong influence on evolutionary dynamics has been widely recognized in eusocial insects and other species (Maynard Smith and Szathmáry 1995; Frank 1995; Michod 1996; Buss 1987), its empirical relevance to human coevolutionary processes has only recently been pursued. Bowles et al. (2002), however, have formally modeled and simulated a gene–culture evolutionary process in which a genetically transmitted altruistic behavior coevolves with a culturally transmitted group-level institution such as resource sharing. They show that such a process could account for the proliferation of the altruistic trait (and group-level sharing institutions) under conditions approximating human society during the Late Pleistocene. Interestingly, in simulations in which they preclude the emergence and proliferation of group-level institutions, the altruistic trait evolves only under implausibly high levels of between-group conflict or low levels of between-group migration and group size. These results suggest that the distinctive capacities of humans for the cultural transmission of institutional conventions may have created niches in which otherwise unlikely evolutionary scenarios could unfold, including the evolution of a genetic predisposition to some forms of altruism.

FUTURE LINES OF THEORETICAL AND EMPIRICAL WORK

Here we discuss the substantial gaps in our empirical and theoretical knowledge. On the empirical side, substantially more data about the lives and conditions of ancestral humans are needed to evaluate the plausibility of these and other evolutionary models. Specifically, data are needed on social group size (band size and ethnolinguistic unit size), deme size, dispersal, migration rates of individuals, genes and material culture, rates of environmental fluctuations, and the frequency of intergroup conflicts (if that can somehow be extracted from the record). From ethnography and child development, we need detailed, micro-level studies of child socialization and human learning: How do children learn about local norms? What combination of teaching, observational learning, trial and error experimentation, and punishment leads a child to adult cultural competency? From both the field and the laboratory, future progress hinges on figuring out how individuals achieve normative competency in their society: How do people infer when they should feel moral outrage and punish a transgression (or, how do they figure out what a transgression is)? When should one feel shame? Also from ethnographic work, we need more in-depth studies of cooperation that focus on the questions: What are the material and fitness payoffs to various forms of cooperation in different societies? Would defection pay? What is the variation within groups in adherence to norms of cooperation and punishment?

What is the variation across groups? What are the sources of variation (e.g., inferential errors, genetic variation, entrepreneurial initiative)?

On the theoretical side, we identified several quite specific lines of theoretical work and some more programmatic concerns. In terms of the former, we need more exploration of *n*-person reputation and reciprocity models — too much of the existing emphasis is on dyadic interaction — and more analysis of costly signaling with repeated interaction. At the more programmatic level, future models of human learning should attempt to take into account what we know, and what we will hopefully learn, about what types of learning and inferential processes (such as analogical extension) humans use to acquire the kind of norm-related information and emotional attachments discussed above. Adding these coevolutionary models will likely provide substantial insight into the nature of human cooperation. By including inferential processes into models of how multiple games are played, we may gain insight into how cooperative norms and learning in one domain can influence cooperative (or noncooperative) behavior in others.

REFERENCES

Atran, S., D. Medin, N. Ross et al. 1999. Folkecology and commons management in the Maya Lowlands. *Proc. Natl. Acad. Sci. USA* **96**:7598–7603.

Atran, S., D. Medin, N. Ross et al. 2002. Folkecology, cultural epidemiology, and the spirit of the commons: A garden experiment in the Maya lowlands, 1991–2001. *Curr. Anthro.* **43**:421–450.

Axelrod, R. 1984. The Evolution of Cooperation. New York: Basic.

Axelrod, R., and W.D. Hamilton. 1981. The evolution of cooperation in biological systems. *Science* **211**:1390–1396.

Baron, R.S., J. Vandello, and B. Brunsman. 1996. The forgotten variable in conformity research: The impact of task importance on social influence. *J. Pers. Soc. Psych.* **71**: 915–927.

Bendor, J., R.M. Kramer, and S. Stout. 1991. When in doubt ...: Cooperation in a noisy prisoner's dilemma. *J. Conflict Resol.* **35**:691–719.

Bliege Bird, R.L., D.W. Bird, G. Kushnick, and E.A. Smith. 2002. Risk and reciprocity in Meriam food sharing. *Evol. Hum. Behav.* **23**:297–321.

Bliege Bird, R.L., E.A. Smith, and D.W. Bird. 2001. The hunting handicap: Costly signaling in human foraging societies. *Behav. Ecol. Sociobiol.* **50**:9–19.

Bloom, P. 2000. How Children Learn the Meaning of Words. Cambridge, MA: MIT Press.

Boehm, C. 1996. Emergency decisions, cultural-selection mechanics, and group selection. *Curr. Anthro.* **37**:763–793.

Bowles, S. 2000. Economic institutions as ecological niches. *Behav. Brain Sci.* **23**.

Bowles, S., J.-K. Choi, and A. Hopfensitz. 2002. The coevolution of individual behaviors and group level institutions. Santa Fe, NM: Santa Fe Institute.

Boyd, R., H. Gintis, S. Bowles, and P.J. Richerson. 2002. The evolution of altruistic punishment. *Proc. Natl. Acad. Sci. USA* **100**:3531–3535.

Boyd, R., and P.J. Richerson. 1985. Culture and the Evolutionary Process. Chicago: Univ. of Chicago Press.

Boyd, R., and P. Richerson. 1988. The evolution of reciprocity in sizable groups. *J. Theor. Biol.* **132**:337–356.

Boyd, R., and P.J. Richerson. 1989. The evolution of indirect reciprocity. *Social Networks* **11**:213–236.

Boyd, R., and P.J. Richerson. 1990. Group selection among alternative evolutionarily stable strategies. *J. Theor. Biol.* **145**:331–342.

Boyd, R., and P.J. Richerson. 1992. Punishment allows the evolution of cooperation (or anything else) in sizable groups. *Ethol. Sociobiol.* **13**:171–195.

Boyd, R., and P. Richerson. 1996. Why culture is common, but cultural evolution is rare. *Proc. Brit. Acad.* **88**:77–93.

Buss, L.W. 1987. The Evolution of Individuality. Princeton, NJ: Princeton Univ. Press.

Cavalli-Sforza, L.L., and M. Feldman. 1981. Cultural Transmission and Evolution. Princeton, NJ: Princeton Univ. Press.

Cavalli-Sforza, L.L., P. Menozzi, and A. Piazza. 1994. The History and Geography of Human Genes. Princeton, NJ: Princeton Univ. Press.

Diamond, J.M. 1997. Guns, Germs, and Steel: The Fates of Human Societies. New York: Norton.

Dunbar, R.I.M. 1998. The social brain hypothesis. *Evol. Anthro.* **6**:178–190.

Ellison, G., and D. Fudenberg. 1993. Rules of thumb for social learning. *J. Pol. Econ.* **101**:612–643.

Fiske, A.P. 1991. Structures of Social Life: The Four Elementary Forms of Human Relations. New York: Free Press.

Fiske, A.P. 1998. Learning a Culture the Way Informants Do: Observing, Imitating, and Participating. Washington, D.C.: American Anthropological Assn.

Frank, S.A. 1995. Mutual policing and repression of competition in the evolution of cooperative groups. *Nature* **377**:520–522.

Gintis, H., E.A. Smith, and S. Bowles. 2001. Costly signaling and cooperation. *J. Theor. Biol.* **213**:103–119.

Gurven, M. 2001. Reservation food sharing among the Aché of Paraguay. *Hum. Nat.* **12**:273–297.

Heinrich, B., and J.M. Marzluff. 1991. Do common ravens yell because they want to attract others? *Behav. Ecol. Sociobiol.* **28**:13–21.

Henrich, J. 2000. Does culture matter in economic behavior? Ultimatum game bargaining among the Machiguenga. *Am. Econ. Rev.* **90**:973–980.

Henrich, J. 2001. Cultural transmission and the diffusion of innovations: Adoption dynamics indicate that biased cultural transmission is the predominate force in behavioral change and much of sociocultural evolution. *Am. Anthro.* **103**:992–1013.

Henrich, J. 2003. Cultural group selection, coevolutionary processes and large-scale cooperation. *J. Econ. Behav. Org.*, in press.

Henrich, J., and R. Boyd. 1998. The evolution of conformist transmission and the emergence of between-group differences. *Evol. Hum. Behav.* **19**:215–242.

Henrich, J., and R. Boyd. 2001. Why people punish defectors: Weak conformist transmission can stabilize costly enforcement of norms in cooperative dilemmas. *J. Theor. Biol.* **208**:79–89.

Henrich, J., R. Boyd, S. Bowles et al. 2001. In search of Homo economicus: Experiments in 15 small-scale societies. *Am. Econ. Rev.* **91**:73–78.

Henrich, J., and F. Gil-White. 2001. The evolution of prestige: Freely conferred deference as a mechanism for enhancing the benefits of cultural transmission. *Evol. Hum. Behav.* **22**:165–196.

Henrich, J., and N. Smith. 2003. Comparative experimental evidence from Peru, Chile and the U.S. shows substantial variation among social groups. In: Cooperation,

Punishment and Self-Interest: Experimental and Ethnographic Evidence from 15 Small Scale Societies, ed. J. Henrich, R. Boyd, S. Bowles et al. New York: Oxford Univ. Press and Ann Arbor: Univ. of Michigan Press, in press.

Insko, C.A., R.H. Smith, M.D. Alicke et al. 1985. Conformity and group size: The concern with being right and the concern with being liked. *Pers. Soc. Psych. Bull.* **11**:41–50.

Johnson, A., and T. Earle. 2000. The Evolution of Human Societies. Stanford, CA: Stanford Univ. Press.

Kaplan, H., and K. Hill. 1985. Food sharing among Aché foragers: Tests of explanatory hypotheses. *Curr. Anthro.* **26**:223–245.

Kelly, R.C. 1985. The Nuer Conquest. Ann Arbor: Univ. of Michigan Press.

Kroll, Y., and H. Levy. 1992. Further tests of the separation theorem and the capital asset pricing model. *Am. Econ. Rev.* **82**:664–670.

Lachmann, M., S. Szamado, and C.T. Bergstrom. 2001. Cost and conflict in animal signals and human language. *Proc. Natl. Acad. Sci. USA* **98**:13,189–13,194.

Laland, K. 1994. Sexual selection with a culturally-transmitted mating preference. *Theor. Pop. Biol.* **45**:1–15.

Laland, K.N., F.J. Odling-Smee, and M. Feldman. 2000. Group selection: A niche construction perspective. *J. Consc. St.* **7**:221–224.

Lancy, D.F. 1996. Playing on Mother Ground: Cultural Routines for Children's Development. London: Guilford.

Lee, R.B. 1979. The !Kung San: Men, Women, and Work in a Foraging Society. Cambridge: Cambridge Univ. Press.

Leimar, O., and P. Hammerstein. 2001. Evolution of cooperation through indirect reciprocity. *Proc. Roy. Soc. Lond. B* **268**:745–753.

Maynard Smith, J., and E. Szathmáry. 1995. The Major Transitions in Evolution. Oxford: Oxford Univ. Press.

Mesterton-Gibbons, M., and L. Dugatkin. 1999. On the evolution of delayed recruitment to food bonanzas. *Behav. Ecol.* **10**:377–390.

Michod, R. 1996. Cooperation and conflict in the evolution of individuality. 2. Conflict mediation. *Proc. Roy. Soc. Lond. B* **263**:813–822.

Milinski, M., D. Semmann, T. Bakker, and H.J. Krambeck. 2001. Cooperation through indirect reciprocity: Image scoring or standing strategy? *Proc. Roy. Soc. Lond. B* **268**:2495–2501.

Nowak, M.A., and K. Sigmund. 1992. Tit-for-tat in heterogeneous populations. *Nature* **355**:250–253.

Nowak, M.A., and K. Sigmund. 1993. A strategy of win-stay, lose-shift that outperforms tit-for-tat in the prisoner's dilemma game. *Nature* **364**:56–58.

Nowak, M.A., and K. Sigmund. 1998. Evolution of indirect reciprocity by image scoring. *Nature* **393**:573–577.

Offerman, T., and J. Sonnemans. 1998. Learning by experience and learning by imitating others. *J. Econ. Behav. Org.* **34**:559–575.

Palmer, C.T., B.E. Fredrickson, and C. Tilley. 1997. Categories and gatherings: Group selection and mythology of cultural anthropology. *Evol. Hum. Behav.* **18**:291–308.

Panchanathan, K., and R. Boyd. 2002. A tale of two defectors: The importance of standing for the evolution of indirect reciprocity. http://www.bol.ucla.edu/~buddha /TakeOfTwoDefectors.pdf..

Pingle, M. 1995. Imitation vs. rationality: An experimental perspective on decision making. *J. Socio-Econ.* **24**:281–315.

Pingle, M., and R.H. Day. 1996. Modes of economizing behavior: Experimental evidence. *J. Econ. Behav. Org.* **29**:191–209.

Richerson, P., and R. Boyd. 1998. The evolution of ultrasociality. In: Indoctrinability, Ideology and Warfare, ed. I. Eibl-Eibesfeldt and F.K. Salter, pp. 71–96. New York: Berghahn.

Richerson, P., and R. Boyd. 2000. Complex societies: The evolutionary dynamics of a crude superorganism. *Hum. Nat.* **10**:253–289.

Rogers, E.M. 1995. Diffusion of Innovations. New York: Free Press.

Ruttan, L., and M. Borgerhoff Mulder. 1999. Are East Africa pastoralists conservationists? *Curr. Anthro.* **40**:621–652.

Simon, H. 1990. A mechanism for social selection and successful altruism. *Science* **250**:1665–1668.

Smith, E.A., and R.L. Bliege Bird. 2000. Turtle hunting and tombstone opening: Public generosity as costly signaling. *Evol. Hum. Behav.* **21**:245–261.

Smith, E.A., R.L. Bliege Bird, and D.W. Bird. 2003. The benefits of costly signaling: Meriam turtle-hunters. *Behav. Ecol.* **14**:116–126.

Smith, J.M., and P.A. Bell. 1994. Conformity as a determinant of behavior in a resource dilemma. *J. Soc. Psych.* **134**:191–200.

Smith, N. 2001. Ethnicity, Reciprocity, Reputation and Punishment: An Ethnoexperimental Study of Cooperation among the Chaldeans and Hmong of Detroit, Michigan. Ph.D. diss. University of California, Los Angeles.

Soltis, J., R. Boyd, and P.J. Richerson. 1995. Can group-functional behaviors evolve by cultural group selection? An empirical test. *Curr. Anthro.* **36**:473–494.

Sugden, R. 1986. The Economics of Rights, Co-operation,and Welfare. Oxford: Basil Blackwell.

Toren, C. 1990. Making Sense of Hierarchy. London: Athlone.

Valente, T. 1995. Network Models of the Diffusion of Innovations. Cresskill, NJ: Hampton Press.

Wit, J. 1999. Social learning in a common interest voting game. *Games Econ. Behav.* **26**:131–156.

Young, D., and R.L. Bettinger. 1992. The Numic spread: A computer simulation. *Am. Antiq.* **57**:85–99.

Young, H.P. 1998. Individual Strategy and Social Structure: An Evolutionary Theory of Institutions. Princeton, NJ: Princeton Univ. Press.

Name Index

Subject Index